Biotechnology Applications of Microalgae

Biotechnology Applications of Microalgae

Editor

Carlos Almeida

MDPI • Basel • Beijing • Wuhan • Barcelona • Belgrade • Manchester • Tokyo • Cluj • Tianjin

Editor
Carlos Almeida
University of Las Palmas de
Gran Canaria
Spain

Editorial Office
MDPI
St. Alban-Anlage 66
4052 Basel, Switzerland

This is a reprint of articles from the Special Issue published online in the open access journal *Marine Drugs* (ISSN 1660-3397) (available at: https://www.mdpi.com/journal/marinedrugs/special_issues/biotechnology_microalgae).

For citation purposes, cite each article independently as indicated on the article page online and as indicated below:

LastName, A.A.; LastName, B.B.; LastName, C.C. Article Title. *Journal Name* **Year**, *Volume Number*, Page Range.

ISBN 978-3-0365-6772-3 (Hbk)
ISBN 978-3-0365-6773-0 (PDF)

Cover image courtesy of Carlos Almeida

Contents

About the Editor

Carlos Almeida

Carlos Almeida has a PhD in marine science and is currently the Head Scientist and a researcher at the Flow Cytometry Unit at the Spanish Bank of Algae (BEA). Dr. Almeida holds scientific training sessions at the Baltic Research Institute in Warnemünde, Germany. He is also the supervisor of 1 PhD thesis, the author of 22 articles in refereed journals and has held 48 presentations in congresses related to oceanography, biotechnology, physiology and algae culture. He has been part of 21 national and European projects, including different research projects related to the cultivation and biotechnological characterization of microalgae and cyanobacteria both at the laboratory scale and the pilot plant (Biodiesel, Miracles, Spiterm, Tropical, Atticua, Rebeca and Sabana projects). He is currently responsible for the Flow Cytometry Unit and co-coordinates the Biotechnology Unit of the Spanish Bank of Algae. These units support the collection of microalgae and cyanobacteria that BEA manages in the isolation of strains and the physiological and biochemical studies related to the intensive cultivation of algae and the transformation and industrial application of the biomass produced.

Preface to "Biotechnology Applications of Microalgae"

Microalgae are currently an inexhaustible source of a great variety of bioactive compounds. For this reason, they have acquired great relevance in various biotechnological industries for use in food, feed, aquaculture, nutraceutical, pharmacological, biomedical, cosmetics, agriculture, energy and environmental processes. The plasticity of these organisms allows their culture conditions to be modified to improve the performance, synthesis and accumulation of the target compounds. However, the low profitability of the production of these compounds due to technical and economic issues associated with the development of cultivation and downstream processes have led to some projects falling by the wayside. New technological advances are thus of the utmost importance for developing this industry so it can one day compete in terms of quality and profitability with other current alternatives, in addition to opening up new possibilities and paths for upcoming biotechnological applications.

This Special Issue, "Biotechnology Applications of Microalgae", is focused on the latest novel advances related to the production of different bioactive compounds from microalgae and their biotechnological use.

Carlos Almeida
Editor

marine drugs

Article

Mixotrophy in a Local Strain of *Nannochloropsis granulata* for Renewable High-Value Biomass Production on the West Coast of Sweden

Valeria Villanova [1,*,†], Christian Galasso [2], Giovanni Andrea Vitale [3], Gerardo Della Sala [3], Johan Engelbrektsson [4,‡], Niklas Strömberg [4,‡], Kashif Mohd Shaikh [1], Mats X. Andersson [1], Fortunato Palma Esposito [3], Susanne Ekendahl [4,‡], Donatella De Pascale [3] and Cornelia Spetea [1,*]

1 Department of Biological and Environmental Sciences, University of Gothenburg, 405 30 Gothenburg, Sweden; mohd.kashif.shaikh@bioenv.gu.se (K.M.S.); mats.andersson@bioenv.gu.se (M.X.A.)
2 Department of Ecosustainable Marine Biotechnology, Stazione Zoologica Anton Dohrn, Calabria Marine Centre, C. da Torre Spaccata, 87071 Amendolara, Italy; christian.galasso@szn.it
3 Department of Ecosustainable Marine Biotechnology, Stazione Zoologica Anton Dohrn, Via A.F. Acton, Molosiglio, 80133 Napoli, Italy; giovanniandrea.vitale@szn.it (G.A.V.); gerardo.dellasala@szn.it (G.D.S.); fortunato.palmaesposito@szn.it (F.P.E.); donatella.depascale@szn.it (D.D.P.)
4 Department of Chemistry, Biomaterials and Textiles, RISE Research Institutes of Sweden AB, 471 56 Gothenburg, Sweden; johan@addscience.se (J.E.); niklas@stromtech.se (N.S.); miljosus@miljosus.se (S.E.)
* Correspondence: valeria.villanova@univr.it (V.V.); cornelia.spetea.wiklund@bioenv.gu.se (C.S.)
† Current address: Department of Biotechnology, University of Verona, 37134 Verona, Italy.
‡ Current address: Industridoktorn®, Kaveldunsvägen 6, 475 37 Bohus-Björkö, Sweden.

Citation: Villanova, V.; Galasso, C.; Vitale, G.A.; Della Sala, G.; Engelbrektsson, J.; Strömberg, N.; Shaikh, K.M.; Andersson, M.X.; Palma Esposito, F.; Ekendahl, S.; et al. Mixotrophy in a Local Strain of *Nannochloropsis granulata* for Renewable High-Value Biomass Production on the West Coast of Sweden. *Mar. Drugs* **2022**, *20*, 424. https://doi.org/10.3390/md20070424

Academic Editor: Carlos Almeida

Received: 20 May 2022
Accepted: 23 June 2022
Published: 28 June 2022

Abstract: A local strain of *Nannochloropsis granulata* (*Ng*) has been reported as the most productive microalgal strain in terms of both biomass yield and lipid content when cultivated in photobioreactors that simulate the light and temperature conditions during the summer on the west coast of Sweden. To further increase the biomass and the biotechnological potential of this strain in these conditions, mixotrophic growth (i.e., the simultaneous use of photosynthesis and respiration) with glycerol as an external carbon source was investigated in this study and compared with phototrophic growth that made use of air enriched with 1–2% CO_2. The addition of either glycerol or CO_2-enriched air stimulated the growth of *Ng* and theproduction of high-value long-chain polyunsaturated fatty acids (EPA) as well as the carotenoid canthaxanthin. Bioassays in human prostate cell lines indicated the highest antitumoral activity for *Ng* extracts and fractions from mixotrophic conditions. Metabolomics detected betaine lipids specifically in the bioactive fractions, suggesting their involvement in the observed antitumoral effect. Genes related to autophagy were found to be upregulated by the most bioactive fraction, suggesting a possible therapeutic target against prostate cancer progression. Taken together, our results suggest that the local *Ng* strain can be cultivated mixotrophically in summer conditions on the west coast of Sweden for the production of high-value biomass containing antiproliferative compounds, carotenoids, and EPA.

Keywords: *Nannochloropsis*; mixotrophy; photobioreactors; CHN analysis; carotenoids; polyunsaturated fatty acids; metabolomics; bioassay; cell death pathway; autophagy; antitumoral activity

1. Introduction

Microalgae are unicellular photosynthetic microorganisms that originated from endosymbiotic events in which a heterotrophic ancestor fused with various photoautotrophic (photosynthetic) organisms [1]. Thanks to this evolutionary history, they possess both photosynthetic and respiratory organelles (chloroplasts and mitochondria, respectively) and hence exhibit trophic flexibility. Although most microalgae are photoautotrophs, some of them are also able to use organic carbon via respiration, either in the dark (heterotrophs) or in a light-dependent manner (mixotrophs). Mixotrophy is the trophic mode in which both CO_2 and organic carbon are assimilated simultaneously, owing to the activation of both respiration and photosynthesis. It can be employed as a method to increase the productivity of microalgae that are cultivated in low light conditions. To minimize the additional cost of organic carbon supplementation, industrial wastewater and biodiesel waste (i.e., glycerol) are often used for algae cultivation and biomass production (as reviewed by Villanova et al. 2021 [2]). The Eustigmatophyceae *Nannochloropsis* genus has been extensively used for biotechnological applications such as the production of biofuel [3,4] and fish feed [5]. Moreover, the *Nannochloropsis* genus is a well-known source of polyunsaturated fatty acids, such as eicosapentaenoic acid (EPA) and carotenoids, that are important for human health [6,7]. Finally, in vitro assays have shown interesting anti-cancer activity of the *Nannochloropsis* biomass on different human cell lines, opening new perspectives for the nutraceutical application of this genus [8–10].

Mixotrophic growth of the *Nannochloropsis* genus (i.e., *Nannochloropsis gaditana* and *Nannochloropsis salina*) been investigated using glucose, glycerol, and acetate as external organic carbon sources or by adding the bacteria-rich medium Lysogeny broth (LB) [11–14]. However, only a handful of research projects have focused on understanding the industrial potential of growing *Nannochloropsis* under this growth mode [12,14].

There is an increasing need for sustainable resources for the production of food, feed, oil-based materials, and energy. Microalgae have a great potential in this respect and can be cultivated outdoors in conditions resembling those in natural habitats. However, to ensure high productivity, it is necessary to exploit the natural diversity of microalgae and select local species and strains that can adjust their physiology to changing environments. This is particularly important in the Nordic countries, which have large variations in light intensity and temperature throughout the year [15]. The overall goal of this work was to employ mixotrophy as a strategy for maximizing the outdoor productivity of a *Nannochloropsis granulata* strain (*Ng*), isolated from the Skagerrak in the northeast Atlantic Ocean [16]. A previous study reported that the local *Ng* was the most productive strain (compared to 166 strains of *Skeletonema marinoi*), in terms of both biomass yield and lipid content when cultivated in photobioreactors that simulate the light and temperature conditions during the summer on the west coast of Sweden [17]. More specifically, in summer, *Ng* reached 3.5 g/L of biomass containing 40% lipids, i.e., 7 and 1.5-fold, respectively, more than the most productive *Skeletonema marinoi* strain in the same conditions [17]. Here, we aimed to further increase the biomass and lipid productivity of *Ng* under simulated summer conditions on the west coast of Sweden by the addition of glycerol, and also to expand the biotechnological potential of the biomass in terms of biofuel, food production, and antitumoral activity. To reach this aim, an interdisciplinary approach was applied for the first time on this strain in order to allow for in-depth physiological and biochemical characterization in mixotrophy as compared to phototrophy; this was combined with the production of high-value molecules and tested on a small scale (i.e., Multi-Cultivator) before being validated on a larger scale (i.e., photobioreactors).

2. Results

2.1. Effect of Mixotrophy on Growth Rates, Photosynthesis, and Biomass Yield in a Multi-Cultivator System

First, we tested the effect of various external carbon sources on *Ng* by monitoring the growth changes in phototrophic and mixotrophic cultures in a Multi-Cultivator MC 1000-OD (Photon System Instruments, Drásov, Czech Republic), with a volume of 80 mL optimised f/2 medium with air bubbling (as described in Section 4.1.1 and shown in Figure S1a). We used a constant light intensity of 300 μmol photons m^{-2} s^{-1} at a temperature of 20 °C, corresponding to the average light intensity and temperature, respectively, during the summer season in Gothenburg. *Ng* was able to grow in mixotrophy using either glycerol or glucose (but not acetate) as an external carbon source, therefore confirming previous results for *Nannochloropsis gaditana* [14] (see Supplementary Materials Figure S2).

Since glycerol is a by-product of biofuel production, and hence a cheap carbon source, it was selected for further experiments with *Ng* in mixotrophic conditions [18].

During the first days of cultivation (0–6 days) in the Multi-Cultivator, the cells grew similarly in mixotrophy and phototrophy. From day 10, the growth in mixotrophy was significantly enhanced as compared to phototrophy (Figure 1a).

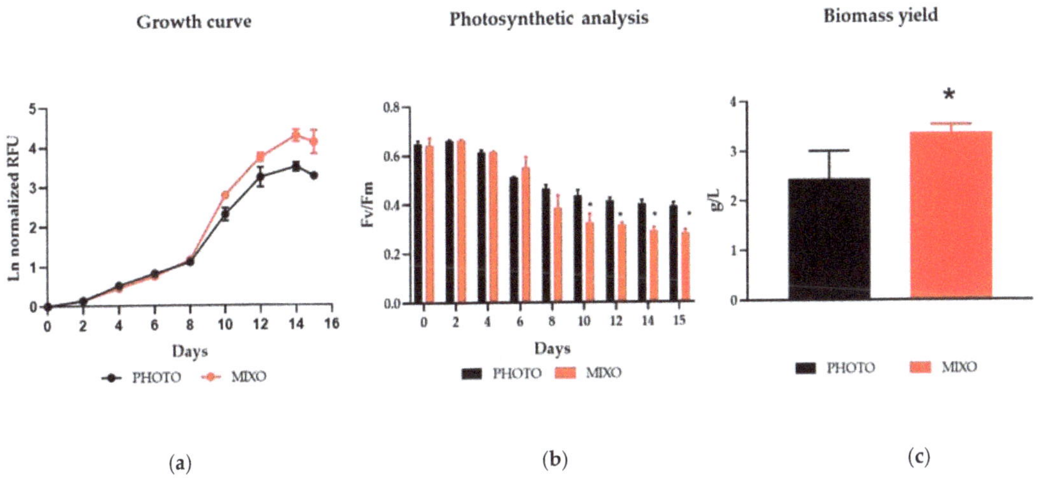

Figure 1. (**a**) Growth profile, (**b**) photosynthetic analysis, and (**c**) final biomass yield of *Nannochloropsis granulata* grown in a Multi-Cultivator at 20 °C, with constant light at 300 µmol photons m^{-2} s^{-1}, under mixotrophy with 10 mM glycerol (MIXO) or under phototrophy with air (PHOTO). The growth was monitored in relative chlorophyll fluorescence units (RFU), and plotted data are values normalized at time zero. Data were considered significant for *p*-values < 0.05 (* $p < 0.05$). Each point is expressed as mean ± standard deviation ($n = 3$). Schemes follow the same formatting.

In the first 4 days of the experiment, *Ng* displayed a maximum quantum efficiency of PSII photochemistry (i.e., F_v/F_m) of about 0.6 in both phototrophy and mixotrophy (Figure 1b), corresponding to a healthy physiological status in the genus [16]. From day 6, a decrease in photosynthetic performance was detected in all tested samples, but this decrease was significantly more pronounced in mixotrophy from day 10. The decrease in Fv/Fm during mixotrophic growth has been previously reported for *Nannochloropsis gaditana* and the diatom *Phaeodactylum tricornutum* [11,19]. The enhanced growth together with the decrease in photosynthesis from day 10 in mixotrophic samples indicate a switch from the phototrophic to the mixotrophic regime. Moreover, the samples grown in mixotrophy showed an increase of about 1.4-fold in the final biomass (Figure 1c). Taken together, these data showed that *Ng* was able to grow under mixotrophy using glycerol as an external carbon source in the tested conditions.

2.2. Effect of Mixotrophy on Growth Rate, Biomass Yield, Energy Production, and Nutrient Removal in Photobioreactors

Next, the cultivation was upscaled to environmental photobioreactors (ePBRs) of 1L (as described in Section 4.1.2 and shown in Figure S1b), using a constant temperature of 20 °C and the simulated photoperiod of summer in Gothenburg, Sweden (Figure 2a). In particular, the mixotrophic growth of samples injected only with air (i.e., 0.04% CO_2) or with 1–2% CO_2-enriched air (MIXO_AIR and MIXO_CO_2, respectively) was compared with a phototrophic control grown with the same aeration system (PHOTO_AIR and PHOTO_CO_2, respectively). When supplied with air in the ePBR, the addition of glycerol (MIXO_AIR) stimulated growth (expressed as relative fluorescence units, RFU) as compared to the phototrophic condition (PHOTO_AIR) starting from day 10 (Figure 2b), which is in line with the observations in the Multi-Cultivator (Figure 1a). Moreover, both samples supplied with CO_2, i.e., MIXO_CO_2 and PHOTO_CO_2, further experienced stimulated growth from day 3 as compared to the samples PHOTO_AIR and MIXO_AIR (Figure 2b). There was no difference in the growth profiles of the conditions PHOTO_CO_2 and MIXO_CO_2; hence, the better growth in these conditions as compared to the others could be attributed to the addition of CO_2 rather than glycerol [20].

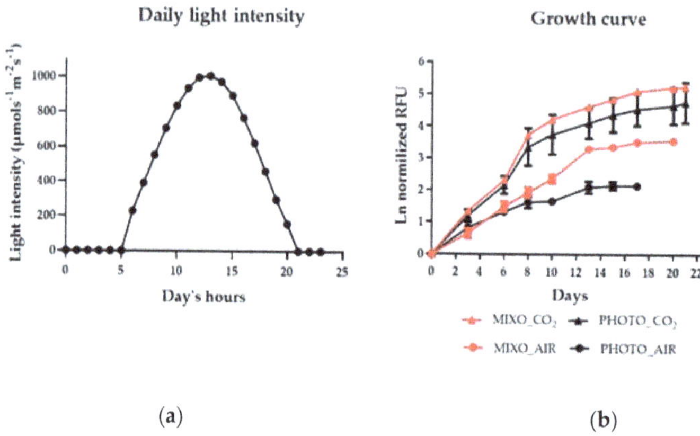

Figure 2. (**a**) Light cycle used in the environmental photobioreactors (ePBRs) for growing *Nannochloropsis granulata*, simulating the summer conditions in Gothenburg; (**b**) Growth profile of *Nannochloropsis granulata* grown with light in a medium injected with air (PHOTO_AIR, black circle); in a medium supplemented with 10 mM glycerol and injected with air (MIXO_AIR, red circle); in a medium injected with 1–2% CO_2-enriched air (PHOTO_CO_2, black square); in a medium supplemented with 10 mM glycerol and injected with 1–2% CO_2-enriched air (MIXO_CO_2, red square). The growth was monitored in relative chlorophyll fluorescence units (RFU), and plotted data are values normalized at time zero. Each time point is the mean ± standard deviation (n = 3–5).

To better compare the four different conditions, both growth (i.e., specific growth rate, biomass yield, and productivity) and energy parameters (calorific value and energetic productivity) were determined. There were no significant differences among PHOTO_CO_2, MIXO_CO_2, and MIXO_AIR in these parameters, and they were all significantly more performant than the condition PHOTO_AIR. More specifically, the specific growth rate, biomass yield, and productivity in these three conditions significantly increased by 1.6-, 3.6-, and 3.6-fold, respectively, when compared to PHOTO_AIR (Table 1). In the same conditions, the calorific value and the energetic productivity of the obtained biomass were enhanced 1.4- and 5-fold, respectively. This indicates that the addition of either/both organic or/and inorganic carbon similarly improved biomass and energy production in the tested conditions. Moreover, both nitrate and phosphate consumption were increased in MIXO_AIR as compared to PHOTO_AIR by about 4-fold (Table 1). In MIXO_CO_2, only phosphate consumption was increased as compared to PHOTO_CO_2, while nitrate consumption was similar in the three conditions MIXO_AIR, PHOTO_CO_2, and MIXO_CO_2. The glycerol consumption was similar among the mixotrophic samples. In addition to reduced growth and nutrient consumption, the samples grown in PHOTO_AIR showed the highest ash content (Table 1), about 3-fold higher than the other three conditions. The PHOTO_CO_2 data for *Ng* are in good agreement with those reported previously for *Ng* in summer conditions [17]. Taken together, the new data show that the addition of glycerol improves the growth and biomass production of *Ng* in summer conditions, reaching productivities similar to those obtained in phototrophy with CO_2.

Table 1. Maximum specific growth rate (*μmax*), biomass yield, biomass productivity, calorific value, energy productivity, nutrient removal rate (N, P, and Gly), and ash content were calculated for *Nannochloropsis granulata* grown in photobioreactors, with light in a medium injected with air (PHOTO_AIR); with light in a medium supplemented with 10 mM glycerol and injected with air (MIXO_AIR); with light in a medium injected with 1–2% CO_2-enriched air (PHOTO_CO_2), and with light in a medium supplemented with 10 μM glycerol and injected with 1–2% CO_2-enriched air (MIXO_CO_2). Each parameter is expressed as mean ± standard deviation (n = 3–5). Different letters (^{a}, ^{b}, ^{c}) denote significant differences among treatments (t-test, $p < 0.05$). P: phosphate, N: nitrate, Gly: glycerol, DW: dry weight, n.a.: not applicable.

Parameters	PHOTO_AIR	MIXO_AIR	PHOTO_CO_2	MIXO_CO_2
Maximum specific growth rate (μ_{max}, d^{-1})	0.26 ± 0.02^a	0.42 ± 0.05^b	0.46 ± 0.01^b	0.41 ± 0.02^b
Biomass yield (g DW/L)	0.90 ± 0.14^a	3.23 ± 0.25^b	3.36 ± 0.38^b	3.38 ± 0.82^b
Biomass productivity (g DW/L/d)	0.05 ± 0.01^a	0.18 ± 0.01^b	0.19 ± 0.02^b	0.19 ± 0.05^b
Calorific value (MJ/kg DW)	18.20 ± 0.80^a	24.90 ± 0.99^b	24.60 ± 1.20^b	24.50 ± 0.71^b
Energy productivity (kJ/L/d)	0.93 ± 0.15^a	4.47 ± 0.39^b	4.14 ± 0.58^b	4.60 ± 1.13^b
P removal rate (mg/L/d)	1.84 ± 0.26^a	6.72 ± 1.03^b	3.07 ± 0.30^c	7.58 ± 1.70^b
N removal rate (mg/L/d)	14.90 ± 0.30^a	63.26 ± 2.76^b	63.61 ± 4.30^b	60.30 ± 5.40^b
Gly removal rate (mg/L/d)	n.a.	61.67 ± 9.93^a	n.a.	68.77 ± 8.40^a
Ash content (% DW)	25.00 ± 1.20^a	8.02 ± 0.20^b	7.95 ± 0.49^b	6.70 ± 1.40^b

2.3. Biomass Composition, Pigment, and Fatty Acid Profiles

The biomass derived from the growth of *Ng* in PBRs in the four different conditions was further analysed in terms of protein-lipid-carbohydrate composition, fatty acid, and pigment profiles. The composition of the ash-free biomass based on CHN analyses was very similar among the four tested conditions, with about 40% proteins, 40% lipids, and 20% carbohydrates (Figure 3a). However, while the fatty acid content of the PHOTO-AIR samples represented 11% of the biomass, it was more than doubled in the samples from the MIXO_AIR, PHOTO_CO_2, and MIXO_CO_2 conditions. Palmitic acid (C16:0), palmitoleic acid (16:1 (n7)), and EPA (20:5 (n3)) were identified as the major fatty acid classes in *Ng* (Figure 3b), as previously reported in the genus [14,21]. The amounts of most detected fatty acids were significantly increased in MIXO_AIR as compared to PHOTO_AIR, particularly the EPA that was triplicated in this condition. MIXO_CO_2 showed the highest fatty acid content in terms of 16:1 (n7) and EPA, which were increased by 1.5- and 3-fold as compared to PHOTO_CO_2 and PHOTO_AIR, respectively. HPLC analysis of pigment extracts from *Ng* revealed the presence of chlorophyll *a*, violaxanthin, and β-carotene as major pigments, and of lutein, zeaxanthin, and canthaxanthin as minor carotenoids (Figure 3c), thus confirming previous results for the strain [16]. In addition, the pigment profile was affected by the growth conditions. The violaxanthin, lutein, and zeaxanthin contents were significantly reduced in MIXO_AIR as compared to PHOTO_AIR. β-carotene content was doubled in PHOTO_CO_2 and MIXO_CO_2 as compared to PHOTO_AIR and MIXO_AIR, whereas the amount of chlorophyll a was not changed among the samples. Furthermore, canthaxanthin was significantly more concentrated in MIXO_CO_2 (about 0.02 μg/mg of dry weight) than in PHOTO_CO_2 (about 0.002 μg/mg of dry weight) and was not detected in the other conditions.

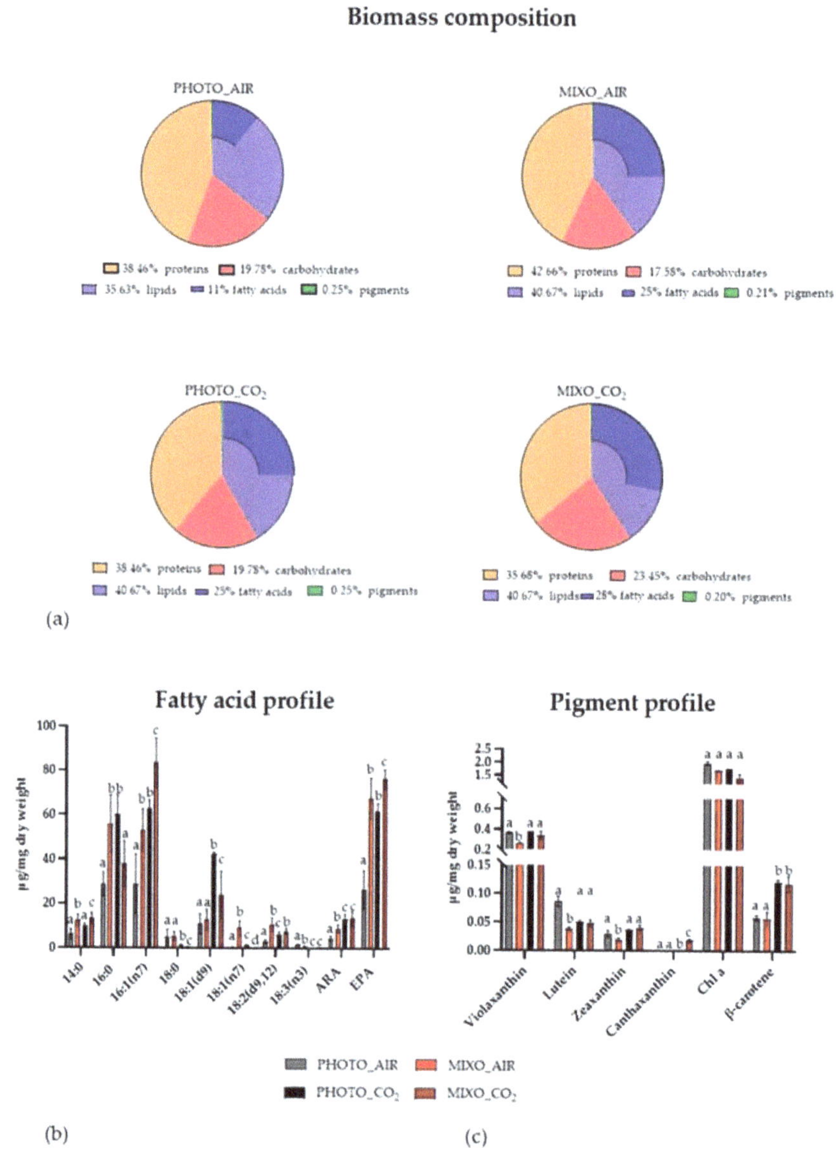

Figure 3. (**a**) Biomass composition, (**b**) Fatty acid, and (**c**) Pigment profile of *Nannochloropsis granulata* grown in photobioreactors with light in a medium injected with air (PHOTO_AIR); with light in a medium supplemented with 10 μM glycerol and injected with air (MIXO_AIR); with light in a medium injected with 1–2% CO_2-enriched air (PHOTO_CO_2); with light in a medium supplemented with 10 μM glycerol and injected with 1–2% CO_2-enriched air (MIXO_CO_2). Each graph shows data as mean ± standard deviation (n = 3–5). Different letters (a, b, c, d) denote significant differences among treatments (t-test, $p < 0.05$). ARA: Arachidonic acid; EPA: Eicosapentaenoic acid; Chl *a*: chlorophyll *a*.

2.4. In Vitro Antiproliferative Effect on Human Cells

Total extracts of *Ng* grown in the four conditions (PHOTO_AIR, MIXO_AIR, MIXO_CO_2, and PHOTO_CO_2) were tested on both normal (PNT2) and cancer (PC3) human prostatic cells. These two cell lines were chosen since prostatic cancer is a very common disease in the male population, representing the second most frequent cancer type in men and the fifth leading cause of death worldwide [22]. Figure 4 shows the results of the viability assay on the two cell lines after 48 h of treatment. Cells grown under phototrophic growth (PHOTO_AIR and PHOTO_CO_2) did not produce metabolites with a strong cytotoxic effect on PC3. Indeed, a reduction of cell viability in this cell line was induced only at the highest concentration (100 μg/mL) (49.5% for PHOTO_AIR and 54.7% for PHOTO_CO_2). A similar effect was observed using the same extracts (PHOTO_AIR and PHOTO_CO_2) on normal cells PNT2; 1 and 10 μg/mL showed comparable viability levels in control cells, while 100 μg/mL reduced the percentage of viable cells comparably to that observed in cancer cells (61.5% for PHOTO_AIR and 67.2% for PHOTO_CO_2). The MIXO conditions (MIXO_AIR and MIXO_CO_2) produced metabolites that strongly reduced the percentage of viable PC3 cells at 100 μg/mL (21.2% for MIXO_AIR and 7.6% for MIXO_CO_2) and only a slight reduction at 10 μg/mL (76.3% for MIXO_AIR and 76.6% for MIXO_CO_2). The same extracts (MIXO_AIR and MIXO_CO_2) did not induce any variations at 1 and 10 μg/mL in normal PNT2 cells, and there was a reduction of cell vitality only at the highest concentration (55.4% for MIXO_AIR and 56.6% for MIXO_CO_2).

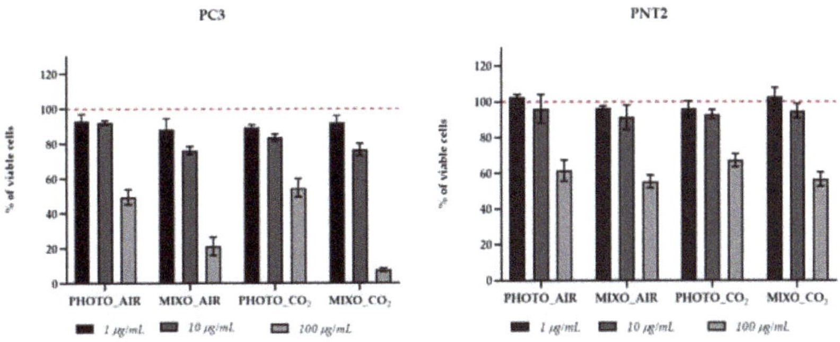

(**a**) (**b**)

Figure 4. MTT cell viability assay on (**a**) PC3 prostate cancer and (**b**) PNT2 normal cell lines. Bar graphs show the percentages of viable cells after 48 h of treatment with 1, 10, and 100 μg/mL of total extracts. Cells treated with DMSO vehicle (0.5%) were used as control and correspond to 100% of cell viability (dotted red lines). Assays were performed in biological triplicates, and the graphs represent means ± standard deviations.

Total extracts were fractionated and tested again on PC3 and PNT2. Both cell lines treated with 1 μg/mL of the fractions did not show reduction in cell viability after 48 h (see Supplementary Materials, Figure S3). The intermediate concentration (10 μg/mL) induced a significant reduction in cell viability only when PC3 cells were treated with the fractions D and E (see Section 4.6) of the MIXO_CO_2 growth condition (Figure 5a). All fractions at 100 μg/mL induced a strong or moderate reduction in viability in both PC3 and PNT2 cells. In particular, all the fractions deriving from the PHOTO_AIR, MIXO_AIR, and PHOTO_CO_2 extracts induced comparable levels of viability in both PC3 and PNT2, while fractions B, C, D, E, and F from MIXO_CO_2 showed higher toxicity in PC3 as compared with PNT2 (Figure 5b).

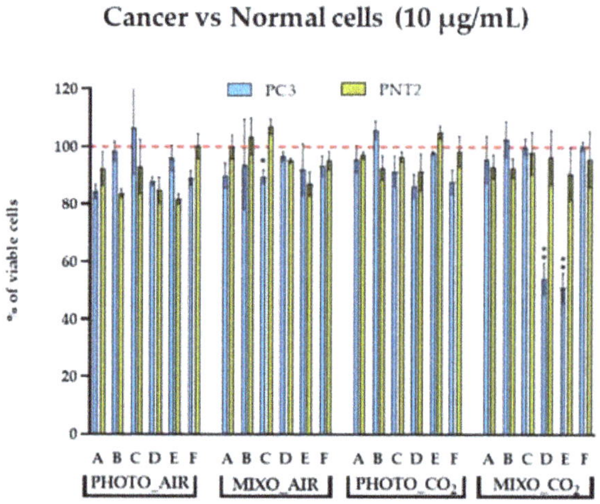

(a)

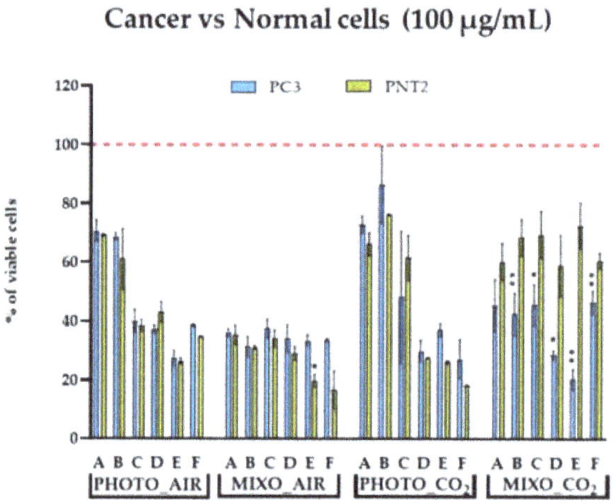

(b)

Figure 5. Cell viability results after treatments of PC3 (blue bar graph) and PNT2 (yellow bar graph) with all fractions (A–F). Histograms represent the percentages of viable cells after 48 h of incubation with (**a**) 10 μg/mL and (**b**) 100 μg/mL of fractions. Cells treated with DMSO vehicle (0.5%) were used as control and correspond to 100% cell viability (dotted red line). Assays were performed in biological triplicate, and graphs represent means ± standard deviations. Differences between viability of PNT2 and PC3 were considered significant for p-values ≤ 0.05 (** $p \leq 0.005$ and * $p \leq 0.05$).

Fractions D and E of MIXO_CO_2 that showed the highest cytotoxicity in PC3 were selected since they exerted the strongest selective antiproliferative effect and were also tested with a wider range of concentrations, at both 24 h (see Supplementary Materials, Figure S4) and 48 h, for the calculation of IC50 (Figure 6). Fraction D exhibited an IC50 of 57 μg/mL while that of fraction E was 54 μg/mL; the latter was selected for cell death pathway analysis due to its lower toxicity in normal cells and its higher fractionation yield with respect to fraction D.

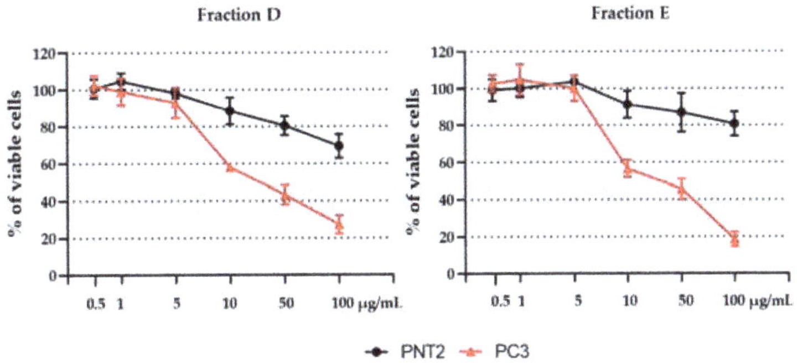

(a) (b)

Figure 6. Cell viability assay on PC3 and PNT2 cells after treatment for 48 h with 0.5, 1, 5, 10, 50, and 100 µg/mL of (**a**) fraction D and (**b**) fraction E. Assays were performed in biological triplicate, and graphs represent means ± standard deviations.

2.5. Chemical Composition of the Bioactive Fractions

Total extracts from the four different culture conditions were analysed by LC-MS in positive ion mode, thus unveiling a similar chemical profile. However, significant changes in the concentrations of the major compounds were clearly observed. Particularly, the most active extract (MIXO_CO_2) displayed higher amounts of the metabolites eluted in the time range between 12.5 and 14.8 min (Figure S5).

Afterwards, we performed an untargeted dereplication of the most active fractions, D and E, of the MIXO_CO_2 crude extract for the rapid identification of compounds responsible for the antiproliferative activity. Fractions D and E were dissolved in mass grade methanol at 1 mg/mL and subjected to liquid chromatography coupled with tandem mass spectrometry (LC-MS/MS) in the data-dependent (DDA) acquisition mode. The obtained MS raw files were processed by MZmine [23] and submitted to GNPS to build a molecular network with the Feature-Based Molecular Networking (FBMN) tool [24]. In the resulting molecular network (Figure 7), each node is representative of an ion detected in one or both fractions, the size of the node is proportional to the area of the extracted ion chromatogram, while the border colour of each node is mapped to the relevant chemical class of the metabolite.

The integration of the molecular networking and MS/MS fragmentation data analysis led to the structural prediction of almost 70 metabolites from the bioactive fractions D and E, which were assigned to five chemical classes, namely betaine diacylglycerols, betaine monoacylglycerols, glycerophospholipids, glycosylmonoacylglycerols, and fatty acids (Figure 7).

Overall, betaine lipids were shown to be the major clusters detected in the bioactive fractions, with betaine diacylglycerols being almost exclusively present in fraction E and betaine monoacylglycerols being the most abundant metabolite in fraction D. Betaine lipids, structural components of cell membranes and chloroplasts, are acylglycerolipids bearing an ether-linked quaternary amine alcohol moiety at the *sn*-3 position, which may be represented by a 2′-(hydroxymethyl)-(*N*,*N*,*N*-trimethyl)-β-alanine, a carboxy-(hydroxymethyl)-choline, or a 4′-(*N*,*N*,*N*-trimethyl)-homoserine. The product ion spectra generated from the $[M + H]^+$ ions of betaine lipids from fractions D and E were all characterized by the presence of the diagnostic fragment ion at m/z 236.1492 ($C_{10}H_{22}NO_5^+$), which is indicative of diacyl- and monoacyl-glycerol-trimethylhomoserine (DGTS and MGTS) or diacyl- and monoacyl-glyceryl-hydroxymethyl-*N*,*N*,*N*-trimethyl-β-alanine (DGTA and MGTA) [25]. Therefore, betaine lipids from the bioactive fractions were predicted to be DGTS and MGTS and/or DGTA and MGTA. Particularly, 39 DGTS/A (Table 2) and 17 MGTS/A (Table 3) were identified in fractions D and E of the MIXO_CO_2 crude extract.

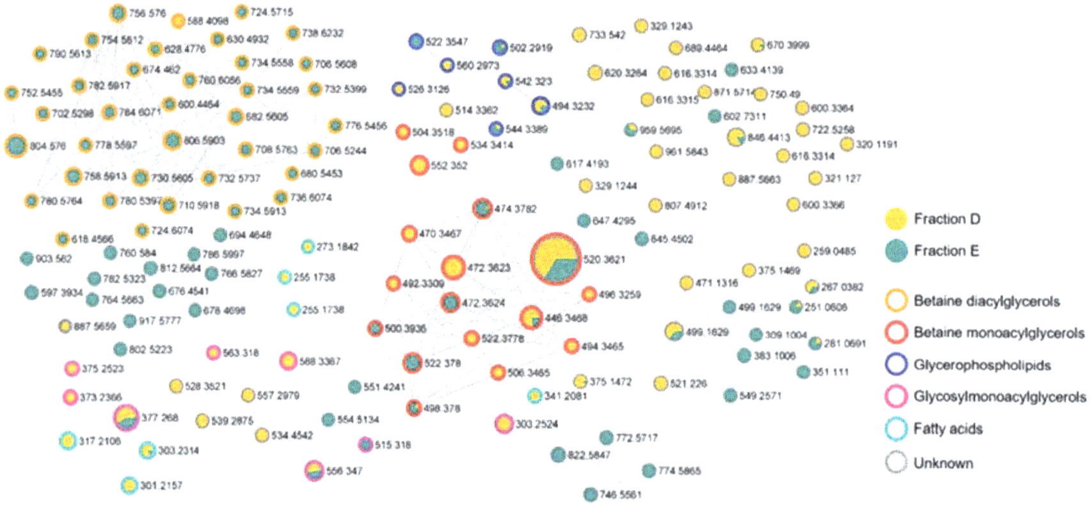

Figure 7. Molecular network obtained combining the LC-MS/MS analyses of fractions D and E of the MIXO_CO_2 extract from *Nannochloropsis granulata*. Nodes are illustrated as a pie chart showing the compound source (yellow: fraction D; green: fraction E). Node size is related to metabolite amount (peak area) while edge thickness reflects cosine score similarity. The colour of the border of each node is mapped to the chemical class assigned by the molecular networking analysis of the LC-MS/MS data.

Table 2. DGTS/A betaine lipids identified by molecular networking analysis of LC-MS/MS data from the bioactive fractions D and E of the MIXO_CO_2 crude extract (mass error < 3 ppm).

m/z	[M + H]$^+$	R_t	Fatty Acyl Chains [1]
674.4620	$C_{39}H_{64}O_8N$	31.3	9:1;O/20:5
600.4464	$C_{33}H_{62}O_8N$	31.3	9:1;O/14:0
588.4098	$C_{31}H_{58}O_9N$	31.3	5:1;O2/16:0
628.4776	$C_{35}H_{66}O_8N$	32.6	9:1;O/16:0
630.4932	$C_{35}H_{68}O_8N$	32.7	9:0;O/16:0
734.5558	$C_{43}H_{76}O_8N$	34.4	16:0/17:4;O
780.5397	$C_{47}H_{74}O_8N$	34.5	17:4;O/20:5
706.5244	$C_{41}H_{72}O_8N$	34.6	16:2/17:4;O
732.5399	$C_{43}H_{74}O_8N$	34.7	16:1/17:4;O
724.5715	$C_{42}H_{78}O_8N$	34.8	16:0/16:2;O
776.5456	$C_{48}H_{74}O_7N$	35.0	18:5/20:5
702.5298	$C_{42}H_{72}O_7N$	35.1	12:0/20:5
752.5455	$C_{43}H_{78}O_9N$	35.2	16:4/20:5
734.5559	$C_{43}H_{76}O_8N$	35.2	16:0/17:4;O
778.5597	$C_{48}H_{76}O_7N$	35.4	18:4/20:5
790.5613	$C_{49}H_{76}O_7N$	35.4	19:5/20:5
754.5612	$C_{46}H_{76}O_7N$	35.5	16:2/20:5
804.5760	$C_{50}H_{78}O_7N$	35.6	20:5/20:5
680.5453	$C_{40}H_{74}O_7N$	35.6	14:0/16:2
780.5764	$C_{48}H_{78}O_7N$	35.7	18:3/20:5
730.5605	$C_{44}H_{76}O_7N$	35.7	14:0/20:5
706.5608	$C_{42}H_{76}O_7N$	35.8	16:2/16:1
756.5760	$C_{46}H_{78}O_7N$	35.8	16:1/20:5
806.5903	$C_{50}H_{80}O_7N$	35.9	20:5/20:4
682.5605	$C_{40}H_{76}O_7N$	36.0	14:0/16:1
782.5917	$C_{48}H_{80}O_7N$	36.0	18:2/20:5
732.5737	$C_{44}H_{78}O_7N$	36.1	14:0/20:4

Table 2. *Cont.*

m/z	[M + H]+	R_t	Fatty Acyl Chains [1]
708.5763	$C_{42}H_{78}O_7N$	36.1	14:0/18:2
708.5763	$C_{42}H_{78}O_7N$	36.1	16:1/16:1
734.5913	$C_{44}H_{80}O_7N$	36.3	16:1/18:2
734.5913	$C_{44}H_{80}O_7N$	36.3	16:0/18:3
734.5913	$C_{44}H_{80}O_7N$	36.3	14:0/20:3
758.5913	$C_{46}H_{80}O_7N$	36.4	16:0/20:5
784.6071	$C_{48}H_{82}O_7N$	36.5	18:1/20:5
710.5918	$C_{42}H_{80}O_7N$	36.7	16:1/16:0
760.6056	$C_{46}H_{82}O_7N$	36.8	16:0/20:4
736.6074	$C_{44}H_{82}O_7N$	37.0	16:1/18:1
736.6074	$C_{44}H_{82}O_7N$	37.0	16:0/18:2
724.6074	$C_{43}H_{82}O_7N$	37.4	16:0/17:1

[1] Fatty acids have been reported using the LIPID MAPS shorthand notation [26]. Fatty acyl chains are indicated as C:N;O, where C is the number of carbon atoms, N is the number of double bond equivalents, and O is the number of additional oxygen atoms linked to the hydrocarbon chain.

Table 3. MGTS/A betaine lipids identified by molecular networking analysis of LC-MS/MS data from the bioactive fractions D and E of the MIXO_CO_2 crude extract (mass error < 3 ppm).

m/z	[M + H]+	R_t	Fatty Acyl Chain [1]
534.3414	$C_{30}H_{48}O_7N$	25.5	20:6;O
504.3518	$C_{26}H_{50}O_8N$	25.7	16:1;O2 [2]
496.3259	$C_{27}H_{46}O_7N$	26.1	17:5;O
552.3520	$C_{30}H_{50}O_8N$	26.5	20:5;O2
492.3309	$C_{28}H_{46}O_6N$	27.8	18:5
506.3465	$C_{29}H_{48}O_6N$	28.8	19:5
494.3465	$C_{28}H_{48}O_6N$	28.8	18:4
470.3467	$C_{26}H_{48}O_6N$	28.9	16:2
446.3468	$C_{24}H_{48}O_6N$	29.4	14:0
472.3624	$C_{26}H_{50}O_6N$	29.4	16:1
520.3621	$C_{30}H_{50}O_6N$	29.7	20:5
472.3623	$C_{26}H_{50}O_6N$	30.0	16:1
522.3780	$C_{30}H_{52}O_6N$	30.1	20:4
498.3780	$C_{28}H_{52}O_6N$	30.6	18:2
522.3778	$C_{30}H_{52}O_6N$	30.7	20:4
474.3782	$C_{26}H_{52}O_6N$	31.3	16:0
500.3936	$C_{28}H_{54}O_6N$	31.7	18:1

[1] Fatty acids have been reported using the LIPID MAPS shorthand notation [26]. Fatty acyl chains are indicated as C:N;O, where C is the number of carbon atoms, N is the number of double bond equivalents, and O is the number of additional oxygen atoms linked to the hydrocarbon chain. [2] MGTS/A with a putative hydroperoxyhexadecenoic acid as a fatty acyl substituent, as revealed by a neutral loss of 34.0055 Da from the $[M + H]^+$ ion, arising from fragmentation of the hydroperoxy group.

As DGTS and DGTA (as well as MGTS and MGTA) are structural isomers, and since no distinctive fragments could be observed in the tandem mass spectra of the $[M + H]^+$ adducts, it was not possible to differentiate unambiguously between the structures of DGTS and DGTA (as well as MGTS and MGTA). In general, the product ion spectra of MGTS/A were dominated by the $[M + H\text{-}H_2O]^+$ daughter ion and allowed us to infer the fatty acyl substituent by the presence of the $C_{10}H_{22}NO_5^+$ fragment ion, which derived from the loss of the fatty acyl chain as ketene (Figure S6). On the other hand, besides the presence of the $[C_{10}H_{22}NO_5]^+$ fragment, the mass tandem spectra of the $[M + H]^+$ ions of DGTS/A displayed fragment ions corresponding to the loss of each fatty acyl substituent at the *sn*-1 and *sn*-2 positions, both as ketene and carboxylic acid, thus giving information about their fatty acyl composition (Figure S7). The DGTS/A and MGTS/A detected in fractions D and E feature a remarkable amount of polyunsaturated fatty acids, including EPA (Tables 2 and 3). The structural prediction of the minor molecular clusters, i.e., glycosylmonoacylglycerols (Table S1), glycerophospholipids (Table S2), and fatty acids (Table S3), is presented in the Supplementary Material.

2.6. Cell Death Pathway

To establish the cell death signalling pathway induced by the bioactive fraction E, a comparative analysis of the gene expression levels was performed between PC3 treated with 54 μg/mL of fraction E (i.e., IC50 value) of MIXO_CO_2 for 3 h and under control conditions (PC3 cells without any treatment, in a complete RPMI medium). Figure 8 shows the relative expression ratios of the analysed genes after treatment with respect to control. Gene expression with a threshold fold regulation value of 2 was used to select genes differentially expressed between the test and control. After a 3 h treatment of PC3, fraction E provoked a strong up-regulation of some autophagy-related genes (orange bar graph), such as ULK1 (11.4-fold change), GAA (9.7-fold change), BECN1 (20.7-fold change), ATG5 (3.0-fold change), ATG16L1 (3.1-fold change), and ATG12 (2.7-fold change). At the same time, five autophagy-related genes were down-regulated: RPS6KB1 (−9.6-fold change), CASP3 (−9.8-fold change), ATG7 (8.4-fold change), ATG3 (−14.5-fold change), and APP (−15.5-fold change). Among the anti-apoptosis genes (green bar graph), one of them was down-regulated (AKT1, −13.3-fold change) while five of them were up-regulated: TNFRSF11 (9.2-fold change), MCL1 (3.2-fold change), CASP2 (6.0-fold change), BIRC3 (3.6-fold change), and BCL2A1 (4.6-fold change). All pro-apoptosis genes found differentially expressed were down-regulated, such as TNFRSF10 (−21.2-fold change), NOL3 (−3.7-fold change), GADD45A (−8.2-fold change), CASP9 (−10.7-fold change), BAX (−28.9-fold change), and APAF1 (−28.1-fold change). Only two genes involved in the necrosis death pathway (grey bar graph) were differentially expressed; SPATA2 was up-regulated (8.0-fold change), and CYLD was down-regulated (−14.4-fold change). NFKB1, CASP3, and AKT1 are genes involved in more than one death pathway investigated (light blue bar graph); NF-κB1 was up-regulated (18.6-fold change), while CASP3 and AKT1 were found to be down-regulated (−9.8 and −13.3-fold change, respectively).

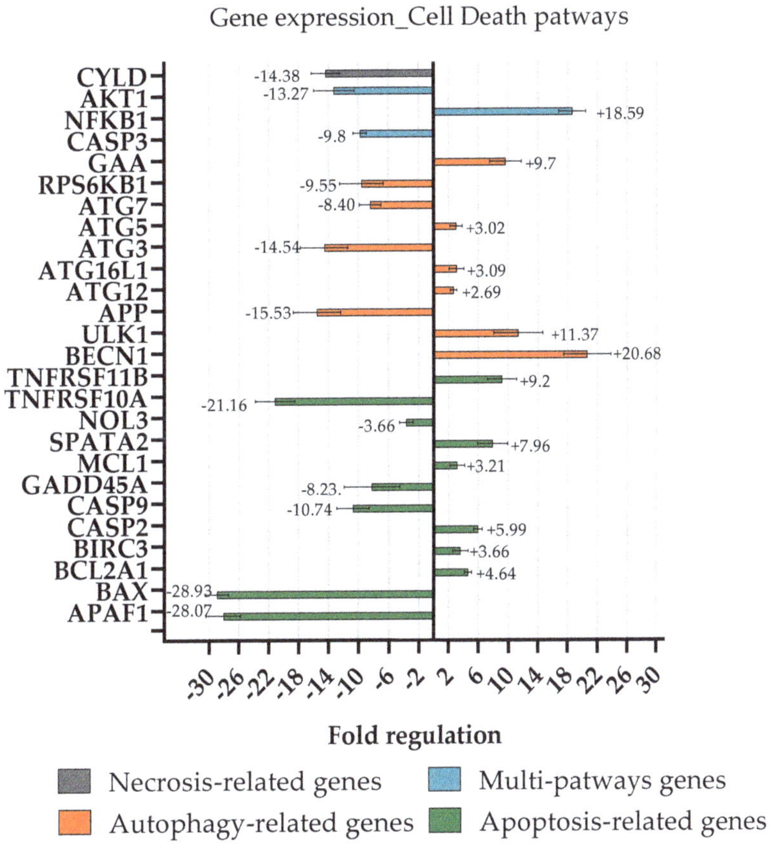

Figure 8. Variation of gene expression levels of death-related genes in PC3 treated with fraction E, with respect to gene expression levels of untreated PC3 cells. The p-values were calculated based on a Student's *t*-test of the replicate $2^{-\Delta CT}$ values for each gene in the control group and the treatment group; all fold regulations reported in this graph show *p*-values ≤ 0.05; thus, they were considered significant.

3. Discussion

Previous studies have investigated the mixotrophic growth of the *Nannochloropsis* genus using various external organic carbon sources [11–14], but only a few have explored the industrial potential of growing *Nannochloropsis* under this growth mode [12,14]. *Ng* has been previously proven as an industrially relevant strain due to its high biomass and its lipid productivity that is compatible with outdoor production during the summer season in Sweden [17]. In this study, we combined physiological, analytical, biological, and metabolic approaches to investigate mixotrophic growth and the biotechnological application of *Ng* grown under this trophic mode. To reduce the energetic cost required for the industrial exploitation of *Ng*, glycerol (i.e., a by-product of biofuel production) was used here as an external, low-cost carbon source. *Ng* was grown in 1L photobioreactors under four different conditions, namely phototrophy, mixotrophy, and bubbled either only with air (PHOTO_AIR and MIXO_AIR) or with CO_2-rich air (PHOTO_CO_2 and MIXO_CO_2). We found that the mixotrophic growth of *Ng* in the presence of both organic and inorganic carbon (MIXO_CO_2) was the best condition for increasing its industrial potential, as recently shown for the diatom *Phaeodactylum tricornutum* [27]. However, our results showed similar biomass productivity for *Ng* grown regardless of the presence of organic, inorganic, or both carbon sources; in contrast, a significant increase of biomass for *Phaeodactylum tricornutum* was found only in the presence of both carbon sources. Different culture conditions and different genus/species can explain these divergent results. Moreover, it was shown that a high concentration of CO_2 inhibits the mixotrophic growth of the *Nannochloropsis* genus [13,28]. Further studies in the presence of different concentrations of CO_2 could clarify the interaction between organic and inorganic carbon during this trophic mode in the *Ng* strain.

To investigate the industrial potential of *Ng*, the biomass composition obtained from the four different conditions was analysed. Our results showed that even if a similar macromolecular composition was maintained, the addition of either organic, inorganic, or both carbon sources (PHOTO_CO_2, MIXO_AIR, and MIXO_CO_2) doubled the fatty acid content as compared with the control PHOTO_AIR. This result was expected as both CO_2 and glycerol are involved in the metabolism of fatty acids in microalgae [19,29]. The increase in fatty acid content explains the increase in calorific value, from 18 to 25 MJ/kg of dry weight in these conditions, as compared to PHOTO_AIR. A calorific value of 25 MJ/kg was also reported for *Ng* in [17] as well as for *Nannochloropsis* sp. grown in Flat-Plate Photobioreactors under N, P starvation [30], thus confirming the genus as promising feedstock for biodiesel production.

Moreover, the fatty acid profile was also affected by culture conditions, showing the increase of both palmitoleic acid (16:1 (n7)) and EPA (20:5) under mixotrophic growth, especially in MIXO_CO_2, as compared to the phototrophic samples. The increase of EPA under mixotrophic growth using glycerol as a carbon source was already shown for *Nannochloropsis gaditana* and *Phaeodactylum tricornutum* [14,19]. EPA belongs to the class of omega-3 polyunsaturated fatty acids (PUFA) that is important for humans and animals that are not able to synthesize them and need to get them from the diet. Some microalgae, including *Nannochloropsis*, are rich in EPA, which suggests several applications in both the food and feed industries (e.g., nutraceuticals and aquaculture) [31]. The EPA content of the biomass was about 8% in the *Ng* grown in MIXO_CO_2, and it reached state-of-the-art EPA-concentration levels as compared to others *Nannochloropsis* species grown in similar conditions [14,32]. The *Nannochloropsis* species is also a good source of valuable carotenoids such as β-carotene and canthaxanthin [16]. The increased amount of these carotenoids in the *Ng* extracts PHOTO_CO_2 and MIXO_CO_2 (0.1 and 0.02 μg/mg dry weight, respectively) was previously reported with the addition of glycerol and of $NaHCO_3$ in other *Nannochloropsis* species [14,33].

Ng extracts displayed antitumoral activity in human prostatic cells (i.e., PC3), as previously found for different *Nannochloropsis* species tested in other human cell lines [8–10]. Interestingly, the PC3 cell viability under the conditions MIXO_AIR and MIXO_CO_2 decreased from 50% to 10–20% when treated with 100 μg/mL of total extracts as compared to PHOTO_AIR and PHOTO_CO_2. In order to further improve the antitumoral activity of *Ng*, the total extracts were fractionated using a polarity gradient elution. The most hydrophobic fractions of MIXO_CO_2 rich in betaine lipids (i.e., D and E, 75% and 100% methanol, respectively) showed increased antiproliferative activity on cancer cells. Indeed, cell viability was already lower than 60% when the cancer cells/PC3 were exposed to 10 μg/mL of each fraction; in contrast, no cytotoxicity was exhibited (cell viability up to 90%) in the normal, prostatic PNT2 cells, as observed for the relevant total extract. In addition, fractions D and E of MIXO_CO_2 were shown to exert highly selective antiproliferative activity towards tumor cells as compared to normal cells at 10 and 100 μg/mL. Similar results were obtained for *Nannochloropsis oculata* by [10], where the sterol-rich fractions showed the strongest antitumoral activity in HL-60 cancer cells at 24 h of treatment (i.e., 50% reduction of cell viability after treatment with 25 μg/mL of the active fractions).

LC-MS/MS analysis of active fractions D and E revealed almost 70 metabolites, which were attributed to five chemical classes: betaine diacylglycerols, betaine monoacylglycerols, glycerophospholipids, glycosylmonoacylglycerols, and fatty acids. In particular, betaine lipids were the most representative classes, with MGTS/A 20:5;O_2 (520.3621) as the most abundant metabolite of the monoacylglycerol group, and DGTS/A 20:5/20:5 (804.478) as the most abundant metabolite of the diacylglycerol group. Both betaine lipids included the polyunsaturated EPA, confirming our previous fatty acid analysis. Betaine lipids, eventually together with others, could be responsible for the antitumoral activity detected in *Ng* fractions, as observed in other organisms [34–36]. Finally, cell death pathways were investigated in order to understand the gene mechanism involved in the antiproliferative effect of these molecules on PC3 cells.

Almost all the genes involved in the activation or downstream process of apoptosis (APAF1 [37], BAX [38], CASP3 and 9 [39], GADD45A [40], NOL3 [41], and TNFRSF10A [42]) were downregulated, suggesting that this cell death pathway is not directly responsible for the high rate of cell death observed after the treatment of PC3. The only exception is SPATA2, a gene encoding for an adaptor protein recruited into the TNF-R1 signalling complex, which is involved in the regulation of RIPK1 [43].

Another evidence supporting the non-involvement of the apoptotic cell death is the upregulation of anti-apoptotic genes such as BCL2A1 [44], BIRC3 [45], and CASP2 [46].

The only gene involved in necrosis found to be differentially expressed was CYLD. CYLD, together with SPATA2, participates at the downstream events of TNF receptor activation and can activate apoptosis (via the CASP 8-FADD complex) or necroptosis (via the RIPK family) [47]. The downregulation of CYLD suggests that these two TNF-dependent processes are not involved in the process of cell death, also considering that only the upregulation of a single gene (SPATA2) has been observed.

Gene expression data related to autophagic factors support the activation of autophagy. GAA [48], ATG5, ATG16L1, ATG12 [49], RPS6KB1 [50], ULK1, and BECN1 [51] are involved in the activation of the autophagic process and in the formation of autophagosomes; these genes were found to be upregulated, except for RPS6KB1.

Additionally, ATG7, ATG3, and APP have been described to be involved in the autophagy cell death pathway; however, after treatment with fraction E, these three genes were downregulated. These can be explained by looking at the involvement of these three factors in the proliferation and migration of cancer cells; ATG7 [52], ATG3 [53], and APP [54] have been found to be upregulated in actively proliferating cancer cells. Thus, the downregulation of these factors together with the induction of the autophagic flux could contribute to the molecular mechanisms underlying the antiproliferative effects exerted by the lipid metabolites contained in fraction E.

To conclude, the simultaneous addition of glycerol and CO_2 in *Ng* could be applied in outdoor systems situated along the west coast of Sweden in order to enhance the industrial potential of this strain for different applications such as biofuel, food, feed, and drug production. The use of wastewater and the optimization of the CO_2 supply (e.g., flue gases) could further reduce the production cost and increase the biomass performance of this strain. However, the addition of organic carbon enhances the chances of competition between microalgae and complex flora [55]. Further pilot outdoor studies are needed to validate the data we obtained in the ePBRs during a simulated season on Sweden's west coast [55].

4. Materials and Methods

4.1. Microalgae Strain and Cultivation Conditions

Nannochloropsis granulata (*Ng*) was initially isolated by Karlson et al. (1996) [15] from the Skagerrak, northeast Atlantic Ocean. For the experiments in this study, it was obtained from the culture collection GUMACC (Gothenburg University Marine Algal Culture Collection, https://www.gu.se/en/marina-vetenskaper/about-us/algal-bank-gumacc, accessed on 1 November 2019). This strain was selected because it was found to be the most productive local strain on the Swedish west coast in summer conditions [17]. *Ng* was not axenic, but 100 μg/L of ampicillin was added at the beginning of the cultivation in order to control the bacterial growth.

Precultures were maintained in 100 mL flasks at 16 °C, with a light intensity of about 20 μmol photons m^{-2} s^{-1} and a 12/12 h L/D (Light/Dark) cycle. The medium used was natural seawater collected from a depth of 30 m at the Tjärnö Research Station, University of Gothenburg, Sweden. The seawater was filtered using two 0.4 μm GF/F glass fibre filters, the salinity was adjusted with deionized water to 26 practical salinity units, and it was sterilized by autoclaving at 121 °C for 20 min. Finally, nutrients from the standard f/2 marine cultivation medium ($NaNO_3$, NaH_2PO_4, microelements, vitamins [56]) were sterilized with cellulose filter paper (with pore size of 0.22 μm) and added to the autoclaved seawater.

4.1.1. Screening in Multi-Cultivator

For the small-scale experiments, *Ng* was grown in a Multi-cultivator MC 1000 OD (Photon System Instruments, Check Republic) using a constant white light with an intensity of 300 μmol photons m^{-2} s^{-1}, at a temperature of 20 °C, and with air bubbling (Figure S1a). These cultivation conditions were used because they correspond to the average light intensity and temperature during the summer season in Gothenburg. The Multi-Cultivator was used for small-scale growth and physiological characterisation thanks to its ability to simultaneously control growth in eight flasks containing 80 mL of liquid culture. The control was grown in phototrophic conditions without the addition of an external carbon source. For the mixotrophic conditions, 10 mM glycerol was added. All the samples were run in triplicate. The medium used was the same as that for the inoculum but using 14-fold concentrated nutrients ($NaNO_3$, NaH_2PO_4, microelements) and vitamins in order to obtain high biomass yields. The enrichment factor was calculated from the required amounts of nutrients to obtain at least 2 g/L of biomass based on Redfield ratio for marine phytoplankton [17].

Algal growth was monitored every two days by measuring chlorophyll *a* fluorescence expressed in relative fluorescence units (RFU), using a Varioscan™ Flash Multimode Reader (Thermo Fisher Scientific, Vantaa, Finland), in 96-well microplates [17]. A total of 250 μL of samples were added into each well of the microplate (in triplicate) and incubated for 10 min in the dark. Dilutions were performed when required (i.e., chlorophyll fluorescence values > 30). Chlorophyll fluorescence was detected using a wavelength of 425 nm for excitation and 680 nm for emission. The growth profiles in the different conditions were normalized as a function of ln (RFU_t/RFU_0), where RFU_t was the chlorophyll a fluorescence at a certain time (t), and RFU_0 was the initial chlorophyll a fluorescence. After the stationary phase was reached, the biomass yield was determined and expressed as g/L of dry weight. A total of 5 mL of final cultures was filtered through pre-weighted dried GF/F (47 mm) Whatman® filters and then washed with 10 mL of 0.5 M ammonium carbonate [57]. Finally, the filters containing the culture were incubated at 100 °C for 24 h and weighted for the determination of the dry weight (biomass yield) according to the following formula:

$$(\text{g of (filter + biomass)}) - (\text{g of filter})/0.005\ \text{L (volume of filtered culture)} \quad (1)$$

4.1.2. Cultivation in Environmental Photobioreactors

Following the first screening in the Multi-Cultivator, larger-scale (i.e., 1 L of liquid culture) cultivation was carried out in photobioreactors in order to collect more biomass for further analysis (i.e., biomass composition, pigment and fatty acid profile, bioassay analysis). The inoculum of *Ng* was grown in 1 L flasks, with 500 mL of culture at room temperature, with a light intensity of 150 μmol photons m^{-2} s^{-1} and a photoperiod of 12 h light/12 h dark, bubbled with air, and stirred at 120 rpm. A total of 100 mL of cells grown for 7 days was inoculated in 900 mL of cultivation medium in environmental photobioreactors (ePBRs) (Figure S1b), corresponding to an initial OD of about 0.1 and using the spectrophotometer Thermo Scientific Evolution 60 at a wavelength of 750 nm. Dilutions were performed for samples with OD 750 nm > 1. OD was monitored throughout the growth experiment; however, only chlorophyll fluorescence (RFU) was shown in order to discriminate microalgal from bacterial growth. The cultivation medium for ePBR experiments was prepared, as described for the screening in the Multi-Cultivator. The ePBRs were programmed for "summer conditions" based on records of air temperature, light intensity, and photoperiod in Gothenburg during the summers of 2014–2016, as designed in [17]. Here, four different conditions were tested: (1) PHOTO_AIR: growth with light in a medium injected with air (i.e., 0.04% CO_2); (2) MIXO_AIR: growth with light in a medium supplemented with 10 mM glycerol and injected with air (i.e., 0.04% CO_2); (3) PHOTO_CO_2: growth with light in a medium injected with 1–2% CO_2-enriched air; (4) MIXO_CO_2: growth with light in a medium injected with glycerol and injected with 1–2% CO_2-enriched air. The concentration of CO_2 in PHOTO_CO_2 and MIXO_CO_2 varied based on the need to maintain the pH values of cultures at 8 through the automatic injection of CO_2. The pH of the other cultures was maintained at the same value by manual injection of 0.4 N H_2SO_4 when needed. The layout of the photobioreactors is modular, and the picture in Figure S1b shows one of three modules in a setup consisting of four PBRs each. The complete setup allows for 12 PBRs in total, of which nine were used for these experiments. Two treatments were used in parallel, and each treatment was allotted from three to five replicate PBRs out of the nine. All three modules were located in a temperature-controlled enclosure. Custom-built microprocessor control modules for pH, light, gas mixing, etc., were located outside the enclosure. Light was controlled separately for each PBR according to the profile in Figure 2a. pH was controlled individually for each PBR through the high-frequency pulsed addition (1 Hz control loop) of a gas mixture, i.e., filtered air or 1–2% CO_2 in filtered air, dispersed through a capillary at the bottom of each PBR. Mixing was accomplished with magnetic stir bars set at 125 rpm. Temperature monitoring was accomplished through a temperature-controlled enclosure and by fine tuning with a water bath, which circulated water through the outer water jacket of each PBR.

Algal growth was monitored every two days by measuring chlorophyll *a* fluorescence expressed in relative fluorescence units (RFU) with the use of a fluorometer (Fluoromax 4, Jobin-Yvon, Horiba Scientific, Palaiseau, France). A total of 3 mL of samples was added in a quartz cuvette and incubated for 10 min in the dark. Dilutions were performed when required (i.e., Chlorophyll fluorescence values > 106). Chlorophyll fluorescence was detected using a wavelength of 425 nm for excitation and 680 nm for emission. The growth profiles in the different conditions were normalized, as described for the screening in Multi-Cultivator. Cultivation was stopped when the stationary phase was reached, and biomass was collected for further analyses.

4.2. Photosynthetic Analysis

Photosynthetic analysis was carried out with a pulse-amplitude-modulated fluorometer DUAL-PAM 100 equipped with a DUAL-DB and a DUAL-E emitter–detector module (Walz, Effeltrich, Germany). The photosynthetic parameter variable fluorescence/maximum fluorescence (F_v/F_m) was determined by measuring 2 mL of a 20 min dark-adapted algae sample with the use of saturated actinic red light (300 μmol photons m^{-2} s^{-1}). F_v/F_m represents the maximum quantum yield of PSII and gives an indication of the physiological state of photosynthetic organisms, where F_v is equal to $F_m - F_0$, and F_m and F_0 are the maximum and minimum fluorescence of the dark-adapted cells, respectively.

4.3. Biomass Analysis

The biomass yield was determined at the end of the growth curve, when approaching the stationary phase, as described for the screening in the Multi-Cultivator. Biomass productivity was calculated by dividing the biomass yield by the number of cultivation days. The maximum specific growth rate (μ_{max}) was calculated during the exponential phase (days 3–6) from RFU data as follows:

$$\mu_{max} = (\ln RFU_{d2} - \ln RFU_{d1})/(d2 - d1), \quad (2)$$

where RFU_{d2} and RFU_{d1} are the relative chlorophyll fluorescence on specific days of cultivation (d2 and d1).

4.3.1. CHN Analysis

On the last day of cultivation, the cells were collected by centrifugation at 7000× *g* for about 20 min, stored at −80 °C, and freeze-dried for 48 h using Alpha 1–2 LD plus,

Martin Christ. About 1 g of freeze-dried biomass was ground using a mortar and pestle for elemental (i.e., Carbon, Hydrogen, Nitrogen, CHN), ash content, and calorific value analyses. CHN and ash content analyses were performed using the standard methods SS-EN-ISO 16,948 with the Elemental Analyzer CHN 628, Leco and the SS-EN-ISO 18,122 with the thermogravimetric analyser TGA 701, Leco, respectively. Finally, the calorific value was determined using the standard method SS-EN-ISO 18,125 and the bomb calorimeter C5003, IKA. The ash-free biomass composition (protein, lipid, and carbohydrate content) was calculated by the equation used in [58].

The energy productivity was calculated by multiplying the biomass productivity with the calorific value.

4.3.2. Fatty Acid Profile

A certain amount (5–10 µg) of freeze-dried biomass was analysed for its fatty acid content by using direct acid transmethylation and Gas Chromatography/Mass Spectrometry (GC–MS) [59]. A total of 5 µg of di-nonadecanoyl-phosphatidylcholine (C19:0) was added to the freeze-dried biomass and used as an internal standard for the quantification. An amount of 2 mL of boiling 2-propanol was then added and incubated at 100 °C for 5 min. The sample was completely evaporated under a stream of N_2, and 1.5 mL of 2.5% H_2SO_4 in methanol (v/v) was added and incubated at 80 °C for 4 h and cooled down for 10 h. A total of 5 mL of 1 M cold NaCl and 1 mL of heptane were added to the samples and then mixed. The samples were centrifugated at 1000 rpm for 10 min to allow for the separation of the phases. The upper phase was collected and analysed using an Agilent 7820 GC (Agilent Technologies Co., Ltd., Shanghai, China) coupled to an Agilent 5975 MS (Agilent Technologies, Wilmington, DE, USA). The obtained fatty acid methyl esters (FAMEs) were separated on a 30 m × 0.25 mm DB-23 capillary column (Agilent), using helium as a carrier gas at a constant flow of 0.6 mL min^{-1} and a temperature of 210 °C. The FAMEs were identified by their comparison with commercial standards from Sigma-Aldrich, Darmstadt, Germany (Me 100, Me81, and individual FAME, Larodan, and Marine PUFA no.3) and quantified by comparison with the internal standard (C19:0). The concentration of FAMEs was then normalized for the freeze-dried biomass and expressed as µg/mg of dried weight.

4.3.3. Pigment Profile

A certain amount of freeze-dried (2–6 µg) biomass was resuspended in 5 mL of 90% (v/v) acetone contained in falcon tubes covered with aluminium foil to prevent the entry of light. The samples were ground in a glass homogenizer and refrigerated at 4 °C for 4 h. After the incubation period, the samples were centrifuged at 3000 rpm for 5 min. Up to 5 mL of the clear supernatant with 90% acetone was taken and used for pigment quantification. The samples were then filtered using a filter with a pore size of 0.2 µm prior to run. The pigment composition of the samples was obtained using HPLC PDA analysis. The samples were analysed in a Shimadzu UFLC system (Shimadzu corporation, Kyoto, Japan) loaded with an Alltima C18 (RP18, ODS, Octadecyl) 150 × 4.6 mm column, using 100 µL injection volume. The carotenoids and chlorophyll a were eluted through a low-pressure gradient system comprised of Solvent A with methanol and 0.5 M ammonium acetate buffer (85:15), solvent B with acetonitrile and milliQ water (90:10), and solvent C with 100% ethyl acetate. The program consisted of solvent 100% B:0% C: (8 min), 90% B:10% C: (8.6 min), 65% B:35% C (13.1 min), 31% B:69% C (21 min), and 100% B:0% C (27 min). Retention time and spectra obtained from standards (DHI, Hørsholm, Denmark) and run under the same conditions were used to identify the carotenoids in the samples. The pigment concentration was obtained using quantification based on the area of each standard. The pigment concentration was then normalized for freeze-dried biomass and expressed as µg/mg of dried weight.

4.4. Nutrient Analysis

Every 2 days, about 2 mL of growing cultures were filtrated using a nylon filter with a pore size of 0.22 μm. The samples were diluted 20 times with MilliQ water and stored at 4 °C until the nitrate (N) and phosphate (P) analysis. The diluted media were then analysed using the ion chromatographic system 882 Compact IC plus, coupled with the 858 Professional Sample Processor Metrohm AG, Herisau, Switzerland, and the anion exchange chromatographic column Metrosep Asup 5–250/4.0, Metrohm AG, Herisau, Switzerlan, using a conductivity detector (Metrohm AG, Herisau, Switzerland, part number 2.850.9010). In this analysis, an injection volume of 100 μL and an eluent of 3.2 mM Na_2CO_3 and 1 mM $NaHCO_3$ pumped at 0.7 mL/min were used.

The glycerol analysis was conducted using the same ion chromatographic system described above after replacing the column and detector with a Metrosep Carb 2-150/4.0 column and an IC amperometry detector (Metrohm AG, Herisau, Switzerland, part number 2.850.9110). The detector was used with palladium/gold electrodes in the pulsed amperometry mode. The injection volume was 20 μL, flowrate was at 0.5 mL min^{-1}, and 100 mM NaOH/10 mM sodium acetate was used as eluent.

N, P and glycerol (Gly) removal rates were calculated with the difference between the initial and final concentrations of these nutrients in the media, divided by the days of the experiment (d), and expressed as mg/mL/d.

4.5. Statistical Analysis

The biomass analysis and nutrient consumption for mixotrophic and phototrophic growth were compared by *t*-test analysis using GraphPad 9.3.1 Software 2365 Northside Dr. Suite 560, San Diego, CA 92108, USA. *p*-values were used to quantify the variability between the four different growth conditions. Data were considered significant for p-values < 0.05.

4.6. Extraction, Fractionation, and Liquid Chromatography—Mass Spectrometry

Freeze-dried biomasses obtained from the four growth conditions were re-suspended in 100% methanol and homogenised by using a glass pestle. Methanol (MeOH) was chosen for the extraction as it is a non-selective solvent and allows for the extraction of the largest possible number of metabolites with a wide polarity range [60]. The samples were kept in agitation and dark conditions for 60 min at room temperature, allowing for the complete extraction of intracellular metabolites. After this step, the samples were centrifuged at 6000× *g* for 10 min at 4 °C in order to discard cellular structures and collect only the supernatants. The supernatants were dried in a rotary evaporator to obtain the dried total extracts to be used for chemical analysis, bioactivity assays, and fractionation. Dried total extracts were re-suspended in methanol at 100 mg/mL and loaded onto the SPE polypropylene column CHROMABOND® C18 ec (column volume 6 mL, filling quantity 1000 mg) assembled on an SPE Vacuum system. The total extracts were separated into six fractions using methanol for a polarity gradient elution: fraction A, 100% H_2O; fraction B, 25% MeOH and 75% H_2O; fraction C, 50% MeOH and 50% H_20; fraction D, 75% MeOH and 25% H_2O; fraction E, 100% MeOH; fraction F, 100% MeOH containing 1% of trifluoroacetic acid. All eluted fractions were dried in a rotary evaporator and stored at −20 °C for further chemical and biological analyses.

In order to outline a metabolomic overview, the crude extracts from the four different culture conditions, i.e., MIXO_AIR, MIXO_CO_2, PHOTO_AIR, and PHOTO_CO_2, were dissolved in mass grade MeOH at a concentration of 1 mg/mL and analysed by LC/MS, with a mass/charge range of 150–1000 m/z on a QTRAP 4500 (SCIEX, Framingham, MA, USA) connected to a Nexera X2 UHPLC (Shimadzu, Kyoto, Japan), which was equipped with a 1.7 m Acquity UPLC BEH C18 column (2.1 × 50 mm). For these experiments, the ESI source was set in positive mode, with the voltage set at 4.5 kV and the capillary temperature at 285 °C. Buffer A (H_2O + 0.1% Formic acid (FA)) and Buffer B (ACN 0.1% + FA) were used for the chromatographic separation, which was executed by adopting the following gradient: from 100% A to 100% B in 20 min, followed by 5 min of 100% B.

The LC-HRMS/MS data-dependent analyses (DDA) of the SPE fractions D and E were carried out on a Thermo LTQ Orbitrap XL (Thermo Fisher Scientific, Waltham, MA, USA) with ESI source coupled to a Thermo U3000 HPLC system equipped with a 5 m Kinetex C18 column (2.1 × 50 mm). The DDA LC-MS/MS experiments were conducted by dissolving the SPE fractions at 1 mg/mL in mass grade MeOH. The gradient program was set as follows: 10% MeOH 1 min, 10−100% MeOH over 30 min, 100% MeOH 10 min. Mass spectra were acquired in the positive ion detection mode. MS parameters were as follows: a spray voltage of 4.8 kV, a capillary temperature of 285 °C, a sheath gas rate of 32 units N_2 (ca. 150 mL/min), and an auxiliary gas rate of 15 units N_2 (ca. 50 mL/min). Data were collected in the DDA mode, in which the five most intense ions of a full-scan mass spectrum were subjected to $HRMS^2$ fragmentation. The m/z range for DDA was set between 150 and 2000 amu. $HRMS^2$ scans were obtained with CID fragmentation, an isolation width of 2.0, normalized collision energy of 35, activation Q of 0.250, and an activation time of 30 ms.

4.7. Antiproliferative Screening and Dose-Response Curve

Antiproliferative experiments were performed on two different cell lines: PC3—human prostate adenocarcinoma (purchased from the American Type Culture Collection, ATCC, product code: CRL-1435™) and PNT2—human normal prostate epithelium immortalized with SV40 (purchased from the Sigma Aldrich, Burlington, MA, USA; product code: 95012613). The selection of a cancer cell line and its normal counterpart is essential as it reveals a selective antiproliferative activity towards cancer cells that has been induced by treatments. PC3 and PNT2 cells were grown in an RPMI (Roswell Park Memorial Institute) medium 1640 completed with 10% FBS (Fetal Bovine serum). Penicillin (100 units/mL) and streptomycin (100 µg/mL) were added to the cell medium. Cells were grown in a 5% CO_2 atmosphere at 37 °C and allowed to reach a maximum confluence of 80% in cell culture flasks with vented filter caps. Before the treatments, the cells were harvested with trypsin (1X), counted, seeded in 96-well plates (2×10^4 cells × $well^{-1}$, with a final volume of 100 µL for each well), and incubated in 5% CO_2 atmosphere at 37 °C overnight. Total extracts (four dried methanolic extracts of microalgal biomasses obtained from the MIXO_AIR, MIXO_CO_2, PHOTO_AIR, and PHOTO_CO_2 growth conditions) and fractions (six dried fractions obtained after SPE fractionation of each total extract) were dissolved in dimethyl sulfoxide (DMSO) and used for all cell treatments. The final concentration of the DMSO used was 0.5% (*v*/*v*) for each treatment. Cells were treated in biological triplicate (three technical replicates were set up for each biological replicate) with all samples (total extracts and fractions) at 1, 10, and 100 µg/mL, for 48 h in a complete cell medium. Two fractions (D and E, Mixo_CO_2) were selected for a further viability assay since they were able to induce the strongest and most selective antiproliferative effect on prostatic cancer cells; thus, they were used to set up a dose–response curve on both cell lines. In this case, the concentrations used were 0.5, 1, 5, 10, 50, and 100 µg/mL, for 24 and 48 h. Control cells were incubated in a complete cell medium with 0.5% of DMSO for all experiments.

Cell Viability

The antiproliferative effect of the samples on cell viability was evaluated using the 3-(4,5-Dimethylthiazol-2-yl)-2,5-diphenyl tetrazolium bromide (MTT) assay. Briefly, at the end of the incubation of the PC3 and PNT2 cells, with the total extracts and all fractions for antiproliferative screening and fractions D and E for the dose–response curve (see Section 4.7), cell culture media (complete RPMI media containing extracts and fractions at different concentrations) were discarded from 96-well plates using a vacuum aspirator system. In each well, fresh media containing 5 µg/mL of the MTT solution were added. Plates were Incubated in a 5% CO_2 atmosphere for 3 h at 37 °C. After incubation, the MTT solution was removed using a vacuum aspirator system, and formazan salts produced by viable cells were dissolved in an isopropanol solution (100 µL) and incubated at room temperature for 30 min on an orbital shaker. The absorbance of each well was read at 570 nm using an Infinite M1000Pro (TECAN, Männedorf, Switzerland) plate reader. The antiproliferative effect of the extracts and fractions at different concentrations was reported as percent of cell viability, calculated as the ratio between the mean absorbance of each treatment and the mean absorbance of the control (cells treated with only 0.5% of DMSO).

4.8. LC-HRMS2 Data Processing and Molecular Networking

LC-HRMS^2 data from the bioactive fractions D and E were processed together to generate a molecular network, using a previously reported method [61,62]. MS raw files were imported into MZmine 2.53 [23]. Mass detection from raw files was performed for mass levels 1 and 2 by setting the noise level at 1000 and 100, respectively. Chromatograms were built by using the ADAP chromatogram algorithm, setting a minimum height of 1000 and an m/z tolerance of 0.05 (or 20 ppm). The baseline cut-off algorithm was employed for chromatogram deconvolution with the following parameters: minimum height peak = 1000, peak duration range = 0.0–10.0 min, baseline level = 100, m/z range for MS^2 scan = 0.05, retention time range = 0.5 min. Chromatogram peaks were aligned by the Join aligner algorithm with the following settings: m/z tolerance at 0.05 or 20 ppm, absolute RT tolerance at 0.5 min. Peaks without associated MS tandem spectra were removed from the peak list. Processed mass data were exported to mgf file and submitted to the Feature Based Molecular Networking (FBMN) tool to generate the molecular network depicted in Figure 7. FBMN parameters were set as follows: precursor ion mass tolerance = 0.02, fragment ion mass tolerances = 0.02 Da, cosine score $\geq$ 0.7, minimum matched fragment ions = 3. The molecular network was visualized in Cytoscape version 3.7.2 (Cytoscape Consortium, San Diego, CA, USA) [63]. Chromatographic data were exported as a csv file from processed LC/MS data by MZmine and then mapped to the relevant nodes in the generated network (Available online: https://gnps.ucsd.edu/ProteoSAFe/status.jsp?task=d0d43c1f93cc4a5db75020e59299334c (accessed on 4 April 2022).

4.9. RNA Extraction and RT^2 Profiler PCR Array

Prior to the RNA extraction, PC3 cells were seeded in 6-well culture plates (2×10^5 cells $\times$ well^{-1}, with a final volume of 3 mL for each well) using a complete RPMI medium and incubated in 5% CO_2 atmosphere at 37 °C overnight. PC3 cells were treated with 52 µg/mL (corresponding to IC50 concentration) of active fraction E; control condition was also set up using PC3 cells in a complete RPMI medium. After 3 h of treatment, media were discarded from control and treated PC3 cells using a vacuum aspirator system, and then PC3 cells were washed directly into the wells by adding PBS and rocking gently. PC3 cells (control and treated) were lysed directly into the wells by adding 0.5 mL $\times$ well of Trisure Reagent (Bioline). RNA was isolated according to the manufacturer's protocol. RNA concentration and purity were assessed using the nanophotomer Nanodrop (Euroclone). A total of 400 ng of RNA was subjected to reverse transcription reaction using the RT^2 first strand kit (Qiagen, cat. 330401, Hilden, Germany) according to the manufacturer's instructions. Real-time quantitative reverse transcription-PCR (qRT-PCR) was performed in biological triplicate using the RT^2 Profiler PCR Array kit (Qiagen, cat.330231) in order to analyse the expression of 84 cell death genes in PC3 after exposure to the active fraction E. Plates were run on a ViiA7 (Applied Biosystems 384 well blocks, Waltham, MA, USA), Standard Fast PCR Cycling protocol with 10 µL reaction volumes. The cycling conditions used were 1 cycle initiation at 95.0 °C for 10 min, followed by amplification for 40 cycles at 95.0 °C for 15 s and 60.0 °C for 1 min. Amplification data were collected with the ViiA 7 RUO Software (Applied Biosystems). Ct values were analysed with the Qiagen data analysis online software (Available online: https://geneglobe.qiagen.com/it/analyze (accessed on 22 December 2021).

Supplementary Materials: The following supporting information can be downloaded at: https://www.mdpi.com/article/10.3390/md20070424/s1, Figure S1: Cultivation systems used in the experiments. Figure S2: Growth profile of *Nannochloropsis granulata* under mixotrophy and phototrophy; Figure S3: Cell viability results after treatments of PC3 and PNT2 with 1 µg/mL of fractions (A–F); Figure S4: Cell viability assay on PC3 and PNT2 cells after treatment for 24 h with fraction D and E; Figure S5: Base peak chromatograms of crude extracts from *Nannochloropsis granulata*. Figure S6: HR ESI-MS^2 spectra of the

[M + H]$^+$ ion of MGTS/A 20:5; Figure S7: HR ESI-MS2 spectra of the [M + H]$^+$ ion of DGTS/A 14:0/20:5; Table S1: Glycosylmonoacylglycerols from the bioactive fractions D and E from *Nannochloropsis granulata*; Table S2: Glycerophospholipids from the bioactive fractions D and E from *Nannochloropsis granulata*; Table S3: Fatty acids from the bioactive fractions D and E from *Nannochloropsis granulata*. Table S4: List of all cell death genes analysed and involved in apoptosis, necrosis, and autophagy. Reference [64] are cited in the supplementary materials.

Author Contributions: V.V., C.G., S.E. and C.S. conceived the project and contributed to the funding acquisition; V.V. performed the algal growth, biomass determination and collection, pigment and fatty acid extraction, and analysed the datasets. C.G. and V.V. performed the bioassay and gene expression analysis and analysed the relative datasets. C.G. and F.P.E. performed the fractionating of the biomass. G.A.V. and G.D.S. performed the metabolomics analysis and analysed the relative datasets. J.E. designed the simulated Nordic seasons, built the PBRs, and performed the nutrient analyses. N.S. performed the glycerol analyses. K.M.S. performed the pigment analysis. M.X.A. performed the fatty acids analyses. C.S., S.E. and D.D.P. supervised parts of the project in the corresponding institutes. V.V., with contributions from C.G., G.A.V., G.D.S. and C.S., wrote the manuscript draft. All authors have read and agreed to the published version of the manuscript.

Funding: This research was funded by supported by the European Union's Horizon 2020 research and innovation programme under the Marie Sklodowska-Curie Grant Agreement No. 844909. KMS was a recipient of a postdoctoral fellowship from the Carl Tryggers Foundation CTS 20:406.

Institutional Review Board Statement: Not applicable.

Informed Consent Statement: Not applicable.

Acknowledgments: We thank Olga Kourtchenko (Department of Marine Sciences, University of Gothenburg) for providing the *Nannochloropsis granulata* strain from the GUMACC collection. We sincerely thank Mathias Berglund for the CHN analysis and the calorific value determination performed at RISE.

Conflicts of Interest: The authors declare no conflict of interest.

References

1. Green, B.R. After the Primary Endosymbiosis: An Update on the Chromalveolate Hypothesis and the Origins of Algae with Chl c. *Photosynth. Res.* **2011**, *107*, 103–115. [CrossRef] [PubMed]
2. Villanova, V.; Spetea, C. Mixotrophy in Diatoms: Molecular Mechanism and Industrial Potential. *Physiol. Plant.* **2021**, *173*, 603–611. [CrossRef] [PubMed]
3. Ma, Y.; Wang, Z.; Yu, C.; Yin, Y.; Zhou, G. Evaluation of the Potential of 9 *Nannochloropsis* Strains for Biodiesel Production. *Bioresour. Technol.* **2014**, *167*, 503–509. [CrossRef] [PubMed]
4. Taleb, A.; Pruvost, J.; Legrand, J.; Marec, H.; Le-Gouic, B.; Mirabella, B.; Legeret, B.; Bouvet, S.; Peltier, G.; Li-Beisson, Y.; et al. Development and Validation of a Screening Procedure of Microalgae for Biodiesel Production: Application to the Genus of Marine Microalgae *Nannochloropsis*. *Bioresour. Technol.* **2015**, *177*, 224–232. [CrossRef]
5. Tibbetts, S.M.; Yasumaru, F.; Lemos, D. In Vitro Prediction of Digestible Protein Content of Marine Microalgae (*Nannochloropsis granulata*) Meals for Pacific White Shrimp (*Litopenaeus vannamei*) and Rainbow Trout (*Oncorhynchus mykiss*). *Algal Res.* **2017**, *21*, 76–80. [CrossRef]
6. Boussiba, S.; Vonshak, A.; Cohen, Z.; Avissar, Y.; Richmond, A. Lipid and Biomass Production by the Halotolerant Microalga *Nannochloropsis Salina*. *Biomass* **1987**, *12*, 37–47. [CrossRef]
7. Marcilla, A.; Gómez-Siurana, A.; Gomis, C.; Chápuli, E.; Catalá, M.C.; Valdés, F.J. Characterization of Microalgal Species through TGA/FTIR Analysis: Application to *Nannochloropsis* sp. *Thermochim. Acta* **2009**, *484*, 41–47. [CrossRef]
8. Hussein, H.A.; Mohamad, H.; Mohd Ghazaly, M.; Laith, A.A.; Abdullah, M.A. Anticancer and Antioxidant Activities of *Nannochloropsis oculata* and *Chlorella* sp. Extracts in Co-Application with Silver Nanoparticle. *J. King Saud Univ.-Sci.* **2020**, *32*, 3486–3494. [CrossRef]
9. Samarakoon, K.W.; Ko, J.Y.; Shah, M.M.R.; Lee, J.H.; Kang, M.C.; O-Nam, K.; Lee, J.B.; Jeon, Y.J. In Vitro Studies of Anti-Inflammatory and Anticancer Activities of Organic Solvent Extracts from Cultured Marine Microalgae. *Algae* **2013**, *28*, 111–119. [CrossRef]
10. Sanjeewa, K.K.A.; Fernando, I.P.S.; Samarakoon, K.W.; Lakmal, H.H.C.; Kim, E.A.; Kwon, O.N.; Dilshara, M.G.; Lee, J.B.; Jeon, Y.J. Anti-Inflammatory and Anti-Cancer Activities of Sterol Rich Fraction of Cultured Marine Microalga *Nannochloropsis oculata*. *Algae* **2016**, *31*, 277–287. [CrossRef]

11. Bo, D.D.; Magneschi, L.; Bedhomme, M.; Billey, E.; Deragon, E.; Storti, M.; Menneteau, M.; Richard, C.; Rak, C.; Lapeyre, M.; et al. Consequences of Mixotrophy on Cell Energetic Metabolism in Microchloropsis Gaditana Revealed by Genetic Engineering and Metabolic Approaches. *Front. Plant Sci.* **2021**, *12*, 628684. [CrossRef] [PubMed]
12. Hu, H.; Gao, K. Optimization of Growth and Fatty Acid Composition of a Unicellular Marine Picoplankton, *Nannochloropsis* sp., with Enriched Carbon Sources. *Biotechnol. Lett.* **2003**, *25*, 421–425. [CrossRef] [PubMed]
13. Sforza, E.; Cipriani, R.; Morosinotto, T.; Bertucco, A.; Giacometti, G.M. Excess CO_2 Supply Inhibits Mixotrophic Growth of *Chlorella protothecoides* and *Nannochloropsis salina*. *Bioresour. Technol.* **2012**, *104*, 523–529. [CrossRef] [PubMed]
14. Menegol, T.; Romero-Villegas, G.I.; López-Rodríguez, M.; Navarro-López, E.; López-Rosales, L.; Chisti, Y.; Cerón-García, M.C.; Molina-Grima, E. Mixotrophic Production of Polyunsaturated Fatty Acids and Carotenoids by the Microalga *Nannochloropsis gaditana*. *J. Appl. Phycol.* **2019**, *31*, 2823–2832. [CrossRef]
15. Cheregi, O.; Ekendahl, S.; Engelbrektsson, J.; Strömberg, N.; Godhe, A.; Spetea, C. Microalgae Biotechnology in Nordic Countries—The Potential of Local Strains. *Physiol. Plant.* **2019**, *166*, 438–450. [CrossRef]
16. Karlson, B.; Potter, D.; Kuylenstierna, M.; Andersen, R.A. Ultrastructure, Pigment Composition, and 18S RRNA Gene Sequence for *Nannochloropsis granulata* sp. nov. (Monodopsidaceae, Eustigmatophyceae), a Marine Ultraplankter Isolated from the Skagerrak, Northeast Atlantic Ocean. *Phycologia* **1996**, *35*, 253–260. [CrossRef]
17. Cheregi, O.; Engelbrektsson, J.; Andersson, M.X.; Strömberg, N.; Ekendahl, S.; Godhe, A.; Spetea, C. Marine Microalgae for Outdoor Biomass Production—A Laboratory Study Simulating Seasonal Light and Temperature for the West Coast of Sweden. *Physiol. Plant.* **2021**, *173*, 543–554. [CrossRef]
18. Monteiro, M.R.; Kugelmeier, C.L.; Pinheiro, R.S.; Batalha, M.O.; da Silva César, A. Glycerol from Biodiesel Production: Technological Paths for Sustainability. *Renew. Sustain. Energy Rev.* **2018**, *88*, 109–122. [CrossRef]
19. Villanova, V.; Fortunato, A.E.; Singh, D.; Bo, D.D.; Conte, M.; Obata, T.; Jouhet, J.; Fernie, A.R.; Marechal, E.; Falciatore, A.; et al. Investigating Mixotrophic Metabolism in the Model Diatom Phaeodactylum Tricornutum. *Philos. Trans. B* **2017**, *372*, 20160404. [CrossRef]
20. Hammer, K.J.; Kragh, T.; Sand-Jensen, K. Inorganic Carbon Promotes Photosynthesis, Growth, and Maximum Biomass of Phytoplankton in Eutrophic Water Bodies. *Freshw. Biol.* **2019**, *64*, 1956–1970. [CrossRef]
21. Conde, T.A.; Neves, B.F.; Couto, D.; Melo, T.; Neves, B.; Costa, M.; Silva, J.; Domingues, P.; Domingues, M.R. Microalgae as Sustainable Bio-Factories of Healthy Lipids: Evaluating Fatty Acid Content and Antioxidant Activity. *Mar. Drugs* **2021**, *19*, 357. [CrossRef] [PubMed]
22. Mucci, L.A.; Wilson, K.M.; Giovannucci, E.L. Epidemiology of Prostate Cancer. *Pathol. Epidemiol. Cancer* **2016**, *10*, 107–125. [CrossRef]
23. Pluskal, T.; Castillo, S.; Villar-Briones, A.; Orešič, M. MZmine 2: Modular Framework for Processing, Visualizing, and Analyzing Mass Spectrometry-Based Molecular Profile Data. *BMC Bioinform.* **2010**, *11*, 395. [CrossRef] [PubMed]
24. Nothias, L.F.; Petras, D.; Schmid, R.; Dührkop, K.; Rainer, J.; Sarvepalli, A.; Protsyuk, I.; Ernst, M.; Tsugawa, H.; Fleischauer, M.; et al. Feature-Based Molecular Networking in the GNPS Analysis Environment. *Nat. Methods* **2020**, *17*, 905–908. [CrossRef]
25. Li, Y.; Lou, Y.; Mu, T.; Xu, J.; Zhou, C.; Yan, X. Simultaneous Structural Identification of Diacylglyceryl-N-Trimethylhomoserine (DGTS) and Diacylglycerylhydroxymethyl-N,N,N-Trimethyl-β-Alanine (DGTA) in Microalgae Using Dual Li+/H+ Adduct Ion Mode by Ultra-Performance Liquid Chromatography/Quadrupole Ti. *Rapid Commun. Mass Spectrom.* **2017**, *31*, 457–468. [CrossRef]
26. Liebisch, G.; Fahy, E.; Aoki, J.; Dennis, E.A.; Durand, T.; Ejsing, C.S.; Fedorova, M.; Feussner, I.; Griffiths, W.J.; Köfeler, H.; et al. Update on LIPID MAPS Classification, Nomenclature, and Shorthand Notation for MS-Derived Lipid Structures. *J. Lipid Res.* **2020**, *61*, 1539–1555. [CrossRef]
27. Villanova, V.; Singh, D.; Pagliardini, J.; Fell, D.; Le Monnier, A.; Finazzi, G.; Poolman, M. Boosting Biomass Quantity and Quality by Improved Mixotrophic Culture of the Diatom Phaeodactylum Tricornutum. *Front. Plant Sci.* **2021**, *12*, 1–14. [CrossRef]
28. Razzak, S.A.; Ilyas, M.; Ali, S.A.M.; Hossain, M.M. Effects of CO_2 Concentration and PH on Mixotrophic Growth of *Nannochloropsis oculata*. *Appl. Biochem. Biotechnol.* **2015**, *176*, 1290–1302. [CrossRef]
29. Sun, Z.; Chen, Y.F.; Du, J. Elevated CO_2 Improves Lipid Accumulation by Increasing Carbon Metabolism in Chlorella Sorokiniana. *Plant Biotechnol. J.* **2016**, *14*, 557–566. [CrossRef]
30. Hulatt, C.J.; Wijffels, R.H.; Bolla, S.; Kiron, V. Production of Fatty Acids and Protein by *Nannochloropsis* in Flat-Plate Photobioreactors. *PLoS ONE* **2017**, *12*, e0170440. [CrossRef]
31. Guschina, I.A.; Harwood, J.L. Lipids and Lipid Metabolism in Eukaryotic Algae. *Prog. Lipid Res.* **2006**, *45*, 160–186. [CrossRef] [PubMed]
32. Shene, C.; Chisti, Y.; Vergara, D.; Burgos-Díaz, C.; Rubilar, M.; Bustamante, M. Production of Eicosapentaenoic Acid by *Nannochloropsis oculata*: Effects of Carbon Dioxide and Glycerol. *J. Biotechnol.* **2016**, *239*, 47–56. [CrossRef] [PubMed]
33. Salbitani, G.; Del Prete, F.; Carfagna, S.; Sansone, G.; Barone, C.M.A. Enhancement of Pigments Production by *Nannochloropsis oculata* Cells in Response to Bicarbonate Supply. *Sustainability* **2021**, *13*, 11904. [CrossRef]
34. Heavisides, E.; Rouger, C.; Reichel, A.F.; Ulrich, C.; Wenzel-Storjohann, A.; Sebens, S.; Tasdemir, D. Seasonal Variations in the Metabolome and Bioactivity Profile of Fucus Vesiculosus Extracted by an Optimised, Pressurised Liquid Extraction Protocol. *Mar. Drugs* **2018**, *16*, 503. [CrossRef] [PubMed]

35. Da Costa, E.; Domingues, P.; Melo, T.; Coelho, E.; Pereira, R.; Calado, R.; Abreu, M.H.; Domingues, M.R. Lipidomic Signatures Reveal Seasonal Shifts on the Relative Abundance of High-Valued Lipids from the Brown Algae Fucus Vesiculosus. *Mar. Drugs* **2019**, *17*, 335. [CrossRef]
36. Da Costa, E.; Melo, T.; Moreira, A.S.P.; Bernardo, C.; Helguero, L.; Ferreira, I.; Cruz, M.T.; Rego, A.M.; Domingues, P.; Calado, R.; et al. Valorization of Lipids from Gracilaria Sp. through Lipidomics and Decoding of Antiproliferative and Anti-Inflammatory Activity. *Mar. Drugs* **2017**, *15*, 62. [CrossRef]
37. Shakeri, R.; Kheirollahi, A.; Davoodi, J. Apaf-1: Regulation and Function in Cell Death. *Biochimie* **2017**, *135*, 111–125. [CrossRef]
38. Peña-Blanco, A.; García-Sáez, A.J. Bax, Bak and beyond—Mitochondrial Performance in Apoptosis. *FEBS J.* **2018**, *285*, 416–431. [CrossRef]
39. Brentnall, M.; Rodriguez-Menocal, L.; de Guevara, R.L.; Cepero, E.; Boise, L.H. Caspase-9, Caspase-3 and Caspase-7 Have Distinct Roles during Intrinsic Apoptosis. *BMC Cell Biol.* **2013**, *14*, 32. [CrossRef]
40. Kleinsimon, S.; Longmuss, E.; Rolff, J.; Jäger, S.; Eggert, A.; Delebinski, C.; Seifert, G. GADD45A and CDKN1A Are Involved in Apoptosis and Cell Cycle Modulatory Effects of ViscumTT with Further Inactivation of the STAT3 Pathway. *Sci. Rep.* **2018**, *8*, 5750. [CrossRef]
41. Gustafsson, Å.B.; Tsai, J.G.; Logue, S.E.; Crow, M.T.; Gottlieb, R.A. Apoptosis Repressor with Caspase Recruitment Domain Protects against Cell Death by Interfering with Bax Activation. *J. Biol. Chem.* **2004**, *279*, 21233–21238. [CrossRef] [PubMed]
42. Rossin, A.; Derouet, M.; Abdel-Sater, F.; Hueber, A.O. Palmitoylation of the TRAIL Receptor DR4 Confers an Efficient TRAIL-Induced Cell Death Signalling. *Biochem. J.* **2009**, *419*, 185–192. [CrossRef] [PubMed]
43. Schlicher, L.; Brauns-Schubert, P.; Schubert, F.; Maurer, U. SPATA2: More than a Missing Link. *Cell Death Differ.* **2017**, *24*, 1142–1147. [CrossRef] [PubMed]
44. Vogler, M. BCL2A1: The Underdog in the BCL2 Family. *Cell Death Differ.* **2012**, *19*, 67–74. [CrossRef]
45. Gressot, L.V.; Doucette, T.; Yang, Y.; Fuller, G.N.; Manyam, G.; Rao, A.; Latha, K.; Rao, G. Analysis of the Inhibitors of Apoptosis Identifies BIRC3 as a Facilitator of Malignant Progression in Glioma. *Oncotarget* **2017**, *8*, 12695–12704. [CrossRef]
46. Brown-Suedel, A.N.; Bouchier-Hayes, L. Caspase-2 Substrates: To Apoptosis, Cell Cycle Control, and Beyond. *Front. Cell Dev. Biol.* **2020**, *8*. [CrossRef]
47. Wagner, S.A.; Satpathy, S.; Beli, P.; Choudhary, C. SPATA 2 Links CYLD to the TNF -α Receptor Signaling Complex and Modulates the Receptor Signaling Outcomes. *EMBO J.* **2016**, *35*, 1868–1884. [CrossRef]
48. Nascimbeni, A.C.; Fanin, M.; Masiero, E.; Angelini, C.; Sandri, M. The Role of Autophagy in the Pathogenesis of Glycogen Storage Disease Type II (GSDII). *Cell Death Differ.* **2012**, *19*, 1698–1708. [CrossRef]
49. Walczak, M.; Martens, S. Dissecting the Role of the Atg12-Atg5-Atg16 Complex during Autophagosome Formation. *Autophagy* **2013**, *9*, 424–425. [CrossRef]
50. Ma, L.; Zhang, D.; Huang, Z.; Zheng, R.; Du, M.; Lv, Q.; Qin, C.; Chu, H.; Yuan, L.; Zhang, Z. Functional Variants of RPS6KB1 and PIK3R1 in the Autophagy Pathway Genes and Risk of Bladder Cancer. *Arch. Toxicol.* **2022**, *96*, 367–375. [CrossRef]
51. Park, J.M.; Seo, M.; Jung, C.H.; Grunwald, D.; Stone, M.; Otto, N.M.; Toso, E.; Ahn, Y.; Kyba, M.; Griffin, T.J.; et al. ULK1 Phosphorylates Ser30 of BECN1 in Association with ATG14 to Stimulate Autophagy Induction. *Autophagy* **2018**, *14*, 584–597. [CrossRef] [PubMed]
52. Zheng, W.; Xie, W.; Yin, D.; Luo, R.; Liu, M.; Guo, F. ATG5 and ATG7 Induced Autophagy Interplays with UPR via PERK Signaling. *Cell Commun. Signal.* **2019**, *17*, 42. [CrossRef] [PubMed]
53. Huang, W.; Zeng, C.; Hu, S.; Wang, L.; Liu, J. ATG3, a Target of MiR-431-5p, Promotes Proliferation and Invasion of Colon Cancer via Promoting Autophagy. *Cancer Manag. Res.* **2019**, *11*, 10275–10285. [CrossRef] [PubMed]
54. Lee, H.N.; Jeong, M.S.; Jang, S.B. Molecular Characteristics of Amyloid Precursor Protein (App) and Its Effects in Cancer. *Int. J. Mol. Sci.* **2021**, *22*, 4999. [CrossRef]
55. Ekendahl, S.; Bark, M.; Engelbrektsson, J.; Karlsson, C.A.; Niyitegeka, D.; Strömberg, N. Energy-Efficient Outdoor Cultivation of Oleaginous Microalgae at Northern Latitudes Using Waste Heat and Flue Gas from a Pulp and Paper Mill. *Algal Res.* **2018**, *31*, 138–146. [CrossRef]
56. Guillard, R.R.; Ryther, J.H. Studies of Marine Planktonic Diatoms. I. Cyclotella Nana Hustedt, and Detonula Confervacea (Cleve) Gran. *Can. J. Microbiol.* **1962**, *8*, 229–239. [CrossRef]
57. Zhu, C.J.; Lee, Y.K.; Chao, T.M. Effects of Temperature and Growth Phase on Lipid and Biochemical Composition of Isochrysis Galbana TK1. *J. Appl. Phycol.* **1997**, *9*, 451–457. [CrossRef]
58. Gnaiger, E.; Bitterlich, G.; Composition, P.B. Original Papers Proximate Biochemical Composition and Caloric Content Calculated from Elemental CHN Analysis: A Stoichiometric Concept. *Int. Assoc. Ecol.* **2012**, *62*, 289–298.
59. Christie, W.W. Preparation of Ester Derivatives of Fatty Acids for Chromatographic Analysis. *Adv. Lipid Methodol.* **1993**, *2*, e111.
60. Mushtaq, M.Y.; Choi, Y.H.; Verpoorte, R.; Wilson, E.G. Extraction for Metabolomics: Access to the Metabolome. *Phytochem. Anal.* **2014**, *25*, 291–306. [CrossRef]
61. Della Sala, G.; Mangoni, A.; Costantino, V.; Teta, R. Identification of the Biosynthetic Gene Cluster of Thermoactinoamides and Discovery of New Congeners by Integrated Genome Mining and MS-Based Molecular Networking. *Front. Chem.* **2020**, *8*, 397. [CrossRef] [PubMed]
62. Caso, A.; Esposito, G.; Della Sala, G.; Pawlik, J.R.; Teta, R.; Mangoni, A.; Costantino, V. Fast Detection of Two Smenamide Family Members Using Molecular Networking. *Mar. Drugs* **2019**, *17*, 618. [CrossRef] [PubMed]

63. Shannon, P.; Markiel, A.; Ozier, O.; Baliga, N.S.; Wang, J.T.; Ramage, D.; Amin, N.; Schwikowski, B.; Ideker, T. Cytoscape: A Software Environment for Integrated Models of Biomolecular Interaction Networks. *Genome Res.* **2003**, *13*, 2498–2504. [CrossRef] [PubMed]
64. Fahy, E.; Subramaniam, S.; Murphy, R.C.; Nishijima, M.; Raetz, C.R.; Shimizu, T.; Spener, F.; van Meer, G.; Wakelam, M.J.; Dennis, E.A. Update of the LIPID MAPS Comprehensive Classification System for Lipids1. *J. Lipid Res.* **2009**, *50*, S9–S14. [CrossRef] [PubMed]

marine drugs

Article

Evaluation of Cellular Uptake and Removal of Chlorpropham in the Treatment of *Dunaliella salina* for Phytoene Production

Laura Mazzucchi, Yanan Xu and Patricia J. Harvey *

School of Science, Faculty of Engineering and Science, University of Greenwich, Central Avenue, Chatham Maritime, Kent ME4 4TB, UK; l.mazzucchi@greenwich.ac.uk (L.M.); y.xu@greenwich.ac.uk (Y.X.)
* Correspondence: p.j.harvey@greenwich.ac.uk; Tel.: +44-20-8331-9972

Abstract: Chlorpropham is a carbamate herbicide that inhibits cell division and has been widely used as a potato sprout suppressant. Recently we showed that the microalga *Dunaliella salina* treated with chlorpropham massively accumulated the colourless carotenoids phytoene and phytofluene. Phytoene and phytofluene are valued for their antioxidant, UV-absorption and skin protectant properties; however, they are present in very low quantities in nature. The low toxicity herbicide chlorpropham seems a promising catalyst to produce phytoene in large quantities from CO_2 and solar energy with *D. salina*. This study explored chlorpropham uptake by the algal cells, the formation of potential intermediate metabolites, and the removal of residual chlorpropham from harvested *D. salina* biomass. Algal biomass rapidly concentrated chlorpropham from culture media. However, washing the harvested biomass with fresh culture medium twice and five times removed ~83 and ~97% of the chlorpropham from the biomass, respectively, and retained algal cell integrity. Furthermore, chloroaniline, a common metabolite of chlorpropham degradation, was not detected in chlorpropham-treated cultures, which were monitored every two days for thirty days. Cells treated with chlorpropham for either 10 min or 24 h continued to over-accumulate phytoene after resuspension in an herbicide-free medium. These data imply that whilst *Dunaliella* cells do not possess the intracellular capacity to degrade chlorpropham to chloroaniline, the effect of chlorpropham is irreversible on cell nuclear division and hence on carotenoid metabolism.

Keywords: *Dunaliella salina*; chlorpropham; herbicide; phytoene; carotenoids

Citation: Mazzucchi, L.; Xu, Y.; Harvey, P.J. Evaluation of Cellular Uptake and Removal of Chlorpropham in the Treatment of *Dunaliella salina* for Phytoene Production. *Mar. Drugs* **2022**, *20*, 367. https://doi.org/10.3390/md20060367

Academic Editor: Carlos Almeida

Received: 30 April 2022
Accepted: 27 May 2022
Published: 30 May 2022

1. Introduction

Chlorpropham (CIPC), isopropyl 3-chlorocarbanilate (IUPAC), is a carbamate herbicide and is very widely used worldwide as a general plant growth regulator to control sprouting and as a herbicide against target weeds [1,2]. Recently, Xu and Harvey [3] showed that the addition of micromolar quantities of chlorpropham to cultures of the microalga, *Dunaliella salina,* inhibited cell division and resulted in the massive overaccumulation of the colourless carotenoids, phytoene and phytofluene. *D. salina* is a halotolerant marine microalga and well known for producing high quantities of β-carotene [4]. Phytoene and phytofluene are precursors of β-carotene and are naturally found in a limited amount in the alga. Chlorpropham was proposed to disrupt synchronised control between nuclear and chloroplast events in cell division, which would normally be associated with carotenogenesis and β-carotene accumulation [4].

Phytoene and phytofluene are similar to β-carotene in being comprised of a C40 backbone of isoprenoid units, which confer antioxidant and anti-inflammatory properties [5]. Unlike β-carotene, however, the absorbance maxima for phytoene and phytoene lie within the UV wavelength range (280–350 nm). Phytoene and phytoene are consequently sought after as UV-protective ingredients in skin protection products [6,7]. Both compounds are found naturally in various stereoisomeric forms, which differ in their physicochemical properties and shape. In simulated gastro-intestinal studies, *cis*-isomers of phytoene and phytofluene from different fruit juices presented with a higher bioaccessibility than their *trans*-isomer counterparts, suggesting that they may have a higher therapeutical value [8]. The isomers from *D. salina* were recently comprehensively characterised and quantified after treatment of *D. salina* cultures with chlorpropham [7]. *Cis-* forms of the isomers are the predominant form in *D. salina*. However, phytoene and phytofluene normally are only found in a limited amount in *D. salina*. This has, until now, limited the potential for industrial-scale developments using this alga [9].

To date, chlorpropham is known to inhibit the process of mitosis in plants and algae by interfering with the spindle microtubule organising centre, causing abnormal or complete suppression of microtubule synthesis and organisation. However, its specific site of action is not yet known [10]. Sterling [11] showed that lipophilic and neutral herbicides, including chlorpropham, could penetrate the cell membrane of some lower and several higher plants by passive diffusion; the kinetics studies of the uptake and efflux of different neutral pesticides have been cited. Little is known about the cellular uptake and accumulation of chlorpropham by microalgae and specifically *D. salina*. Likewise, the biotransformation of chlorpropham has been investigated in higher plants and mainly in potatoes due to its extensive use on this crop. In studies performed with potatoes, chlorpropham was shown to be decarboxylated to 3-chloroaniline (3-CA) [12], which is an organochlorine compound listed on the European Community priority pollutant Circular No 90–55 (1990) [13], and classified as highly toxic for the environment and humans. Studies on the metabolism of chlorpropham in algae are limited. To our knowledge, only John et al. [14] investigated its degradation by a selection of green algae (such as *Chlorella pyrenoidosa* and *Chlamydomonas, Ulothrix fimbriata*) and blue green algae (*Anacistis nidulans*). Using colorimetric methods for detection of 3-CA, they observed that only *A. nidulans* transformed chlorpropham to 3-CA.

The aim of the present work was to gain insights into the interaction between the herbicide chlorpropham and *D. salina* cells by studying its uptake and possible metabolism to 3-CA in cell cultures of *D. salina*. The objectives involved the detection of chlorpropham and its metabolites in the algal biomass, the study of its concentration over time and its removal from the biomass.

2. Results

2.1. Determination of Chlorpropham and Its Metabolites in Algal Biomass and Extracts

2.1.1. HPLC Detection of Chlorpropham and 3-Chloroaniline

Harvested biomass from *D. salina* cultures treated with 20 µM–1 mM chlorpropham over 30 days was extracted with methanol for chlorpropham and 3-chloroaniline, and the extracts were analysed by HPLC at an interval of 2 days for 30 days. HPLC chromatograms of the extracts revealed the presence of chlorpropham in the treated biomass, based on the spectral properties of pure standard and its retention time (5 min, see Figure 1), but 3-chloroaniline could not be detected (detection limit 1 ng/mL). The recovery efficiency of 3-chloroaniline from *D. salina* biomass was assessed with the 3-chloroaniline standard using methanol and determined to be 91%. These data suggested that chlorpropham was not metabolised to 3-chloroaniline by *D. salina* cells.

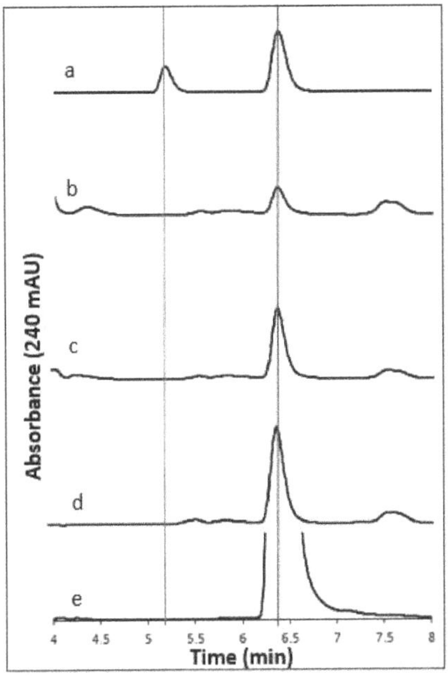

Figure 1. HPLC chromatograms of (a) chlorpropham (7.5 μg/mL) and 3-chloroaniline (1 μg/mL) standard spiked in *D. salina*, and extracts from *D. salina* treated with (b) 20 μM, (c) 50 μM, (d) 200 μM and (e) 1 mM chlorpropham for one month; red line (RT = 6.39 min) highlights the peaks corresponding to chlorpropham; blue line (RT = 5.24 min) indicates peak of 3-chloroaniline. n.d. not detected.

2.1.2. Determination of Chlorpropham and Phytoene in *D. salina* Biomass

Figure 2 shows the quantification of chlorpropham and phytoene in *D. salina* extracts from unwashed biomass, which was extracted with either MeOH, MeOH/MTBE, MTBE or EtOH. As shown in Figure 2a, the chlorpropham contents in the extracts with different solvents were not significantly different ($p > 0.05$, Anova), whereas the phytoene contents of the extracts (Figure 2b) were significantly different ($p < 0.001$, Anova); extraction with either EtOH, MeOH or MeOH/MTBE yielded similar levels of phytoene ($p > 0.05$), whilst MTBE gave the lowest yield of phytoene. In all subsequent experiments, MeOH was therefore used to assess the contents of phytoene and chlorpropham in harvested biomass.

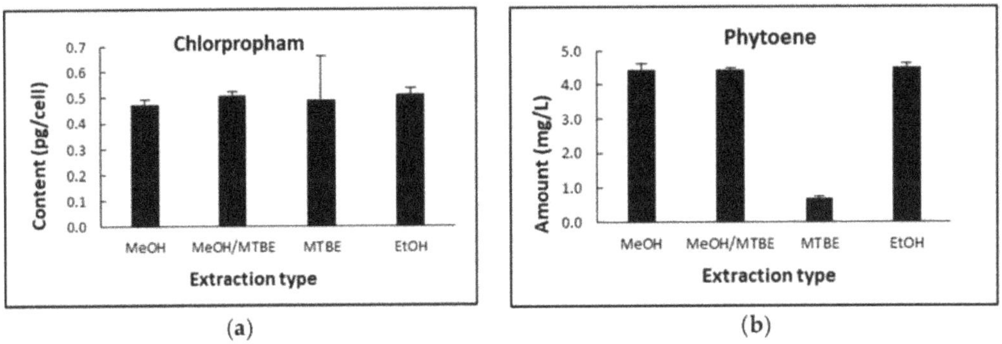

Figure 2. (**a**) Chlorpropham content with different extraction solvents; (**b**) phytoene yield with different extraction solvents.

2.2. *Cellular Uptake and Accumulation of Chlorpropham during D. salina Cultivation*

2.2.1. Chlorpropham Uptake over Time

The kinetics of binding and uptake of chlorpropham by *D. salina* cells is shown in Figure 3, which displays chlorpropham amounts in the harvested biomass and culture medium over a period of 30 days. After 5 min following the addition of chlorpropham to cultures, unwashed biomass contained 1.92 ± 0.23 mg chlorpropham/g biomass (AFDW). Over the next 28 days, there was no further significant increase in cell density ($p > 0.05$). However, the amount of chlorpropham continued to associate with the biomass at a linear rate of 0.04 mg/day and, after 28 days, reached 3.22 mg/g AFDW. In concert, a corresponding significant decrease ($p < 0.05$) of chlorpropham amount in the culture medium between day 0 and day 28 (from 4.31 to 2.59 mg/L) was recorded (see Figure 3). Chlorpropham solution in uninoculated control had no significant change ($p > 0.05$) in chlorpropham concentration over the same period.

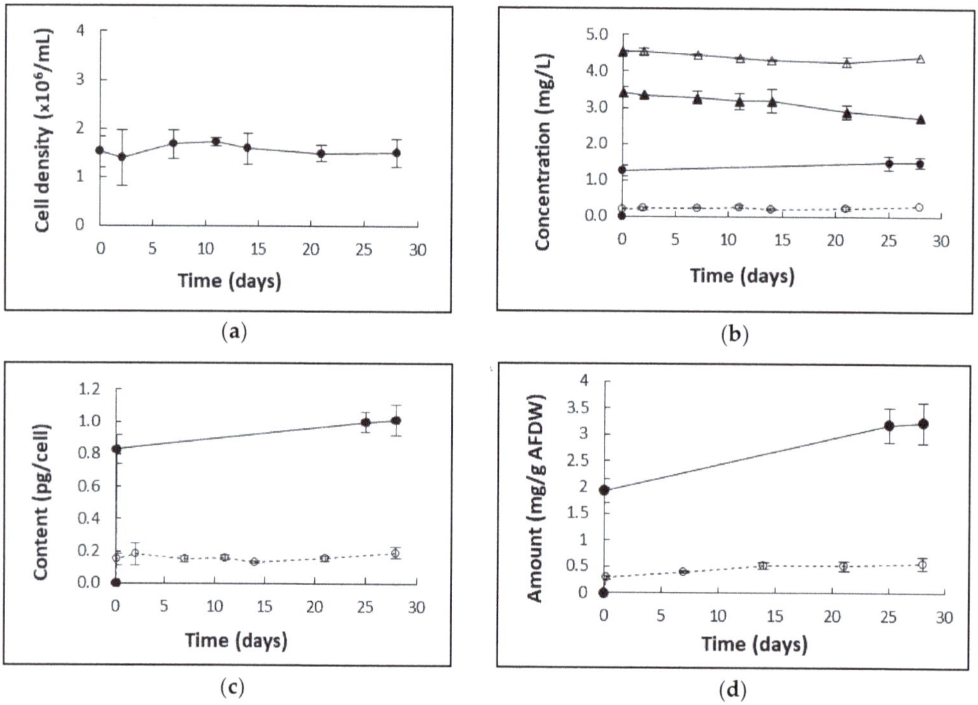

Figure 3. (**a**) The cell density of the *D. salina* cultures treated with 20 μM chlorpropham; day 0 represents the start of the treatment. (**b**) Changes in chlorpropham amount (mg/L culture) over time in different fractions of *D. salina* cultures treated with 20 μM chlorpropham; unwashed biomass after centrifugation (circle, full line), biomass washed twice with fresh culture medium (broken line), supernatant from centrifugation (full triangle) and chlorpropham solution control that was not inoculated (open triangle); (**c**) cellular content of chlorpropham in unwashed biomass of *D. salina* cultures treated with 20 μM herbicide (full line) and washed twice (broken line); (**d**) chlorpropham (mg/g AFDW) in *D. salina* total biomass treated with 20 μM herbicide when cells were not washed (full line) and washed twice (broken line).

On the other hand, the amount of chlorpropham associated with washed biomass compared to unwashed never reached more than the initial concentration of 0.2 pg/cell throughout the 28-day test period, indicating that only a small fraction of the original amount of chlorpropham added to cultures might be needed to solicit phytoene accumulation (Figure 3c,d).

2.2.2. Effect of Chlorpropham Concentration

Different concentrations of chlorpropham were added to *D. salina* cultures, and the amount of chlorpropham taken up by the cells increased linearly with the increasing concentration of the chlorpropham added into the cultures (Figure 4). In 50 µM chlorpropham-treated cultures, the amount of chlorpropham in the cells was three times higher than that in 20 µM chlorpropham-treated cells. Additionally, in 200 µM chlorpropham-treated cultures, the amount was 15 times higher than that in 20 µM treated cultures.

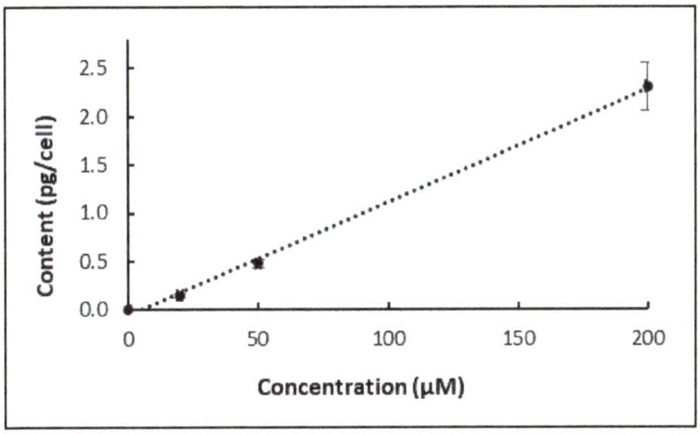

Figure 4. Correlation between the concentration of chlorpropham added to *D. salina* cultures and the chlorpropham content in *D. salina* cells (the concentration was measured after 24 h of treatment).

2.2.3. Effect of Cell Density

Two sets of cultures with different cell densities (1.36×10^6/mL and 0.4×10^6/mL) were treated with 20 µM chlorpropham to study the effect of different cell densities on chlorpropham accumulation. The data in Figure 5 show that cell volume overall increased (significantly) from day 0 onwards in all the cultures ($p < 0.05$), in line with the progressive increase in the phytoene content and chlorpropham amount in the cells. See also Figure A1, Appendix A. The increasing cellular content of phytoene over time (Figure 5d) may contribute to cell swelling reflected in a slightly increased rate of accumulation of chlorpropham during the treatment period (Figure 5c). The volume increase was significantly higher in cultures of lower density. Notably, from day 22 of chlorpropham treatment, cells of cultures with low density became pale and irregular in shape, and cultures turned light grey, intracellular phytoene significantly decreased ($p < 0.05$) and chlorpropham associated with this biomass sharply increased ($p > 0.05$) between day 15 and 22.

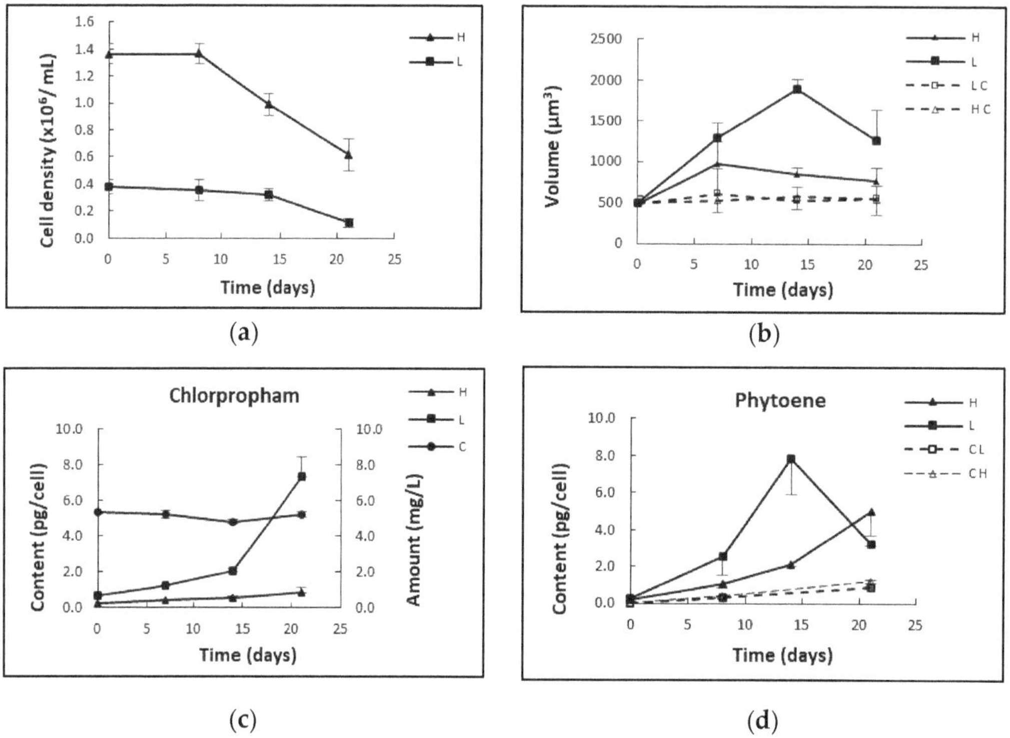

Figure 5. (**a**) Cell densities of the cultures treated with 20 µM chlorpropham. (**b**) Cell volume measurements of cultures treated at the cell densities of 1.36 × 10^6 cells/mL (H) and 0.4 × 10^4 cells/mL (L); (**c**) Chlorpropham content (pg/cell) in cultures at different cell densities (H and L in the figure) and in not inoculated control cultures; (**d**) Phytoene content in the biomass of low- and high-density-treated cultures or untreated controls (CH/CL).

2.3. *Process Optimisation to Reduce the Concentration of Chlorpropham Loosely-Associated with Algal Biomass*

2.3.1. Different Numbers of Washing Cycles

Harvested *D. salina* biomass was washed with a fresh culture medium, and the effectiveness was assessed so that the amount of chlorpropham associated with the biomass and thereafter the phytoene extracts could be minimised. Treated *D. salina* cells were centrifuged, and pellets were transferred into a fresh herbicide-free medium. The change in the chlorpropham content in the harvested biomass was monitored, as shown in Figure 6. The results show that almost 90% of the chlorpropham associated with unwashed *D. salina* cells was released into the washing medium after two minutes of transferring the treated biomass into the fresh medium, indicating the release of chlorpropham from the cells to the fresh medium occurred very rapidly.

Table 1 and Figure 7 show chlorpropham removal and phytoene recovery from harvested biomass with different numbers of washing cycles. A significant difference between the content of chlorpropham derived from washed and unwashed biomass was found ($p < 0.05$). Chlorpropham contents in the biomass washed either 2, 5 or 10 times were 6.2, 45.5 and 307 times lower, respectively, than the unwashed biomass. More than 99% of chlorpropham in the harvested biomass was removed after washing with a fresh medium 10 times. Given that the harvested pellets had a volume of ~50 µL in each test and that the concentration of chlorpropham in the culture was 20 µM, the calculated herbicide carried over by water in the pellet volume was ~0.21 µg. This value was 85 times and 18 times lower in respect of the total amount of chlorpropham in both unwashed (18 μg) and washed biomass (3.75 μg) extracted from the pellet, suggesting that the biomass had concentrated chlorpropham from the extracellular medium.

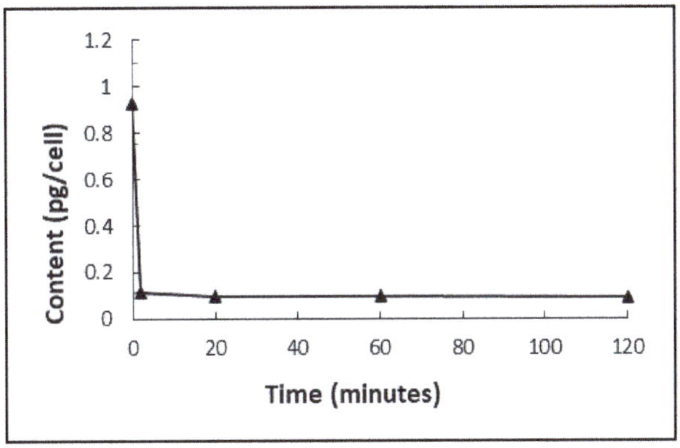

Figure 6. Loss of cellular chlorpropham amount in the harvested *D. salina* cells treated with 20 μM of chlorpropham. Unwashed cell pellets were obtained from 30 mL of treated cultures and then transferred to 60 mL of fresh herbicide-free medium.

Table 1. Amount of chlorpropham (expressed in μg) removed from the biomass of *D. salina* was washed with a different number of cycles by fresh culture medium or dH_2O; 10 mL of culture was harvested and washed with 10 mL of washing solution.

Wash Type	Chlorpropham Removed [μg]	Loss, % Chlorpropham	Tot Intact Cell Number ($\times 10^5$)
No wash	0	0	266.10 ± 2.90
Wash (×2)	7.446 ± 0.353	83.89	262.11± 44.5
Wash (×5)	8.681 ± 0.314	97.77	250 ± 25.70
Wash (×10)	8.848 ± 0.337	99.69	162 ± 19.50
dH_20 wash (×2)	8.7109 ± 0.274		

Since *D. salina* cells lack a cell wall and are therefore relatively easily ruptured, quantification of phytoene in harvested biomass served as an internal marker and the ratio of chlorpropham:phytoene was recorded. The results show that the total cells that remained intact after washing 2, 5 or 10 times with fresh medium were 98.5%, 94% and 61.1%, respectively, in comparison to the biomass harvested without washing. Additionally, the recovery of phytoene was also investigated from samples when the volume ratio of washing solution:harvested cultures was 2:1 (Figure 7a). More than 90% of phytoene was recovered when *D. salina* biomass was washed less or equal to 5 times, and more than 80% recovered when biomass was washed 10 times.

Washing *D. salina* cells with water caused a substantial loss of phytoene (90% of total phytoene). The ratio of phytoene/chlorpropham increased with increasing washing cycles (Figure 7b), but it remained small when cells were washed with water, which caused complete cell disruption. Washing with water ruptured cells as indicated both visually and by the low value obtained for the ratio of chlorpropham:phytoene. These results confirm that washing with a culture medium successfully removes chlorpropham from *D. salina* biomass without bursting the cells.

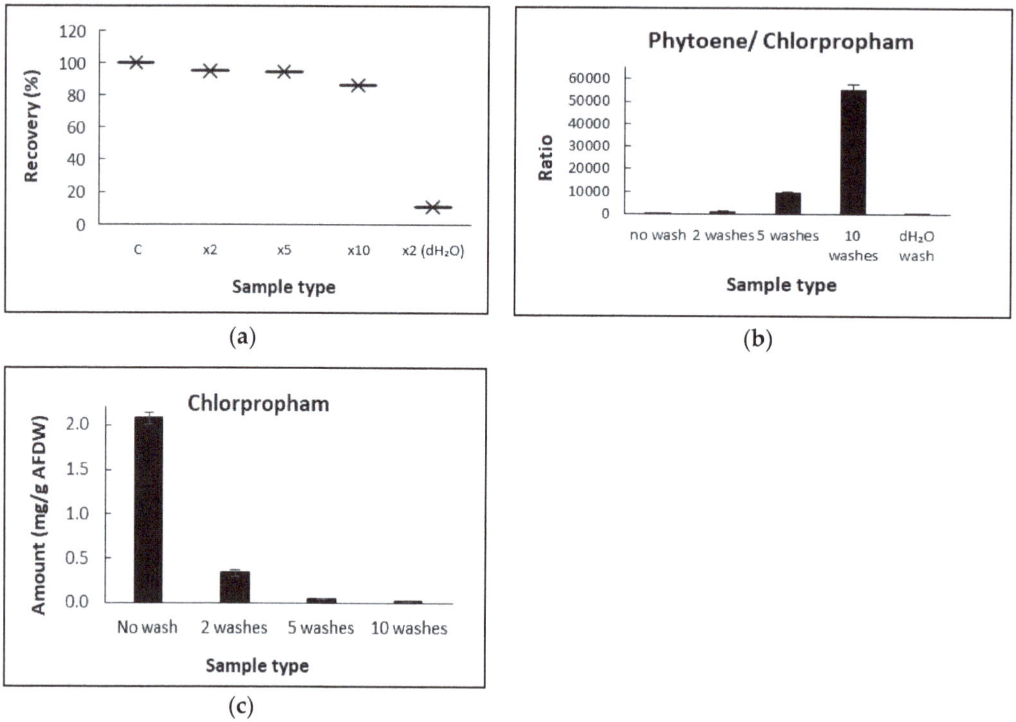

Figure 7. (**a**) Mean recovery (%) of phytoene from biomass after washing with fresh medium or water for different times. (**b**) phytoene/chlorpropham ratios when cells are not washed (1), washed twice (2), washed 5 times (3), 10 times (4) and washed with water (5). (**c**) Amount of chlorpropham in the final *D. salina* biomass (mg/g).

2.3.2. Different Washing Volumes

Table 2 shows that the decrease in chlorpropham in harvested biomass was positively correlated to the increase in wash solution: harvest volume ratio. Chlorpropham amount in the biomass was 8.9 times lower than that recorded in unwashed biomass when the volume ratio of washing solution: harvest culture was 1:1; 17.8 times lower when the ratio was 2:1, and 71.6 times lower when the ratio was 4:1 (Table 2).

Table 2. Values of the amount of chlorpropham (pg/cell) with different wash solution/ harvest volumes (mL) tested.

v/*v* Wash Solution/Harvest (mL)	Chlorpropham Content	Chlorpropham Bound to Biomass (%) (pg/Cell)
	1/1	
washed	0.066 ± 0.007	11.2
unwashed	0.59± 0.008	
	2/1	
washed	0.035 ± 0.004	5.93
unwashed	0.063 ± 0.093	
	4/1	
washed	0.015 ± 0.003	1.6
unwashed	1.10 ± 0.021	

2.3.3. Phytoene Production in Washed D. salina Cells after Chlorpropham Treatment

D. salina cultures treated with chlorpropham for either 10 min or 24 h were harvested, and then pellets were resuspended in herbicide-free fresh medium to investigate the remained effects of chlorpropham on algal cells in terms of carotenoid accumulation (Figure 8; Figure A2, Appendix A). Cells harvested from both treated cultures continued to accumulate phytoene as well as β-carotene at significantly faster rates than in controlled, treated cultures. This is probably because the cells were diluted to lower densities in the fresh medium and gained access to higher light energy as well as higher nutrient levels. There is no significant difference between the cells treated for 10 min and those treated for 24 h ($p > 0.05$), confirming that the effects occurred in the cells rapidly within the first few minutes.

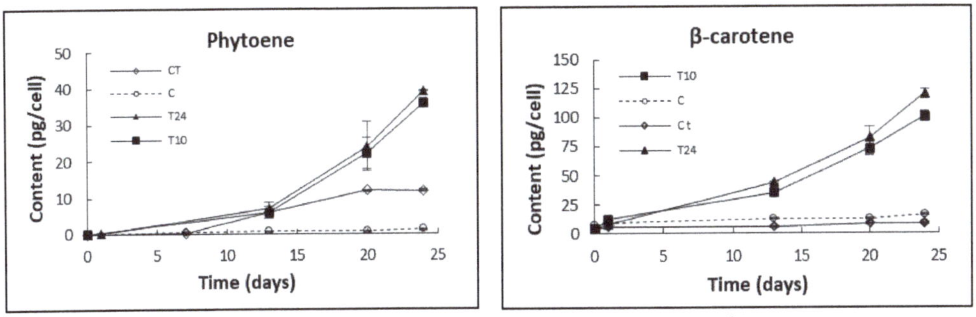

(a) (b)

Figure 8. (**a**) Phytoene and (**b**) β-carotene content (pg/cell) in the different cultures (T10 = 10 min treatment; T24 = 24 h treatment; C = control; C t = control treated).

3. Discussion

Chlorpropham is a plant herbicide and a mitotic inhibitor which has been used extensively as a potato sprout suppressant. When *D. salina* cultures were treated with a micromolar concentration of chlorpropham, the two colourless carotenoid precursors, phytoene and phytofluene, massively accumulated [3]. The present work was undertaken to explore the interaction between chlorpropham and *D. salina* cells and the accumulation and removal of chlorpropham in the algal biomass.

D. salina biomass concentrated chlorpropham from the culture medium, but most could be removed by washing the biomass. A similar phenomenon has been previously reported using higher plant tissues [15]: in the study of the absorption of C-labelled herbicides washed for different cycles and time periods, herbicides such as fluorodifen (log K ow of 1.84 at 25 °C), which were concentrated by higher plant tissues, were removed after 10 min washing period. In the present study, the fresh culture medium was chosen as the washing solution to maintain the osmotic pressure and reduce the likelihood of cell rupture during washing since *D. salina* cells have no rigid cell wall [16]. Under these circumstances, the intracellular phytoene content remained within the cells.

Monitoring of the chlorpropham uptake by *D. salina* over time showed that the association of chlorpropham with the harvested biomass occurred within the first few minutes of treatment. Lipophilic herbicides have been shown to freely transfer across the cell membrane of cells via passive diffusion until the chemical equilibrium between the internal and external concentration is reached [11] and at a rate dependent on their lipophilicity. Chlorpropham is relatively lipophilic with a partition coefficient of octanol/water = 5.75×10^3. *D. salina* cells, moreover, are bounded from the extracellular medium by a lipophilic membrane comprising a glycocalyx-like cell layer of varying thickness [16]. Although the precise site of action of chlorpropham is not clear, it is likely to be internal, as chlorpropham is known to disrupt mitosis by interfering with the spindle microtubule organising centres in lower and higher plants species [10,17] or by interacting with the microtubules directly [18].

In *D. salina* cultures, chlorpropham may be taken up by cells by internal passive diffusion through the cell membrane; this needs further exploration. In the present study, *D. salina* cells treated with chlorpropham for either 10 min or 24 h continued to accumulate phytoene even after their transfer into herbicide-free fresh medium and at the same rate as those cells that remained in chlorpropham-containing culture medium.

Chlorpropham accumulation during cell growth may parallel the increase in cell volume; increases in cell volume in cells treated with chlorpropham have been shown previously [19]. Our data show that the content of phytoene significantly accumulated over the period tested (21 days), and cell volumes of all treated cultures were observed to increase over time after the treatment while phytoene and phytofluene were accumulated, in accord also with [3]. Additionally, the sharp increase in volume in cultures of low cell number paralleled higher carotenoid accumulation, suggesting a correlation between *D. salina* cell number and the amount of chlorpropham added. Therefore, the physiological changes observed on day 21 of treatment in cultures with lower cell density may be linked to a higher toxicity effect of intracellular herbicide. It was shown previously that cell swelling occurred in cells grown once a specific level of chlorpropham was reached [20]. On the other hand, when cultures were treated for only 10 min or 24 h and resuspended in fresh medium, the effect of chlorpropham on the overaccumulation of phytoene was shown even after 24 days, proving the irreversibility of the herbicide after short exposure without showing cell toxicity caused by prolonged exposures to the herbicide.

The formation of the metabolite, 3-chloroaniline, has been a concern for chlorpropham applications. In this study, no 3-chloroaniline was detected in cultures treated with up to 1 mM herbicide with the method used, suggesting that *D. salina* might not be capable of breaking down chlorpropham. There is little information available regarding the degradation of chlorpropham in microalgal species. John et al. [14] investigated the presence of 3-chloroaniline in different microalgae and a cyanobacterium and showed that *Ulothrix fimbriata*, which belongs to the same phylum *Chlorophyta* as *D. salina*, did not produce 3-chloroaniline, in contrast to the blue-green alga *Anacystis nidulans*, which possessed the enzyme acylamidase. On the other hand, studies on potato tissues attributed the presence of 3-chloroaniline after chlorpropham treatment to the thermal degradation of chlorpropham during the fogging application [21], when extremely high temperatures (>300 °C) are used; high temperatures have been reported to trigger the degradation to 3-chloroaniline [22]. Additionally, 3-chloroanline may form during potato storage due to the activity of microorganisms capable of degrading the parent compound to this metabolite [23].

The present work aimed to elucidate details of the use of chlorpropham in *D. salina* cultivation for phytoene production. Chlorpropham added to cultures was concentrated in the biomass. The effects caused by chlorpropham on cell metabolism are irreversible. The constant increase in the chlorpropham content associated with the biomass over time could be linked to increased cell volume during the accumulation of the colourless carotenoids, phytoene and phytofluene. Washing repeatedly with fresh medium will remove most chlorpropham associated with harvested biomass, up to more than 99% of its initial amount. Phytoene and phytofluene are high valued compounds which are sought to have high beneficial health properties; however, their availability in nature is low. The use of chlorpropham with *D. salina* cultivation may represent a facile, low-cost method for producing large quantities of phytoene.

4. Materials and Methods

4.1. Alga Strain and Cultivation

D. salina strain DF15 (CCAP 19/41) was obtained from the Marine Biological Association, UK (MBA). Cultures were cultivated in 500 mL Modified Johnsons Medium [24] containing 1.5 M NaCl and 10 mM $NaHCO_3$ in an illuminated incubator (Varicon Aqua, Worcester, UK) under white light of 500 μmol m^{-2} s^{-1} at 25 ± 2 °C. For the different kinetic studies of treatments with herbicides, triplicate sets of cultures were grown to mid-late log phase, and each set was treated with chlorpropham at different concentrations from 20 μM up to 1 mM. Flasks containing only fresh culture media added with the same amount of herbicide served as blank controls to monitor the natural degradation rate of chlorpropham over time without algal cells.

4.2. Standards and Solvents

Phytoene standard (LGC Limited, Teddington, UK), chlorpropham (PESTANAL, analytical standard) and 3-chloroaniline (99% purity) were purchased from Sigma-Aldrich (Merck KGaA, Darmstadt, Germany). Methanol (MeOH) and Methyl Tert Butyl Ether (MTBE), both HPLC grade, were purchased from Fischer Scientific UK Ltd. (Loughborough, Leicestershire, UK).

4.3. Extraction and Analysis of Phytoene

To determine the yield of phytoene and other carotenoids in chlorpropham-treated and control cultures of *D. salina* at different time periods, 5 to 10 mL of *D. salina* cultures were harvested at 3000× *g* for 5 min at 20 °C with an Eppendorf centrifuge 5810R (Eppendorf, UK). Carotenoids were extracted and analysed by HPLC as described in our previous work [7]: 10 mL of MeOH: MTBE (80:20) was added to the samples, which were first sonicated for 60 s and then vortexed for 20 s. Samples were clarified at the centrifuge, and the top solvent phase was collected. The extracts were filtered (0.20 μm filter) into amber HPLC vials before analysis. The carotenoid extracts were routinely flushed with nitrogen after extraction. To determine the kinetics for phytoene accumulation, *D. salina* cultures (cell density = 85.5 × 10^4/mL) were treated with chlorpropham for either 10 min or for 24 h, and then aliquots of the cultures (30 mL) were transferred to fresh, herbicide-free, culture medium (60 mL), and the rate of phytoene accumulation (pg/cell) determined for 24 days. The phytoene yield in those cultures was compared to cultures not treated and cultures treated for all the 24 days analysis period. To determine the yield of phytoene, a calibration curve was generated for its quantification.

4.4. Extraction and Analysis of Chlorpropham and Its Potential Metabolites

4.4.1. Sample Preparation

Analytical methods for chlorpropham and its metabolite 3-chloroaniline include high-performance liquid chromatography or gas chromatography. However, gas chromatography often requires the derivatisation of the compounds before analysis and, therefore, HPLC was selected as the main tool for chlorpropham and 3-CA analysis in this study. To monitor the amount of chlorpropham in *D. salina* cells on different days of treatment (day 0 to day 28), 10 to 15 mL cultures were harvested at 3000× *g* for 5 min at 20 °C. The extraction of chlorpropham from *D. salina* biomass was evaluated with the following solvent extracts: MeOH (100%), MTBE (100%), Ethanol (100%), MTBE/MeOH (20/80). 2.5 mL MeOH (100%) was added to the biomass, and the solvent suspension was sonicated for 20 s, vortexed for 20 s and centrifuged at 3,000× *g* for 5 min. The upper phase with the solvent was collected, filtered with 0.2 μm size filters into amber HPLC vials and analysed by HPLC.

To enhance the compatibility with the HPLC mobile phase, as a stronger sample phase solvent than the mobile phase may lead to inaccuracies in the results, a final sample phase consisting of 90% sample (in MeOH) and 10% of HPLC grade water was prepared. To analyse samples deriving from the algal supernatant, 5 mL of each collected supernatant was diluted with 15 mL of MeOH (HPLC grade) to reach a ratio (25:75, H_2O/MeOH).

4.4.2. HPLC Analysis of Chlorpropham and 3-Chloroaniline

The presence of chlorpropham and its metabolite, 3-chloroaniline, was evaluated with HPLC by matching the retention time and the UV-vis spectra features of authentic standards; these values are shown in Table 3. The instrument was equipped with Diode-Array Detection (HPLC-DAD; Agilent Technologies 1200 series, Agilent, Santa Clara, CA, USA), an online degasser and a quaternary pump system. Samples were separated using a C18 250 × 4.6 mm Waters Spherisorb 5 µm ODS2. MeOH/ dH_2O mixtures with different ratios of water (from 0 to 10%) were tested as mobile phase solvents for analysis. The method parameters chosen were as follows: the column temperature was set at 20 °C, and the gradient solvent system was MeOH/ dH_2O (90/10) (A) running at 100% A for the first 8 min at a flow rate of 0.6 mL/min, then running at 100% MeOH (B) for one minute, before adjusting to 80% B and 20% C (C = 100% MTBE) for the next 20 min at a flow rate of 1.0 mL/min (to elute the carotenoids) before going back to the initial conditions at 30 min. The total run time was 45 min. The absorbance at different wavelengths (210, 240, 250, 282, 355, 450, 480 nm) was monitored. All data were acquired and analysed with Chemstation for LC System software. The method was validated for precision linearity, selectivity, limit of detection and quantification as in [25]. Calibration curves were obtained for chlorpropham and chloroaniline quantification by spiking known concentrations of chlorpropham and chloroaniline standards into crude algal extracts and supernatant and preparing serial dilution (at least 7 points) within the range needed for the experiment. At least three fresh solutions for each concentration were used, and the mean and standard deviation was obtained. Each area value was plotted against each corresponding concentration to obtain the best-fitted line. To detect the presence of 3-chloroaniline in the biomass of *D. salina*, cultures were treated with an increasing concentration of chlorpropham (up to 1 mM) and aliquots of cultures (from 10 to 100 mL) were harvested and extracted. Then, the solvent was evaporated with the vacuum evaporator Genevac and resuspended to the compatible mobile phase.

Table 3. Values of retention time (min) and UV absorption λ_{max} (nm) of chlorpropham and 3-chloroaniline standards and extracts from the biomass of *D. salina*.

Property	Chlorpropham Standard	Chlorpropham Biomass	3-CA Standard	3-CA Biomass
HPLC retention time (min)	6.39 ± 0.014	6.37 ± 0.009	5.241	n.d.
UV absorption λ_{max} (nm)	238, 278	238, 278	242, 292	n.d.

4.5. Confocal Microscopy Analysis

The cell volume of *D. salina* in treated cultures was measured with an LSM880 Bruker confocal microscope. Two triplicate sets of cultures with different cell densities were set up, with the culture of lower density diluted from the one of higher density. Both cultures were treated with 20 µM chlorpropham. The laser channel was set at 488 nm, and 2D images were acquired as follows: the speed was set at 5, and the averaging number was 4. Images size was: 425.1 µM × 425.1 µM, and the pixel size was 0.42 µm. *D. salina* cells were fixed with formalin (2%) before analysis. Different images for each culture flask were acquired to have at least 50 to 100 cells per replicate flask to measure, and the volume of the cells was calculated by measuring their diameter and assuming that *D. salina* has a spherical shape.

4.6. Chlorpropham Removal by Washing Process

To eliminate the traces of chlorpropham surrounding the biomass of *D. salina*, samples were washed with water or the culture medium. Furthermore, the amounts of chlorpropham derived from the biomass washed with different washing cycles (0, 2, 5 and 10) or different washing solutions/harvest ratios were determined. Samples of each washing experiment were collected in triplicate from the same culture flask. The washing experiments conducted with different washing cycles were performed as follows: 10 mL culture was harvested via centrifugation at 3000× *g* for 5 min, the supernatant discharged, and the biomass pellet was resuspended completely in 10 mL washing solution (either culture medium or dH_2O) by vortexing for 5 s; the suspension was centrifuged at 3,000× *g* for 5 min. For the experiment conducted with different harvest cultures/ washing solution volumes, the ratios were as follows: (1) 10 mL culture/10 mL washing solution; (2) 10 mL culture/20 mL washing solution; (3) 10 mL culture/40 mL washing solution.

4.7. Statistical Analysis

Data are presented as the mean and standard deviation of triplicates. IBM SPSS statistics 64-bit was used to perform ANOVA analysis with a significant level of $p < 0.05$ to compare the significance between data at different treatment times and between control and treatment cultures. Microsoft Excel for Office MSO 64-bit was used for graphs representations.

Author Contributions: Conceptualisation, L.M., Y.X. and P.J.H.; Data curation, L.M.; Formal analysis, L.M.; Funding acquisition, P.J.H.; Investigation, L.M. and Y.X.; Methodology, L.M. and Y.X.; Project administration, P.J.H.; Supervision, Y.X. and P.J.H.; Writing–original draft, L.M. and Y.X.; Writing–review and editing, L.M., Y.X. and P.J.H. All authors have read and agreed to the published version of the manuscript.

Funding: L.M. was funded by a VC Scholarship, the University of Greenwich. Y.X. and P.J.H. were funded by the University of Greenwich and by the Interreg 2 Seas programme 2014-2020, co-funded by the European Regional Development Fund under subsidy contract No. ValgOrize 2S05017.

Institutional Review Board Statement: Not applicable.

Informed Consent Statement: Not applicable.

Conflicts of Interest: The authors declare no conflict of interest.

Appendix A

Figure A1 displays the change in cell volume over time for two sets of *D. salina* cultures with different cell density (high = 1.36×10^6/mL and low = 0.4×10^6/mL) that had been treated with 20 μM chlorpropham. Figure A2 displays the difference in cell volume of cultures treated with the herbicide for a short period of time and suspended in herbicide-free fresh medium (a and b), cultures treated for all the analysis period (24 days), (c) and not treated cultures (d).

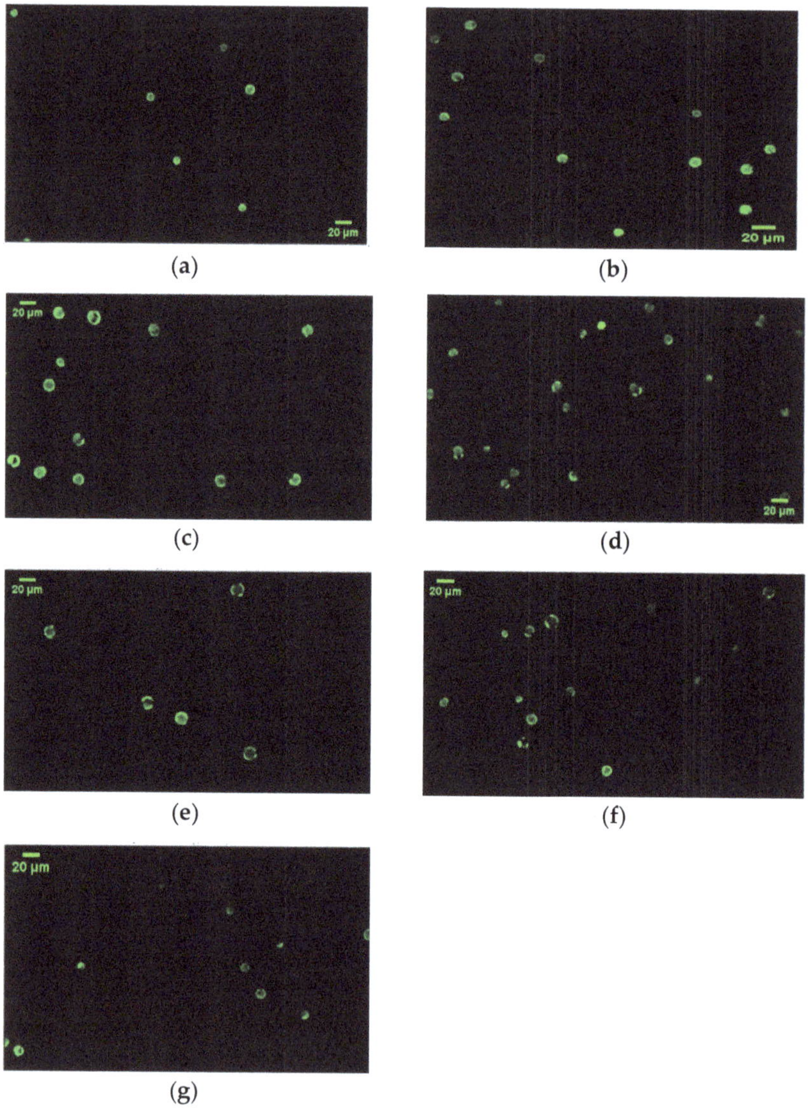

Figure A1. Confocal microscopy images of cultures treated with 20 µM of chlorpropham (**a**) low density, time 0; (**b**) high density, time 0; (**c**) low density, time 7; (**d**) high density, time 7; (**e**) low density, time 15; (**f**) high density, time 15; (**g**) high density, time 22.

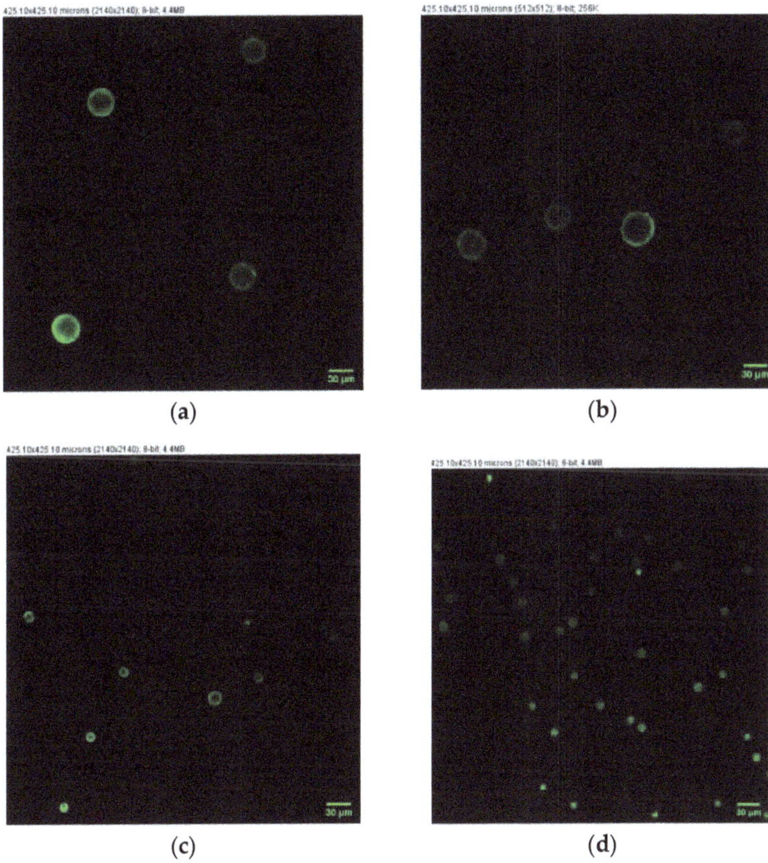

Figure A2. Confocal microscopy images of cultures treated with 20 µM of chlorpropham for (**a**) 10 min, (**b**) 24 h, (**c**) all the experimental period, and (**d**) not treated cultures at day 20 of the experiment.

References

1. Pujar, N.K.; Premakshi, H.G.; Laad, S.; Pattar, S.V.; Mirjankar, M.; Kamanavalli, C.M. Biodegradation of chlorpropham and its major products by *Bacillus licheniformis* NKC-1. *World J. Microbiol. Biotechnol.* **2018**, *34*, 1–11. [CrossRef] [PubMed]
2. European Food Safety Authority (EFSA); Arena, M.; Auteri, D.; Barmaz, S.; Bellisai, G.; Brancato, A.; Brocca, D.; Bura, L.; Byers, H.; Chiusolo, A.; et al. Peer review of the pesticide risk assessment of the active substance chlorpropham. *EFSA J.* **2017**, *15*, e04903. [PubMed]
3. Xu, Y.; Harvey, P.J. Phytoene and phytofluene overproduction by *Dunaliella salina* using the mitosis inhibitor chlorpropham. *Alg. Res.* **2020**, *52*, 102–126. [CrossRef]
4. Xu, Y.; Harvey, P.J. Carotenoid production by *Dunaliella salina* under red light. *Antioxidants* **2019**, *8*, 123. [CrossRef] [PubMed]
5. Rodriguez-Concepcion, M.; Avalos, J.; Bonet, M.L.; Boronat, A.; Gomez-Gomez, L.; Hornero-Mendez, D.; Limon, M.C.; Meléndez-Martínez, A.J.; Olmedilla-Alonso, B.; Palou, A.; et al. A global perspective on carotenoids: Metabolism, biotechnology, and benefits for nutrition and health. *Prog. Lipid Res.* **2018**, *70*, 62–93. [CrossRef] [PubMed]
6. Meléndez-Martínez, A.J.; Mapelli-Brahm, P.; Benítez-González, A.; Stinco, C.M. A comprehensive review on the colorless carotenoids phytoene and phytofluene. *Arch. Biochem. Biophys.* **2015**, *572*, 188–200. [CrossRef] [PubMed]
7. Mazzucchi, L.; Xu, Y.; Harvey, P.J. Stereoisomers of colourless carotenoids from the marine microalga *Dunaliella salina*. *Molecules* **2020**, *25*, 1880. [CrossRef] [PubMed]
8. Mapelli-Brahm, P.; Corte-Real, J.; Meléndez-Martínez, A.J.; Bohn, T. Bioaccessibility of phytoene and phytofluene is superior to other carotenoids from selected fruit and vegetable juices. *Food Chem.* **2017**, *229*, 304–311. [CrossRef] [PubMed]
9. Werman, M.J.; Mokady, S.; Ben-Amotz, A. Bioavailability of the isomer mixture of phytoene and phytofluene-rich alga *Dunaliella bardawil* in rat plasma and tissues. *J. Nutr. Biochem.* **2002**, *13*, 585–591. [CrossRef]
10. Vaughn, K.C.; Lehnen, L.P. Mitotic disrupter herbicides. *Weed Sci.* **1991**, *39*, 450–457. [CrossRef]

11. Sterling, T.M. Mechanisms of herbicide absorption across plant membranes and accumulation in plant cells. *Weed Sci.* **1994**, *42*, 263–276. [CrossRef]
12. Kim-Kang, H. *Metabolism of C-14 Chlorpropham in Stored Potatoes-Nature of the Residue in Potatoes, RPT0066*; XenoBiotic Laboratories, Inc.: Plainsboro, NJ, USA, 1991.
13. David, B.; Lhote, M.; Faure, V.; Boule, P. Ultrasonic and photochemical degradation of chlorpropham and 3-chloroaniline in aqueous solution. *Water Res.* **1998**, *32*, 2451–2461. [CrossRef]
14. John, S.; Wright, L.; Maule, A. Transformation of the herbicides propanil and chlorpropham by micro-algae. *Pestic. Sci.* **1982**, *13*, 253–256. [CrossRef]
15. Boulware, M.A.; Camper, N.D. Sorption of some 14C-herbicides by isolated plant cells and protoplasts. *Weed Sci.* **1973**, *21*, 145–149. [CrossRef]
16. Ben-Amotz, A.; Avron, M. The biotechnology of cultivating the halotolerant alga *Dunaliella*. *Trends Biotechnol.* **1990**, *8*, 121–126. [CrossRef]
17. Hidalgo, A. Abnormal mitosis and growth inhibition in *Allium cepa* roots induced by propham and chlorpropham. *Cytobios* **1989**, *57*, 7–14.
18. Eleftheriou, E.P.; Bekiari, E. Ultrastructural effects of the herbicide chlorpropham (CIPC) in root tip cells of wheat. *Plant Soil* **2000**, *226*, 11–19. [CrossRef]
19. Sumida, S.; Yoshida, R.; Ueda, M. Studies of pesticides effects on Chlorella metabolism III. Effect of isopropyl 3-chlorocarbanilate (chlorpropham) on cell cycle and biosynthesis. *Plant Cell Physiol.* **1977**, *18*, 9–16.
20. Fedtke, C. *Biochemistry and Physiology of Herbicide Action*; Springer: Berlin/Heidelberg, Germany, 2012.
21. Park, L.J. Chlorpropham Distribution in Potato Stores and Evaluation of Environmental Issues Relating to Its Use. Doctoral Dissertation, University of Glasgow, Glasgow, UK, 2004.
22. Nagayama, T.; Kikugawa, K. Influence of frying and baking on chlorpropham residue. *Eisei Kagaku* **1992**, *38*, 78–83. [CrossRef]
23. Park, L.; Duncan, H.; Briddon, A.; Jina, A.; Cunnington, A.; Saunders, S.; Unit, S.B. *Review and Development of the CIPC Application Process and Evaluation of Environmental Issues*; Potato Council Ltd.: Oxford, UK, 2009.
24. Borowitzka, M.A. *Algal Growth Media and Sources of Algal Cultures*; Cambridge University Press: Cambridge, UK, 1988.
25. Baber, N. International conference on harmonisation of technical requirements for registration of pharmaceuticals for human use (ICH). *Br. J. Clin. Pharmacol.* **1994**, *37*, 401. [CrossRef] [PubMed]

marine drugs

MDPI

Article

Evaluation of the Potential of Lipid-Extracted *Chlorella vulgaris* Residue for *Yarrowia lipolytica* Growth at Different pH Levels

Guillaume Delfau-Bonnet [1,2], Nabila Imatoukene [2], Tiphaine Clément [2], Michel Lopez [2], Florent Allais [2] and Anne-Lise Hantson [1,*]

[1] Chemical and Biochemical Process Engineering Unit, Faculty of Engineering, University of Mons, 7000 Mons, Belgium; delfaug@yahoo.com
[2] Unite Recherche et Developpement Agro-Biotechnologies Industrielles (URD ABI), Centre Europeen de Biotechnologie et Bieconomie (CEBB), AgroParisTech, 51110 Pomacle, France; nabila.imatoukene@agroparistech.fr (N.I.); tiphaine.clement@inrae.fr (T.C.); michel.lopez@agroparistech.fr (M.L.); florent.allais@agroparistech.fr (F.A.)
* Correspondence: anne-lise.hantson@umons.ac.be; Tel.: +32-65374419

Abstract: Projections show that the cultivation of microalgae will extend to the production of bio-based compounds, such as biofuels, cosmetics, and medicines. This will generate co-products or residues that will need to be valorized to reduce the environmental impact and the cost of the process. This study explored the ability of lipid-extracted *Chlorella vulgaris* residue as a sole carbon and nitrogen source for growing oleaginous yeasts without any pretreatment. Both wild-type *Yarrowia lipolytica* W29 and mutant JMY3501 (which was designed to accumulate more lipids without their remobilization or degradation) showed a similar growth rate of 0.28 h^{-1} at different pH levels (3.5, 5.5, and 7.5). However, the W29 cell growth had the best cell number on microalgal residue at a pH of 7.5, while three times fewer cells were produced at all pH levels when JMY3501 was grown on microalgal residue. The JMY3501 growth curves were similar at pH 3.5, 5.5, and 7.5, while the fatty-acid composition differed significantly, with an accumulation of α-linolenic acid on microalgal residue at a pH of 7.5. Our results demonstrate the potential valorization of *Chlorella vulgaris* residue for *Yarrowia lipolytica* growth and the positive effect of a pH of 7.5 on the fatty acid profile.

Keywords: *Yarrowia lipolytica*; *Chlorella vulgaris*; growth; fatty acids

Citation: Delfau-Bonnet, G.; Imatoukene, N.; Clément, T.; Lopez, M.; Allais, F.; Hantson, A.-L. Evaluation of the Potential of Lipid-Extracted *Chlorella vulgaris* Residue for *Yarrowia lipolytica* Growth at Different pH Levels. *Mar. Drugs* **2022**, *20*, 264. https://doi.org/10.3390/md20040264

Academic Editor: Carlos Almeida

Received: 24 February 2022
Accepted: 11 April 2022
Published: 13 April 2022

1. Introduction

Microalgae are a diverse phylogenetic group of unicellular photosynthetic organisms. They represent species that live in a wide range of fresh and marine environments. Due to their diversity and ability to capture and convert carbon dioxide, they are promising candidates for the production of bio-based building blocks and molecules. These molecules can be used in animal feed, food supplements, nutraceuticals, cosmetics, and pharmaceuticals [1–4]. Microalgae have also been proposed as a promising candidate for lipid accumulation due to their adaptability to environmental stresses and rapid growth rate, as well as the large number of biochemical compounds they produce [5–7]. Among microalgae, three main genera have been widely used: *Chlamydomonas* sp., *Scenedesmus* sp., and *Chlorella* sp. [8]. Increased use of *Chlorella vulgaris* to produce polyunsaturated fatty acids (PUFA) has raised concerns regarding the environmental impact of its by-products. Indeed, delipidated algal co-products account for approximately 65% of the harvested biomass. It has been reported that the global production of microalgae was 15,000 tons/year [9]. In 2020, the production of *Chlorella vulgaris* within European Union was 82 tons [10]. However, there is lack of studies regarding effective utilization of the microalgal residues, which is left after lipid extraction but still enriched with significant amount of proteins, carbohydrates, or cellulose, so we aimed to study *Chlorella vulgaris* residue as substrate [11].

To decrease the quantity of by-products, residual microalgae biomass obtained after lipids extraction has been considered as a promising substrate for oleaginous yeasts, as residual microalgae biomass is a renewable source of high levels of organic carbon and nitrogen. Many studies on growth and lipid production from oleaginous microorganisms have used residual microalgae biomass after lipids extraction as the sole carbon and nitrogen source [12,13]. However, delipidated microalgae are not usable as a substrate because the remaining components of the cell wall and membrane are not bioavailable for microorganisms [14]. Due to the recalcitrance of algal by-products, pretreatments have been applied to lipid-extracted microalgae before their use [4,15]. As an example of a pretreatment, chemical hydrolysis often generates inhibitory substances that can hinder or abolish the growth of microorganisms cultivated in the resulting hydrolysates [12,16–18]. Seo et al. used physical means, such as ultra-sonication or hydrothermal cavitation, to solubilize residual algal biomass from biodiesel production; however, high energy consumption and scalability are not easy issues to solve [13]. Some researchers reported the use of *C. vulgaris* hydrolysates generated after lipids extraction and enzymatic treatment for the production of bioethanol with a wide diversity of microorganisms, including *E. coli*, *P. stipitis*, and *S. cerevisiae* [19–22]. Moreover, other researchers engineered a strand of *E. coli* that transformed *Chlorella emersonii* hydrolysate to 2-pyrone 4,6-dicarboxylic acid [23]. Shahi et al. attempted to produce bio-oil from a residual biomass of microalgae *Dunaliella sp.* after lipids extraction using hydrothermal liquefaction and achieved a bio-oil yield of 11.8% in dry matter [17].

To simplify the fermentation process and avoid the residual toxic compounds resulting from pretreatment, this study did not submit the *C. vulgaris* residue substrate to any mechanical, chemical, or enzymatic hydrolysis to enhance assimilation of nutrients by oleaginous yeasts.

Y. lipolytica is one of the oleaginous yeasts that can accumulate more than 30% of its dry weight in lipids [24]. It represents an attractive source of valuable compounds that can be used in biotechnology, pharmaceuticals, and the food industry [25,26]. *Y. lipolytica* can grow in a wide range of pH levels and temperatures and can metabolize many substrates, including alkane derivatives and hydrophobic substrates, such as lipids [27,28].

Additionally, *Y. lipolytica* can tolerate some organic solvents, such as those used for lipids extraction, such as hexane or chloroform/methanol [29]. Carbon and nitrogen sources, C/N ratio, temperature, dissolved oxygen, agitation, and pH have been shown to influence the quantity and quality of lipids accumulation significantly, particularly the saturated to unsaturated fatty acids ratio [30–33]. Most reported experiments have used wild-type *Y. lipolytica* strains, such as W29. However, mutant *Y. lipolytica* strains, such as JMY3501, were designed to accumulate more than 45% lipids through the deletion of genes involved in lipids catabolism and triacylglycerol (TAG) intracellular remobilization [34]. The fatty acid profile measured in mutant yeast was compared with that synthesized on the W29 strain, both cultivated on microalgal substrate.

In this study, we examined the potential use of lipid-extracted *C. vulgaris* residue as a sole carbon and nitrogen source for growing two strains of *Y. lipolytica* (W29 and JMY3501). The main objective of this study was to evaluate the impact of this original substrate on the growth rate and fatty acid profile. In addition to sterilization, the microalgal substrate did not undergo any physical, enzymatic, or chemical pretreatment, not only to prevent the generation of inhibitors but also to develop an eco-friendly fermentation process. Moreover, to achieve complete exploitation of microalgae biomass in a microalgae biorefinery, no carbon source was added. The effect of pH on both cell growth and the fatty acid profile was the major criterion considered in this work.

2. Results

This study aimed to evaluate the possibility of using lipid-extracted *C. vulgaris* without pretreatment as a sole carbon and nitrogen source for the growth and fatty acid profile of *Y. lipolytica* strains. The study also assessed the impact of different pH levels on the growth and fatty acid profiles of wild-type W29 and mutant JMY3501 yeasts cultivated in *C. vulgaris* residue or on yeast-extract peptone dextrose (YPD) media. We used different studies to determine the pH levels to be tested. Although the best conditions for the growth of *Y. lipolytica* W29 have been reported as a pH of 5.5 and temperature of 28 °C [35], Zhang et al. concluded that the best pH for producing lipids was 2.0 [33]. Neutral and alkaline conditions have also been shown to enhance protein solubilization and dispersion of the microalgal cell wall [36]. Therefore, we performed the tests in this study at three pH levels: (i) a pH of 3.5 allowed for a better production of lipids, (ii) a pH of 5.5 provided better growth conditions for the yeasts, and (iii) a pH of 7.5 showed better dispersion of the substrate and therefore better accessibility.

2.1. Biochemical and Elemental Analysis of *C. vulgaris* Residue

In this study, we first determined the biomass composition of *C. vulgaris* after lipids extraction (Table 1). As expected, the main constituent in dry *C. vulgaris* residue was proteins at 40.1 ± 5.3% (*w/w* of dry biomass). Other components included residual fatty acids and ash, which were measured as 0.34 ± 0.02% and 11.2 ± 0.1% (*w/w* of dry biomass), respectively. Reducing carbohydrates composed about 15.6 ± 2.1% of the dry biomass.

Table 1. Biochemical and elementary composition of lipid-extracted *C. vulgaris* residue.

Biochemical Compound	Content (Percentage of Dry Weight)
Proteins	40.1 ± 5.3
Reducing carbohydrates	15.6 ± 2.1
Fatty acids	0.34 ± 0.02
Elementary Composition	**Content (Percentage of Dry Weight)**
Total carbon	34.9 ± 0.4
Total nitrogen	7.8 ± 0.2
Ash	11.2 ± 0.1

The green algal lipid-extracted biomass showed high amounts of total carbon and nitrogen, representing 34.9 ± 0.4% and 7.8 ± 0.2%, respectively (Table 1). The acquired biomass data thus suggested that *C. vulgaris* could serve as a sole substrate source of carbon and nitrogen for microbial cultivation. It had an average carbon/nitrogen (C/N ratio) of 4.5, close to that of the YPD medium (C/N = 5.3) used as a control in this study.

2.2. Growth of Wild-Type and Mutant *Y. lipolytica* Strains in Lipid-Extracted *C. vulgaris* Residue and Control YPD Media

We conducted all yeast cultures with the same batch of lipid-extracted *C. vulgaris* residue for 50 h to limit the impact of microalgal biochemical composition. Additionally, we performed all cultures in duplicate. Then, we evaluated the effect of pH on the cell growth of *Y. lipolytica* W29 and JMY3501 cultivated in YPD and *C. vulgaris* residue media.

2.2.1. The Effect of pH on the Cell Growth of *Y. lipolytica* W29

Figure 1a compares the growth profiles of *Y. lipolytica* W29 in YPD medium at a pH of 5.5 and in lipid-extracted *C. vulgaris* residue medium at different pH levels. Table 2 shows the main parameters used to assess the growth of wild-type W29 and JMY3501 yeasts, in both the control YPD and *C. vulgaris* residue media.

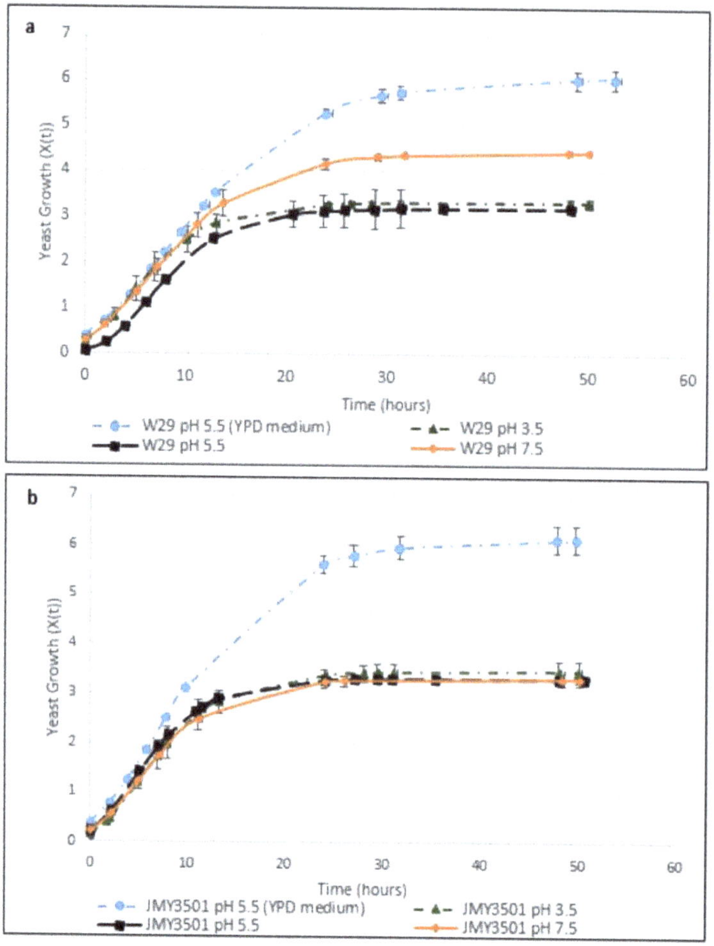

Figure 1. (**a**) Growth curves of *Y. lipolytica* W29 on YPD medium at a pH of 5.5 (blue circle) and on *C. vulgaris* residue at pH levels of 3.5 (green triangle), 5.5 (black square), and 7.5 (orange diamond). (**b**) Growth curves of *Y. lipolytica* JMY3501 on YPD medium at a pH of 5.5 (blue circle) and on *C. vulgaris* residue at pH levels of 3.5 (green triangle), 5.5 (black square), and 7.5 (orange diamond). Growth curves were obtained using the Gompertz model.

The specific growth rate of the wild-type strain W29 in the YPD and *C. vulgaris* residue media appeared similar regardless of pH level (3.5, 5.5, and 7.5). We obtained a maximum specific growth rate (μ_{max}) of 0.28 h^{-1} (Table 2).

However, the wild-type W29 cell concentration showed major differences depending on the culture conditions. When the wild-type strain was cultivated on the microalgal residue medium at a pH of 7.5, the stationary phase was reached after 25 h. In other pH levels, the stationary phase was reached after only 15 h (Table 2, Figure 1a). The final cell concentration of the W29 strain in the *C. vulgaris* medium at a pH of 5.5 or 7.5 was above 70% of the final cell concentration at a pH of 3.5 (Table 2).

Table 2. Summary of the growth parameters of *Y. lipolytica* W29 and JMY3501 on YPD (pH 5.5) and on *C. vulgaris* residue media at different pH levels (3.5, 5.5, and 7.5) after 50 h of cultivation.

Strains Medium	pH	Specific Growth Rate (h^{-1})	Cell Concentration (10^6 Cells·mL^{-1})	Generation Time (h^{-1})	Stationary Phase (h)
Y. lipolytica W29 YPD medium	5.5	0.28 ± 0.03	1082.8 ± 4.61	2.50 ± 0.26	35
Y. lipolytica W29 Lipid-extracted *C. vulgaris* residue	3.5	0.28 ± 0.05	59.6 ± 8.1	2.56 ± 0.49	15
	5.5	0.27 ± 0.06	200.8 ± 94.2	2.67 ± 0.59	15
	7.5	0.26 ± 0.01	235.4 ± 81.5	2.64 ± 0.09	25
Y. lipolytica JMY3501 YPD medium	5.5	0.3238 ± 0.0004	1501.7 ± 33.4	2.14 ± 0.00	35
Y. lipolytica JMY3501 Lipid-extracted *C. vulgaris* residue	3.5	0.28 ± 0.08	77.1 ± 13.7	2.60 ± 0.76	15
	5.5	0.28 ± 0.01	62.8 ± 18.3	2.49 ± 0.11	15
	7.5	0.26 ± 0.03	73.9 ± 6.6	2.69 ± 0.30	15

2.2.2. The Effect of pH on the Cell Growth of JMY3501

We then examined the effect of pH (3.5, 5.5, and 7.5) on the growth of the JMY3501 strain in the *C. vulgaris* residue medium compared with that at a pH of 5.5 in the YPD medium. JMY3501 is a genetically modified strain in which the β-oxidation pathway, intracellular lipid mobilization, and alkaline protease have been blocked, rendering overexpression of the genes that push and pull TAG biosynthesis [34]. We compared the results with those of the W29 strain cultivated in the same media. The mutant strain showed almost equivalent microbial growth compared to the W29 strain (0.28 ± 0.01 h^{-1}, Table 2) at different pH levels when the same residue was used (Figure 1b, Table 2). However, we obtained a maximum specific growth rate (μ_{max}) of 0.3238 ± 0.0004 h^{-1} when we cultivated the JMY3501 strain in the YPD medium at a pH of 5.5 (Table 2). We did not observe a significant difference between the two strains under the same conditions in terms of maximum specific growth rates, nor did we observe a lag time, regardless of the pH level or medium used (*C. vulgaris* residue and YPD at pH 5.5). The cell concentration of the JMY3501 strain was similar for the three pH levels, ranging from 62.8 ± 18.3 to 77.1 ± 13.7 10^6 cells.ml^{-1} after 50 h of cultivation (Table 2). Thus, the growth of JMY3501 cultivated in the *C. vulgaris* residue medium did not seem to be influenced by the pH variation in the medium.

These results clearly show that the growth profile of the W29 and JMY3501 strains differed in YPD and *C. vulgaris* residue media. The wild-type strain had more difficulty exploiting the microalgal substrate at a pH of 3.5. Although its cell growth was three times higher at pH levels of 5.5 and 7.5 in *C. vulgaris* residue, this growth was still 4–5 times lower compared with that in the YPD medium. Regarding cultures of mutant yeast in *C. vulgaris* residue, performance of W29 was improved by 30% regardless of pH compared with the JMY3501 yeast at pH levels of 5.5 and 7.5 in the microalgal residue used as the sole source of carbon and nitrogen. Though we did not observe changes in growth rate analysis at different pH levels, we expected to see changes in the fatty acids profile.

2.3. *Fatty Acid Profiles of Wild-Type and Mutant Y. lipolytica Strains*

Several factors, such as the physiological state, dissolved oxygen, agitation, osmotic pressure, nutrients, and carbon and nitrogen sources, influence the biosynthesis of fatty acid methyl esters (FAMEs). In this study, we measured the effect of the substrate and pH levels on the fatty acid profiles of the yeast strains.

2.3.1. Effect of Different pH Levels on the Fatty Acid Profile of *Y. lipolytica* W29

We compared the fatty acid profiles of the W29 strain after 50 h of cultivation in both media: YPD at a pH of 5.5 (yeast control) and *C. vulgaris* residue at different pH levels. We incubated blank *C. vulgaris* residue at a pH of 7.5 for 50 h to evaluate the temporal evolution of fatty acids without yeast growth. We selected this pH because of the improved substrate homogenization. Fatty acid concentrations were quantified using a gas chromatography–flame ionization detector (GC-FID). The results are shown in Figure 2a,b.

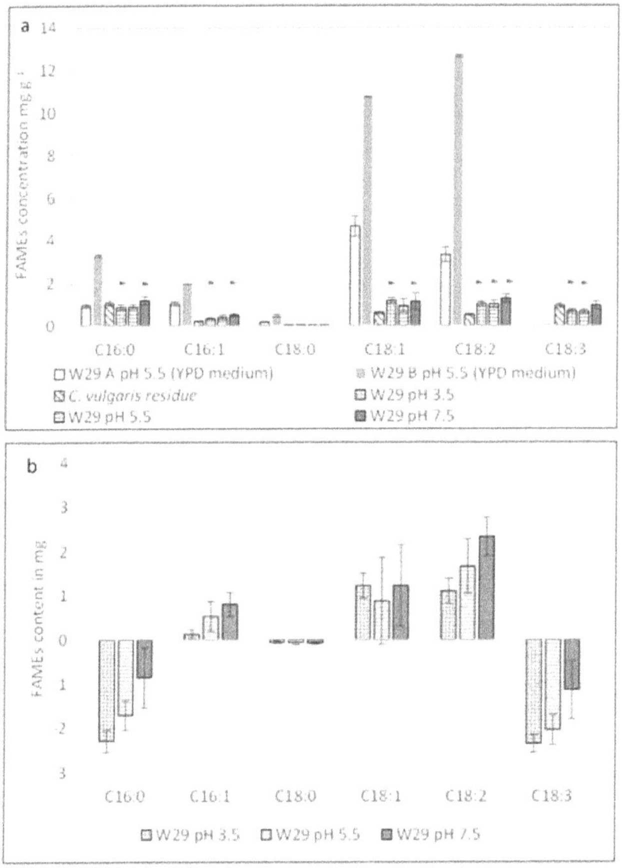

Figure 2. (**a**) Raw data regarding fatty acid content in *Y. lipolytica* W29 cells cultivated in the YPD medium at a pH of 5.5 (A and B represent two different fermenters in the YPD medium), algal raw material control (*C. vulgaris* residue incubated at a pH of 7.5), and the W29 strain cultivated in *C. vulgaris* residue for 50 h at different pH levels of 3.5, 5.5, and 7.5. (A statistical difference was found between the initial concentration in the *C. vulgaris* residue and final concentration of fatty acids after fermentation in the same residue; Kruskal–Wallis test, n = 4, ddl = 1, * significant at a *p*-value < 0.05.) (**b**) Evolution of fatty acid quantities in the fermenter after 50 h cultivation of the W29 strain in *C. vulgaris* residue. These values were obtained by subtraction of the initial fatty acid quantities present in the *C. vulgaris* residue from the final fatty acid quantities present in the fermenter. (C16:0: palmitic acid; C16:1: palmitoleic acid; C18:0: stearic acid; C18:1: oleic acid; C18:2: linoleic acid; C18:3: α-linolenic acid.)

In the YPD medium, oleic acid (C18:1) was the predominant fatty acid found in the wild-type W29 strain (12.8 mg·g^{-1} of biomass dry weight), while linolenic acid (C18:2) reached 7.2 mg.g^{-1} of the dry biomass. Palmitic acid (C16:0) and palmitoleic acid (C16:1) were identified, but no linolenic acid (C18:3) was detected in the W29 strain cultivated in the YPD medium. Regardless of pH, the wild-type strain fermented in the *C. vulgaris* residue medium showed no significant changes regarding the fatty acid profile present in the initial algal raw material, as shown in Figure 2a.

FAMEs analyses were performed on the biomass produced after 50 h of yeast growth without any separation of the yeasts and remaining *C. vulgaris* residue. To demonstrate a potential biosynthesis of FAMEs by the wild-type W29, we subtracted the fatty acid quantities present in the initial *C. vulgaris* residue from the final fatty acid content measured in the recovered biomass. Thus, the fatty acid content shown in Figures 2b and 3b was likely underestimated.

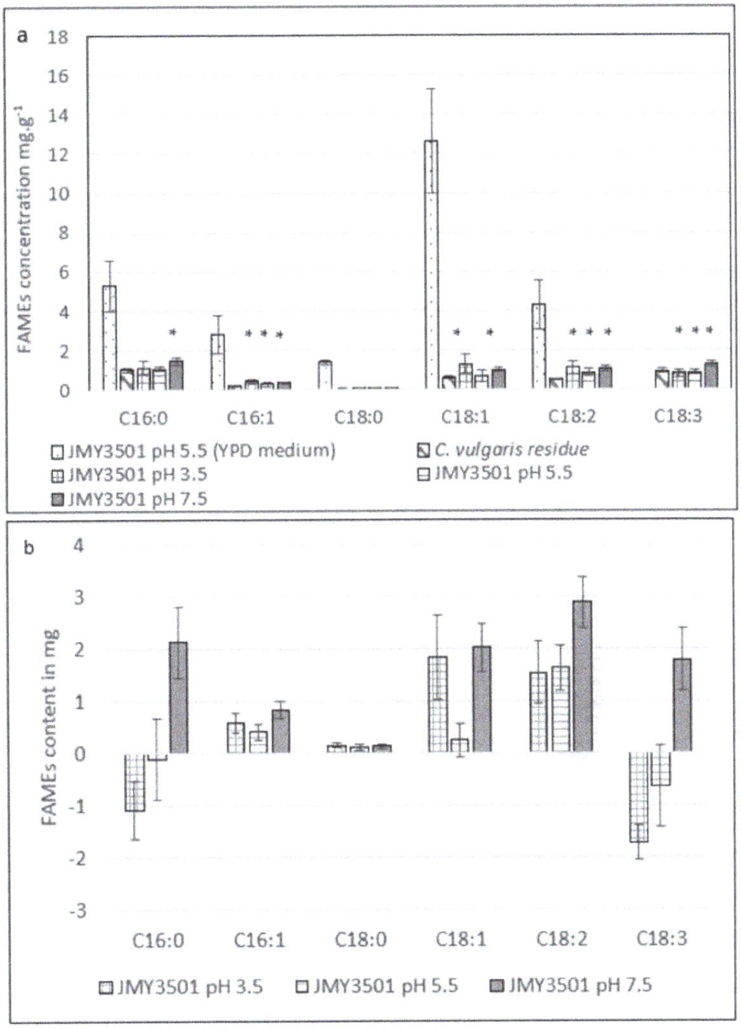

Figure 3. (**a**) Raw data on the fatty acid content of the control (*C. vulgaris* cultivated at a pH of 7.5) and the JMY3501 strain cultivated on YPD at a pH of 5.5 and on *C. vulgaris* residue media for 50 h at different pH levels. (A statistical difference was found between the initial and final concentration of fatty acids in the *C. vulgaris* residue; Kruskal–Wallis test, n = 4, ddl = 1, * significant at p-value < 0.05.) (**b**) Evolution of fatty acid quantities in the fermenter after cultivation of the JMY3501 strain in *C. vulgaris* residue for 50 h. These values were obtained by subtracting the initial fatty acid quantities present in *C. vulgaris* residue from the final fatty acid quantities present in the fermenter. (C16:0: palmitic acid; C16:1: palmitoleic acid; C18:0: stearic acid; C18:1: oleic acid; C18:2: linoleic acid; C18:3: α-linolenic acid).

As shown in Figure 2b, the wild-type strain catabolized C16:0 and C18:3 at different pH levels (3.5, 5.5, and 7.5; Kruskal–Wallis test, p-value < 0.05), while C16:1, C18:1, and C18:2 seemed to be anabolized by the *Y. lipolytica* W29 at pH levels of 3.5, 5.5, and 7.5 (Kruskal–Wallis test, p-value < 0.02). As seen in Table 3, the total quantity of FAMEs and biomass remain unchanged in the wild-type strain after 50 h of cultivation in the *C. vulgaris* residue medium (regardless of pH).

Table 3. Total FAMEs and biomass at different pH levels for the W29 and JMY3501 strains cultivated in lipid-extracted *C. vulgaris* residue medium and YPD at a pH of 5.5 after 50 h of cultivation.

Strains	pH	Media	Total Concentration of FAMEs mg.g^{-1}	Total Dry Biomass (g)	Total FAMEs (mg)
Lipid-extracted *C. vulgaris* residue	7.5		3.3 ± 0.2	5.4 ± 0.9	17.8 ± 4.1
Y. lipolytica W29	5.5	YPD	25.4 ± 18.4	4.0 ± 0.4	101.6 ± 83.8
Y. lipolytica W29	3.5	Lipid-extracted *C. vulgaris* residue	4.1 ± 0.5	3.8 ± 0.4	15.6 ± 3.3
Y. lipolytica W29	5.5	Lipid-extracted *C. vulgaris* residue	3.9 ± 0.7	4.4 ± 0.3	17.1 ± 4.4
Y. lipolytica W29	7.5	Lipid-extracted *C. vulgaris* residue	5.1 ± 0.7	4.0 ± 0.2	20.1 ± 3.9
Y. lipolytica JMY3501	5.5	YPD	26.5 ± 6.1	4.4 ± 0.5	115.8 ± 39.8
Y. lipolytica JMY3501	3.5	Lipid-extracted *C. vulgaris* residue	4.9 ± 1.5	4.1 ± 0.7	19.1 ± 9.4
Y. lipolytica JMY3501	5.5	Lipid-extracted *C. vulgaris* residue	3.9 ± 0.6	5.0 ± 0.1	19.5 ± 3.5
Y. lipolytica JMY3501	7.5	Lipid-extracted *C. vulgaris* residue	5.3 ± 0.6	5.26 ± 0.02	27.6 ± 1.2

2.3.2. Effect of Different pH Levels on the Fatty Acid Profile of *Y. lipolytica* JMY3501

When comparing the wild-type and modified yeast strains cultivated on the YPD medium (Figures 2a and 3a), we noticed similar fatty acid profiles. In contrast, the JMY3501 strain was genetically engineered to be incapable of degrading fatty acids and remobilizing TAGs. On the YPD medium, the average fatty acid accumulation in the JMY3501 strain (26.5 ± 6.1 mg.g^{-1}) was equivalent to that of the W29 strain (25.4 ± 18.4 mg.g^{-1}). This showed the importance of a high C/N ratio for fatty acid accumulation [37].

The fatty acid content of the JMY3501 strain cultivated in the *C. vulgaris* residue medium at different pH levels indicated the presence mainly of long-chain fatty acids with 16 and 18 carbon atoms. To obtain an accurate understanding of the fatty acid metabolism of the mutant yeast, we calculated the final quantity of each identified fatty acid after deduction of the FAMEs quantity present in the microalgal raw material control bioreactor (Figure 3b). In addition to the fatty acid profile of the *C. vulgaris* residue, JMY3501 cultivation appears to induce an increase in C18:2 synthesis regardless of pH (Kruskal–Wallis test, p-value < 0.02), while C18:1 rose only at pH levels of 3.5 and 7.5 (Kruskal–Wallis test, p-value < 0.02).

JMY3501 growth at a pH of 7.5 in *C. vulgaris* appeared to be the only culture showing an increase in the total FAMEs associated with an unchanged quantity of biomass after 50 h of cultivation (Table 3). Under these specific conditions, fatty acid quantities differed from the other tested conditions. Indeed, at a pH of 7.5, the JMY3501 strain cultivated on *C. vulgaris* residue synthesized significant quantities of C16:0, C18:2, and C18:3 (Kruskal–Wallis test, p-value < 0.02; Figure 3b).

3. Discussion

The broad objective of this study was to assess the potential use of lipid-extracted *C. vulgaris* residue as a sole source of carbon and nitrogen without pretreatment for growing two strains of *Y. lipolytica* (W29 and JMY3501). We examined the effect of substrate and pH on both cell growth and the fatty acid profile.

First, we compared the cell growth of wild-type and mutant *Y. lipolytica* strains on lipid-extracted *C. vulgaris* residue and control YPD media. Different pH levels (3.5, 5.5, and 7.5) were evaluated. The specific growth rates of the wild-type strain W29 in the YPD and *C. vulgaris* residue media were similar regardless of pH (3.5, 5.5, and 7.5). A maximum specific growth rate (μ_{max}) of 0.28 h^{-1} was obtained. This finding matches those reported in studies in which *Y. lipolytica* W29 was cultivated in glucose [37–39]. However, yeast cellular growth (cell counts) was directly impacted by the relevant strain/substrate combination, especially with complex raw material. This explains the differences observed regarding the maximal growth of the W29 and JMY3501 strains cultivated on *C. vulgaris* residue.

The final cell concentration of the W29 strain in the *C. vulgaris* medium at a pH of 5.5 or 7.5 was above 70% of the final cell concentration at a pH of 3.5 (Table 2). This discrepancy may have been due to the reduced accessibility of the nutrients contained in *C. vulgaris* residue because of the pH and lack of pretreatments. Since proteins and polysaccharides are major microalgal macroconstituents, their metabolization requires the secretion of hydrolytic enzymes, such as proteases, from the wild-type W29 strain. This strain had the capacity to secrete alkaline proteases at a pH of 7.5; the lack of alkaline protease secretion by the JMY3501 strain prevented protein degradation and thus the release of amino acids and peptides from the microalgal substrate. Therefore, cell growth in the microalgal medium appeared to be three times lower in the deleted *xpr2* gene JMY3501 mutant at pH levels of 5.5 and 7.5 than in the wild-type W29 strain at the same pH levels [40].

To simplify the process and avoid the residual toxic compounds that result from pretreatment, we did not submit the *C. vulgaris* residue substrate to any mechanical, chemical, or enzymatic hydrolysis to enhance the yeasts' assimilation of nutrients [19]. Therefore, both *Y. lipolytica* strains had less readily available carbon and nitrogen sources for their growth. Several studies have explored the feasibility of using microalgae hydrolysate for microbial growth [12,13,17]. However, complex detoxification would be necessary prior to fermentation; hence, these researchers did not undertake the additional costs and eco-friendly aspects that were adopted in our study.

Regarding the lipids accumulation within the JMY3501 cultivated on *C. vulgaris* residue at different pH levels, the cultivation of JMY3501 on microalgal residue with a C/N ratio of 4.5 achieved a moderate cellular lipid content (5.3 ± 0.6 mg.g^{-1} of biomass dry weight) at a pH of 7.5, similar to that achieved by the W29 strain under the same conditions (5.1 ± 0.7 mg.g^{-1} of biomass dry weight). A pH of 7.5 appeared to have a positive effect on lipid accumulation when the mutant strain was cultivated on microalgal residue as its sole source of nutrients (27.6 ± 1.2 mg of total FAMEs in the JMY3501 biomass vs. 17.8 ± 4.1 mg in the *C. vulgaris* residue). The overall production efficiency of FAMEs was similar for the two strains (W29 and JMY3501). However, the cell growth of the JMY3501 strain was weaker. Therefore, the conversion yield of *C. vulgaris* residue into fatty acids was better in the JMY3501 strain. Previous research has shown that lipid accumulation is highly dependent on the C/N ratio and is induced by nitrogen starvation [24,41] and has reported that the optimal C/N ratio for lipids accumulation is around 30 [42,43]. *C. vulgaris* residue has a high nitrogen content and a low C/N ratio, which is not favorable for lipid production based on the nitrogen-limitation strategy. In this study, the JMY3501 strain deleted in the alkaline protease was chosen to promote lipid synthesis requiring a high C/N ratio. This choice was made at the expense of cell cultures requiring metabolizable nitrogen to direct lipid anabolism.

The FAMEs profiles and mass balances of the W29 and JMY3501 strains showed that pH and culture media influence the balance between lipid anabolism and catabolism. In both yeasts, lipid synthesis tended to increase when the pH in the microalgal residue media increased. This phenomenon was even more pronounced in the JMY3501 strain, with significant metabolization of C18:3 at a pH of 3.5; lipid profile changes were limited at a pH of 5.5, while a clear trend toward lipid anabolism occurred at a pH of 7.5 with the noteworthy appearance of C18:3. The synthesis of α-linolenic acid (C18:3 [c9, c12, c15]) in the JMY3501 strain cultivated on *C. vulgaris* residue after 50 h (Figure 3b) was confirmed by GC-MS (data not shown). *Y. lipolytica* is known not to synthesize this fatty acid naturally [44–47].

When the JMY3501 strain was cultivated on *C. vulgaris* residue, the amount of fatty acids varied depending on the pH level (Figure 3a). The best production of C18:1 and C18:2 was observed at a pH of 3.5, while production of C18:3 was higher at a pH of 7.5. Previous studies have found that when *Y. lipolytica* is cultivated in a canola oil substrate, about 22% of the total FAMEs are C18:3. This C18:3 is not synthesized but originates from the plant after being accumulated by the yeast [30,31]. Few researchers have reported significant quantities of C18:3 in different *Y. lipolytica* strains grown on glucose-based medium. In 2020, Carsanba et al. showed that C18:3 content in wild-type strains W29 and H917 reached 4.1% and 14.2%, respectively, based on the total lipids and 19.9% in the mutant strain Pol1dL. In 2014, Mattanna et al. described the *Y. lipolytica* strain QU22 isolated in artisanal Brazilian cheeses whose total lipid fraction contained 13% of C18:3 [44,48]. However, no information was provided about the C18:3 synthetic pathways in *Y. lipolytica* strains. In oilseeds and microorganisms, the synthesis of α-linolenic acid occurs through desaturation of linoleic acid (C18:2 c9, c12) with Δ15 desaturase. Synthesis can also be carried out by a Δ12-15 desaturase in several species of fungi [45,49,50].

In this study, no C18:3 was measured in the wild-type W29 or JMY3501 strain cultivated in YPD medium. This PUFA was only synthesized when the mutant JMY3501 was grown at a pH of 7.5 on *C. vulgaris* residue (Kruskal–Wallis test, p-value = 0.02). This C18:3 synthesis could be attributed to an unknown factor present in *C. vulgaris* residue released at a pH of 7.5 that induced the synthesis in the JMY3501 strain specifically engineered to inhibit lipid catabolism. These results provide a basis for further exploration of C18:3 production by *Y. lipolytica*: genetic modifications can be combined with i) the overexpression of cellulases and proteases to release polysaccharides and proteins from microalgal residue; ii) addition of a carbon source in the medium, such as glycerol; and iii) chemical, enzymatic, or physical pretreatments of the residue.

4. Materials and Methods

4.1. Strains and Plasmid Genotype

Y. lipolytica strain W29 (CLIB89, *MATa* WT) was obtained from the Centre International de Ressources Microbiennes (CIRM), while the JMY3501 strain was provided by the laboratory Biologie Intégrative du Métabolisme Lipidique INRAE, UMR1319, Jouy-en-Josas, France. *Y. lipolytica* W29 is a wild-type strain [41]. The prototrophic strain JMY3501 (*MATa ura3-302 leu2-270 xpr2-322 Δpox1-6 Δtgl4 + pTEF-DGA2 LEU2ex + pTEF-GPD1 URA3ex*) was previously described by Lazar et al. and is unable to degrade fatty acids or remobilize TAG due to deletion of the POX genes and the TGL4 gene, respectively [34]. The deletion of POX1-6 completely blocks β-oxidation in the peroxisome [51], whereas deletion of TGL4 prevents the release of fatty acids from the lipid body [52]. Additionally, this strain overexpresses *YlDGA2* and *YlGPD1*, which push and pull TAG biosynthesis. GPD1 is involved in glycerol-3-phosphate formation, a precursor of TAG [53], while the DGA2 gene encodes the acyltransferase involved in the final step of TAG formation [54]. The *xpr2* gene, which directs the synthesis of a thermosensitive alkaline protease, was also deleted.

4.2. Culture Conditions

YPD medium was used to preculture yeasts at 28 °C and 150 rpm for 15 h. The medium contained 10 g.L^{-1} yeast extract (VWR Chemicals ®), 20 g.L^{-1} peptone (Universal Peptone M66, Merck®), and 20 g.L^{-1} glucose (VWR chemicals ®). The W29 and JMY3501 strains were cultivated in a 500 mL bioreactor (INFORS®HT, Switzerland) with a final operating

Author Contributions: Conceptualization, A.-L.H. and F.A.; methodology, A.-L.H., T.C. and G.D.-B.; formal analysis, G.D.-B., N.I., T.C., M.L. and A.-L.H.; investigation, G.D.-B.; data curation, G.D.-B.; writing—original draft preparation, G.D.-B., N.I. and M.L.; writing—review and editing, N.I., M.L., F.A. and A.-L.H.; supervision A.-L.H. All authors have read and agreed to the published version of the manuscript.

Funding: This research was funded by the financial support of the European Regional Development Fund (ERDF) under the Interreg France-Wallonia-Vlaanderen program and in the framework of the ALPO project Polymer Materials from Microalgae Biomass.

Institutional Review Board Statement: Not applicable.

Data Availability Statement: Not applicable.

Acknowledgments: The authors are grateful to Interreg V project ALPO for financial support. The authors acknowledge Jean-Marc Nicaud for providing *Yarrowia lipolytica* JMY3501 strain.

Conflicts of Interest: The authors declare no conflict of interest. The funders had no role in the design of the study; in the collection, analyses, or interpretation of data; in the writing of the manuscript; or in the decision to publish the results.

References

1. Safi, C.; Zebib, B.; Merah, O.; Pontalier, P.Y.; Vaca-Garcia, C. Morphology, composition, production, processing and applications of *Chlorella vulgaris*: A review. *Renew. Sustain. Energy Rev.* **2014**, *35*, 265–278. [CrossRef]
2. De Morais, M.G.; da Silva Vaz, B.; de Morais, E.G.; Costa, J.A.V. Biologically Active Metabolites Synthesized by Microalgae. *Biomed. Res. Int.* **2015**, *2015*, 835761. [CrossRef]
3. Falaise, C.; François, C.; Travers, M.-A.; Morga, B.; Haure, J.; Tremblay, R.; Turcotte, F.; Pasetto, P.; Gastineau, R.; Hardivillier, Y.; et al. Antimicrobial compounds from eukaryotic microalgae against human pathogens and diseases in aquaculture. *Mar. Drugs* **2016**, *14*, 159. [CrossRef] [PubMed]
4. Tang, D.Y.Y.; Khoo, K.S.; Chew, K.W.; Tao, Y.; Ho, S.H.; Show, P.L. Potential utilization of bioproducts from microalgae for the quality enhancement of natural products. *Bioresour. Technol.* **2020**, *304*, 835761. [CrossRef] [PubMed]
5. Cho, H.U.; Cho, H.U.; Park, J.M.; Park, J.M.; Kim, Y.M. Enhanced microalgal biomass and lipid production from a consortium of indigenous microalgae and bacteria present in municipal wastewater under gradually mixotrophic culture conditions. *Bioresour. Technol.* **2017**, *228*, 290–297. [CrossRef] [PubMed]
6. Zabed, H.M.; Akter, S.; Yun, J.; Zhang, G.; Awad, F.N.; Qi, X.; Sahu, J.N. Recent advances in biological pretreatment of microalgae and lignocellulosic biomass for biofuel production. *Renew. Sustain. Energy Rev.* **2019**, *105*, 105–128. [CrossRef]
7. Ferreira, G.F.; Ríos Pinto, L.F.; Carvalho, P.O.; Coelho, M.B.; Eberlin, M.N.; Maciel Filho, R.; Fregolente, L.V. Biomass and lipid characterization of microalgae genera Botryococcus, Chlorella, and Desmodesmus aiming high-value fatty acid production. *Biomass Convers. Biorefinery* **2019**, *11*, 1675–1689. [CrossRef]
8. Rumin, J.; Nicolau, E.; de Oliveira, R.G.; Fuentes-Grünewald, C.; Flynn, K.J.; Picot, L. A bibliometric analysis of microalgae research in the world, Europe, and the European Atlantic area. *Mar. Drugs* **2020**, *18*, 79. [CrossRef]
9. Benemann, J. Microalgae for Biofuels and Animal Feeds. *Energies* **2013**, *6*, 5869–5886. [CrossRef]
10. Araújo, R.; Calderón, F.V.; Sánchez López, J.; Azevedo, I.C.; Bruhn, A.; Fluch, S.; Tasende, M.G.; Ghaderiardakani, F.; Ilmjärv, T.; Laurans, M.; et al. Current Status of the Algae Production Industry in Europe: An Emerging Sector of the Blue Bioeconomy. *Front. Mar. Sci.* **2021**. [CrossRef]
11. Bhattacharya, M.; Goswami, S. Microalgae—A green multi-product biorefinery for future industrial prospects. *Biocatal. Agric. Biotechnol.* **2020**, *25*, 101580. [CrossRef]
12. Younes, S.; Bracharz, F.; Awad, D.; Qoura, F.; Mehlmer, N.; Brueck, T. Microbial lipid production by oleaginous yeasts grown on *Scenedesmus obtusiusculus* microalgae biomass hydrolysate. *Bioprocess Biosyst. Eng.* **2020**, *43*, 1629–1638. [CrossRef] [PubMed]
13. Seo, Y.H.; Han, S.; Han, J.I. Economic biodiesel production using algal residue as substrate of lipid producing yeast *Cryptococcus curvatus*. *Renew. Energy* **2014**, *69*, 473–478. [CrossRef]
14. Yen, H.W.; Brune, D.E. Anaerobic co-digestion of algal sludge and waste paper to produce methane. *Bioresour. Technol.* **2007**, *98*, 130–134. [CrossRef]
15. Jankowska, E.; Sahu, A.K.; Oleskowicz-Popiel, P. Biogas from microalgae: Review on microalgae's cultivation, harvesting and pretreatment for anaerobic digestion. *Renew. Sustain. Energy Rev.* **2017**, *75*, 692–709. [CrossRef]
16. Sun, M.L.; Madzak, C.; Liu, H.H.; Song, P.; Ren, L.J.; Huang, H.; Ji, X.J. Engineering *Yarrowia lipolytica* for efficient γ-linolenic acid production. *Biochem. Eng. J.* **2017**, *117*, 172–180. [CrossRef]
17. Shahi, T.; Beheshti, B.; Zenouzi, A.; Almasi, M. Bio-oil production from residual biomass of microalgae after lipid extraction: The case of *Dunaliella* sp. *Biocatal. Agric. Biotechnol.* **2020**, *23*, 101494. [CrossRef]

18. Huang, X.; Bai, S.; Liu, Z.; Hasunuma, T.; Kondo, A.; Ho, S.H. Fermentation of pigment-extracted microalgal residue using yeast cell-surface display: Direct high-density ethanol production with competitive life cycle impacts. *Green Chem.* **2020**, *22*, 153–162. [CrossRef]
19. Kim, K.H.; Choi, I.S.; Kim, H.M.; Wi, S.G.; Bae, H.-J.J. Bioethanol production from the nutrient stress-induced microalga *Chlorella vulgaris* by enzymatic hydrolysis and immobilized yeast fermentation. *Bioresour. Technol.* **2014**, *153*, 47–54. [CrossRef]
20. Lee, S.; Oh, Y.; Kim, D.; Kwon, D.; Lee, C.; Lee, J. Converting carbohydrates extracted from marine algae into ethanol using various ethanolic *Escherichia coli* strains. *Appl. Biochem. Biotechnol.* **2011**, *164*, 878–888. [CrossRef]
21. Lee, O.K.; Oh, Y.-K.; Lee, E.Y. Bioethanol production from carbohydrate-enriched residual biomass obtained after lipid extraction of *Chlorella* sp. KR-1. *Bioresour. Technol.* **2015**, *196*, 22–27. [CrossRef]
22. De Farias Silva, C.E.; Meneghello, D.; Bertucco, A. A systematic study regarding hydrolysis and ethanol fermentation from microalgal biomass. *Biocatal. Agric. Biotechnol.* **2018**, *14*, 172–182. [CrossRef]
23. Htet, A.N.; Noguchi, M.; Ninomiya, K.; Tsuge, Y.; Kuroda, K.; Kajita, S.; Masai, E.; Katayama, Y.; Shikinaka, K.; Otsuka, Y.; et al. Application of microalgae hydrolysate as a fermentation medium for microbial production of 2-pyrone 4,6-dicarboxylic acid. *J. Biosci. Bioeng.* **2018**, *125*, 717–722. [CrossRef] [PubMed]
24. Ratledge, C.; Wynn, J.P. The biochemistry and molecular biology of lipid accumulation in oleaginous microorganisms. *Adv. Appl. Microbiol.* **2002**, *51*, 1–52. [CrossRef] [PubMed]
25. Groenewald, M.; Boekhout, T.; Neuvéglise, C.; Gaillardin, C.; van Dijck, P.W.M.; Wyss, M. *Yarrowia lipolytica*: Safety assessment of an oleaginous yeast with a great industrial potential. *Crit. Rev. Microbiol.* **2014**, *40*, 187–206. [CrossRef] [PubMed]
26. Spagnuolo, M.; Hussain, M.S.; Gambill, L.; Blenner, M. Alternative substrate metabolism in *Yarrowia lipolytica*. *Front. Microbiol.* **2018**, *9*, 1–14. [CrossRef] [PubMed]
27. Thevenieau, F.; Beopoulos, A.; Desfougeres, T.; Sabirova, J.; Albertin, K.; Zinjarde, S.; Nicaud, J.-M. Uptake and Assimilation of Hydrophobic Substrates by the Oleaginous Yeast *Yarrowia lipolytica*. In *Handbook of Hydrocarbon and Lipid Microbiology*; Springer: Berlin/Heidelberg, Germany, 2010; pp. 1513–1527.
28. Tezaki, S.; Iwama, R.; Kobayashi, S.; Shiwa, Y.; Yoshikawa, H.; Ohta, A.; Horiuchi, H.; Fukuda, R. Δ12-fatty acid desaturase is involved in growth at low temperature in yeast *Yarrowia lipolytica*. *Biochem. Biophys. Res. Commun.* **2017**, *488*, 165–170. [CrossRef]
29. Krzyczkowska, J.; Kozłowska, M. Effect of Oils Extracted from Plant Seeds on the Growth and Lipolytic Activity of *Yarrowia lipolytica* Yeast. *J. Am. Oil Chem. Soc.* **2017**, *94*, 661–671. [CrossRef]
30. Sestric, R.; Munch, G.; Cicek, N.; Sparling, R.; Levin, D.B. Growth and neutral lipid synthesis by *Yarrowia lipolytica* on various carbon substrates under nutrient-sufficient and nutrient-limited conditions. *Bioresour. Technol.* **2014**, *164*, 41–46. [CrossRef]
31. Yang, S.; Wang, W.; Wei, H.; van Wychen, S.; Pienkos, P.T.; Zhang, M.; Himmel, M.E. Comparison of nitrogen depletion and repletion on lipid production in yeast and fungal species. *Energies* **2016**, *9*, 685. [CrossRef]
32. Lopes, M.; Gomes, A.S.; Silva, C.M.; Belo, I. Microbial lipids and added value metabolites production by *Yarrowia lipolytica* from pork lard. *J. Biotechnol.* **2018**, *265*, 76–85. [CrossRef] [PubMed]
33. Zhang, S.; Jagtap, S.S.; Deewan, A.; Rao, C.V. pH selectively regulates citric acid and lipid production in *Yarrowia lipolytica* W29 during nitrogen-limited growth on glucose. *J. Biotechnol.* **2019**, *290*, 10–15. [CrossRef] [PubMed]
34. Lazar, Z.; Dulermo, T.; Neuvéglise, C.; Crutz-Le Coq, A.M.; Nicaud, J.M. Hexokinase-A limiting factor in lipid production from fructose in *Yarrowia lipolytica*. *Metab. Eng.* **2014**, *26*, 89–99. [CrossRef] [PubMed]
35. Sekova, V.Y.; Dergacheva, D.I.; Isakova, E.P.; Gessler, N.N.; Tereshina, V.M.; Deryabina, Y.I. Soluble Sugar and Lipid Readjustments in the *Yarrowia lipolytica* Yeast at Various Temperatures and pH. *Metabolites* **2019**, *9*, 307. [CrossRef]
36. Rashidi, B.; Trindade, L.M. Detailed biochemical and morphologic characteristics of the green microalga *Neochloris oleoabundans* cell wall. *Algal Res.* **2018**, *35*, 152–159. [CrossRef]
37. Ochoa-Estopier, A.; Guillouet, S.E. D-stat culture for studying the metabolic shifts from oxidative metabolism to lipid accumulation and citric acid production in *Yarrowia lipolytica*. *J. Biotechnol.* **2014**, *170*, 35–41. [CrossRef]
38. Timoumi, A.; Cléret, M.; Bideaux, C.; Guillouet, S.E.; Allouche, Y.; Molina-Jouve, C.; Fillaudeau, L.; Gorret, N. Dynamic behavior of *Yarrowia lipolytica* in response to pH perturbations: Dependence of the stress response on the culture mode. *Appl. Microbiol. Biotechnol.* **2017**, *101*, 351–366. [CrossRef]
39. Azambuja, S.P.H.; Bonturi, N.; Miranda, E.A.; Gombert, A.K. Physiology and lipid accumulation capacity of different *Yarrowia lipolytica* and *Rhodosporidium toruloides* strains on glycerol. *bioRxiv* **2018**, 1–18. [CrossRef]
40. Gonzalez-Lopez, C.I.; Szabo, R.; Blanchin-Roland, S.; Gaillardin, C. Genetic control of extracellular protease synthesis in the yeast *Yarrowia lipolytica*. *Genetics* **2002**, *160*, 417–427. [CrossRef]
41. Barth, G.; Gaillardin, C. Yarrowia lipolytica. In *Nonconventional Yeasts in Biotechnology*; Wolf, K., Ed.; Springer: Berlin/Heidelberg, Germany, 1996; pp. 313–388.
42. Beopoulos, A.; Chardot, T.; Nicaud, J.M. *Yarrowia lipolytica*: A model and a tool to understand the mechanisms implicated in lipid accumulation. *Biochimie* **2009**, *91*, 692–696. [CrossRef]
43. Wu, S.; Zhao, X.; Shen, H.; Wang, Q.; Zhao, Z.K. Microbial lipid production by *Rhodosporidium toruloides* under sulfate-limited conditions. *Bioresour. Technol.* **2011**, *102*, 1803–1807. [CrossRef] [PubMed]
44. Mattanna, P.; Dallé da Rosa, P.; Gusso, A.P.; Richards, N.; Valente, P. Enhancement of microbial oil production by alpha-linolenic acid producing *Yarrowia lipolytica* strains QU22 and QU137. *Food Sci. Biotechnol.* **2014**, *23*, 1929–1934. [CrossRef]

45. Cordova, L.T.; Alper, H.S. Production of α-linolenic acid in *Yarrowia lipolytica* using low-temperature fermentation. *Appl. Microbiol. Biotechnol.* **2018**, *102*, 8809–8816. [CrossRef] [PubMed]
46. Gemperlein, K.; Dietrich, D.; Kohlstedt, M.; Zipf, G.; Bernauer, H.S.; Wittmann, C.; Wenzel, S.C.; Müller, R. Polyunsaturated fatty acid production by *Yarrowia lipolytica* employing designed myxobacterial PUFA synthases. *Nat. Commun.* **2019**, *10*, 4055. [CrossRef]
47. Yan, F.X.; Dong, G.R.; Qiang, S.; Niu, Y.J.; Hu, C.Y.; Meng, Y.H. Overexpression of △12, △15-Desaturases for Enhanced Lipids Synthesis in *Yarrowia lipolytica*. *Front. Microbiol.* **2020**, *11*, 289. [CrossRef]
48. Carsanba, E.; Papanikolaou, S.; Fickers, P.; Erten, H. Lipids by *Yarrowia lipolytica* Strains Cultivated on Glucose in Batch Cultures. *Microorganisms* **2020**, *8*, 1054. [CrossRef]
49. Cui, J.; He, S.; Ji, X.; Lin, L.; Wei, Y.; Zhang, Q. Identification and characterization of a novel bifunctional Δ12/Δ15-fatty acid desaturase gene from *Rhodosporidium kratochvilovae*. *Biotechnol. Lett.* **2016**, *38*, 1155–1164. [CrossRef]
50. Sayanova, O.; Haslam, R.; Guschina, I.; Lloyd, D.; Christie, W.W.; Harwood, J.L.; Napier, J.A. A bifunctional Δ12,Δ15-desaturase from *Acanthamoeba castellanii* directs the synthesis of highly unusual n-1 series unsaturated fatty acids. *J. Biol. Chem.* **2006**, *281*, 36533–36541. [CrossRef]
51. Beopoulos, A.; Mrozova, Z.; Thevenieau, F.; Le Dall, M.T.; Hapala, I.; Papanikolaou, S.; Chardot, T.; Nicaud, J.M. Control of lipid accumulation in the yeast *Yarrowia lipolytica*. *Appl. Environ. Microbiol.* **2008**, *74*, 7779–7789. [CrossRef]
52. Dulermo, T.; Tréton, B.; Beopoulos, A.; Gnankon, A.P.K.; Haddouche, R.; Nicaud, J.M. Characterization of the two intracellular lipases of *Y. lipolytica* encoded by TGL3 and TGL4 genes: New insights into the role of intracellular lipases and lipid body organisation. *Biochim. Biophys. Acta—Mol. Cell Biol. Lipids* **2013**, *1831*, 1486–1495. [CrossRef]
53. Dulermo, T.; Nicaud, J.M. Involvement of the G3P shuttle and B-oxidation pathway in the control of TAG synthesis and lipid accumulation in *Yarrowia lipolytica*. *Metab. Eng.* **2011**, *13*, 482–491. [CrossRef]
54. Beopoulos, A.; Haddouche, R.; Kabran, P.; Dulermo, T.; Chardot, T.; Nicaud, J.M. Identification and characterization of DGA2, an acyltransferase of the DGAT1 acyl-CoA:diacylglycerol acyltransferase family in the oleaginous yeast *Yarrowia lipolytica*. New insights into the storage lipid metabolism of oleaginous yeasts. *Appl. Microbiol. Biotechnol.* **2012**, *93*, 1523–1537. [CrossRef] [PubMed]
55. Ryckebosch, E.; Muylaert, K.; Foubert, I. Optimization of an analytical procedure for extraction of lipids from microalgae. *J. Am. Oil Chem. Soc.* **2012**, *89*, 189–198. [CrossRef]
56. Tjørve, K.M.C.; Tjørve, E. The use of Gompertz models in growth analyses, and new Gompertz-model approach: An addition to the Unified-Richards family. *PLoS ONE* **2017**, *12*, e0178691. [CrossRef] [PubMed]
57. Miller, G.L. Use of Dinitrosalicylic Acid Reagent for Determination of Reducing Sugar. *Anal. Chem.* **1959**, *31*, 426–428. [CrossRef]
58. Rice, E.W.; Baird, R.B.; Eaton, A.D. *Standard Methods for the Examination of Water and Wastewater*, 23th ed.; American Public Health Association: Washington, DC, USA; American Water Works Association: Washington, DC, USA, 2017.
59. Sluiter, A.; Hames, B.; Ruiz, R.; Scarlata, C.; Sluiter, J.; Templeton, D. *Determination of Ash in Biomass*; NREL Laboratory Analytical Procedure (LAP): Golden, CO, USA, 2008; pp. 1–17.
60. Bligh, E.G.; Dyer, W.J. A rapid method of total lipid extraction and purification. *Can. J. Biochem. Physiol.* **1959**, *37*, 911–917. [CrossRef]

marine drugs

MDPI

Article

Total Phenolic Levels, In Vitro Antioxidant Properties, and Fatty Acid Profile of Two Microalgae, *Tetraselmis marina* Strain IMA043 and Naviculoid Diatom Strain IMA053, Isolated from the North Adriatic Sea

Riccardo Trentin [1,2], Luísa Custódio [1,*], Maria João Rodrigues [1], Emanuela Moschin [2], Katia Sciuto [3], José Paulo da Silva [1] and Isabella Moro [2,*]

1 Centre of Marine Sciences, Faculty of Sciences and Technology, University of Algarve, Ed. 7, Campus of Gambelas, 8005-139 Faro, Portugal; riccardo.trentin.2@studenti.unipd.it (R.T.); mary_p@sapo.pt (M.J.R.); jpsilva@ualg.pt (J.P.d.S.)

2 Department of Biology, University of Padova, Via U. Bassi 58/B, 35131 Padova, Italy; emanuela.moschin@unipd.it

3 Department of Chemical, Pharmaceutical and Agricultural Sciences, University of Ferrara, Via Luigi Borsari 46, 44121 Ferrara, Italy; katia.sciuto@unife.it

* Correspondence: lcustodio@ualg.pt (L.C.); isabella.moro@unipd.it (I.M.)

Citation: Trentin, R.; Custódio, L.; Rodrigues, M.J.; Moschin, E.; Sciuto, K.; da Silva, J.P.; Moro, I. Total Phenolic Levels, In Vitro Antioxidant Properties, and Fatty Acid Profile of Two Microalgae, *Tetraselmis marina* Strain IMA043 and Naviculoid Diatom Strain IMA053, Isolated from the North Adriatic Sea. *Mar. Drugs* **2022**, *20*, 207. https://doi.org/10.3390/md20030207

Academic Editor: Carlos Almeida

Received: 20 January 2022
Accepted: 11 March 2022
Published: 12 March 2022

Publisher's Note: MDPI stays neutral with regard to jurisdictional claims in published maps and institutional affiliations.

Abstract: This work studied the potential biotechnological applications of a naviculoid diatom (IMA053) and a green microalga (*Tetraselmis marina* IMA043) isolated from the North Adriatic Sea. Water, methanol, and dichloromethane (DCM) extracts were prepared from microalgae biomass and evaluated for total phenolic content (TPC) and in vitro antioxidant properties. Biomass was profiled for fatty acid methyl esters (FAME) composition. The DCM extracts had the highest levels of total phenolics, with values of 40.58 and 86.14 mg GAE/g dry weight (DW in IMA053 and IMA043, respectively). The DCM extracts had a higher radical scavenging activity (RSA) than the water and methanol ones, especially those from IMA043, with RSAs of 99.65% toward 2,2′-azino-bis(3-ethylbenzothiazoline-6-sulfonic acid)diammonium salt (ABTS) at 10 mg/mL, and of 103.43% against 2,2-diphenyl-1-picrylhydrazyl (DPPH) at 5 mg/mL. The DCM extract of IMA053 displayed relevant copper chelating properties (67.48% at 10 mg/mL), while the highest iron chelating activity was observed in the water extract of the same species (92.05% at 10 mg/mL). Both strains presented a high proportion of saturated (SFA) and monounsaturated (MUFA) fatty acids. The results suggested that these microalgae could be further explored as sources of natural antioxidants for the pharmaceutical and food industry and as feedstock for biofuel production.

Keywords: biodiesel; microalgal biotechnology; natural antioxidants

1. Introduction

The large amount of resources provided by the marine environment constitute the basis of many economic activities and, according to the Green Paper—Towards a Future Maritime Policy for the Union, the exploitation of the marine biodiversity will contribute to many industrial sectors, including those related to aquaculture, healthcare, cosmetics, food/feed, and energy [1].

In particular, algal extracts have been gaining increasing interest due to their unique chemical composition and possibility of wide industrial applications over the last few years [2–4]. Recent developments in genetic engineering, coupled with the enormous biodiversity of microalgae, make this group of organisms one of the most promising sources for new products and applications [4]. With the development of new culture and screening techniques, microalgal biotechnology can already meet the high demands of both the food and pharmaceutical industries [5] and is considered fundamental for the development of sustainable processes that contribute to the global bioeconomy [6]. However, marine microalgae remain largely unexplored and still represent an almost untapped reservoir of novel products and metabolites, such as lipids, with possible applications in different sectors, including aquaculture, human health, nutrition, and biofuel production [1,4,7].

Green algae (Chlorophyta) and diatoms (Bacillariophyta) are acknowledged as important sources of high-added-value natural products. However, a reliable molecular identification is critical for the bioprospection of these organisms [8]. Mistakes in sample identification can have direct implications for the interpretation of biochemical and physiological results, undermining our understanding of the biology of these species, as well as the possibility of a commercial exploitation [9]. Phylogenetically related species to *Tetraselmis marina* (IMA043) and a naviculoid diatom (IMA053) have important commercial applications. For example, *Tetraselmis* (Stein) species (green marine microalgae) are commonly used in aquaculture, due to their high nutritional value in terms of proteins and fatty acids (FA) [10]. Specifically, different strains of the species *T. marina* (Cienkowski) R.E.Norris, Hori, and Chihara are used as sources of carotenoids and FA for feed, food, and biofuel production (strain CTM 20,015) [10] and in the bioremediation of water contaminated with copper, iron, and manganese (strain AC16-MESO) [11]. Furthermore, Naviculales (Bessey) is an order of benthic diatoms comprising more than 5600 species having a similar boat-shape, and produces several bioactive compounds of commercial interest [12,13]. For example, phytosterols from *Navicula incerta* Grunow strain KMMCC B-001 induced apoptosis in hepatocarcinoma (HepG2) cells by upregulating the expression of pro-apoptotic gene (Bax, p53) while downregulating the anti-apoptotic gene Bcl-2 [14]. Other members of this order, including *Navicula* sp. strain JPCC DA0580 [15] and *Fistulifera solaris* strain JPCC DA0580, are considered a promising source of lipids for biofuel production [16].

The main goal of this work was to evaluate the potential of two microalgae strains, isolated from the North Adriatic Sea and identified combining morphological and DNA sequence data, as new sources of natural antioxidants, FA, and phenolic compounds. For this purpose, the naviculoid diatom strain IMA053 and *T. marina* strain IMA043 were isolated from water samples collected in the North Adriatic Sea (Chioggia, Italy) cultivated under laboratory conditions and identified by light microscopy and genetic analyses (amplification of the 18S rDNA gene). The obtained dried biomass was characterized in terms of FA methyl ester contents (FAME) by gas chromatography–mass spectrometry (GC–MS), and used for the preparation of methanol, water, and dichloromethane (DCM) extracts. The extracts were evaluated for total content of phenolics (TPC) by a spectrophotometric method, and for in vitro antioxidant activity by radical-based methods (scavenging on the 1,1-diphenyl-2-picrylhydrazyl (DPPH) and 2,2′-azino-bis 3-ethylbenzothiazoline-6-sulfonic acid (ABTS) radicals) and metal-based methods (metal chelation of copper and iron, and ferric reducing activity).

With this study, we have contributed to the knowledge about the production capability of different potentially useful compounds by the genus *Tetraselmis* (phylum Chlorophyta) and the family Stauroneidaceae (phylum Bacillariophyta). In particular, besides adding data to the little information so far available for the TPC of the naviculoid family Stauroneidaceae, this was the first time that the TPC of the species *T. marina* was analyzed. In addition, the antioxidant properties of different extracts from the two investigated microalgal strains were considered by applying five complementary methods, including radical and metal-related assays. In some cases, the obtained results confirmed what was previously found for other species of the genus *Tetraselmis* and for other naviculoid species, while, in other cases, our findings were not in agreement with the evidence already reported for other phylogenetically related taxa. The same could be said for the FAMEs profiles that were obtained for both *Tetraselmis marina* (IMA043) and the naviculoid diatom (IMA053). Possible explanations for the observed discrepancies with previous studies could be explained by the choice of the growing conditions, by the polarity of the extraction solvents used in this study, and by interspecific variations. Moreover, as our results regard new and unexplored taxa, the possession of unique properties by different microalgal taxonomic entities is emphasized.

2. Results and Discussion

2.1. Molecular Analyses

To establish the taxonomical position of the strains IMA043 and IMA053, isolated from the seawater samples collected from the North Adriatic Sea, a molecular analysis was carried out, through the 18S rRNA gene. Regarding the IMA043 strain, the comparison of its 18S rRNA gene sequence with other ones available in GenBank showed the highest percentage identities with sequences of *T. marina* CCMP898 (99.90%) and *Tetraselmis* sp. KMMCC 1293 (100%). This was evident also in the ML tree generated by IQTREE, where strain IMA043 was part of the *T. marina* clade (Figure 1), confirming the identification based on morphological observations.

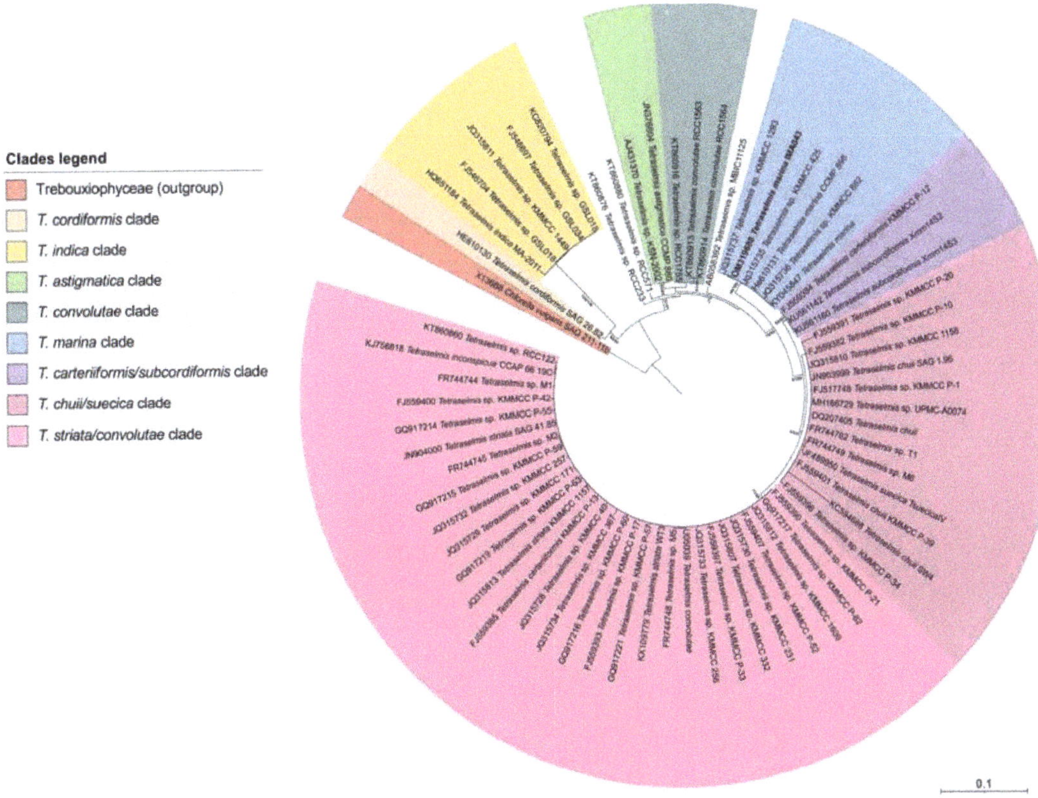

Figure 1. Maximum likelihood (ML) phylogeny based on the 18S rRNA for IMA043-aligned dataset. Approximate Likelihood Ratio Test (SH-aLRT) and ML bootstrap support values (BP) are indicated at nodes. Approximate Likelihood Ratio Test and bootstrap supports ≥70% are shown. Scale number indicates substitutions/site.

The comparison of the 18S rRNA sequence of strain IMA053 with the other sequences from GenBank showed the highest percentage identities with sequences of Bacillariophyta genera (*Sternimirus*, *Prestauroneis* and *Stauroneis*), all belonging to Stauroneidaceae family. In particular, the highest percentage identity (99.65%) was with a sequence of 'Bacillariophyta sp. MBIC10102′, a strain not identified at the species or genus level. Maximum-likelihood (ML) phylogeny of 18S rRNA gene sequences of Naviculales confirmed our identification of strain IMA053 as a species belonging to the Stauroneidaceae family (Figure 2), but to better identify this strain at the genus or species level, it will be necessary to carry out further molecular analyses by using different markers (*rbc*L, *psb*C, and ITS).

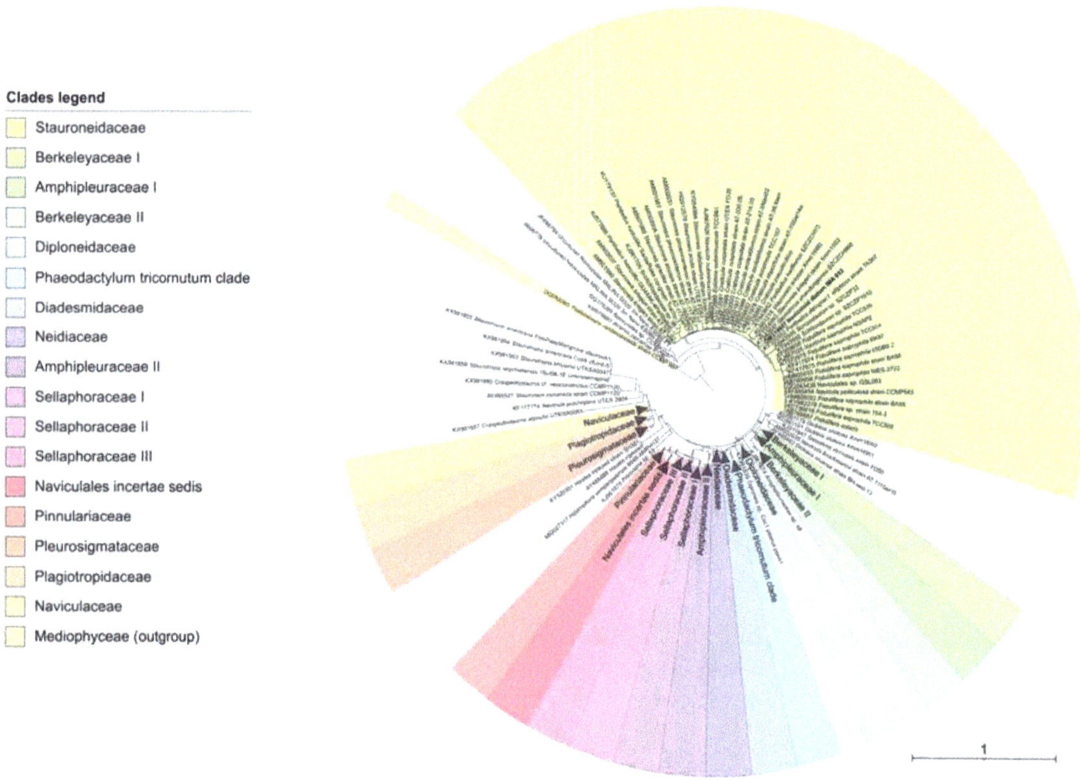

Figure 2. Maximum likelihood (ML) phylogeny based on the 18S rRNA for IMA053 aligned dataset. Approximate Likelihood Ratio Test (SH-aLRT) and ML bootstrap support values (BP) are indicated at nodes. Approximate Likelihood Ratio Test and bootstrap supports ≥70% are shown. Scale number indicates substitutions/site.

2.2. Light Microscopy

Light microscope observations of the isolated *T. marina* IMA043 showed slightly compressed oval cells, about 12 µm long, with the posterior part wider than the apical one. Cells were shown in the anterior end of a depression, where four equal-length flagella were located, emerging from the cell in two pairs (Figure 3a). Cells were green in color with a single large cup-shaped chloroplast, sometimes located in the posterior end of the cell. A single pyrenoid was present within the chloroplast. The pyrenoid occupied a central position inside the chloroplast (Figure 3a). Moreover, in the anterior part of the cell, there was an orange-red eyespot (Figure 3a). *Tetraselmis marina* cells divided in a nonmotile vegetative stage (Figure 3b).

Naviculoid diatom IMA053 cells showed a simple shape and a bilateral symmetry. Valves were lanceolate with rounded apices in the valve view (Figure 3c). In the valve view, the central raphe system was evident (Figure 3c). Even if the valves generally offer more features useful for the identification of the species, during our observation, there were many lipid inclusions preventing the valve analysis (Figure 3d).

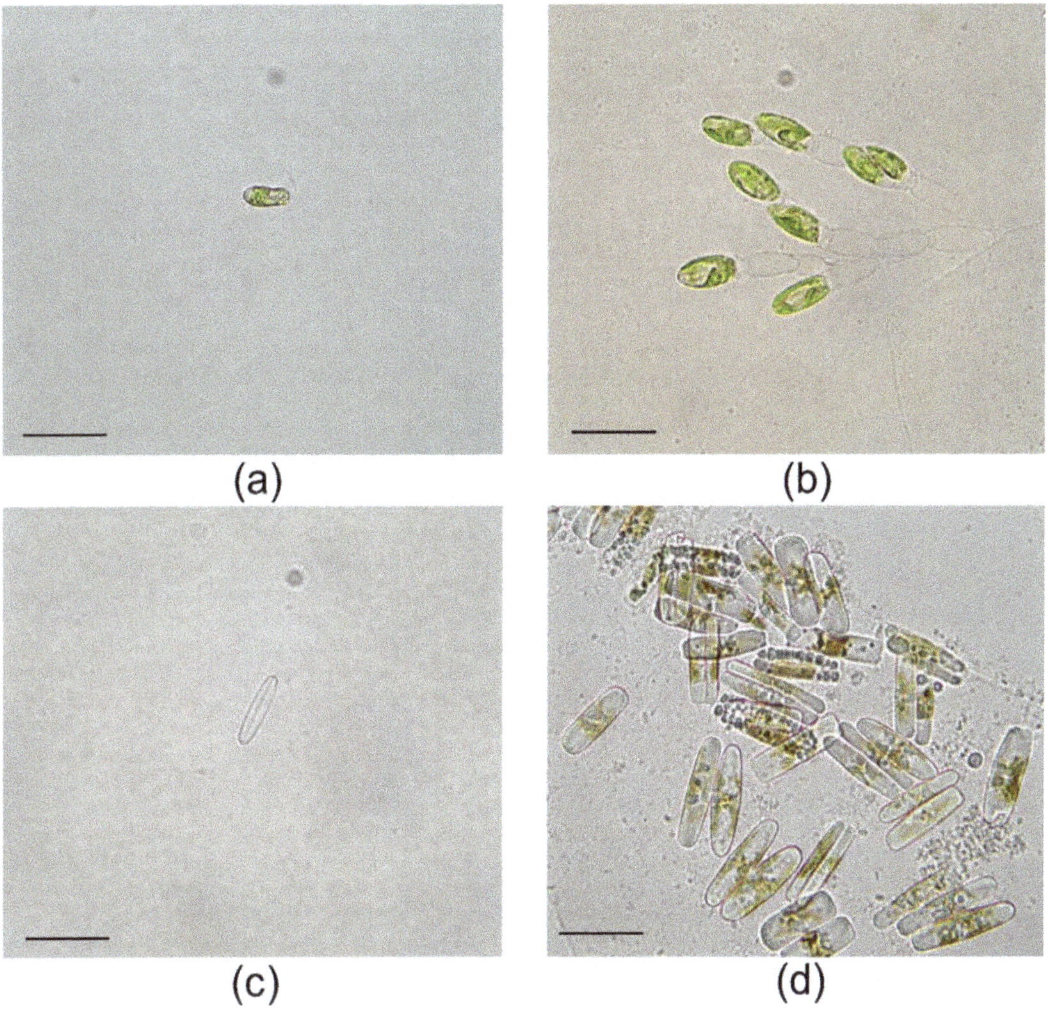

Figure 3. Light microscopy image of the strains IMA043 and IMA053: (**a**) single cell of *T. marina* IMA043, scale bar = 20 μm; (**b**) nonmotile vegetative stage of *T. marina* IMA043, scale bar = 20 μm; (**c**) valve view of a single cell of the naviculoid diatom IMA053, scale bar = 20 μm; (**d**) lipid-rich cells of naviculoid diatoms IMA053, scale bar = 20 μm.

2.3. Total Phenolics Content (TPC) of the Extracts

Polyphenolic compounds are the main contributors to the antioxidant defense mechanisms of plants and algae and exhibit several interesting biological activities relevant for commercial exploitation, related to their radical scavenging properties, such as anti-inflammatory, anti-atherosclerotic, and anti-carcinogenic [17–19]. Therefore, this work explored the TPC of extracts from the two strains, *T. marina* IMA043 and naviculoid diatom IMA053 (Table 1). The TPC of the different samples varied from 17.80 mg GAE/g dry weight (DW) in the water extract of naviculoid diatom IMA053 to 86.14 GAE/g (DW) in the DCM extract of *T. marina* IMA043. The extraction of different phenolics relies on their polar properties and, especially, polyphenols are often more soluble in organic solvents less polar than water, for example, DCM [19–21].

Table 1. Total phenolic content (TPC, mg GAE/100 g DW, GAE: gallic acid equivalents) of different extracts of *T. marina* (IMA043) and naviculoid diatom (IMA053).

	Extract	TPC
T. marina (IMA043)	Water	20.45 ± 1.87 [d]
	Methanol	25.19 ± 1.26 [cd]
	DCM	86.14 ± 4.52 [a]
Naviculoid diatom (IMA053)	Water	17.80 ± 0.88 [d]
	Methanol	31.85 ± 0.92 [c]
	DCM	40.58 ± 0.66 [b]

Data represent the mean ± standard error of mean (SEM) ($n \geq 6$). In each column, different letters mean significant differences (Multiple Comparisons of Means: Tukey Contrast, 95% family-wise confidence level).

To our best knowledge, there is no published information on the total levels of phenolics of *T. marina* and little is known regarding the phenolic content of species of the Stauroneidaceae family. This work highlighted that the TPCs of both *T. marina* IMA43 and naviculoid diatom IMA053 are higher than those described in the literature for related species, obtained with the same methodology (F–C method). In particular, methanol and hexane extracts from *T. chuii* [22], ethanol/water extracts from *T. suecica* (Kylin) Butcher [23], methanol extract from *Navicula* sp. [24], and 50% ethanol extracts from *Stauroneis* sp. LACW24 [25] showed lower levels of phenolics compared to IMA043 and IMA053. These differences could be explained by the use of solvents with different polarities for the extractions, such as hexane and chloroform, or by choosing different growing conditions that can affect the final composition of microalgae [20–23]. In fact, several environmental conditions could be stimulants for the accumulation of different phenolics, including nitrogen and phosphorus limitation, and salinity, on the microalgae culture medium [10,24]. This is because phenolic compounds act as defense against a wide range of stresses and are thus accumulated in response to these environmental conditions [25]. Moreover, variations in the TPC of extracts obtained by extraction with different solvents are ascribed to different polarities of the compounds present in the biomass. Phenols include hydroxyl groups (polar part) attached to an aromatic ring (nonpolar part) [26]. This stereochemistry distinguishes phenols according to their polarity, which has been used for the recovery of phenols from natural sources by extraction with different solvents. Therefore, the extraction yield in terms of polyphenolic components strongly depends on the nature of the solvent and the extraction is affected by the diffusion coefficient and the dissolution rate of compounds until they reach the equilibrium concentration inside the solvent [27]. Phenolic compounds usually have a more polar nature, as reported by different authors. For example, Hajimahmoodi et al. [28] and Goiris et al. [22] obtained the highest phenolic content in the most polar extract (hot water extracts) from different microalgae species, such as *Chlorella* and *Tetraselmis* sp. However, Li et al. [18] found the highest content of phenolic substances in apolar extracts, namely hexane fractions from *Chlorella* and *Nostoc* species. The recovery of certain phenolic terpenes has been referred to occur preferably with less polar solvents, while more polar solvents have been reported to extract flavonoid glycosides and higher-molecular-weight phenols [29]. In our work, the results suggest that the species being studied are rich in polyphenolic compounds of a more nonpolar nature extractable with solvents such as DCM.

2.4. *In Vitro Antioxidant Properties*

The accumulation of reactive oxygen species (ROS) generates oxidative stress, causing, for example, photosynthesis failure [30]. To cope with oxidative stress, microalgae produce different classes of antioxidant compounds (e.g., phenols, sulphated polysaccharides, carotenoids) that could be exploited as food supplements, food preservatives, nutraceuticals, or pharmaceuticals [19,22,29–32]. In this work, the extracts from *T. marina* (IMA043) and the naviculoid diatom (IMA053) were evaluated for antioxidant activity by five complementary methods, including radical and metal (copper and iron)-related assays. The less polar (DCM) extracts had the highest capability to scavenge the DPPH radical, 178.75% (*T. marina*) and 91.38% (naviculoid diatom), when compared to the methanol and water extracts (Table 2). The high percentage of DPPH scavenging activity of the DCM extracts from *T. marina* is consistent with other findings on *T. suecica* [33] and *T. chui* [34]. The water and methanol extracts from *T. marina* had no capacity to scavenge DPPH, even though other authors detected a high RSA in water extract from *Tetraselmis* sp. and in methanol extracts from *T. tetrathele* (West) Butcher and *T. chuii* Butcher [21,35,36]. However, interspecific variations and different growing conditions, such as nutrient availability and light intensity, can account for the differences observed for different species belonging to the *Tetraselmis* genus [22,33,36]. In turn, the methanol and water extracts from *T. marina* IMA043 had no capacity to scavenge the ABTS radical; however, the DCM extract showed a higher RSA toward the ABTS radical when compared with all the extracts from the naviculoid diatom (IMA053), at all tested concentrations. These results are consistent with those found in the literature, as a relatively high ABTS scavenging activity is reported in ethanol/water extracts from algae of the genus *Tetraselmis*, such as *Tetraselmis* sp. and *T. suecica* [22], and diatoms, including *Chaetoceros calcitrans* (Paulsen) H.Takano and *Phaeodactylum tricornutum* Bholin [22,37]. Moreover, studies on the antioxidant properties of other phylogenetically related diatoms, such as *Navicula incerta* [38] and *Stauroneis* sp. LACW24 [23], reported a high RSA toward DPPH and ABTS, respectively, which is in accordance with our results on the naviculoid diatom (IMA053). At the concentration of 10 mg/mL, the methanol and DCM extracts presented a moderate RSA toward the ABTS radical (35.54% and 43.83%, respectively), while the water extract had a lower antioxidant capacity (23.51%).

Table 2. Radical scavenging on DPPH and ABTS radicals of different extracts from naviculoid diatom (IMA053) and *T. marina* (IMA043). Results are expressed as antioxidant activity (% activity).

	Extract	ABTS			DPPH		
		1 mg/mL	5 mg/mL	10 mg/mL	1 mg/mL	5 mg/mL	10 mg/mL
Naviculoid diatom (IMA053)	Methanol	n.a.	23.37 ± 0.58 ᶜ	35.54 ± 1.28 ᵇ	n.a.	12.53 ± 0.44 ᶜ	18.91 ± 0.46 ᶜ
	Water	n.a.	11.06 ± 1.27 ᵈ	23.51 ± 1.81 ᶜ	n.a.	n.a.	n.a.
	DCM	10.09 ± 0.53 ᵇ	46.91 ± 0.42 ᵇ	43.83 ± 0.84 ᵇ	18.25 ± 6.08 ᵇ	41.55 ± 2.47 ᵇ	91.38 ± 11.80 ᵇ
T. marina (IMA043)	Methanol	n.a.	n.a.	n.a.	n.a.	17.29 ± 0.35 ᶜ	24.94 ± 0.71 ᶜ
	Water	n.a.	n.a.	n.a.	n.a.	n.a.	n.a.
	DCM	24.66 ± 0.32 ᵃ	88.19 ± 2.41 ᵃ	99.65 ± 3.76 ᵃ	40.96 ± 0.94 ᵃ	103.43 ± 7.12 ᵃ	178.75 ± 6.13 ᵃ
	BHT *	94.77 ± 0.04			80.07 ± 0.72		

Values represent the mean ± standard error of mean (SEM) ($n \geq 6$). n.a: no activity; * positive control tested at 1 mg/mL; DCM: dichloromethane. For the same column, different letters are significantly different (Multiple Comparisons of Means: Tukey Contrast, 95% family-wise confidence level).

The FRAP of the extracts at the concentration of 10 mg/mL ranged from 21.47% and 26.75% in the water extracts from both *T. marina* (IMA043) and the naviculoid diatom (IMA053) to 150.10% in the DCM extract from the naviculoid diatom (IMA053) and 103.17% in the methanol extract of *T. marina* (Table 3). Other microalgae exhibited a strong ferrous reducing power, as reported for acetone extracts of *Navicula* sp. [39] and hydroethanolic extracts from *Tetraselmis* sp. [22]. Molecules capable of reducing iron are electron donors that reduce the oxidized intermediates of lipid peroxidation processes, acting as primary and secondary antioxidants [40]. Our results indicate that both microalgal strains contain molecules that act as electron donors stabilizing radical species and counteracting their harmful effect [41], thus preventing damage to DNA, lipids, proteins, and other biomolecules [42].

Transition metal ions, such as copper (Cu^{2+}) and iron (Fe^{2+}), can lead to the formation of free radicals that may cause modifications to DNA bases, enhance lipid peroxidation, and affect calcium and sulphydryl homeostasis [43]. Therefore, the use of natural products with the capacity to chelate those metals is considered a useful strategy to manage disorders related to the accumulation of such ions [44]. According to our findings, the water and DCM extracts of the naviculoid diatom (IMA053) showed moderate to high metal chelating activity, while the methanol extract had no metal chelating properties, unlike the results reported for *Navicula incerta* [38] (Table 4). However, these differences can be explained by different causes, including interspecific variations and extrinsic factors, such as light intensity, salinity, and nutrient levels, or by the different tested extracts [45]. The water extract from *T. marina* showed a high capacity to chelate Fe^{2+}; a similar metal chelating activity is reported in the literature for the genus *Tetraselmis* [36].

Table 3. Ferric reducing antioxidant power (FRAP) of different extracts from naviculoid diatom (IMA053) and *T. marina* (IMA043). Results are expressed as antioxidant activity (% activity).

	Extract	FRAP		
		1 mg/mL	**5 mg/mL**	**10 mg/mL**
Naviculoid diatom (IMA053)	Methanol	22.01 ± 0.68 c	60.68 ± 3.54 ab	70.23 ± 5.47 b
	Water	n.a.	23.29 ± 4.24 b	26.75 ± 1.73 c
	DCM	46.35 ± 8.26 ab	135.10 ± 45.63 a	150.10 ± 17.85 a
T. marina (IMA043)	Methanol	36.17 ± 2.04 bc	92.28 ± 4.13 ab	103.17 ± 3.26 b
	Water	n.a.	19.56 ± 2.15 b	21.47 ± 2.14 c
	DCM	67.16 ± 3.46 a	40.02 ± 5.62 ab	n.a.

Values represent the mean ± standard error of mean (SEM) ($n \geq 6$). n.a: no activity; DCM: dichloromethane. For the same column, different letters are significantly different (Multiple Comparisons of Means: Tukey Contrast, 95% family-wise confidence level).

Table 4. Metal chelating activities on copper (CCA) and iron (ICA) of different extracts from naviculoid diatom (IMA053) and *T. marina* (IMA043).

	Extract	ICA			CCA		
		1 mg/mL	**5 mg/mL**	**10 mg/mL**	**1 mg/mL**	**5 mg/mL**	**10 mg/mL**
Naviculoid diatom (IMA053)	Methanol	n.a.	n.a.	n.a.	n.a.	n.a.	n.a.
	Water	n.a.	83.43 ± 2.22 a	92.05 ± 3.54 a	13.92 ± 2.39 b	29.23 ± 5.53 a	35.31 ± 3.89 b
	DCM	n.a.	49.18 ± 3.77 b	63.57 ± 5.34 b	n.a.	21.11 ± 8.01 a	67.48 ± 14.68a
T. marina (IMA043)	Methanol	34.12 ± 3.00	n.a.	n.a.	n.a.	n.a.	n.a.
	Water	n.a.	59.25 ± 4.23 b	68.00 ± 6.13 b	21.00 ± 2.88 a	25.92 ± 3.41 a	28.87 ± 1.57 b
	DCM	n.a.	n.a.	n.a.	n.a.	n.a.	11.58 ± 1.49 c
	EDTA *	72.28 ± 2.79			89.69 ± 0.55		

Values represent the mean ± standard error of mean (SEM) ($n \geq 6$). n.a: no activity; * positive control tested at 1 mg/mL; DCM: dichloromethane. For the same column, different letters are significantly different (Multiple Comparisons of Means: Tukey Contrast, 95% family-wise confidence level).

2.5. Relation between Phenolic Contents of the Extracts and the In Vitro Antioxidant Properties

As phenolic compounds are considered one of the most important classes of natural antioxidants, a relation between the TPC of the two strains of Adriatic microalgae and their RSA was expected. In this sense, the linear association between TPC and the antioxidant activity obtained with the different assays (FRAP, DPPH, and ABTS) was expressed as the correlation coefficient R^2. The correlation coefficient between FRAP and the TPC of the studied samples was R^2 = 0.75, whereas, for the correlation between antioxidant activity (DPPH and ABTS) and TPC, it was higher (R^2 = 0.85 for DPPH and R^2 = 0.95 for ABTS) (Figure 4). These positive correlations suggest that 75% of the capacity of the naviculoid diatom (IMA053) and *T. marina* (IMA043) strains to reduce iron was due to the contribution of phenolic compounds, while from 85% to 95% of the antioxidant capability of these microalgae was related to the presence of phenols (Figure 4). The remaining 25% (for FRAP), 15% (for DPPH), and 5% (for ABTS) of the antioxidant activity that is not explained by the presence of phenols could be ascribed to other molecules present in the extracts with antioxidant properties [46]. Overall, our results suggest that phenolic compounds were a major contributor to the antioxidant capacities of the tested extracts.

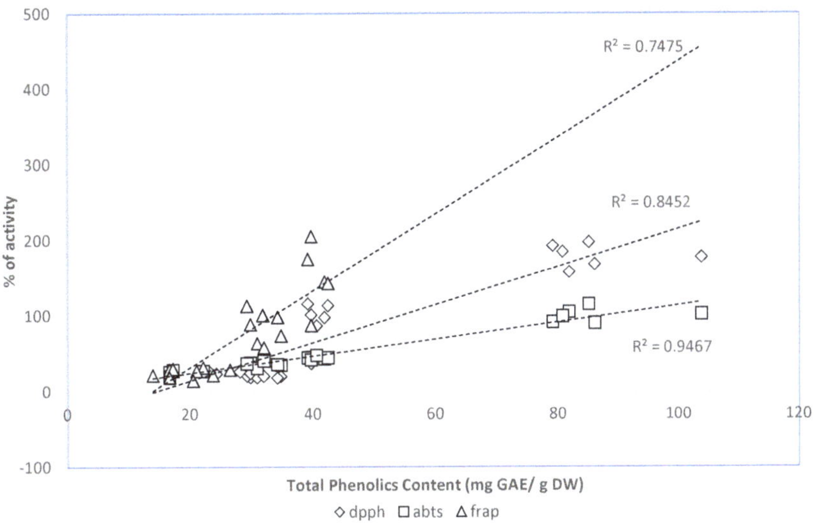

Figure 4. Correlation between the antioxidant capacity and the total phenolic content of the ethanol extracts of naviculoid diatom (IMA053) and *T. marina* (IMA043).

2.6. FAMEs Profile

Tetraselmis marina (IMA043) and the naviculoid diatom (IMA053) were analyzed for FAMEs profile, by GC–MS, after direct transesterification of the dried biomass, and the results are summarized on Table 5. Biomass from both species had a prevalence of MUFA (49.81% for IMA053 and 45.41% for IMA043), followed by SFA (45.70% for IMA053 and 38.28% for IMA043) and PUFA (4.57% for IMA053 and 16.31% for IMA043). A similar FAME profile was observed for *Tetraselmis* sp. [47], while higher concentrations of PUFAs were found in *Tetraselmis* sp. IMP3 and *Tetraselmis* sp. CTP4 [48]. The FAMEs profile of IMA053 is similar to that of *Stauroneis* sp. LACW24 [49]; however, differences in PUFAs content were observed in *Navicula* sp. maintained in the MAScIR's microalgae collection [47]. These variations in the relative proportions of the FAMEs composition of microalgae has a high phenotypic plasticity and can occur in response to different culture conditions [49,50]. In particular, an increase in PUFA levels at low culture temperatures is a common trend reported for microalgae [51]. The FA profile of *T. marina* consisted predominantly of palmitic (C16:0) and oleic (C18:1n9c) acids, in accordance with other findings for the same species [10] and for other species of the same genus, such as *Tetraselmis* sp. [47]. The main FA identified in the naviculoid diatom (IMA053) were palmitoleic acid (C16:1) and palmitic acid (C16:0), as previously reported for other closely related species, including *Navicula* UMACC 231 [52] and *Stauroneis* sp. LACW24 [49]. Among these abundant FA detected in IMA043 and IMA053, palmitic acid (C16:0) from microalgae is widely recognized as an antimicrobial agent [53,54], while palmitoleic acid (C16:1) accumulation is reported to be a remarkable characteristic for the production of biodiesel [55], and preliminary clinical trials suggested that this compound could promote weight loss, reduce cholesterol levels, and manage inflammation [56]. Oleic acid (C18:1n9c), abundant in *T. marina* IMA043, is the most common MUFA in human nutrition and emerging studies have shown that diets enriched in oleic acids can contribute to the management and prevention of obesity, and one of its derivatives (oleoylethanolamide) has been demonstrated to reduce hunger and subsequent food consumption [57].

The lipid content is an important prerequisite determining the suitability of microalgae for commercial biofuel production [58] and the high proportion of SFAs and MUFAs in the naviculoid diatom (IMA053) and *T. marina* (IMA043), suggesting their potential use for that purpose [59,60]. Other closely related species such as *N. pelliculosa* (Kützing) Hilse and

T. suecica have already proved to be desirable for biodiesel production as they can accumulate a large mass of oils per unit volume of the microalgal broth per day [61,62]. High SFAs and MUFAs percentages, together with the absence of C18:3 in the lipid profile of the naviculoid diatom (IMA053) and *T. marina* (IMA043), ensure that both species meet European standard specifications (SFA and MUFA $\geq$ 12%, C18:3 $\leq$ 1) for biodiesel [63,64]. Particularly, the naviculoid diatom (IMA053) showed the highest oleic acid content, which is an important characteristic to produce good-quality biodiesel. It is reported that oils with high oleic acid content have a reasonable balance of fuel, including its ignition quality, combustion heat, cold filter plugging point (CFPP), oxidative stability, viscosity, and lubricity, which are determined by the structure of its component fatty esters [65].

Table 5. Fatty acid profile of the microalgae *T. marina* (IMA043) and the naviculoid diatom (IMA053). Values are given as means of total FAME percentage ± standard deviation (*n* = 3). n.d., not detected.

Fatty Acid	Common Name	*T. marina* (IMA043)	Naviculoid Diatom (IMA053)
∑SFA		**38.28 ± 2.00**	**45.70 ± 1.28**
(C14:0)	Methyl myristate	0.03 ± 0.08	3.15 ± 0.08
(C15:0)	Methyl pentadecanoate	n.d.	1.50 ± 0.96
(C16:0)	Methyl palmitate	37.74 ± 1.99	40.68 ± 0.83
(C18:0)	Methyl stearate	0.42 ± 0.12	0.37 ± 0.11
(C24:0)	Methyl lignocerate	0.09 ± 0.09	n.d.
∑MUFA		**45.41 ± 1.75**	**49.81 ± 1.00**
(C16:1)	Methyl palmitoleate	0.71 ± 0.20	47.25 ± 0.92
(C18:1*n*9c)	*cis*-9-Oleic acid methyl ester	37.52 ± 0.49	0.51 ± 0.15
(C18:1*n*9t)	*trans*-9-Elaidic acid methyl ester	3.17 ± 1.23	1.03 ± 0.29
(C20:1)	Methyl cis-11-eicosenoate	4.01 ± 1.12	n.d.
(C22:1*n*9)	Methyl erucate	n.d.	0.60 ± 0.18
(C24:1*n*9)	Methyl nervonate	n.d.	0.42 ± 0.13
∑PUFA		**16.31 ± 1.64**	**4.57 ± 0.47**
(C19:3*n*3)	Methyl linolenate	2.38 ± 1.43	n.d.
(C18:2*n*6c)	Methyl linoleate	11.01 ± 0.29	0.08 ± 0.08
(C20:4*n*6)	*cis*-5,8,11,14-Eicosatetraenoic acid methyl ester	n.d.	1.06 ± 0.34
(C20:5*n*3)	*cis*-5,8,11,14,17-Eicosapentaenoic acid methyl ester	2.50 ± 0.72	3.43 ± 0.32
(C20:3*n*3)	*cis*-11,14,17-Eicosatrienoic acid methyl ester	0.41 ± 0.22	n.d.
(C20:2)	*cis*-11,14-Eicosadienoic acid methyl ester	0.02 ± 0.03	n.d.
∑*n*-3		**5.28 ± 1.61**	**3.43 ± 0.32**
∑*n*-6		**11.01 ± 0.29**	**1.14 ± 0.35**
∑*n*-6/∑*n*-3		**2.08**	**0.33**
PUFA/SFA		**0.43**	**0.10**

3. Materials and Methods

3.1. Chemicals

The compounds DPPH, ABTS, and FAME standards (Supelco® 37 Component FAME Mix) were purchased from Sigma (Steinheim am Albuch, Germany). Merck (Darmstadt, Germany) supplied the Folin–Ciocalteau (F–C) phenol reagent and all solvents used for chemical analyses. Additional reagents and solvents were obtained from VWR International (Leuven, Belgium).

3.2. Sampling, Strain Isolation and Culture Set Up

Seawater samples were collected in the North Adriatic Sea, near Chioggia, through a plankton net with a nylon mesh of 45 µm. The single-cell isolation of *Tetraselmis marina* (IMA043) and the naviculoid diatom (IMA053) was performed with a heated flame, extended, and broken Pasteur glass pipette. After isolation, cells were maintained in liquid cultures with F/2 medium [66] in a growth chamber at 16 °C, with a photosynthetic photon flux density of 35 µmol photons m^{-2} s^{-1} under a 12-h:12-h light/dark cycle. Microalgal biomass was harvested by centrifugation and immediately frozen at −20 °C. Freeze-dried biomass was obtained upon lyophilization for 24 h and stored at room temperature (RT) in the dark.

3.3. Genetic Analyses

Total genomic DNA of IMA043 and IMA053 was extracted from microalgal pellets, using the Genomic DNA purification kit (Fermentas, Burlington, ON, Canada) according to the manufacturer's protocol. Amplification of the 18S rDNA gene was carried out with the primers Euk528F and EukB [67,68] under the following PCR conditions: initial denaturation at 95 °C for 5 min, followed by 30 cycles (95 °C for 50 s, 52 °C for 50 s, and 72 °C for 1 min 30 s) and final elongation at 72 °C for 8 min. The obtained PCR products were visualized with GelRed (Biotium) staining after electrophoresis in a 1% agarose gel, purified with the ExoSAP-ITTM kit (Amersham Biosciences, Piscataway, New York, NY, USA), and directly sequenced with the same primers used in the amplification reaction. Sequencing was performed at the BMR-Genomics Sequencing Service (Padua University) on both strands to ensure accuracy of the results. The final consensus sequences were assembled using the SeqMan II program from the Lasergene software package (DNAStar, Madison, WI, USA) and analyzed by similarity search using the BLAST program [69], available at the NCBI web server (www.ncbi.nlm.nih.gov/blast, accessed on 23 December 2021). The 18S rRNA sequences generated in this study were deposited in GenBank with the following accession numbers: OM319687 for naviculoid diatom IMA053 and OM319688 for *Tetraselmis marina* IMA043.

For the phylogenetic analysis, two separate datasets were created for the 18S rRNA partial sequences of the strains IMA043 and IMA053, including the new sequences generated in this study and other related sequences obtained from the NCBI archives. Each dataset was aligned with the online version of MAFFT (https://mafft.cbrc.jp/alignment/server, accessed on 27 December 2021). Maximum likelihood (ML) phylogenetic trees were computed for both datasets using the IQ-TREE webserver (http://iqtree.cibiv.univie.ac.at/, accessed on 27 December 2021) [70] using the best-fitting models (TIM3 + F + I + G4 for the IMA043-aligned dataset and GTR + F + I + G4 for the IMA053-aligned dataset) selected by the Model Finder algorithm [71] implemented in IQ-TREE. The statistical support for tree topologies was computed by performing 10,000 ultrafast bootstrap replicates [72] and the SH-aLRT branch test [73]. Trees were visualized in iTOL v5 (https://itol.embl.de, accessed on 27 December 2021) [74].

3.4. Light Microscopy

Cultures of the strains IMA043 and IMA053 were observed with a light microscope (LM) Leitz Dialux 22 (Wetzlar, Germany), equipped with a digital image acquisition system.

3.5. Preparation of the Extracts

The extracts were prepared by an ultrasound-assisted sequential method, as follows. One gram of dried biomass was mixed with water (200 mL) and the algal cell walls were disrupted in an ultrasonic bath (USC-TH, VWR, Portugal; capacity of 5.4 L, frequency of 45 kHz, a supply of 230 V, tub heater of 400 W, temperature control made by a LED display) for 15 min. Samples were filtered (Whatman no. 4) and the residue was then sequentially extracted with methanol and DCM, as described previously. The extracts were dried under a reduced vacuum pressure at 40 °C, weighed, dissolved in the corresponding solvent (water, methanol, and DCM) at the concentration of 50 mg/mL, and stored (4 °C for the methanol and DCM extracts, −20 °C for the water extract).

3.6. Total Phenolics Content (TPC) of the Extracts

TPC was determined by the F–C method [75] adapted to 96-well plates according to Custódio et al. [21]. Extracts (5 µL at the concentration of 10 mg/mL) were mixed with 100 µL of 10-fold-diluted F–C reagent in distilled water. After 5 min of incubation at RT, 100 µL of sodium carbonate (75 g/L, *w*/*v*, in water) was added and incubated for 90 min at RT. The absorbance was measured at 725 nm on a microplate reader (Biochrom™ EZ Read 400, Biochrom Ltd., Cambridge, UK) and the results were expressed as gallic acid equivalents in milligrams per gram of extract (mg GAE/g dry weight, DW). A calibration curve was built using gallic acid standard solutions.

3.7. *In Vitro Antioxidant Properties*

The in vitro antioxidant properties of the extracts were evaluated by five complementary methods, including radical scavenging activity (RSA) on the free radicals (DPPH and ABTS), metal chelation of iron and copper ions, and ferric reducing power (FRAP). The extracts were tested at three different concentrations (1, 5, and 10 mg/mL). The absorbances were measured using a microplate reader (Biochrom™ EZ Read 400, Biochrom Ltd., Cambridge, United Kingdom) and the RSA and metal chelating activities were calculated as the percentage of inhibition relative to a blank-containing solvent instead of the extracts. FRAP results were expressed as inhibition (%) relative to the positive control (butylated hydroxytoluene: BHT), tested at 1 mg/mL.

3.7.1. Radical-Based Assays: RSA on DPPH and ABTS

The RSA against DPPH was evaluated by the method of Brand Williams et al. [76], adapted to 96-well microplates as reported by Custódio et al. [77]. Briefly, 22 µL of the extracts was mixed with 200 µL of an ethanol DPPH solution (120 µM) in 96-well microplates. After 30 min of incubation at RT in the dark, the absorbance was measured at 515 nm. The RSA on the ABTS radical was determined according to the method described by Re et al. [78]. A stock solution of ABTS (7.4 mM) was prepared in potassium persulfate (2.6 mM), left in the dark for 12–16 h at RT, and diluted with ethanol until it reached an absorbance of 0.7 at 734 nm. For the assay, 10 µL of the extracts was mixed with 190 µL of ABTS in 96-well microplates and incubated in darkness at RT for 6 min. The absorbance was read at 734 nm. In both assays, BHT was used as the positive control at the concentration of 1 mg/mL.

3.7.2. Metal-Based Assays: Ferric Reducing Activity Power (FRAP) and Metal Chelation of Iron and Copper

FRAP was evaluated according to Megías et al. [79]. In brief, 50 µL of the samples was mixed in 96-well microplates with 50 µL of potassium ferricyanide (1% in water) and 50 µL of distilled water. After 20 min of incubation in the dark at 50 °C, 50 µL of 10% TCA (trichloroacetic acid in water) and 10 µL of ferric chloride solution (0.1% in water) were added. The absorbance was measured at 700 nm after 10 min of incubation at RT, and BHT was used as the standard.

Copper chelating activity (CCA) was assessed according to Megías et al. [79]. Samples (30 µL) were mixed in 96-well microplates with 200 µL of 50 mM sodium acetate buffer (pH 6), 6 µL of pyrocatechol violet (PV, 4 mM in the acetate buffer), and 100 µL of copper sulphate (50 µg/mL in water). The absorbance was measured at 632 nm. Iron chelating activity (ICA) was determined as described by Megías et al. [79] by measuring the formation of the Fe^{2+} ferrozine complex. Briefly, 30 µL of the extracts was mixed in 96-well microplates with 200 µL of distilled water and 30 µL of an iron (II) chloride solution (0.1 mg/mL in water) and incubated for 30 min at RT. Afterward, 12.5 µL of ferrozine solution (40 mM in water) was added and the change in color was measured at 562 nm. In both assays, ethylenediamine tetraacetic acid (EDTA 1 mg/mL), a synthetic metal chelator, was used as a positive control.

3.8. *Evaluation of the FA Profile of the Biomass*

3.8.1. FAME Preparation

Lipids were converted into the corresponding FAMEs according to the method described by Lepage and Roy [80]. In brief, 1.5 mL of the derivatization solution (methanol/acetyl chloride, 20:1, *v*/*v*) were mixed with 100 mg of dried algae. After 15 min at RT in an ultrasound-water bath, 1 mL of hexane was added to the samples and then heated for one hour at 100 °C. One milliliter of distilled water was added to the transesterification solution after cooling in an ice bath, and the organic phase was removed and dried with anhydrous sodium sulphate. The extracts were then filtered (0.2 µm) and the FAMEs profile determined by GC/MS [81].

3.8.2. Determination of FAMEs Profile by GC–MS

An Agilent GC–MS (Agilent Technologies 6890 Network GC System, 5973 Inert Mass Selective Detector) equipped with a ZB-5MS capillary column (30 m × 0.25 mm internal diameter, 0.25 µm film thickness, Agilent Tech) with helium as the carrier gas was used to determine the FAMEs profile. The injection of samples occurred at 300 °C with a temperature profile of the GC oven at 60 °C (1 min), 30 °C min^{-1} to 120 °C, 4 °C min^{-1} to 250 °C, and 20 °C min^{-1} to 300 °C (4 min). Total ion mode was used for the identification and quantification of FAMEs, and the A—Supelco® 37 Component FAME Mix||(Sigma-Aldrich, Sintra, Portugal) was used as a standard. Different calibration curves were generated for each of the FAME in this standard and results were expressed as mg/g of dry weight (DW).

3.9. Statistical Analysis

All the experiments were conducted at least in triplicate and results were expressed as mean ± standard deviation (SD). Significant differences ($p < 0.05$) were assessed by one-way analysis of variance (ANOVA). If significant, Tukey's pairwise multiple comparison tests were applied. *R*-statistics® 3.5.3 version was used for the statistical analyses.

4. Conclusions

In this work, we explored two strains of microalgae: naviculoid diatom (IMA053) and *T. marina* (IMA043), as sources of natural antioxidants, phenolic compounds, and FAs. Both strains were identified by molecular and morphological analyses before the bioprospecting of these organisms. Overall, the obtained results suggest that DCM extracts from both species have a relevant antioxidant activity against DPPH and ABTS radicals. The naviculoid diatom (IMA053) had a higher iron and copper chelating ability. The extracts of both microalgae were rich in phenolics and a positive correlation between the TCP and the RSA was found, indicating that these algae could be a promising source for nutraceutical and pharmaceutical industries. Moreover, our results suggest that the high proportion of SFAs and MUFAs could render IMA053 and IMA043 desirable feedstocks for biofuel production. The obtained results could be a springboard on these microalgal strains for developing further studies, such as biochemical variation on the phenolics and lipid production and on the FAs profile under different growing conditions. Future steps in accomplishing the study of the strains IMA043 and IMA053 will be the evaluation of the chemical profiles of the extracts by LC/MS/MS and the assessment of their bioactivity through antimicrobial, enzyme inhibitory, and cytotoxicity assays.

Author Contributions: Conceptualization, R.T., L.C. and I.M.; methodology, L.C., K.S., M.J.R. and J.P.d.S.; formal analysis, R.T. and L.C.; investigation, R.T., J.P.d.S., M.J.R., K.S. and E.M.; data curation, R.T.; writing—original draft preparation, R.T.; writing—review and editing, K.S., J.P.d.S., E.M., L.C. and I.M.; supervision, L.C. and I.M.; project administration, L.C.; funding acquisition, L.C. All authors have read and agreed to the published version of the manuscript.

Funding: This research was funded by the Foundation for Science and Technology (FCT) and the Portuguese National Budget, through the UIDB/04326/2020 project. Luísa Custódio was supported by the FCT Scientific Employment Stimulus (CEECIND/00425/2017).

Institutional Review Board Statement: Not applicable.

Informed Consent Statement: Not applicable.

Data Availability Statement: Not applicable.

Conflicts of Interest: The authors declare no conflict of interest.

References

1. Guedes, A.C.; Amaro, H.M.; Malcata, F.X. Microalgae as sources of carotenoids. *Mar. Drugs* **2011**, *9*, 625–644. [CrossRef] [PubMed]
2. Tang, D.Y.Y.; Khoo, K.S.; Chew, K.W.; Tao, Y.; Ho, S.H.; Show, P.L. Potential utilization of bioproducts from microalgae for the quality enhancement of natural products. *Bioresour. Technol.* **2020**, *304*, 122997. [CrossRef] [PubMed]
3. Jimenez-Lopez, C.; Pereira, A.G.; Lourenço-Lopes, C.; Garcia-Oliveira, P.; Cassani, L.; Fraga-Corral, M.; Prieto, M.A.; Simal-Gandara, J. Main bioactive phenolic compounds in marine algae and their mechanisms of action supporting potential health benefits. *Food Chem.* **2021**, *341*, 128262. [CrossRef] [PubMed]
4. Wu, J.; Gu, X.; Yang, D.; Xu, S.; Wang, S.; Chen, X.; Wang, Z. Bioactive substances and potentiality of marine microalgae. *Food Sci. Nutr.* **2021**, *9*, 5279–5292. [CrossRef] [PubMed]
5. Pulz, O.; Gross, W. Valuable products from biotechnology of microalgae. *Appl. Microbiol. Biotechnol.* **2004**, *65*, 635–648. [CrossRef] [PubMed]
6. Fernández, F.G.A.; Reis, A.; Wijffels, R.H.; Barbosa, M.; Verdelho, V.; Llamas, B. The role of microalgae in the bioeconomy. *New Biotechnol.* **2021**, *61*, 99–107. [CrossRef]
7. Priyadarshani, I.; Rath, B. Commercial and industrial applications of micro algae—A review. *J. Algal Biomass Util.* **2012**, *3*, 89–100.
8. Bohutskyi, P.; Liu, K.; Nasr, L.K.; Byers, N.; Rosenberg, J.N.; Oyler, G.A.; Betenbaugh, M.J.; Bouwer, E.J. Bioprospecting of microalgae for integrated biomass production and phytoremediation of unsterilized wastewater and anaerobic digestion centrate. *Appl. Microbiol. Biotechnol.* **2015**, *99*, 6139–6154. [CrossRef]
9. Rindi, F.; Braga, J.C.; Martin, S.; Peña, V.; Le Gall, L.; Caragnano, A.; Aguirre, J. Coralline Algae in a Changing Mediterranean Sea: How Can We Predict Their Future, if We Do Not Know Their Present? *Front. Mar. Sci.* **2019**, *6*, 723. [CrossRef]
10. Dahmen-Ben Moussa, I.; Chtourou, H.; Karray, F.; Sayadi, S.; Dhouib, A. Nitrogen or phosphorus repletion strategies for enhancing lipid or carotenoid production from Tetraselmis marina. *Bioresour. Technol.* **2017**, *238*, 325–332. [CrossRef]
11. Cameron, H.; Mata, M.T.; Riquelme, C. The effect of heavy metals on the viability of Tetraselmis marina AC16-MESO and an evaluation of the potential use of this microalga in bioremediation. *PeerJ* **2018**, *6*, e5295. [CrossRef] [PubMed]
12. Fimbres-Olivarría, D.; López-Elías, J.A.; Carvajal-Millán, E.; Márquez-Escalante, J.A.; Martínez-Córdova, L.R.; Miranda-Baeza, A.; Enríquez-Ocaña, F.; Valdéz-Holguín, J.E.; Brown-Bojórquez, F. Navicula sp. Sulfated Polysaccharide Gels Induced by Fe(III): Rheology and Microstructure. *Int. J. Mol. Sci.* **2016**, *17*, 1238. [CrossRef] [PubMed]
13. Guiry, M.D.; Guiry, G.M. Algaebase. Available online: https://www.algaebase.org/about/ (accessed on 3 January 2022).
14. Kim, Y.S.; Li, X.F.; Kang, K.H.; Ryu, B.M.; Kim, S.K. Stigmasterol isolated from marine microalgae Navicula incerta induces apoptosis in human hepatoma HepG2 cells. *BMB Rep.* **2014**, *47*, 433–438. [CrossRef] [PubMed]
15. Matsumoto, M.; Sugiyama, H.; Maeda, Y.; Sato, R.; Tanaka, T.; Matsunaga, T. Marine diatom, Navicula sp. strain JPCC DA0580 and marine green alga, Chlorella sp. strain NKG400014 as potential sources for biodiesel production. *Appl. Biochem. Biotechnol.* **2010**, *161*, 483–490. [CrossRef] [PubMed]
16. Muto, M.; Tanaka, M.; Liang, Y.; Yoshino, T.; Matsumoto, M.; Tanaka, T. Enhancement of glycerol metabolism in the oleaginous marine diatom Fistulifera solaris JPCC DA0580 to improve triacylglycerol productivity. *Biotechnol. Biofuels* **2015**, *8*, 4. [CrossRef] [PubMed]
17. Chung, K.T.; Wong, T.Y.; Wei, C.I.; Huang, Y.W.; Lin, Y. Tannins and human health: A review. *Crit. Rev. Food Sci. Nutr.* **1998**, *38*, 421–464. [CrossRef]
18. Li, H.B.; Cheng, K.W.; Wong, C.C.; Fan, K.W.; Chen, F.; Jiang, Y. Evaluation of antioxidant capacity and total phenolic content of different fractions of selected microalgae. *Food Chem.* **2007**, *102*, 771–776. [CrossRef]
19. Haminiuk, C.W.I.; Plata-Oviedo, M.S.V.; de Mattos, G.; Carpes, S.T.; Branco, I.G. Extraction and quantification of phenolic acids and flavonols from Eugenia pyriformis using different solvents. *J. Food Sci. Technol.* **2012**, *51*, 2862–2866. [CrossRef]
20. Lee, S.-H.; Karawita, R.; Affan, A.; Lee, J.-B.; Lee, K.-W.; Lee, B.-J.; Kim, D.-W.; Jeon, Y.-J. Potential of Benthic Diatoms Achnanthes longipes, Amphora coffeaeformisand Navicula sp. (Bacillariophyceae) as Antioxidant Sources. *ALGAE* **2009**, *24*, 47–55. [CrossRef]
21. Custódio, L.; Justo, T.; Silvestre, L.; Barradas, A.; Duarte, C.V.; Pereira, H.; Barreira, L.; Rauter, A.P.; Albericio, F.; Varela, J. Microalgae of different phyla display antioxidant, metal chelating and acetylcholinesterase inhibitory activities. *Food Chem.* **2012**, *131*, 134–140. [CrossRef]
22. Goiris, K.; Muylaert, K.; Fraeye, I.; Foubert, I.; De Brabanter, J.; De Cooman, L. Antioxidant potential of microalgae in relation to their phenolic and carotenoid content. *J. Appl. Phycol.* **2012**, *24*, 1477–1486. [CrossRef]
23. Archer, L.; McGee, D.; Parkes, R.; Paskuliakova, A.; McCoy, G.R.; Adamo, G.; Cusimano, A.; Bongiovanni, A.; Gillespie, E.; Touzet, N. Antioxidant Bioprospecting in Microalgae: Characterisation of the Potential of Two Marine Heterokonts from Irish Waters. *Appl. Biochem. Biotechnol.* **2021**, *193*, 981–997. [CrossRef] [PubMed]
24. Bhosale, P. Environmental and cultural stimulants in the production of carotenoids from microorganisms. *Appl. Microbiol. Biotechnol.* **2004**, *63*, 351–361. [CrossRef]
25. Kepekçi, R.A.; Saygideger, S.D. Enhancement of phenolic compound production in Spirulina platensis by two-step batch mode cultivation. *J. Appl. Phycol.* **2012**, *24*, 897–905. [CrossRef]
26. Queimada, A.J.; Mota, F.L.; Pinho, S.P.; Macedo, E.A. Solubilities of Biologically Active Phenolic Compounds: Measurements and Modeling. *J. Phys. Chem. B* **2009**, *113*, 3469–3476. [CrossRef]
27. Pan, W.; Xu, H.; Cui, Y.; Song, D.; Feng, Y.Q. Improved liquid–liquid–liquid microextraction method and its application to analysis of four phenolic compounds in water samples. *J. Chromatogr. A* **2008**, *1203*, 7–12. [CrossRef] [PubMed]

28. Hajimahmoodi, M.; Faramarzi, M.A.; Mohammadi, N.; Soltani, N.; Oveisi, M.R.; Nafissi-Varcheh, N. Evaluation of antioxidant properties and total phenolic contents of some strains of microalgae. *J. Appl. Phycol.* **2010**, *22*, 43–50. [CrossRef]
29. Oreopoulou, V.; Tzia, C. Utilization of Plant By-Products for the Recovery of Proteins, Dietary Fibers, Antioxidants, and Colorants. In *Utilization of by-Products and Treatment of Waste in the Food Industry*; Springer: Boston, MA, USA, 2007; pp. 209–232. [CrossRef]
30. Dring, M.J. Stress Resistance and Disease Resistance in Seaweeds: The Role of Reactive Oxygen Metabolism. *Adv. Bot. Res.* **2005**, *43*, 175–207. [CrossRef]
31. Cirulis, J.T.; Scott, J.A.; Ross, G.M. Management of oxidative stress by microalgae. *Can. J. Physiol. Pharmacol.* **2013**, *91*, 15–21. [CrossRef] [PubMed]
32. Scaglioni, P.T.; Badiale-Furlong, E. Can Microalgae Act as Source of Preservatives in Food Chain? *J. Food Sci. Eng.* **2017**, *7*, 283–296. [CrossRef]
33. Ulloa, G.; Otero, A.; Sánchez, M.; Sineiro, J.; Núñez, M.J.; Fábregas, J. Effect of Mg, Si, and Sr on growth and antioxidant activity of the marine microalga Tetraselmis suecica. *J. Appl. Phycol.* **2011**, *24*, 1229–1236. [CrossRef]
34. Banskota, A.H.; Sperker, S.; Stefanova, R.; McGinn, P.J.; O'Leary, S.J.B. Antioxidant properties and lipid composition of selected microalgae. *J. Appl. Phycol.* **2018**, *31*, 309–318. [CrossRef]
35. Natrah, F.M.I.; Yusoff, F.M.; Shariff, M.; Abas, F.; Mariana, N.S. Screening of Malaysian indigenous microalgae for antioxidant properties and nutritional value. *J. Appl. Phycol.* **2007**, *19*, 711–718. [CrossRef]
36. Custódio, L.; Soares, F.; Pereira, H.; Barreira, L.; Vizetto-Duarte, C.; Rodrigues, M.J.; Rauter, A.P.; Albericio, F.; Varela, J. Fatty acid composition and biological activities of *Isochrysis galbana* T-ISO, *Tetraselmis* sp. and *Scenedesmus* sp.: Possible application in the pharmaceutical and functional food industries. *J. Appl. Phycol.* **2014**, *26*, 151–161. [CrossRef]
37. Foo, S.C.; Yusoff, F.M.; Ismail, M.; Basri, M.; Khong, N.M.H.; Chan, K.W.; Yau, S.K. Efficient solvent extraction of antioxidant-rich extract from a tropical diatom, Chaetoceros calcitrans (Paulsen) Takano 1968. *Asian Pac. J. Trop. Biomed.* **2015**, *5*, 834–840. [CrossRef]
38. Affan, A.; Karawita, R.; Jeon, Y.J.; Lee, J.B. Growth characteristics and antioxidant properties of the benthic diatom *Navicula incerta* (Bacillariophyceae) from Jeju Island, Korea[1]. *J. Phycol.* **2007**, *43*, 823–832. [CrossRef]
39. Karthikeyan, P. In vitro Antioxidant Activity of Marine Diatoms. *IOSR J. Environ. Sci. Toxicol. Food Technol.* **2013**, *5*, 32–37. [CrossRef]
40. Ordoñez, A.A.L.; Gomez, J.D.; Vattuone, M.A.; Isla, M.I. Antioxidant activities of Sechium edule (Jacq.) Swartz extracts. *Food Chem.* **2006**, *97*, 452–458. [CrossRef]
41. Zengin, G.; Uysal, A.; Aktumsek, A.; Mocan, A.; Mollica, A.; Locatelli, M.; Custodio, L.; Neng, N.R.; Nogueira, J.M.F.; Aumeeruddy-Elalfi, Z.; et al. Euphorbia denticulata Lam.: A promising source of phyto-pharmaceuticals for the development of novel functional formulations. *Biomed. Pharmacother.* **2017**, *87*, 27–36. [CrossRef]
42. Halliwell, B. Antioxidants in human health and disease. *Annu. Rev. Nutr.* **1996**, *16*, 33–50. [CrossRef]
43. Valko, M.; Rhodes, C.J.; Moncol, J.; Izakovic, M.; Mazur, M. Free radicals, metals and antioxidants in oxidative stress-induced cancer. *Chem. Biol. Interact.* **2006**, *160*, 1–40. [CrossRef] [PubMed]
44. Plaza, M.; Santoyo, S.; Jaime, L.; García-Blairsy Reina, G.; Herrero, M.; Señoráns, F.J.; Ibáñez, E. Screening for bioactive compounds from algae. *J. Pharm. Biomed. Anal.* **2010**, *51*, 450–455. [CrossRef] [PubMed]
45. Kelman, D.; Posner, E.K.; McDermid, K.J.; Tabandera, N.K.; Wright, P.R.; Wright, A.D. Antioxidant Activity of Hawaiian Marine Algae. *Mar. Drugs* **2012**, *10*, 403–416. [CrossRef] [PubMed]
46. Djeridane, A.; Yousfi, M.; Nadjemi, B.; Boutassouna, D.; Stocker, P.; Vidal, N. Antioxidant activity of some algerian medicinal plants extracts containing phenolic compounds. *Food Chem.* **2006**, *97*, 654–660. [CrossRef]
47. Maadane, A.; Merghoub, N.; Ainane, T.; El Arroussi, H.; Benhima, R.; Amzazi, S.; Bakri, Y.; Wahby, I. Antioxidant activity of some Moroccan marine microalgae: Pufa profiles, carotenoids and phenolic content. *J. Biotechnol.* **2015**, *215*, 13–19. [CrossRef] [PubMed]
48. Cardoso, C.; Pereira, H.; Franca, J.; Matos, J.; Monteiro, I.; Pousão-Ferreira, P.; Gomes, A.; Barreira, L.; Varela, J.; Neng, N.; et al. Lipid composition and some bioactivities of 3 newly isolated microalgae (*Tetraselmis* sp. IMP3, *Tetraselmis* sp. CTP4, and *Skeletonema* sp.). *Aquac. Int.* **2020**, *28*, 711–727. [CrossRef]
49. Parkes, R.; Archer, L.; Gee, D.M.; Smyth, T.J.; Gillespie, E.; Touzet, N. Differential responses in EPA and fucoxanthin production by the marine diatom *Stauroneis* sp. under varying cultivation conditions. *Biotechnol. Prog.* **2021**, *37*, e3197. [CrossRef]
50. Guzmán, H.M.; de la Valido, A.J.; Duarte, L.C.; Presmanes, K.F. Analysis of interspecific variation in relative fatty acid composition: Use of flow cytometry to estimate unsaturation index and relative polyunsaturated fatty acid content in microalgae. *J. Appl. Phycol.* **2010**, *23*, 7–15. [CrossRef]
51. Renaud, S.M.; Thinh, L.V.; Lambrinidis, G.; Parry, D.L. Effect of temperature on growth, chemical composition and fatty acid composition of tropical Australian microalgae grown in batch cultures. *Aquaculture* **2002**, *211*, 195–214. [CrossRef]
52. Teoh, M.L.; Chu, W.L.; Marchant, H.; Phang, S.M. Influence of culture temperature on the growth, biochemical composition and fatty acid profiles of six Antarctic microalgae. *J. Appl. Phycol.* **2004**, *16*, 421–430. [CrossRef]
53. Najdenski, H.M.; Gigova, L.G.; Iliev, I.I.; Pilarski, P.S.; Lukavský, J.; Tsvetkova, I.V.; Ninova, M.S.; Kussovski, V.K. Antibacterial and antifungal activities of selected microalgae and cyanobacteria. *Int. J. Food Sci. Technol.* **2013**, *48*, 1533–1540. [CrossRef]
54. Davoodbasha, M.A.; Edachery, B.; Nooruddin, T.; Lee, S.Y.; Kim, J.W. An evidence of C16 fatty acid methyl esters extracted from microalga for effective antimicrobial and antioxidant property. *Microb. Pathog.* **2018**, *115*, 233–238. [CrossRef] [PubMed]

55. Wang, F.; Gao, B.; Huang, L.; Su, M.; Dai, C.; Zhang, C. Evaluation of oleaginous eustigmatophycean microalgae as potential biorefinery feedstock for the production of palmitoleic acid and biodiesel. *Bioresour. Technol.* **2018**, *270*, 30–37. [CrossRef] [PubMed]
56. De Souza, C.O.; Vannice, G.K.; Neto, J.C.R.; Calder, P.C. Is Palmitoleic Acid a Plausible Nonpharmacological Strategy to Prevent or Control Chronic Metabolic and Inflammatory Disorders? *Mol. Nutr. Food Res.* **2018**, *62*, 1700504. [CrossRef]
57. Tutunchi, H.; Ostadrahimi, A.; Saghafi-Asl, M. The Effects of Diets Enriched in Monounsaturated Oleic Acid on the Management and Prevention of Obesity: A Systematic Review of Human Intervention Studies. *Adv. Nutr.* **2020**, *11*, 864–877. [CrossRef] [PubMed]
58. Deshmukh, S.; Kumar, R.; Bala, K. Microalgae biodiesel: A review on oil extraction, fatty acid composition, properties and effect on engine performance and emissions. *Fuel Process. Technol.* **2019**, *191*, 232–247. [CrossRef]
59. Xu, H.; Miao, X.; Wu, Q. High quality biodiesel production from a microalga *Chlorella protothecoides* by heterotrophic growth in fermenters. *J. Biotechnol.* **2006**, *126*, 499–507. [CrossRef]
60. Trentin, R.; Custódio, L.; Rodrigues, M.J.; Moschin, E.; Sciuto, K.; da Silva, J.P.; Moro, I. Exploring *Ulva australis* Areschoug for possible biotechnological applications: In vitro antioxidant and enzymatic inhibitory properties, and fatty acids contents. *Algal Res.* **2020**, *50*, 101980. [CrossRef]
61. Li, Q.; Du, W.; Liu, D. Perspectives of microbial oils for biodiesel production. *Appl. Microbiol. Biotechnol.* **2008**, *80*, 749–756. [CrossRef]
62. Chisti, Y. Biodiesel from microalgae. *Biotechnol. Adv.* **2007**, *25*, 294–306. [CrossRef]
63. Huang, M.; Xin, X.; Shi, H.; Lin, Y.; Ling, H.; Ge, J. Identification of six novel microalgal strains and characterization of its potential for development of high-value compounds. *S. Afr. J. Bot.* **2022**, *147*, 1–7. [CrossRef]
64. Rodolfi, L.; Zittelli, G.C.; Bassi, N.; Padovani, G.; Biondi, N.; Bonini, G.; Tredici, M.R. Microalgae for oil: Strain selection, induction of lipid synthesis and outdoor mass cultivation in a low-cost photobioreactor. *Biotechnol. Bioeng.* **2009**, *102*, 100–112. [CrossRef] [PubMed]
65. Abou-Shanab, R.A.I.; Hwang, J.H.; Cho, Y.; Min, B.; Jeon, B.H. Characterization of microalgal species isolated from fresh-water bodies as a potential source for biodiesel production. *Appl. Energy* **2011**, *88*, 3300–3306. [CrossRef]
66. Guillard, R.R.L. Culture of Phytoplankton for Feeding Marine Invertebrates. In *Culture of Marine Invertebrate Animals*; Springer: Boston, MA, USA, 1975; pp. 29–60. [CrossRef]
67. Edgcomb, V.P.; Kysela, D.T.; Teske, A.; de Vera Gomez, A.; Sogin, M.L.; Edgcomb, V.P.; Kysela, D.T.; Teske, A.; de Vera Gomez, A.; Sogin, M.L. Benthic eukaryotic diversity in the Guaymas Basin hydrothermal vent environment. *Proc. Natl. Acad. Sci. USA* **2002**, *99*, 7658–7662. [CrossRef]
68. Medlin, L.; Elwood, H.J.; Stickel, S.; Sogin, M.L. The characterization of enzymatically amplified eukaryotic 16S-like rRNA-coding regions. *Gene* **1988**, *71*, 491–499. [CrossRef]
69. Altschul, S.F.; Gish, W.; Miller, W.; Myers, E.W.; Lipman, D.J. Basic local alignment search tool. *J. Mol. Biol.* **1990**, *215*, 403–410. [CrossRef]
70. Trifinopoulos, J.; Nguyen, L.T.; von Haeseler, A.; Minh, B.Q. W-IQ-TREE: A fast online phylogenetic tool for maximum likelihood analysis. *Nucleic Acids Res.* **2016**, *44*, W232–W235. [CrossRef] [PubMed]
71. Kalyaanamoorthy, S.; Minh, B.Q.; Wong, T.K.F.; Von Haeseler, A.; Jermiin, L.S. ModelFinder: Fast model selection for accurate phylogenetic estimates. *Nat. Methods* **2017**, *14*, 587–589. [CrossRef]
72. Minh, B.Q.; Nguyen, M.A.T.; Von Haeseler, A. Ultrafast Approximation for Phylogenetic Bootstrap. *Mol. Biol. Evol.* **2013**, *30*, 1188–1195. [CrossRef]
73. Guindon, S.; Dufayard, J.F.; Lefort, V.; Anisimova, M.; Hordijk, W.; Gascuel, O. New Algorithms and Methods to Estimate Maximum-Likelihood Phylogenies: Assessing the Performance of PhyML 3.0. *Syst. Biol.* **2010**, *59*, 307–321. [CrossRef]
74. Letunic, I.; Bork, P. Interactive Tree Of Life (iTOL) v5: An online tool for phylogenetic tree display and annotation. *Nucleic Acids Res.* **2021**, *49*, W293–W296. [CrossRef] [PubMed]
75. Velioglu, Y.S.; Mazza, G.; Gao, L.; Oomah, B.D. Antioxidant Activity and Total Phenolics in Selected Fruits, Vegetables, and Grain Products. *J. Agric. Food Chem.* **1998**, *46*, 4113–4117. [CrossRef]
76. Brand-Williams, W.; Cuvelier, M.E.; Berset, C. Use of a Free Radical Method to Evaluate Antioxidant Activity. *LWT-Food Sci. Technol.* **1995**, *28*, 25–30. [CrossRef]
77. Custódio, L.; Patarra, J.; Albericio, F.; da Rosa Neng, N.; Nogueira, J.M.F.; Romano, A. Phenolic composition, antioxidant potential and in vitro inhibitory activity of leaves and acorns of *Quercus suber* on key enzymes relevant for hyperglycemia and Alzheimer's disease. *Ind. Crops Prod.* **2015**, *64*, 45–51. [CrossRef]
78. Re, R.; Pellegrini, N.; Proteggente, A.; Pannala, A.; Yang, M.; Rice-Evans, C. Antioxidant activity applying an improved ABTS radical cation decolorization assay. *Free Radic. Biol. Med.* **1999**, *26*, 1231–1237. [CrossRef]
79. Megías, C.; Pastor-Cavada, E.; Torres-Fuentes, C.; Girón-Calle, J.; Alaiz, M.; Juan, R.; Pastor, J.; Vioque, J. Chelating, antioxidant and antiproliferative activity of *Vicia sativa* polyphenol extracts. *Eur. Food Res. Technol.* **2009**, *230*, 353–359. [CrossRef]

80. Lepage, G.; Roy, C.C. Improved recovery of fatty acid through direct transesterification without prior extraction or purification. *J. Lipid Res.* **1984**, *25*, 1391–1396. [CrossRef]
81. Pereira, H.; Barreira, L.; Figueiredo, F.; Custódio, L.; Vizetto-Duarte, C.; Polo, C.; Rešek, E.; Engelen, A.; Varela, J. Polyunsaturated Fatty Acids of Marine Macroalgae: Potential for Nutritional and Pharmaceutical Applications. *Mar. Drugs* **2012**, *10*, 1920–1935. [CrossRef] [PubMed]

marine drugs

Article

Developing a *Chromochloris zofingiensis* Mutant for Enhanced Production of Lutein under CO_2 Aeration

Yuanyuan Ren [1,2,3], Jinquan Deng [2,3], Yan Lin [2,3], Junchao Huang [2,3,*] and Feng Chen [2,3,*]

1 Institute for Food and Bioresource Engineering, College of Engineering, Peking University, Beijing 100871, China; 1701111648@pku.edu.cn
2 Shenzhen Key Laboratory of Marine Microbiome Engineering, Institute for Advanced Study, Shenzhen University, Shenzhen 518060, China; 1900392002@email.szu.edu.cn (J.D.); yanhely@163.com (Y.L.)
3 Institute for Innovative Development of Food Industry, Shenzhen University, Shenzhen 518060, China
* Correspondence: huangjc65@szu.edu.cn (J.H.); sfchen@szu.edu.cn (F.C.)

Abstract: Microalgae are competitive and commercial sources for health-benefit carotenoids. In this study, a *Chromochloris zofingiensis* mutant (*Cz-pkg*), which does not shut off its photosystem and stays green upon glucose treatment, was generated and characterized. *Cz-pkg* was developed by treating the algal cells with a chemical mutagen as *N*-methyl-*N'*-nitro-*N*-nitrosoguanidine and followed by a color-based colony screening approach. *Cz-pkg* was found to contain a dysfunctional cGMP-dependent protein kinase (PKG). By cultivated with CO_2 aeration under mixotrophy, the mutant accumulated lutein up to 31.93 ± 1.91 mg L^{-1} with a productivity of 10.57 ± 0.73 mg L^{-1} day^{-1}, which were about 2.5- and 8.5-fold of its mother strain. Besides, the lutein content of *Cz-pkg* could reach 7.73 ± 0.52 mg g^{-1} of dry weight, which is much higher than that of marigold flower, the most common commercial source of lutein. Transcriptomic analysis revealed that in the mutant *Cz-pkg*, most of the genes involved in the biosynthesis of lutein and chlorophylls were not down-regulated upon glucose addition, suggesting that PKG may regulate the metabolisms of photosynthetic pigments. This study demonstrated that *Cz-pkg* could serve as a promising strain for both lutein production and glucose sensing study.

Keywords: microalgae; *Chromochloris zofingiensis*; lutein; CO_2 aeration; cGMP-dependent kinase

Citation: Ren, Y.; Deng, J.; Lin, Y.; Huang, J.; Chen, F. Developing a *Chromochloris zofingiensis* Mutant for Enhanced Production of Lutein under CO_2 Aeration. *Mar. Drugs* **2022**, *20*, 194. https://doi.org/10.3390/md20030194

Academic Editor: Carlos Almeida

Received: 11 February 2022
Accepted: 5 March 2022
Published: 7 March 2022

1. Introduction

Lutein is a natural carotenoid that has drawn great interest for its health-promoting functions, such as scavenging free radicals, preventing age-related macular degeneration (AMD) and Alzheimer's Disease (AD), and beneficial for skin health [1,2]. At present, commercial lutein mostly derives from marigold petals, while harvesting only in specific seasons coupled with time-consuming petal collection hinders the large-scale production [3]. Though lutein is also common in vegetables, its daily dietary uptake is still insufficient for all populations. Thus, searching for better lutein sources as nutritional supplement is of significance.

Microalgae are potent sources of carotenoids that served as either primary carotenoids for photosynthesis or secondary ones in response to adverse conditions [4]. Serving as an essential photosynthesis pigment, the production of lutein in microalgae is related to photosynthetic activity. Compared with terrestrial plants, microalgae have higher photosynthetic efficiency and growth rates [4]. A number of microalgal species, including *Chlorella protothecoides* [5], *Parachlorella* sp. JD-076 [6], *Scenedesmus* sp. [7], and *Chlorella vulgaris* UTEX 265 [8], have been investigated for lutein production with limiting successes.

Chromochloris zofingiensis is a green microalga in the class *Chlorophyceae* that can grow fast under autotrophy, mixotrophy, and heterotrophy [9]. *C. zofingiensis* switched off photosynthesis in the presence of glucose, resulting in degrading of chlorophylls and accumulation of the secondary astaxanthin [10]. We had previously characterized a *Cz-bkt1* mutant that failed to accumulate astaxanthin, but instead accumulated high amounts of zeaxanthin when induced with high light and glucose [11].

In this study, we developed and characterized a novel mutant strain of *C. zofingiensis* that did not shut off its photosynthetic system under mixotrophic cultivation, and therefore could maintain high photosynthetic activity, cell growth and accumulated much higher amounts of pigments including lutein under various culture conditions. The objectives of this study are (1) to uncover the mutated gene and explain the phenotype differences between the mutant and the wild type; (2) to investigate the transcriptome differences between the mutant and the wild type; (3) to find the optimal CO_2 concentration for high yield and productivity of lutein from the mutant; and (4) to put forward a novel hypothesis of glucose-sensing in microalgae.

2. Results

2.1. Isolation of a "Stay-Green" Mutant of *C. zofingiensis*

MNNG (*N*-methyl-*N′*-nitro-*N*-nitrosoguanidine) has been proved to be an effective chemical for creating microalgal mutants with enhanced production of carotenoids [11]. In this study, MNNG was applied to generate mutants of *C. zofingiensis* followed by growing the treated cells on plates with Kuhl medium containing 15 g L^{-1} for three weeks. Generally, red colonies appeared due to the accumulation of red ketocarotenoids in the algal cells induced by glucose [12]. However, we found a green colony (here we named *Cz-pkg*) that might fail to accumulate ketocarotenoids whereas maintain stable photosynthesis pigments.

To characterize this apparent difference, the *Cz-pkg* was picked out and went on cultivation in liquid Kuhl medium with or without glucose. When cultivated in medium without glucose, both *Cz-pkg* and WT (wild type) cell cultures appeared green (Figure 1a). However, in the culture of medium containing glucose, *Cz-pkg* displayed green color while WT showed yellow to orange color (Figure 1a).

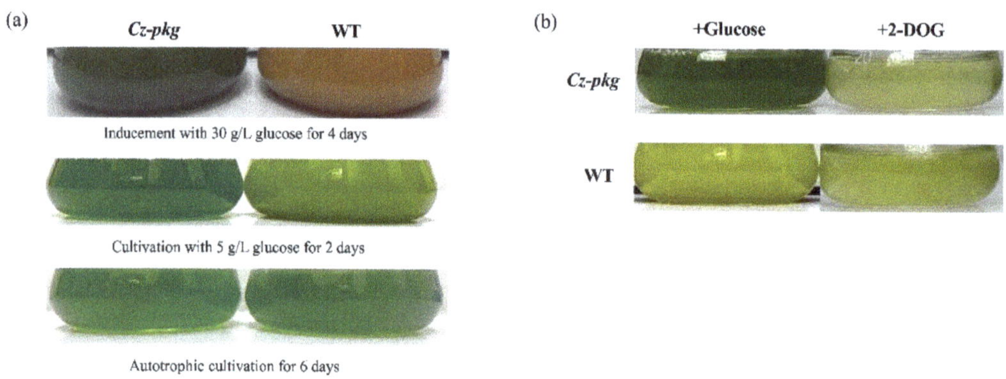

Figure 1. (**a**) Phenotypes of *Cz-pkg* and WT cultures under autotrophy, 5 g L^{-1} and 30 g L^{-1} glucose-addition mixotrophy. (**b**) Growth status of *Cz-pkg* and WT cultures under 5 g L^{-1} glucose or 2-DOG cultivation on Day 3.

HPLC analysis showed that under photoautotrophic condition, both WT and *Cz-pkg* shared similar pigment profiles under autotrophy, mainly as chlorophyll a, chlorophyll b, and lutein (Figure 2). However, when induced by 30 g L^{-1} glucose, WT accumulated mostly ketocarotenoids, mainly astaxanthin (1.06 ± 0.12 mg g^{-1} DW) (Figure 2). In contrast, *Cz-pkg* mainly accumulated chlorophylls and lutein (4.08 ± 0.19 mg g^{-1} DW), which was over 10-fold of WT (0.37 ± 0.06 mg g^{-1} DW). Thus, *Cz-pkg* might be a potential strain for lutein accumulation and production under various culture conditions.

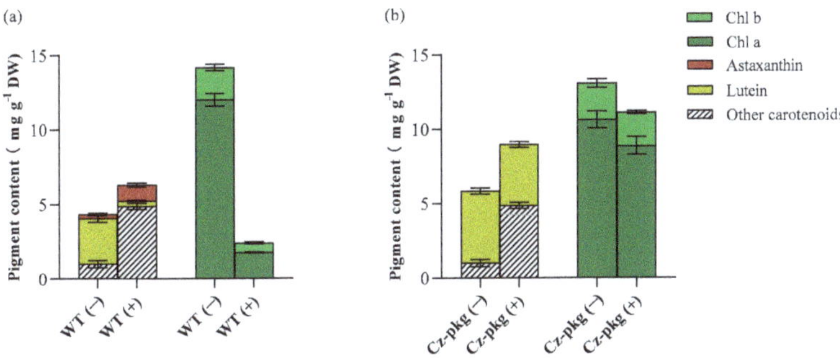

Figure 2. Pigment profiles of (**a**) WT and (**b**) *Cz-pkg* under autotrophy or glucose inducement on Day 6. (−) and (+) represent cultures without glucose or with 30 g L^{-1} addition. Data in the figure were presented in the form of means (n = 3) ± the standard deviation.

2.2. A Nonsense Mutation Occurred in PKG Gene of Cz-pkg

Since glucose failed to shut off the photosynthesis of *Cz-pkg* nor induce astaxanthin production, it is possible that a very regulating gene may loss its function. Hexokinase (HXK) is a conserved enzyme in generating glucose-6-phosphate from glucose in sugar metabolism. This reacting step was revealed to involve in switching off the photosynthesis of *C. zofingiensis* [10]. Thus, we first proposed that our stay-green *Cz-pkg* might loss its HXK function. *C. zofingiensis* consists of only one *HXK* gene [10], and we cloned and sequenced the *HXK* gene of *Cz-pkg*; however, no difference was found in the genes between WT and the mutant (data not shown).

2-DOG is a glucose analog commonly used to investigate sugar sensing in cells. 2-DOG can be uptake into microalgal cells and phosphorylated by hexokinase, however, the product cannot be further metabolized [10]. When treated with 2-DOG, both WT and *Cz-pkg* died due to the photosynthetic switching off (Figure 1b), supporting that HXK was normal in *Cz-pkg*. For a comprehensive knowledge of the mutation in *Cz-pkg*, transcriptome analysis was applied to find out the possible mutation point.

As proposed, we focused on nonsense mutation in regulation genes. Single-nucleotide polymorphisms (SNP) analysis located a SNP occurred in a cGMP-dependent protein kinase (PKG) gene that an A to T substitution led to the change of TTG (encoding for leucine) to UAG (stop codon), resulting in nonsense mutation of the PKG gene (Figure S1). PKG plays an essential role in sensing guanosine-3′, 5′-cyclic monophosphate (cGMP) in diverse physiological processes in animals and plants in the NO-cGMP-PKG pathway [13,14]. Moreover, we determined the transcriptional levels of the key genes involved in the biosynthesis of photosynthetic pigments.

2.3. Glucose Differentially Regulates the Biosynthesis of Photosynthetic Pigments

qRT-PCR was used to detect the transcription of genes encoding for the components of photosystem II (PS II) and photosystem I (PS I), which participate in the initial steps of photosynthesis, driving solar energy into chemical energy for the biosynthesis of organic compounds in oxygenic photosynthetic organisms [15]. As shown in Figure 3a, most genes of photosystem I and II were significantly downregulated in WT under 30 g L^{-1} glucose inducement, consistent with the shut off photosynthesis (Figure 1a). In contrast, except for PSBQ1 and PSBW of photosystem II, most of the genes were slightly upregulated in *Cz-pkg*. In addition, most genes involved in chlorophyll formation were significantly downregulated in WT, while there were no significant changes in *Cz-pkg* when induced with glucose (Figure 3b), some were even slightly upregulated. Typically, light-dependent protochlorophyllide oxidoreductase (POR), a key enzyme in chlorophyll synthesis, responsible for the successive reduction to form chlorophyllide a [16], was significantly downregulated in WT

(12.14-fold decrease). Similarly, Mg-protoporphyrin IX methyltransferase (CHLM) was also significantly downregulated in WT (over 10-fold decrease) (Figure 3b).

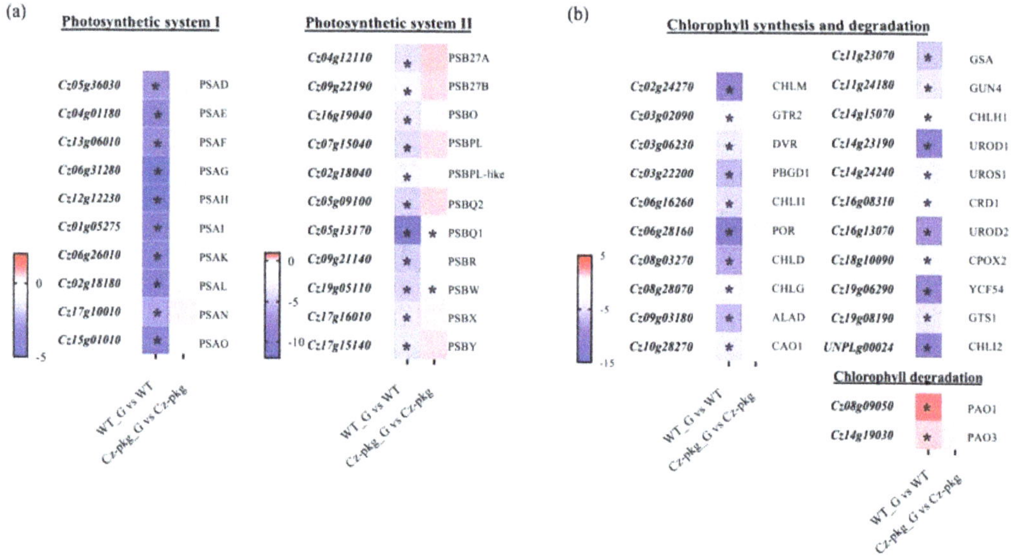

Figure 3. Different expressions involved in photosynthesis between WT and *Cz-pkg* mixotrophic cells with 30 g L^{-1} glucose addition. Heat map illustrating differences of mRNA levels of the genes related with (**a**) photosystem, (**b**) chlorophyll synthesis and degradation. WT_G and Cz-pkg_G represent cultures with 30 g L^{-1} glucose. Fold change of the mRNA levels was calculated as Log_2FC and displayed in the heat map. Significant difference (at least a two-fold change and FDR adjusted $p < 0.05$) is indicated with an asterisk. PSA, Photosystem I Subunit; PSB, Photosystem II Subunit; CHLM, Mg-Protoporphyrin IX Methyltransferase; GTR2, Glutamyl tRNA Reductase; DVR, Divinyl Chlorophyllide a 8-Vinyl Reductase; PBGD1, Porphobilinogen Deaminase; CHLI1, Mg-Chelatase Subunit I; POR, Light-Dependent Protochlorophyllide Oxidoreductase; CHLD, Mg-Chelatase Subunit D; CHLG, Chlorophyll Synthetase; ALAD, Delta-Aminolaevulinic Acid Dehydratase; CAO1, Chlorophyllide a Oxygenase; GSA, Glutamate-Semialdehyde Aminotransferase; GUN4, Tetrapyrrole Binding Protein; CHLH1, Mg-Chelatase Subunit H; UROD1, Uroporphyrinogen III Decarboxylase; UROS1, Uroporphyrinogen III Synthase; CRD1, Mg-Protoporphyrin Monomethyl Ester Cyclase; UROD2, Uroporphyrinogen III Decarboxylase; CPOX2, Coproporphyrinogen-III Oxidase; YCF54, Ycf54 Conserved Hypothetical Protein; GTS1, Glutamyl-Glutaminyl Non-Discriminatory tRNA Synthetase; CHLI2, Mg-Chelatase Subunit I; PAO1, Pheophorbide a Oxygenase; PAO3, Pheophorbide a Oxygenase.

Pheophorbide a oxygenase (PAO) is a key enzyme for chlorophyll degradation [17]. As shown in Figure 3b, under 30 g/L glucose inducement, apart from downregulation of chlorophyll synthesis, WT cells went through upregulation of chlorophyll degradation, leading to the orange color of its suspension culture (Figure 1). In contrast, *Cz-pkg* exhibited stable expression of the related genes (Figure 3) and maintained stable contents of chlorophylls (Figure 2) under 30 g L^{-1} glucose induction. As a result, *Cz-pkg* displayed stay-green phenotype (Figure 1) and contained much higher amounts of lutein than WT (Figure 2).

To further understand the different regulation of carotenoid biosynthesis between WT and *Cz-pkg*, we detected the expression of genes related to carotenoid biosynthesis (Figure 4). Under 30 g L^{-1} glucose, the expression of essential enzymes involved in the MEP (methylerythritol phosphate) pathway was significantly upregulated in WT, such as DXR (1-deoxy-d-xylulose 5-phosphate reductoisomerase), HDS (4-hydroxy-3-methylbut-2-en-1-yl diphosphate synthase) and HDR (4-hydroxy-3-methylbut-2-en-1-yl diphosphate reductase). Moreover, the expression of LCYB (Lycopene Beta-Cyclase) to the β-branch carotenoids was significantly upregulated, leading to more isoprenoid skeletons for astaxanthin and canthaxanthin; while the expression of enzymes leading to lutein were significantly downregulated, such as LCYE (Lycopene Epsilon-Cyclase), CYP97A2 (Cytochrome P450-Type Carotene Hydroxylase), and CYP97C. In addition, the downregulated expression of ZEP 1 (Zeaxanthin Epoxidase) and NSY (Neoxanthin Synthase) restricted the formation of vioxanthin and neoxanthin, leading to more zeaxanthin for astaxanthin synthesis in WT. In contrast, there are no significant expression changes in *Cz-pkg*. Thus, *Cz-pkg* could have a potential for sustainable production of lutein.

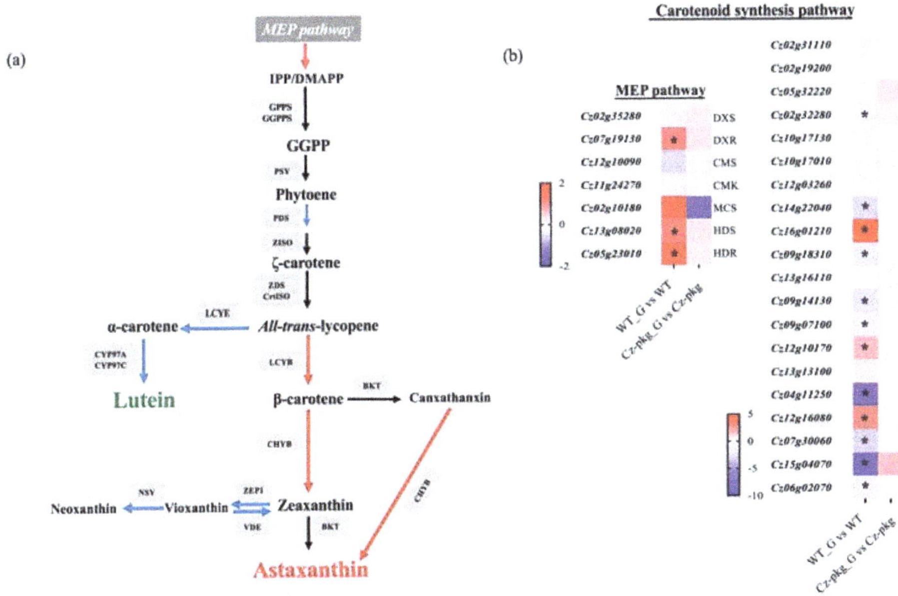

Figure 4. (**a**) Carotenoid biosynthetic pathways in *C. zofingiensis*. Red arrows indicate pathways with significant up-regulations, and blue arrows indicate pathways with significant down-regulations. (**b**) Heat map illustrating differences of gene expressions related with carotenogenesis between WT and *Cz-pkg* under 30 g L^{-1} glucose. Fold change of the mRNA levels was calculated as Log_2FC and displayed in the heat map. Significant difference (at least a two-fold change and FDR adjusted $p < 0.05$) is indicated with an asterisk. DXS, 1-deoxy-d-xylulose 5-phosphate synthase; DXR, 1-deoxy-d-xylulose 5-phosphate reductoisomerase; CMS, 2-C-methyl-d-erythritol 4-phosphate cytidylyltransferase; CMK, 2-C-methyl-d-erythritol 4-phosphate cytidylyltransferase; MCS, 2-C-methyl-d-erythritol 2,4-cyclodiphosphate synthase; HDS, 4-hydroxy-3-methylbut-2-en-1-yl diphosphate synthase; HDR, 4-hydroxy-3-methylbut-2-en-1-yl diphosphate reductase; GPPS, geranyl diphosphate synthase; GGPPS, geranylgeranyl pyrophosphate synthase; PSY, Phytoene Synthase; PDS, Phytoene Desaturase; ZISO, Zeta-Carotene Isomerase; ZDS, Zeta-Carotene Desaturase; CRTISO, Carotene Isomerase; LCYE, Lycopene Epsilon-Cyclase; CYP97, Cytochrome P450-Type Carotene Hydroxylase; LCYB, Lycopene Beta-Cyclase; BKT, Beta-Ketolase; CHYB, Beta-Carotene Hydroxylase; ZEP1, Zeaxanthin Epoxidase; NSY, Neoxanthin Synthase (ABA4); VDE, Violaxanthin de-epoxidase.

2.4. The Growth of Cz-pkg under Different Trophic Modes

To assess if *Cz-pkg* has potential for lutein production, mixotrophic cultivation with gradient concentrations of glucose (from 5 g L^{-1} to 50 g L^{-1}) was applied to find out the best growing condition. Though high glucose concentration increased biomass with longer cultivation time, the maximum specific growth rates showed a descending trend with the increase of sugar concentration, and it was the highest at 5 g L^{-1} (u_{max}= 0.0375 ± 0.0025 h^{-1},

Figure S2). This result is consistent with previous research that 5 g L^{-1} glucose addition is the optimal condition for cultivation of *C. zofingiensis*. [18]. Thus, for mixotrophy of *Cz-pkg*, 5 g L^{-1} was chosen as the optimal glucose concentration for further experiments.

The biomass concentration of *Cz-pkg* under mixotrophy was higher than the sum of the concentrations under autotrophy and heterotrophy (Figure 5a). In contrast, as WT shut off photosynthesis in the presence of glucose, its biomass under mixotrophy was lower than the sum of those under autotrophy and heterotrophy (Figure 5a). As shown in Figure 5b, the growth curves of *Cz-pkg* and WT under 5 g L^{-1} glucose were determined and fitted to a logistic growth model by Prism (R^2 > 95%). According to the fitted results, *Cz-pkg* could accumulate 1.23-fold biomass of WT, although it reached the plateau stage later and its fit-calculated maximum specific growth rate (u_{max} = 0.05008 h^{-1}) was lower than that of WT (u_{max} = 0.05784 h^{-1}). The slow growth rate may be due to its slower utilization of glucose. As *Cz-pkg* keeps green in the presence of glucose, its photosystem can work efficiently for uptaking both inorganic (CO_2) and organic carbon sources (glucose) under mixotrophic cultivation.

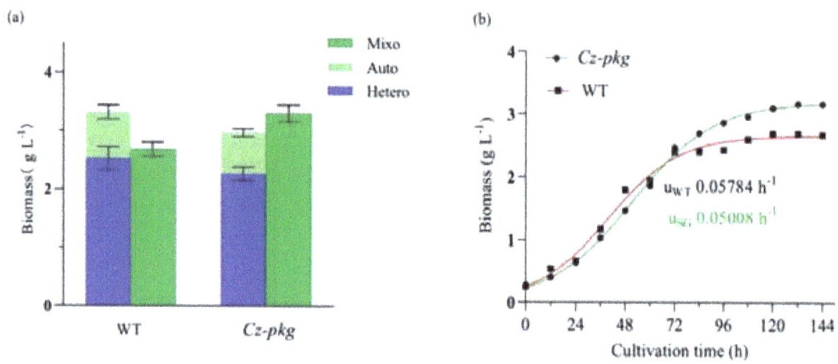

Figure 5. (**a**) Biomass of WT and *Cz-pkg* under different trophic modes on Day 5 in flasks; (**b**) growth curve of WT and *Cz-pkg* under 5 g L^{-1} glucose in flasks. Solid symbols represent actual sampling points, while red and green curves as fitted growth curves (R^2 > 0.99). Data were presented in the form of mean ± the standard deviation (n = 3).

2.5. Supplemented CO_2 Promotes Cell Growth and Lutein Production

In contrast to marigold flower that accumulates esterified lutein, microalgae produce lutein mostly in free form [1]. Up until now, only a limited number of microalgae have been exploited for lutein production, and typically their lutein contents range about 340 to 760 mg/100 g DW [19]. CO_2 was shown to increase algal photosynthesis [20]. Since *Cz-pkg* does not shut off photosynthesis in the presence of glucose, five cultivation conditions (flasks without aeration, air, 2.5% CO_2, 4% CO_2, and 5% CO_2) were performed under both autotrophy and mixotrophy to evaluate the effects of trophic mode and CO_2 concentrations on cell growth and lutein accumulation.

As shown in Figure 6, all cultures with aeration showed higher biomass than those in flasks, whether under autotrophy or mixotrophy, which may be due to better gas exchange under aeration and light exposure of cells. However, when the CO_2 concentration reached 5%, the biomass concentrations and lutein yields showed reverse tendency.

As shown in Table 1, 5% CO_2 led to a decrease in the content of photosynthetic pigments (both chlorophylls and lutein), and the proportion of lutein under mixotrophy with 5% CO_2 decreased to 44.4% of total carotenoids, nearly two-fold lower than other cultures. HPLC analysis revealed that *Cz-pkg* also accumulated zeaxanthin up to 1.70 mg g^{-1} DW, which occupied about 30% of the total carotenoids. Higher concentrations of CO_2 had previously been found to cause lower photosynthetic efficiency and cell growth of *Chlorella minutissima* [21] and *Desmodesmus* sp. [22], possibly resulting from excessive changes of pH values in the medium led by high soluble CO_2 concentrations.

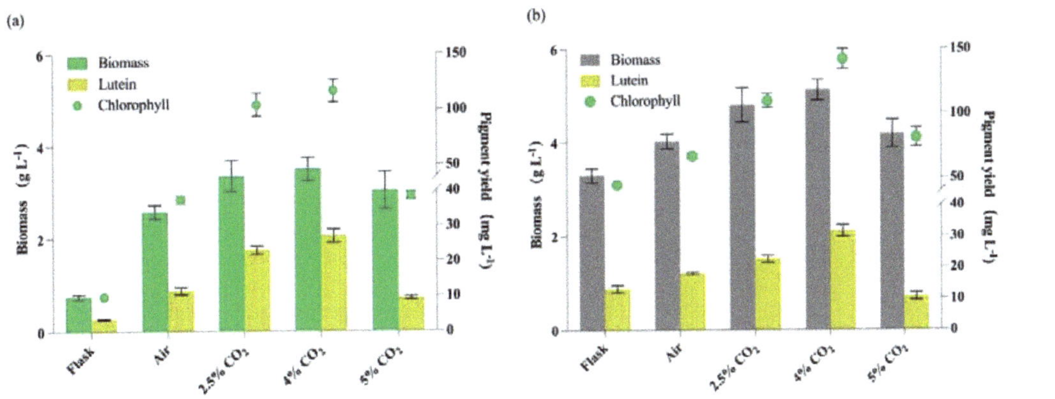

Figure 6. The growth and pigment production of *Cz-pkg* under different culture conditions. (**a**) Autotrophic cultures in flasks or aerations with different CO_2 concentrations. (**b**) Mixotrophic cultures with 5 g L^{-1} glucose addition in flasks or aerations with different CO_2 concentrations. Data were presented in the form of mean ± the standard deviation (n = 3).

Table 1. Pigment profiles of *Cz-pkg* under different cultivation conditions in autotrophy and mixotrophy. Data in the table were presented in the form of means (n = 3) ± the standard deviation.

Cultivation Conditions	Pigment Composition (mg g^{-1} DW)			
	Total Carotenoids	Lutein	Chl a	Chl b
		Autotrophy		
flasks	5.32 ± 0.31	4.63 ± 0.22	10.63 ± 0.97	2.43 ± 0.30
Bubble tubes + Air	5.28 ± 0.22	4.40 ± 0.37	11.06 ± 1.01	3.41 ± 0.11
Bubble tubes + 2.5% CO_2	7.12 ± 0.80	5.70 ± 0.20	23.44 ± 2.05	7.56 ± 1.08
Bubble tubes + 4.0% CO_2	8.40 ± 0.81	7.73 ± 0.52	25.50 ± 2.44	6.86 ± 0.41
Bubble tubes + 5.0% CO_2	3.78 ± 0.61	3.06 ± 0.19	10.38 ± 1.20	2.25 ± 0.06
		Mixotrophy		
flasks	5.94 ± 0.33	3.97 ± 0.33	11.37 ± 0.24	2.18 ± 0.50
Bubble tubes + Air	5.83 ± 0.34	4.47 ± 0.10	12.94 ± 0.41	3.67 ± 0.23
Bubble tubes + 2.5% CO_2	6.72 ± 0.35	4.71 ± 0.25	17.83 ± 1.00	5.02 ± 0.14
Bubble tubes + 4.0% CO_2	7.82 ± 0.49	6.28 ± 0.57	20.80 ± 1.32	6.93 ± 0.20
Bubble tubes + 5.0% CO_2	5.67 ± 0.25	2.52 ± 0.28	15.97 ± 1.30	3.46 ± 0.51

As shown in Figure 6a, 4% CO_2 under autotrophy showed the best on both biomass and lutein accumulation, and the lutein yield under autotrophy can reach 27.83 ± 1.87 mg L^{-1}, which was 8.02-fold of that in flask and 2.46-fold of that with air aeration (0.04% CO_2). As shown in Figure 6b, under mixotrophy, the lutein yields were relatively higher than those with the same CO_2 aeration under autotrophy. This result is consistent with a previous study that mixotrophy was more favorable for lutein production [6]. The maximal lutein yield under mixotrophy was 31.93 ± 1.91 mg L^{-1} with 4% CO_2, which was 2.43-fold of that in flask and 1.78-fold of that with air aeration. Besides, the content of lutein was in positive correlation with the content of chlorophylls (Figure 6).

To further find out the pattern of lutein accumulation in *Cz-pkg*, we determined the time-course lutein content during cultivation. As shown in Table 2, lutein began to increase rapidly when the cells entered the plateau phase (from Day 6 to Day 8), which may be due to the increased cell density, so more light-harvesting pigments are needed to meet the energy required for cell growth. Yeh et al. [23] suggested that in the later growth stage of *Desmodesmus* sp., lutein accumulation was essential for maintaining the structural integrity of LHCs and promoted photosynthesis under low light conditions due to the self-shading effect.

Table 2. Time-course of lutein contents (mg g^{-1} DW) and the maximal productivity (mg L^{-1} day^{-1}) of *Cz-pkg* with different CO_2 concentration aeration. Data in the table were presented in the form of means (n = 3) ± the standard deviation.

Conditions	Lutein Contents (mg g^{-1} DW)				Pmax (mg L^{-1} day^{-1})
	Day 2	Day 4	Day 6	Day 8	
Auto + 2.5% CO_2	2.16 ± 0.06	2.68 ± 0.14	3.25 ± 0.33	5.70 ± 0.35	5.45 ± 0.52
Auto + 4.0% CO_2	2.50 ± 0.25	2.77 ± 0.08	3.81 ± 0.40	7.73 ± 0.52	8.80 ± 0.60
Mixo + 2.5% CO_2	2.02 ± 0.13	2.40 ± 0.22	2.98 ± 0.21	4.71 ± 0.25	5.61 ± 0.44
Mixo + 4.0% CO_2	1.98 ± 0.16	2.85 ± 0.16	3.45 ± 0.17	6.28 ± 0.57	10.57 ± 0.73

Auto, autotrophy; Mixo, mixotrophy with 5 g L^{-1} glucose; P_{max}, the maximal productivity of lutein.

At present, marigold flower is still the major source of natural lutein, and it contains lutein from 0.17 to 5.70 mg g^{-1} [3]. Compared with marigold, the concentrations of lutein in microalgae are much higher [19]. As most microalgae contain lutein less than 5 mg g^{-1} DW [11], microalgal species containing lutein more than 5 mg g^{-1} have been acknowledged as potential sources for lutein production [19].

With 4% CO_2 aeration under mixotrophy, *Cz-pkg* can achieve 6.28 ± 0.57 mg g^{-1} DW of lutein with a yield of 31.93 ± 1.91 mg L^{-1}. Under autotrophy, *Cz-pkg* can also accumulate large content of lutein as 7.73 ± 0.52 mg g^{-1} DW with a relatively high yield of 27.83 ± 1.87 mg L^{-1}. As shown in Table 2, the maximal productivity of lutein was achieved also under mixotrophy with 4.0% CO_2 as 10.57 ± 0.73 mg L^{-1} day^{-1}, 1.20-fold of that under autotrophy (8.80 ± 0.60 mg L^{-1} day^{-1}), which is the highest among the known species for lutein production as listed in Table 3. Considering that the production of carotenoids by *C. zofingiensis* can be greatly enhanced by optimization of culture conditions as reported by previous studies [11,19,24], *Cz-pkg* may serve as a potential source for lutein production on industrial scales.

Table 3. Natural sources and several microalgae species potential for lutein production. - represents no related data.

Sources	Lutein Content	Productivity (mg L^{-1} day^{-1})	References
Tagetes erecta (Marigold flower)	0.17–5.70 mg g^{-1}	-	[3]
Chicken egg yolk	16.22 μg g^{-1}	-	[25]
Brassica oleracea (Broccoli)	39 μg g^{-1}	-	[25]
Tetracystis intermedium	3.5 mg g^{-1}	-	[25]
Chlorella sorokiniana FZU60	11.22 mg g^{-1}	8.25	[1]
Chlorella vulgaris CS-41	9.0 mg g^{-1}	1.56	[1]
Chlorella sp. GY-H4	8.9 mg g^{-1}	10.50	[1]
Chlorella sorokiniana MB-1-M12	7.39 mg g^{-1}	3.43	[19]
Chlorella minutissima MCC-27	7.05 mg g^{-1}	6.34	[19]
Chlorella vulgaris	5–9 mg g^{-1}	1.60	[19]
Chlorella sorokiniana AK-1	4.56 mg g^{-1}	3.56	[19]
Chromochloris zofingiensis WT	3.07 mg g^{-1}	1.24	This study
Cz-pkg	7.73 mg g^{-1}	10.57	This study

3. Discussion

As sugars play vital roles in regulations of cell growth, physiology, metabolism, and gene expression in plants and microalgae, it is of great significance to investigate the genes and enzymes associated with glucose sensing and responding. Roth et al. previously found that glucose-treated *hxk1* mutants did not shut off photosynthesis or accumulate astaxanthin in the presence of glucose under light, same as *Cz-pkg* in this research [10]. They hypothesized that HXK was necessary for glucose-mediated photosynthesis repression, and G-6-P (glucose-6-phosphate), the downstream product of glucose phosphorylated by hexokinase (HXK) was also closely related [10]. Though the functions of PKG have been shown important in plants, its definite properties and functions still remain unclear [26,27]. As *PKG* mutant *Cz-pkg* does not shut off photosynthesis or accumulate astaxanthin in the presence of glucose, the kinase PKG might be significant in regulating downstream metabolism after microalgae sensing glucose. Further study should focus on exploiting the functions of PKG and determine its role in glucose sensing of microalgae with molecular methods, such as *CRISPR/Cas9*, to achieve rigor results.

4. Materials and Methods

4.1. Microalgae Strain and Cultivation

C. zofingiensis (ATCC 30412) was obtained from American Type Culture Collection (ATCC, Rockville, MD, USA) and cultivated in Kuhl medium as reported previously [11]. The Kuhl medium contains KNO_3 1.01 g L^{-1}, $NaH_2PO_4{\cdot}H_2O$ 0.62 g L^{-1}, $Na_2HPO_4{\cdot}2H_2O$ 0.089 g L^{-1}, $MgSO_4{\cdot}7H_2O$ 0.247 g L^{-1}, $CaCl_2{\cdot}2H_2O$ 14.7 mg L^{-1}, $Na_2EDTA{\cdot}H_2O$ 6.95 mg L^{-1}, $FeSO_4{\cdot}7H_2O$ 6.95 mg L^{-1}, H_3BO_3 0.061 mg L^{-1}, $MnSO_4{\cdot}H_2O$ 0.169 mg L^{-1}, $ZnSO_4{\cdot}7H_2O$ 0.287 mg L^{-1}, $CuSO_4{\cdot}5H_2O$ 0.0025 mg L^{-1}, $(NH_4)_6Mo_7O_{24}{\cdot}4H_2O$ 0.01235 mg L^{-1}. Briefly, the algal cells were cultured in 100 mL fresh Kuhl medium containing 5 g L^{-1} glucose (in 250-mL flasks) with orbital shaking at 150 rpm, and illuminated with continuous light of 30 μmol photons m^{-2} s^{-1} (cool-white fluorescent tube light). Cells grown to late exponential phase were used as seed cells for further experiments. For cultivation in bubble tubes, the seed cells were inoculated into glass columns at 0.5 g L^{-1} under illumination of 150 μmol photons m^{-2} s^{-1} from one side, and separately aerated by air, 2.5% (*v*/*v*), 4% (*v*/*v*), and 5% (*v*/*v*) CO_2-enriched air in different experimental groups. All experiments were operated in triplicate. Cell biomass was determined according to our previous study [28]. By sampling at a 12 h time interval, the biomass concentrations were fitted according to a logistic growth model as the following equation:

$$Y = Y_M \times Y_0 / \left((Y_M - Y_0) \times e^{(-kx)} + Y_0 \right) \quad (1)$$

where Y_M is the maximum biomass (g L^{-1}), Y_0 is the initial inoculum biomass (g L^{-1}), k is the rate constant (h^{-1}), and x is the cultivation time (h).

4.2. Mutant Generation, Selection, and Identification

The mutagenesis procedure by *N*-methyl-*N'*-nitro-*N*-nitrosoguanidine (MNNG) was performed successfully according to our previous study [29]. Surviving colonies were then isolated and transferred to fresh 1/2 Kuhl medium with or without 15 g L^{-1} glucose. In detail, the selection procedure was conducted as follows: (a) each of single colonies were suspended in two liquid cultures (with or without glucose); (b) a clone showing green color in the presence of glucose was isolated. The culture of the stay-green mutant and the wild type were then transferred to Kuhl medium with 5 mM 2-deoxy-D-glucose (2-DOG) to identify if the mutant strain shuts off photosynthesis in the presence of glucose according to [10]. To eliminate the mutant with transport deficiency of glucose, the mutant was also cultivated in dark in the presence of glucose.

4.3. Pigment Extraction and Analysis

Algal cells were harvested after centrifugation, and the pellets were lyophilized and grinded in 2 mL tubes with 3 stainless steel beads for 10 × 30 s at 1/30 frequency with the TissueLyser II (QIAGEN, Hilden, Germany). Then, the debris were extracted with pre-chilled acetone (HPLC grade) for three times until they were almost colorless. The supernatants were collected by centrifugation (13,000× *g* for 10 min at 4 °C) and filtered through a 0.22 μm Millipore organic membrane. Absorbance values at 470, 652.4, and 665.2 nm were measured with a spectrophotometer for pigment quantifications according to our previous study [28]. Pigment profile analysis was performed by high performance liquid chromatography (HPLC, Waters) according to Huang et al. with modifications [11]. Briefly, HPLC was equipped with a Waters YMC Carotenoid C30 column (5 μm, 4.6 × 250 mm), and the mobile phase consisted of solvent A (methanol: isopropanol, 68:32, *v*/*v*) and solvent B (acetonitrile: methanol: water, 84:2:14, *v*/*v*/*v*). A total of 10 μL of each sample was analyzed at a flow rate of 0.80 mL min^{-1} with a gradient mode (0–15 min: linear gradient of 0–100%

of A; 15−30 min: linear gradient of 0−100% of B). Compounds were detected at 450 and 480 nm. The peaks of each compound were identified by their absorption spectra, and the retention times were compared with the biological reference standards for recognition and quantification. The lutein productivity was calculated with the following equation:

$$\text{Productivity}\ \left(\text{mg L}^{-1}\text{day}^{-1}\right) = \frac{\text{Biomass concentration}\ \left(\text{g L}^{-1}\right) \times \text{Lutein content}\ \left(\text{mg g}^{-1}\right)}{\text{Cultivation time (day)}}. \quad (2)$$

4.4. DNA Extraction and Molecular Characterization of Mutant

Genomic DNA of *C. zofingiensis* was extracted with Chelex-100 chelating resin (Bio-Rad Laboratories, Hercules, CA, USA) according to Kang et al. [30]. Briefly, microalgal cells were centrifuged at 14,000 rpm for 1 min, and the supernatant was discarded. Then, 1 mL PBS solution was added, vortexed and centrifuged to wash the cells twice. A total of 100 µL of autoclaved 5% Chelex-100 chelating resin (suspended in 0.1M Tris and 0.5 mM EDTA, pH = 8.0) was added, vortexed and boiled at 100 °C for 20 min. The suspension was then and centrifuged at 14,000 rpm for 1 min. The supernatant was collected and used as DNA samples for the characterization experiments.

To reveal the molecular base of the mutant, DNA sequences of putative mutated genes were compared with that of wild type by PCR amplification and sequencing. According to the whole-genome of *C. zofingiensis* [31], primers were designed to amplify the full length of *CzHXK1* and *CzPKG* genes as follows: HXK1F: 5′ ATGAAACTTGACGCAGACACTCAACG 3′ and HXK1R: 5′ TTAGGCAGTAGTGCTTGGCAGGGGGTC 3′ for HXK1; and PKGF: 5′ ATGGGGAACTCGCACAGCCAG 3′ and PKGR: 5′ TGAGCAGTGATGTAGCACTGGCAG 3′ for PKG. The PCR procedure was set as: 98 °C for 3 min, 36 cycles of 98 °C for 10 s, 60 °C for 5 s, and 68 °C for 6 min (for *CzHXK1*) or 8 min (for *CzPKG*). An elongation procedure was added at 68 °C for 5 min. PCR products were gel purified and sequenced. Sequence alignments were completed through BLAST online (https://blast.ncbi.nlm.nih.gov/Blast.cgi, accessed on 10 October 2021).

4.5. Transcriptome Sequencing and Analysis of Differentially Expressed Genes

Algal cultures (both Cz-WT and *Cz-pkg*) in the presence of glucose were harvested at exponential phase for RNA isolation and sequencing with an Illumina Novaseq 6000 system (Illumina Inc., San Diego, CA, USA) by Majorbio Bio-pharm Technology Co., Ltd. (Shanghai, China). The transcriptome sequences are accessible in the NCBI Sequence Read Archive database (http://www.ncbi.nlm.nih.gov/sra/, accessed on 10 October 2021) under the accession number PRJNA664005. Expression of the annotated genes was profiled by the values of TPM (transcripts per million reads) through RSEM v1.3.1. Fold change of a gene between two samples was considered significant when $|\log_2(\text{TPM}_{\text{Sample 1}}/\text{TPM}_{\text{Sample 2}})| \geq 1$ with Padjust < 0.001.

4.6. RNA Isolation and Quantitative Real-Time PCR

Validation of the interesting key genes was also performed by quantitative real-time PCR (qRT-PCR) analysis using the primers listed in Table S1. Total RNA was isolated with TRIzol reagent (Invitrogen, Shanghai, China) according to the manufacturer's protocol. To remove possible contaminating DNA, RNase-free DNase I (TaKaRa, Beijing, China) was used to treat raw RNA samples. Nanodrop 2000 (Thermo Scientific, Shanghai, China) was used to determine the concentration and quality of RNA was checked by electrophoresis. Total RNA (~1 µg) was then reversely transcribed to cDNA using Prime ScriptTM RT reagent kit (TaKaRa, Beijing, China) according to the manufacturer's protocol.

qRT-PCR was performed on a CFX Connect Real-Time System (Bio-Rad) with a 20 µL reaction volume, containing 10 µL of TB GREEN ® Premix Ex TaqTM II (Tli RNaseH Plus) (TaKaRa), 0.8 µL of each primer (10 µM), 2 µL of template cDNA, and 6.4 µL sterile distilled water (DNase free). The qRT-PCR protocol was set as follow: 30 s at 95 °C followed by 40 cycles of 5 s at 95 °C and 30 s at 60 °C. A procedure of 0.5 °C increment at 5 s/step from

65 °C to 95 °C was added after for the melt curve analysis. All experiments were operated in triplicate and data were analyzed by the CFX Manager™ Software v3.1 (Bio-Rad, Hercules, California, USA). The relative gene expression level was calculated based on the $2^{-\Delta\Delta CT}$ method [32] and the actin gene was set as reference.

4.7. Statistical Analysis

All the experiments above were conducted in at least triplets to guarantee the reproducibility. Statistical analysis was carried out by using GraphPad Prism 9.0 and Microsoft Excel. A one-way analysis of variance (ANOVA) was applied for the determination of the significant differences from the control groups for each experimental condition separately ($p < 0.05$). All the data are presented in the form as means value ($n = 3$) ± the standard deviation.

5. Conclusions

In this study, the *C. zofingiensis* mutant *Cz-pkg* was generated and characterized. Under 30 g L^{-1} glucose inducement, *Cz-pkg* consisted of high amount of lutein and stayed green, while WT accumulated astaxanthin with red phenotype. *Cz-pkg* consists of a dysfunctional PKG, leading to not shutting off its photosystem and staying green, with higher biomass and lutein production under mixotrophy. Specifically, coupled with 4.0% CO_2 aeration, the mixotrophic *Cz-pkg* with 5 g L^{-1} glucose produced lutein 31.93 ± 1.91 mg L^{-1} with a productivity of 10.57 ± 0.73 mg L^{-1} day^{-1}. This study demonstrated that Cz-*pkg* could serve as a promising strain for lutein production.

Supplementary Materials: The following are available online at https://www.mdpi.com/article/10.3390/md20030194/s1, Figure S1: The sequence alignment of *PKG* gene of WT (the upper) and *Cz-pkg* and the peptide sequence of PKG in WT, Figure S2: Biomass concentration and the specific growth rate of *Cz-pkg* under cultivation with different glucose concentrations. Table S1: Primers for qRT-PCR analysis. Table S2: Fold change of transcriptional differences of genes related to photosynthetic systems, chlorophyll synthesis and degradation.

Author Contributions: Conceptualization, resources, and supervision, F.C. and J.H.; writing—original draft preparation, Y.R. and J.D.; writing—review and editing, Y.L., J.H. and F.C. All authors have read and agreed to the published version of the manuscript.

Funding: This research was funded by National Natural Science Foundation of China (Project 31471717), Key Realm R&D Program of Guangdong Province (No. 2018B020206001), Guangdong Province Zhujiang Talent Program (No. 2019ZT08H476) and Science and Technology Innovation Commission of Shenzhen (No. KQTD20180412181334790).

Institutional Review Board Statement: Not applicable.

Informed Consent Statement: Not applicable.

Data Availability Statement: Not applicable.

Conflicts of Interest: The authors declare no conflict of interest.

References

1. Ren, Y.; Sun, H.; Deng, J.; Huang, J.; Chen, F. Carotenoid Production from Microalgae: Biosynthesis, Salinity Responses and Novel Biotechnologies. *Mar. Drugs* **2021**, *19*, 713. [CrossRef]
2. Roberts, R.L.; Green, J.; Lewis, B. Lutein and zeaxanthin in eye and skin health. *Clin. Dermatol.* **2009**, *27*, 195–201. [CrossRef] [PubMed]
3. Lin, J.H.; Lee, D.J.; Chang, J.S. Lutein production from biomass: Marigold flowers versus microalgae. *Bioresour. Technol.* **2015**, *184*, 421–428. [CrossRef] [PubMed]
4. Liu, C.; Hu, B.; Cheng, Y.; Guo, Y.; Yao, W.; Qian, H. Carotenoids from fungi and microalgae: A review on their recent production, extraction, and developments. *Bioresour. Technol.* **2021**, *337*, 125398. [CrossRef] [PubMed]
5. Shi, X.; Zhang, X.; Chen, F. Heterotrophic production of biomass and lutein by *Chlorella protothecoides* on various nitrogen sources. *Enzym. Microb. Technol.* **2000**, *27*, 312–318. [CrossRef]

6. Heo, J.; Shin, D.-S.; Cho, K.; Cho, D.-H.; Lee, Y.J.; Kim, H.-S. Indigenous microalga *Parachlorella* sp. JD-076 as a potential source for lutein production: Optimization of lutein productivity via regulation of light intensity and carbon source. *Algal Res.* **2018**, *33*, 1–7. [CrossRef]
7. Přibyl, P.; Pilný, J.; Cepák, V.; Kaštánek, P. The role of light and nitrogen in growth and carotenoid accumulation in *Scenedesmus* sp. *Algal Res.* **2016**, *16*, 69–75. [CrossRef]
8. Gong, M.; Bassi, A. Investigation of Chlorella vulgaris UTEX 265 Cultivation under Light and Low Temperature Stressed Conditions for Lutein Production in Flasks and the Coiled Tree Photo-Bioreactor (CTPBR). *Appl. Biochem. Biotechnol.* **2017**, *183*, 652–671. [CrossRef]
9. Mao, X.; Zhang, Y.; Wang, X.; Liu, J. Novel insights into salinity-induced lipogenesis and carotenogenesis in the oleaginous astaxanthin-producing alga *Chromochloris zofingiensis*: A multi-omics study. *Biotechnol. Biofuels* **2020**, *13*, 73. [CrossRef]
10. Roth, M.S.; Westcott, D.J.; Iwai, M.; Niyogi, K.K. Hexokinase is necessary for glucose-mediated photosynthesis repression and lipid accumulation in a green alga. *Commun. Biol.* **2019**, *2*, 347. [CrossRef]
11. Huang, W.; Lin, Y.; He, M.; Gong, Y.; Huang, J. Induced High-Yield Production of Zeaxanthin, Lutein, and beta-Carotene by a Mutant of *Chlorella zofingiensis*. *J. Agric. Food Chem.* **2018**, *66*, 891–897. [CrossRef] [PubMed]
12. Roth, M.S.; Gallaher, S.D.; Westcott, D.J.; Iwai, M.; Louie, K.B.; Mueller, M.; Walter, A.; Foflonker, F.; Bowen, B.P.; Ataii, N.N.; et al. Regulation of Oxygenic Photosynthesis during Trophic Transitions in the Green Alga *Chromochloris zofingiensis*. *Plant Cell* **2019**, *31*, 579–601. [CrossRef] [PubMed]
13. Shen, Q.; Zhan, X.; Yang, P.; Li, J.; Chen, J.; Tang, B.; Wang, X.; Hong, Y. Dual Activities of Plant cGMP-Dependent Protein Kinase and Its Roles in Gibberellin Signaling and Salt Stress. *Plant Cell* **2019**, *31*, 3073–3091. [CrossRef] [PubMed]
14. Bassot, A.; Chauvin, M.A.; Bendridi, N.; Ji-Cao, J.; Vial, G.; Monnier, L.; Bartosch, B.; Alves, A.; Cottet-Rousselle, C.; Gouriou, Y.; et al. Regulation of Mitochondria-Associated Membranes (MAMs) by NO/sGC/PKG Participates in the Control of Hepatic Insulin Response. *Cells* **2019**, *8*, 1319. [CrossRef] [PubMed]
15. Goldschmidt-Clermont, M.; Bassi, R. Sharing light between two photosystems: Mechanism of state transitions. *Curr. Opin. Plant Biol.* **2015**, *25*, 71–78. [CrossRef]
16. Reinbothe, C.; El Bakkouri, M.; Buhr, F.; Muraki, N.; Nomata, J.; Kurisu, G.; Fujita, Y.; Reinbothe, S. Chlorophyll biosynthesis: Spotlight on protochlorophyllide reduction. *Trends Plant Sci.* **2010**, *15*, 614–624. [CrossRef]
17. Pruzinska, A.; Tanner, G.; Anders, I.; Roca, M.; Hortensteiner, S. Chlorophyll breakdown: Pheophorbide a oxygenase is a Rieske-type iron-sulfur protein, encoded by the accelerated cell death 1 gene. *Proc. Natl. Acad. Sci. USA* **2003**, *100*, 15259–15264. [CrossRef]
18. Zhang, Z.; Huang, J.J.; Sun, D.; Lee, Y.; Chen, F. Two-step cultivation for production of astaxanthin in *Chlorella zofingiensis* using a patented energy-free rotating floating photobioreactor (RFP). *Bioresour. Technol.* **2017**, *224*, 515–522. [CrossRef]
19. Zheng, H.; Wang, Y.; Li, S.; Nagarajan, D.; Varjani, S.; Lee, D.-J.; Chang, J.-S. Recent advances in lutein production from microalgae. *Renew. Sustain. Energy Rev.* **2022**, *153*, 111795. [CrossRef]
20. Fu, J.; Huang, Y.; Liao, Q.; Xia, A.; Fu, Q.; Zhu, X. Photo-bioreactor design for microalgae: A review from the aspect of CO_2 transfer and conversion. *Bioresour. Technol.* **2019**, *292*, 121947. [CrossRef]
21. Dineshkumar, R.; Dash, S.K.; Sen, R. Process integration for microalgal lutein and biodiesel production with concomitant flue gas CO_2 sequestration: A biorefinery model for healthcare, energy and environment. *RSC Adv.* **2015**, *5*, 73381–73394. [CrossRef]
22. Xie, Y.; Zhao, X.; Chen, J.; Yang, X.; Ho, S.H.; Wang, B.; Chang, J.S.; Shen, Y. Enhancing cell growth and lutein productivity of *Desmodesmus* sp. F51 by optimal utilization of inorganic carbon sources and ammonium salt. *Bioresour. Technol.* **2017**, *244 Pt 1*, 664–671. [CrossRef]
23. Yeh, T.-J.; Tseng, Y.-F.; Chen, Y.-C.; Hsiao, Y.; Lee, P.-C.; Chen, T.-J.; Chen, C.-Y.; Kao, C.-Y.; Chang, J.-S.; Chen, J.-C.; et al. Transcriptome and physiological analysis of a lutein-producing alga *Desmodesmus* sp. reveals the molecular mechanisms for high lutein productivity. *Algal Res.* **2017**, *21*, 103–119. [CrossRef]
24. Sun, N.; Wang, Y.; Li, Y.-T.; Huang, J.-C.; Chen, F. Sugar-based growth, astaxanthin accumulation and carotenogenic transcription of heterotrophic *Chlorella zofingiensis* (Chlorophyta). *Process. Biochem.* **2008**, *43*, 1288–1292. [CrossRef]
25. Mario, O.B.; Luis, M.C.; Ming, H.L.; Juan, M.D.; Gustavo, C.H. Lutein as a functional food ingredient: Stability and bioavailability. *J. Funct. Foods* **2020**, *66*, 103771. [CrossRef]
26. Marondedze, C.; Groen, A.J.; Thomas, L.; Lilley, K.S.; Gehring, C. A Quantitative Phosphoproteome Analysis of cGMP-Dependent Cellular Responses in *Arabidopsis thaliana*. *Mol. Plant* **2016**, *9*, 621–623. [CrossRef] [PubMed]
27. Nan, W.; Wang, X.; Yang, L.; Hu, Y.; Wei, Y.; Liang, X.; Mao, L.; Bi, Y. Cyclic GMP is involved in auxin signalling during Arabidopsis root growth and development. *J. Exp. Bot.* **2014**, *65*, 1571–1583. [CrossRef] [PubMed]
28. Ren, Y.; Sun, H.; Deng, J.; Zhang, Y.; Li, Y.; Huang, J.; Chen, F. Coordinating Carbon Metabolism and Cell Cycle of *Chlamydomonas-reinhardtii* with Light Strategies under Nitrogen Recovery. *Microorganisms* **2021**, *9*, 2480. [CrossRef]
29. Ye, Y.; Huang, J.C. Defining the biosynthesis of ketocarotenoids in *Chromochloris zofingiensis*. *Plant Divers.* **2020**, *42*, 61–66. [CrossRef]
30. Kang, M.; Yang, J.S.; Kim, Y.; Kim, K.; Choi, H.; Lee, S.H. Comparison of DNA extraction methods for drug susceptibility testing by allele-specific primer extension on a microsphere-based platform: Chelex-100 (in-house and commercialized) and MagPurix TB DNA Extraction Kit. *J. Microbiol. Methods* **2018**, *152*, 105–108. [CrossRef]

31. Roth, M.S.; Cokus, S.J.; Gallaher, S.D.; Walter, A.; Lopez, D.; Erickson, E.; Endelman, B.; Westcott, D.; Larabell, C.A.; Merchant, S.S.; et al. Chromosome-level genome assembly and transcriptome of the green alga *Chromochloris zofingiensis* illuminates astaxanthin production. *Proc. Natl. Acad. Sci. USA* **2017**, *114*, E4269–E4305. [CrossRef] [PubMed]
32. Livak, K.J.; Schmittgen, T.D. Analysis of relative gene expression data using real-time quantitative PCR and the $2^{-\Delta\Delta CT}$ Method. *Methods* **2001**, *25*, 402–408. [CrossRef] [PubMed]

marine drugs

MDPI

Article

Characterization of Novel Selected Microalgae for Antioxidant Activity and Polyphenols, Amino Acids, and Carbohydrates

Paula Santiago-Díaz [1,2], Argimiro Rivero [1,2], Milagros Rico [1,2,*] and Juan Luis Gómez-Pinchetti [2,3]

1 Departamento de Química, Universidad de Las Palmas de Gran Canaria, Campus de Tafira, 35017 Las Palmas de Gran Canaria, Spain; paula.santiago@ulpgc.es (P.S.-D.); argimiro.rivero@ulpgc.es (A.R.)

2 Instituto de Oceanografía y Cambio Global (IOCAG), Campus de Taliarte, Universidad de Las Palmas de Gran Canaria, Unidad Asociada ULPGC-CSIC, 35214 Telde, Spain; juan.gomez@ulpgc.es

3 Banco Español de Algas (BEA), Universidad de Las Palmas de Gran Canaria, Muelle de Taliarte s/n, 35214 Telde, Spain

* Correspondence: milagros.ricosantos@ulpgc.es; Tel.: +34-928-454418

Abstract: The biochemical composition of three novel selected microalgae strains (Chlorophyta) was evaluated to confirm their potential possibilities as new sustainably produced biomass with nutritional, functional, and/or biomedical properties. Extracts from cultured *Pseudopediastrum boryanum*, *Chloromonas cf. reticulata*, and *Chloroidium saccharophilum* exhibited higher radical scavenging activity of DPPH (1,1-diphenyl-2-picrylhydrazyl) when compared to butylated hydroxytoluene (BHT), but lower than butylated hydroxyanisole (BHA). Total phenolic compounds and amino acids were determined by newly developed RP-HPLC methods. Total phenolic contents, as $\mu g\ g^{-1}$ of dry biomass, reached 27.1 for *C. cf. reticulata*, 26.4 for *P. boryanum*, and 55.8 for *C. saccharophilum*. Percentages of total analysed amino acids were 24.3, 32.1, and 18.5% of dry biomass, respectively, presenting high values for essential amino acids reaching 54.1, 72.6, and 61.2%, respectively. Glutamic acid was the most abundant free amino acid in all microalgae samples, followed by proline and lysine in *C. saccharophilum* and *P. boryanum*, and methionine and lysine in *C. reticulata*. Soluble carbohydrates in aqueous extracts ranged from 39.6 for *C. saccharophilum* to 49.3% for *C. reticulata*, increasing values to 45.1 for *C. saccharophilum* and 52.7% for *P. boryanum* in acid hydrolysates of dried biomass. Results confirmed the potential possibilities of these microalgae strains.

Keywords: amino acids; carbohydrates; microalgae; phenolic compounds; radical scavenging activity (RSA); RP-HPLC

Citation: Santiago-Díaz, P.; Rivero, A.; Rico, M.; Gómez-Pinchetti, J.L. Characterization of Novel Selected Microalgae for Antioxidant Activity and Polyphenols, Amino Acids, and Carbohydrates. *Mar. Drugs* **2022**, *20*, 40. https://doi.org/10.3390/md20010040

Academic Editor: Carlos Almeida

Received: 17 December 2021
Accepted: 29 December 2021
Published: 30 December 2021

1. Introduction

Changes in human and animal nutrition are essential, among other actions, to achieve several of the UN Sustainable Development Goals [1]. Diets rich in meat and processed foods are detrimental to health and are also associated with higher environmental costs and greenhouse gas emissions. Therefore, it is a priority to reduce their consumption in order to mitigate their negative impact on health [2].

Algae in general and microalgae in particular are described as a novel rich source of nutrients and contain natural products with several properties and applications in many industrial fields, including food, feed, cosmetics, pharmaceuticals, and biofuel production [3–5].

Polysaccharides from macro- and microalgae are considered a source of dietary fibre with bioactive properties improving the levels of blood glucose and cholesterol [6,7]. These algal polysaccharides also show other potent biological activities such as antioxidant, antifungal, antiviral, antibacterial, and antitumoral properties; tyrosinase inhibitory activity; and anti-inflammatory and immunomodulatory characteristics [6,8].

Balanced diets in amino acids of natural origin and safe sources are strongly recommended [9]. Proteins are one of the main components of microalgae, reaching up to 70%

of dry biomass in some species [10,11] and containing up to 50% essential amino acids and higher antioxidant activities than those of common proteins in the human diet [12,13]. Several microalgal peptides also exhibit antihypertensive, immunomodulatory, anticancer, hepatoprotective, antiatherosclerosis, anticoagulant, anti-UV radiation, antiosteoporosis, and antimicrobial activities [14]. Studies focused on drugs combined with glutamic acid (glutamate) confirmed an increase in their efficacy [15]. In addition, glutamic and aspartic acids contribute to enhancing (i) the flavour of meat, soy sauce, seafood, and some processed foods [16] and to (ii) protein solubility for pharmaceutical uses [17]. Dietary supplementation with proline may also be advantageous under certain physiological and pathological conditions [18].

In addition, different types of antioxidants such as phenolic compounds from seaweeds and microalgae have also been reported [19–21]. Food enrichment with microalgae is a simple and well-known method for improving the physicochemical, nutritional, and sensory properties [22]. The antioxidant capacity and phenolic content of broccoli soup increased when freeze-dried *Spirulina* sp., *Chlorella* sp., or *Tetraselmis* sp. was added at concentrations ranging from 0.5 to 2.0% (*w/v*) [23]. According to Žugčić et al. [24], beef patties prepared with microalgal proteins (1% *Chlorella* or 1% *Spirulina* of 60 and 70% purity, respectively) increased the concentrations of all amino acids, especially aspartic and glutamic acids, concluding that microalgae proteins could be useful candidates for new meat products in the food and feed industries [25].

Considering all this application potential and despite the rich biodiversity of microalgae, only a few species are exploited from a biotechnological point of view, and only 18 species of the phylum Chlorophyta are being produced in Europe [26]. The objective of this study was to evaluate the biochemical composition of three selected, not previously studied, freshwater microalgae strains *Chloromonas cf. reticulata, Pseudopediastrum boryanum*, and *Chloroidium saccharophilum* for their potential in developing food and feed products with high nutritional and therapeutic/functional values. For this purpose, methanol extracts obtained from laboratory-controlled cultured biomass were screened for their capacity to scavenge the 1,1-diphenyl-2-picrylhydrazyl (DPPH) radical and compared with food additives (butylated hydroxyanisole (BHA) and butylated hydroxytoluene (BHT)) with known antioxidant activities [27]. Ten different phenolic compounds and ten selected amino acids were identified and quantified by newly developed RP-HPLC methods, and total soluble carbohydrate contents obtained by two different extraction protocols were also determined.

2. Results

2.1. Radical Scavenging Activity

As it is observed in Figure 1, extracts from *Pseudopediastrum boryanum* exhibited the highest capacity to scavenge free radical DPPH (30.19%), followed by *Chloroidium saccharophilum* (26.95%) and *Chloromonas cf. reticulata* (19.33%). All microalgae samples showed a higher RSA than BHT (17.37%) and lower activity than BHA (48.69%).

2.2. Identification and Quantification of Phenolic Compounds

The identification and quantification of 10 polyphenols were achieved by an updated RP-HPLC method in less than 30 min (retention times ranged from 4.69 to 25.87 min). Linearity was evaluated using the method of least squares of a plot of integrated peak area versus mean concentration from three area measurements. The correlation coefficients were not less than 0.9995. Precision was assessed using six determinations at 1 μg mL^{-1} and expressed as relative standard deviation (RSD), which ranged from 1.80 to 3.85%. The limits of detection (LOD) and the limits of quantification (LOQ) were calculated assuming a minimum detectable signal-to-noise level of 3 and 10, respectively. LOD were found to be in the range of 0.0221–0.2003 μg mL^{-1}, and the LOQ were observed in the range of 0.0736–0.6676 μg mL^{-1}. The recoveries were found in the range from 91.8 to 109.2%.

All ten analysed phenolic compounds were identified for the strain *Chloroidium saccharophilum* (Table 1). Rutin and protocatechuic, coumaric, ferulic, and gentisic acids were not detected in the extracts obtained from *Chloromonas reticulata. Pseudopediastrum boryanum* showed a lack of coumaric and gentisic acids. *C. saccharophilum* exhibited the highest content of phenolic compounds (55.83 μg g^{-1} of dry weight) followed by *C. reticulata* and *P. boryanum* (27.10 and 26.40 μg g^{-1} of dry weight), respectively.

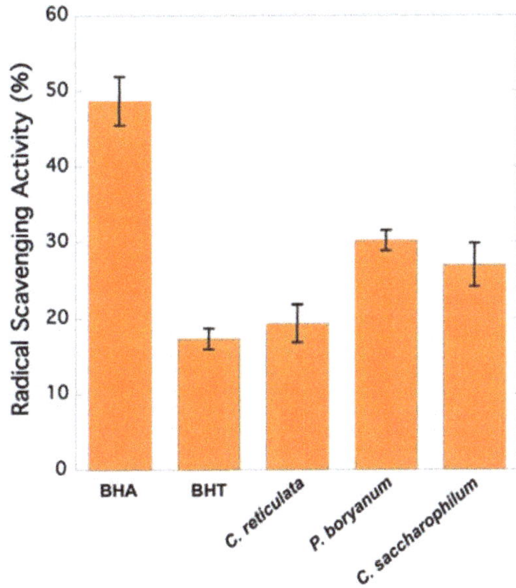

Figure 1. Radical scavenging activities (RSA) of synthetic compounds (BHA and BHT) and microalgae extracts expressed as DPPH inhibition percentage: 100 × (1 − Abs in the presence of sample/Abs in the absence of sample).

Table 1. Total polyphenol contents (μg g^{-1} of dry biomass) of the analysed microalgae strains.

Polyphenol	*Chloromonas cf. reticulata*	*Pseudopediastrum boryanum*	*Chloroidium saccharophilum*
Gallic acid	5.08 ± 0.25	1.44 ± 0.99	2.88 ± 1.22
Protocatechuic acid	n.d. *	1.23 ± 0.11	3.08 ± 1.34
Catechin	5.68 ± 0.74	4.14 ± 0.13	12.34 ± 1.13
Vanillic acid	1.61 ± 0.21	2.63 ± 0.23	14.15 ± 2.84
Epicatechin	10.43 ± 0.21	5.27 ± 0.84	1.90 ± 0.33
Syringic acid	3.12 ± 0.32	1.12 ± 0.10	5.84 ± 0.66
Rutin	n.d. *	6.40 ± 1.03	6.87 ± 3.42
Gentisic acid	n.d. *	n.d. *	2.42 ± 2.21
Coumaric acid	n.d. *	n.d. *	3.95 ± 2.09
Ferulic acid	n.d. *	2.44 ± 0.34	2.40 ± 1.67
Total	27.10 ± 2.03	26.40 ± 4.02	55.83 ± 16.90

* n.d.: not detected.

2.3. Identification and Quantification of Free and Total Analysed Amino Acids

A new RP-HPLC method was developed for detecting and quantifying 10 amino acids in microalgae extracts in less than 35 min (the retention times ranged from 6.19 to 34.87 min). Correlation coefficients were not lower than 0.9976. Precision expressed as RSD, the LODs, and the LOQs were calculated as above. RSD ranged from 1.73 to 3.88%. The percentage recoveries were from 92.4 to 101.8%. The LOD ranged from 0.0006 to 0.01 μg mL^{-1}, and the LOQ from 0.0015 to 0.0335 μg mL^{-1}. This methodology was fast, precise, and accurate and allowed the simultaneous quantification of 10 amino acids with a high sensitivity and reproducibility.

All 10 free amino acids studied were present in all the microalgae strains (Table 2). Glutamic acid was the most abundant in all microalgae samples, ranging from 461.82 to 5630.37 μg g^{-1} of dry weight. *C. saccharophilum* exhibited a remarkably higher amount of each single amino acid compared to the other microalgae strains, reaching 20.46 mg of total free amino acids per gram of dry biomass.

Table 2. Free analysed amino acid contents (μg g^{-1} of dry weight) of the three microalgae strains.

Amino Acid	*Chloromonas cf. reticulata*	*Pseudopediastrum boryanum*	*Chloroidium saccharophilum*
Arginine	107.88 ± 1.67	448.00 ± 90.00	668.34 ± 48.16
Glutamic acid	461.82 ± 1.70	1937.00 ± 28.00	5630.37 ± 135.89
Aspartic acid	89.31 ± 0.55	157.57 ± 16.49	805.95 ± 18.82
Proline	112.99 ± 0.51	1282.00 ± 38.00	5546.42 ± 141.14
Methionine	146.00 ± 6.30	201.00 ± 77.00	805.95 ± 90.52
Valine	130.00 ± 0.38	285.00 ± 6.14	1379.65 ± 159.98
Lysine	196.84 ± 0.87	697.65 ± 52.14	2847.93 ± 157.67
Isoleucine	65.34 ± 0.08	86.03 ± 4.00	866.81 ± 15.26
Phenylalanine	107.42 ± 0.19	108.00 ± 2.00	1379.65 ± 30.57
Histidine	116.67 ± 1.22	154.00 ± 21.00	530.92 ± 38.35
	Sum of amino acids in μg g^{-1} of dry weight		
∑NEEA	772.0 ± 4.43	3824.6 ± 172.5	12,651.1 ± 344.0
∑EEA	762.3 ± 9.04	1531.7 ± 162.3	7810.9 ± 492.5
∑FAA	1534.33 ± 13.5	5356.3 ± 334.8	20,462.0 ± 836.5

∑NEEA, sum of non-essential amino acids; ∑EEA, sum of essential amino acids; ∑FAA, sum of free amino acids.

Apart from values for glutamic acid, Table 2 shows that the most abundant free amino acids in *C. saccharophilum* were proline, lysine, valine, and phenylalanine; in *P. boryanum* were proline, lysine, and arginine; and in *C. reticulata* were lysine and methionine. The percentages of essential amino acids were 49.7% in *C. reticulata*, 38.2% in *C. saccharophilum*, and 28.6% in *P. boryanum* (Table 2).

Table 3 shows the total content of the ten amino acids analysed in the acid-hydrolysed microalgae extracts. Total contents of these amino acids in *P. boryanum*, *C. reticulata*, and *C. saccharophilum* were 32.06, 24.27, and 18.45% on a dry weight basis, with high percentages of the analysed essential amino acids: 72.6, 54.1, and 61.2%, respectively. Lysine was the most abundant amino acid in *C. saccharophilum* (36.49 mg g^{-1} of dry weight), proline in *C. reticulata* (51.69 mg g^{-1} of dry weight), and methionine in *P. boryanum* (137.20 mg g^{-1} of dry weight).

2.4. Carbohydrate Contents

Aqueous extraction of *C. saccharophilum, P. boryanum, and C. reticulata* biomass showed 39.64, 43.24, and 49.25% in soluble carbohydrates, respectively (Figure 2). The highest carbohydrate contents reaching 45.06, 52.67, and 51.24% were obtained when extractions were carried out under acid hydrolysis conditions.

Table 3. Contents of 10 analysed amino acids (mg g^{-1} of dry weight) of three microalgae strains.

Amino Acid	*Chloromonas cf. reticulata*	*Pseudopediastrum boryanum*	*Chloroidium saccharophilum*
Arginine	21.29 ± 3.47	7.12 ± 0.75	5.98 ± 0.66
Glutamic acid	23.84 ± 6.25	19.48 ± 1.41	30.50 ± 1.28
Aspartic acid	14.67 ± 3.59	11.38 ± 0.84	3.44 ± 0.21
Proline	51.69 ± 1.75	49.90 ± 4.25	31.68 ± 1.42
Methionine	40.81 ± 9.78	137.20 ± 24.41	29.89 ± 8.96
Valine	19.72 ± 1.79	20.21 ± 1.59	18.48 ± 0.45
Lysine	31.56 ± 1.66	36.20 ± 2.32	36.49 ± 1.99
Isoleucine	13.28 ± 1.35	13.43 ± 1.32	11.36 ± 0.71
Phenylalanine	19.76 ± 1.37	20.01 ± 1.20	14.92 ± 0.21
Histidine	6.07 ± 0.13	5.63 ± 0.69	1.81 ± 0.01
	Sum of amino acids in mg g^{-1} of dry weight (%)		
∑NEEA	111.5 ± 15.1 (45.9)	87.9 ± 7.25 (27.4)	71.6 ± 3.57 (38.8)
∑EEA	131.2 ± 16.1 (54.1)	232.7 ± 30 (72.6)	112.9 ± 11.3 (61.2)
∑TAA	242.7 ± 31.1	320. 6 ± 38.1	184.5 ± 14.9

∑NEEA, sum of non-essential amino acids; ∑EEA, sum of essential amino acids; ∑TAA, sum of total amino acids.

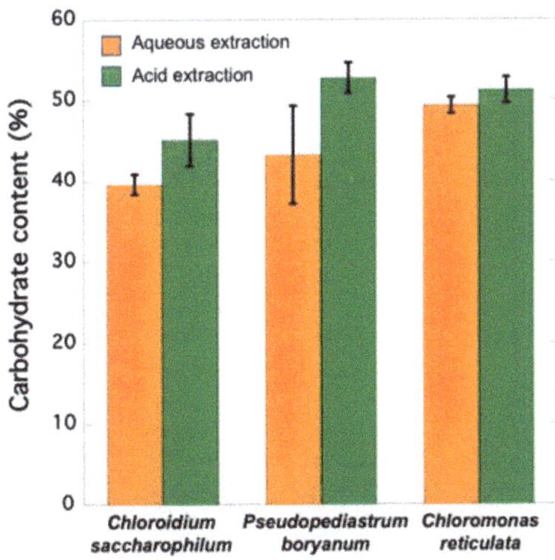

Figure 2. Carbohydrate content (as percent of glucose equivalents per dry weight) of the analysed microalgae strains.

3. Discussion

3.1. Algal Material and Extraction Procedures

The Canary Islands are mountainous with a sub-tropical volcanic origin, supporting high levels of solar radiation all year round. These environmental conditions generate highly diverse habitats and ecosystems, forcing microorganisms to adapt and accumulate metabolites that might be interesting from a biotechnological approach [28].

The selected microalgae strains analysed in the present study were bioprospected from different locations and environments and, after clonal isolation, adapted to laboratory growth conditions and BG-11 culture media before scale-up for the evaluation of growth characteristics and biomass production.

Several factors should be considered in the extraction and quantification of metabolites, such as the previous treatment of microalgae cells and storage, the hydrolysis conditions (time, temperature, and acid concentration), and the mechanical/chemical extraction methods. Cell wall disruption is a necessary preliminary step to make the cell contents accessible and digestible and prevent incorrect measurements. The efficiency of cell disruption methods depends on the species being investigated (including cell wall type) and their physiological state. Therefore, the total rupture of the cell membrane should be confirmed by making observations with a microscope to prevent underestimation of the metabolite contents [11].

Kröger et al. [29] compared several methods for effective extraction from the microalgae *Scenedesmus rubescens*, concluding that freeze-drying produces cell wall damage and therefore improves the extraction yields. Moreover, several drying methods have also been studied by de Farias Neves et al. [30], who confirmed that freeze-drying is the most suitable microalgae drying method without bioactive compound loss. For all these reasons, in our work, biomass was freeze-dried and cells examined under a microscope to check complete cell wall breaking.

3.2. Antioxidant (RSA) Activity and Phenolic Contents

The synthetic compounds BHA and BHT are widely used as antioxidant food additives at a maximum level of 0.02% [31]. However, these compounds are classified as cancer promoters capable of inducing cytotoxicity and apoptosis [32]. Therefore, many studies have extensively studied their replacement by safer and inexpensive natural antioxidants. Phenolic compounds of natural origin exhibit antioxidant properties capable of extending food shelf life by preventing rancidity due to oxidation. They also exhibit protective effects against oxidative stress in biological systems [19].

Previous studies showed that gentisic acid, gallic acid, catequin, epicatechin, protocatechuic, and syringic acids exhibited higher antioxidant activity than BHT and BHA [33]. Catechin, epicatechin, gallic acid, and vanillic acid, among others, are better radical scavengers than many monomeric flavones and even flavonols because they act particularly well as H-atom donors [34]. The presence of these compounds in microalgae increases their potential health benefits. The microalgae species *C. saccharophilum, P. boryanum*, and *C. reticulata* showed a higher RSA than that of BHT (Figure 1). Gentisic and coumaric acids were not detected in the extracts derived from *P. boryanum* (Table 1), which showed the highest RSA (30.19%) and the lowest phenolic content (26.40 µg g^{-1} of dry biomass). All ten phenolic compounds, gallic acid, protocatechuic acid, catechin, vanillic acid, rutin, epicatechin, syringic acid, gentisic acid, coumaric acid, and ferulic acid, were detected in *C. saccharophilum*, which exhibited the highest content of phenolic compounds, followed by *C. reticulata* and *P. boryanum* (55.83, 27.10, and 26.40 µg g^{-1} of dry biomass, respectively). The lack of five analysed phenolics (rutin and coumaric, ferulic, protocatechuic, and gentisic acids) was observed in the extracts prepared with *C. reticulata*, which gave the lowest relative RSA (19.33%), but a phenolic content similar to that of *P. boryanum* (Table 1). Similar findings were reported by Corrêa da Silva et al. [35] who studied the total phenolic content and antioxidant activity of microalgae *P. boryanum* grown in different culture media. Their results also showed extracts with higher phenolic content (measured through the Folin–Ciocalteu assay) but lower antioxidant activity (assay ABTS (2,2′-azino-bis(3-ethylbenzothiazoline-6-sulphonic acid)). No correlation between DPPH inhibition and phenolic contents was found by Blagojević et al. [36], who evaluated the phenolic profiles and antioxidant activities of 10 cyanobacterial strains cultured in BG11 medium in the presence and absence of nitrogen. In our study, a small group of phenolic compounds was quantified. The antioxidant response of these compounds varies remarkably depending on their chemical structure [34], and other possible antioxidants present in the mixture have not been considered, as phenolics are not the only contributors to the antioxidant activities in algae. In addition, the extracts are complex mixtures including active components at low levels, and their activities depend on the relative concentrations of these components and the interfering compounds as well as the synergistic, additive, or antagonistic effects between them [37].

Samples in this study showed a higher content of phenolic compounds than five different microalgae and cyanobacterial species evaluated by Klejdus et al. [38], who identified eight phenolic compounds and quantified the highest content in the green microalgae *Spongiochloris spongiosa* (6.656 µg g^{-1} of dry biomass). Blagojević et al. [36] only detected 8 of 45 polyphenols investigated in several cyanobacteria species. These authors observed that cyanobacteria *Arthrospira* S1 and *Anabaena* C2 showed a lower total phenolic content (24.05 and 18.72 µg g^{-1} of biomass, respectively) than those obtained in our samples. Onofrejová et al. [39] identified 12 phenolic compounds in the freshwater microalgae *Spongiochloris spongiosa* and cyanobacterium *Anabela doliolum*, also quantifying lower contents (5.1 and 3.6 µg g^{-1}, respectively) than those determined in the present study.

3.3. Free Amino Acid Contents

In this study, 10 amino acids were selected because of their antioxidant properties and their important role in cellular metabolism as key precursors for synthesis of several metabolites [9,40–43]. Therefore, a well-balanced diet can ensure the intake needs of essential and non-essential amino acids of the body to function properly.

All microalgal samples presented the 10 free amino acids evaluated—histidine, methionine, valine, lysine, isoleucine, phenylalanine, arginine, proline, and glutamic and aspartic acids—with quantitative differences for individual compounds between the three different strains (Table 2). Glutamic acid was the most abundant, ranging from 461.8 to 5630.4 µg g^{-1} of dry biomass. These results are in accordance with those reported by Vendruscolo et al. [44], who quantified 15 amino acids in two Chlorophyceae (*Chlorella vulgaris* and *Scenedesmus obliquus*) and two cyanobacteria (*Aphanothece microscopica* and *Phormidium autumnale*), concluding that glutamic acid was the most abundant detected amino acid in three of them (*Scenedesmus obliquus* showed a higher content of alanine). Our findings also agree partially with previous studies focused on the free amino acid profile determination of seaweeds and microalgae, which confirmed that glutamic and aspartic acids were the most abundant free amino acids (up to 26% of the free amino acid fraction) [45,46].

C. saccharophilum showed the highest amount of total analysed free amino acids, reaching 20.5 mg g^{-1} of dry biomass (13.34 and 3.82 times higher than those of *C. reticulata* and *P. boryanum*, respectively). Vendruscolo et al. [44] reported lower amounts of the total sum of 15 amino acids quantified in two Chlorophyceae and two cyanobacteria (ranging from 0.371 to 1.525 mg g^{-1} of dry biomass). Machado et al. [47] determined the total sum of 20 free amino acids in four seaweed species (*Porphyra dioica, Porphyra umbilicalis, Gracilaria vermiculophylla*, and *Ulva rigida)*, which ranged from 3.36 to 16.17 mg g^{-1} of dry biomass. Their values were higher than those for *C. reticulata*, but lower than those found for *C. saccharophilum* in this study (1.53 mg and 20.46 mg g^{-1} of dry biomass, respectively).

Apart from glutamic acid, Table 2 shows that the most abundant free amino acids were proline, lysine, valine, and phenylalanine in *C. saccharophilum*; proline, lysine, and arginine in *P. boryanum*; and lysine and methionine in *C. reticulata*. On the contrary, Vendruscolo et al. [44] only detected methionine and lysine in one of the four analysed microalgae (30.55 and 8.05 µg g^{-1} of dry biomass, respectively, in *Scenedesmus obliquus*), and Machado et al. [47] only identified methionine in one of the above cited red seaweed (160 µg g^{-1} of dry biomass in *Porphyra dioica*).

Under natural conditions, the composition of free amino acids in microalgae can vary dramatically during the growing season. Kolmakova et al. [45] concluded that the composition of free amino acids of diatoms and green microalgae and cyanobacteria is extremely sensitive to external factors such as available nitrogen and light intensity and photoperiod and also depends on the growth phase of the microalgae culture [48]. According to Granum et al. [49], extracellular free amino acid contents exuded by the marine diatom *Skeletonema costatum* changed drastically from the exponential to stationary growth phase, and the cellular free amino acid levels reached values between 8% (end of scotophase) and 22% (end of photophase) of cellular organic N and decreased by 90% within 24 h of N depletion. In fact, intracellular amino acids have been proposed to be used as an index of the physiological status of the diatom *Rhizosolenia delicatula* [50].

3.4. Total Contents of Analysed Amino Acids

Kolmakova et al. [45] reported that the percentages of total essential and non-essential amino acids in diatoms and green microalgae and cyanobacteria are stable and show a common profile, with leucine as the most abundant essential amino acid, methionine and histidine as the least abundant, and glutamic and aspartic acids as the most abundant non-essential amino acids (up to 20% of the sum of amino acids). However, methionine was the most abundant amino acid in *P. boryanum* in the present study (137.2 mg g^{-1} of dry weight). Several authors have confirmed that non-essential glutamic and aspartic acids in the cell hydrolysates constituted 22–44% of the total amino acids in algae [51]. Cobos et al. [52] reported relatively similar amino acid profiles in four freshwater Chlorophyta microalgae, *Ankistrodesmus* sp., *Haematococcus* sp., *Scenedesmus* sp., and *Chlorella* sp., where aspartic acid ranged from 20.94 to 38.21 mg g^{-1} of dry weight and leucine from 20.08 to 40.99 mg g^{-1} of dry weight, and the least abundant amino acid was histidine, ranging from 4.10 to 7.24 mg g^{-1} of dry weight. The percentages of glutamic and aspartic acids in this work were 9.6, 15.9, and 18.4% in *P. boryanum*, *C. reticulata*, and *C. saccharophilum*, respectively, which would presumably decrease if a larger number of amino acids were analysed. In accordance with Cobos et al. [52] and Kolmakova et al. [45], the least abundant amino acid was histidine, ranging from 1.81 to 6.07 mg g^{-1} of dry weight (Table 3). However, methionine, proline, and lysine were the most abundant amino acids in *C. reticulata* and *P. boryanum*, while *C. saccharophilum* gave a higher content of lysine, proline, and glutamic acid. Lysine was the most abundant amino acid in *C. saccharophilum* (36.49 mg g^{-1} of dry weight) and proline in *C. reticulata* (51.69 mg g^{-1} of dry weight). Several studies have also exhibited relevant amounts of lysine and proline [11,53]. *C. saccharophilum*, which showed the highest free amino acid content, presented the lowest total content of analysed amino acids after acid hydrolysis, followed by *C. reticulata* and *P. boryanum* (184.5, 242.7, and 320.6 mg g^{-1} of dry weight, respectively). The analysis of total amino acids does not distinguish between free amino acids, which represent less than 10% of the total amino acids, and those which are bound in proteins [54]. This fact can lead to significant differences between free and total amino acid profiles. In addition, it is important to note that high total analysed amino acid values were obtained, considering that only 10 amino acids were quantified in the present study. Machado et al. [47] evaluated the presence of 20 amino acids in four macroalgae (*Porphyra dioica*, *Porphyra umbilicalis*, *Gracilaria vermiculophylla*, and *Ulva rigida*) and reported total contents of amino acids ranging from 96.22 to 286.56 mg g^{-1} of dry weight, and from 57.63 to 173.08 mg g^{-1} of dry weight if only the 10 amino acids analysed in our study were considered. These authors found fractions of free amino acids ranging from 3.15 to 7.18 g per 100 g of total amino acids and from 3.56 to 5.57% considering only our 10 amino acids. Vieira et al. [13] reported higher fractions of free amino acids (grams per 100 g of total amino acids) ranging from 6.47 to 24.0% in brown seaweed species and from 3.40 to 14.0% in red and green seaweeds. These differences may be due to the fact that the extraction of free amino acids was carried out with 0.2M perchloric acid, which could have hydrolysed peptides and increased the amount of free amino acids versus the aqueous extraction performed by Machado et al. [47]. Our results showed lower fractions of the 10 analysed free amino acids in *C. reticulata* and *P. boryanum* (0.63 and 1.67%, respectively) and a higher fraction in *C. saccharophilum* (11.1%) than those reported by Machado et al. [47].

Higher percentages of the analysed essential amino acids were found in *P. boryanum* and *C. saccharophilum* (72.6 and 61.2%, respectively) than those found by Machado et al. [46] when only the 10 amino acids evaluated in this study were considered (between 54% in *Gracilaria vermiculophylla* and 57.87% in *Ulva rigida*). *C. reticulata* showed a similar percentage of the analysed essential amino acids (54.1%). In accordance with Sui et al. [55], these high percentages could be due to the illumination cycle applied during the culture (18:6 h L:D) and the late-exponential growth phase for harvesting cells. Araya et al. [56] quantified the content of seventeen amino acids in five species of microalgae (*Haematococcus pluvialis*, *Skeletonema costatum*, *Arthrospira* sp., *Acutodesmus acuminatus*, and *Botryococcus braunii*). Their results showed that the highest amount of amino acids (267.6 mg g^{-1}) was found in *Arthrospira* sp., and the other four species contained lower amounts than those found in this study (below 141.3 mg g^{-1} found in *Botryococcus braunii*).

Lourenço et al. [54] found different amino acid contents (as percentage of dry matter) in *Chlorella minutissima* and *Prorocentrum minimum* at the following growth phases: mid-exponential (24.79 and 24.18%, respectively), late-exponential (36.96 and 30.40%, respectively), early stationary (36.12 and 27.44%, respectively), and late-stationary (22.46 and 26.25, respectively).

On the other hand, different microalgae species show specific needs of L:D cycles and light intensity for productive photosynthesis [57]. Long illumination periods have been previously reported as a stress condition to induce the accumulation of lipids and carotenoids such as astaxanthin [58]. Sui et al. [55] studied the impact of two L:D cycles (12:12 and 24:0 h) on *Dunaliella salina* protein production, concluding that continuous illumination led to higher protein content (0.62 g L^{-1} on day 16 in the exponential phase), which decreased in the stationary phase (0.49 g L^{-1} on day 28). On the contrary, microalgae cultured under a 12:12 h L:D cycle gave a constant accumulation of proteins that reached 0.43 g L^{-1} in the stationary phase. The contents of all individual essential amino acids increased between 5% and 58% in cells cultured under an L:D cycle, reaching 30% of the total protein, and increased dramatically by 17-125% in cells cultured under continuous illumination, reaching 44% of the total protein content. Seyfabadi et al. [59] studied the behaviour of *Chorella vulgaris* incubated at 37.5, 62.5, and 100 µmol photons m^{-2} s^{-1} irradiance and 8:16, 12:12, and 16:8 h L:D photoperiods. It was confirmed that a longer illumination period increases protein contents. In fact, the cycle 16:8 h L:D yielded the highest protein accumulation under each irradiance assayed, reaching the maximum at 100 µmol photons $m^{-2}s^{-1}$ irradiance and a 16:8 h L:D cycle. The photoperiod 16:8 h L:D used in the present study also might stimulate the production of several essential amino acids.

Gorissen et al. [60] analysed the amino acid contents after the acid hydrolysis of 35 protein samples commercially available as isolated protein powder suitable for application in human nutrition or animal feeds. *P. boryanum* showed a higher total amount of six essential amino acids than several of these dietary protein samples, whose content of eight essential amino acids was quantified (oat (137), lupin (131), wheat (180), hemp (116), microalgae (157), soy (199), brown rice (221), corn (210), and egg (165), where values in parentheses mean mg g^{-1} of raw material). *C. saccharophilum* and *C. reticulata* also showed comparable contents to several protein sources analysed by Gorissen et al. [60]. The high levels of the analysed amino acids with high percentages of essential amino acids make microalgae strains in our study a novel source of essential amino acids with potential use in food and functional products.

3.5. Carbohydrate Contents

C. saccharophilum, *C. reticulata*, and *P. boryanum* showed 45.1, 51.2, and 52.7% in carbohydrates, respectively (Figure 2). These results partially agree with those reported by Schulze et al. [61] who analysed freeze-dried biomass from 46 microalgae after hydrolysis with 2N HCl for 1 h and revealed carbohydrate contents ranging from 16.5% (*Mychonastes* sp.) to 71.6% (*Porphyridium purpureum*). Templenton et al. [62] published lower amounts of carbohydrates in strains of *Phaeodactylum tricornutum*, *Nannochloropsis* sp., and *Chlorella vulgaris* (19.6, 8.6, and 20.5% of dry weight, respectively) than those observed in our study by applying the same hydrolysis conditions. This might be due to the fact that biomass was air dried before the acid hydrolysis process in their study, and we used freeze-dried biomass. Our results align well with those of Visca et al. [63], who compared two drying methods before extracting *Scenedesmus* sp. and *Chlorella* sp. biomass: (1) cells were dried at 105°C for 12 h, and (2) cells were freeze-dried, and then both were subjected to similar hydrolysis conditions as those described in the present work. *Scenedesmus* sp. yielded 30.5% carbohydrates and reached a maximum carbohydrate purity when freeze-drying pretreatment was used and lipids were removed (58.7% and 51.8%, depending on the extracting solvent). The authors concluded that *Chlorella* sp. carbohydrate content (17.7%) was not affected by the freeze-drying process because its cell wall is weaker. However, Safi et al. [11] reported that *Chlorella* sp. has a robust cell wall, and its disruption is a necessary preliminary step to quantify the total/maximum content of each metabolite. These findings agree with those reported by Stirk et al. [64], who observed that a simple freeze-drying step is not enough to break the tough cell wall of *Chlorella* sp., which requires methods combining freeze-drying with sonication or ball-milling.

Carbohydrate contents evaluated without methods involving previous acid hydrolysis in 12 species of seaweeds washed with tap water and air dried on blotting paper to remove excess water ranged from 20.47 to 23.9% carbohydrates [65]. Our results obtained without acid pretreatment showed higher contents than those described by Manivannan et al. [65]: *C. saccharophilum, P. boryanum*, and *C. reticulata* yielded 39.6, 43.2, and 49.3% carbohydrates, respectively. These results agree with the total carbohydrate content quantified in microalgae *Neochloris oleoabundans* (~40%) by Suarez Garcia et al. [66].

Further analytical studies of oligosaccharide structure and composition in the samples of novel microalgae, including the strains analysed in the present study, are necessary to unlock the potential applications of these microalgae and their components. In particular, algal polysaccharides have shown numerous industrial applications including antioxidant and antitumor effects, immunostimulating functions, cosmetics and cosmeceuticals, or prebiotic properties as functional foods or nutraceuticals [67,68].

4. Materials and Methods

4.1. Chemicals

Methanol (HPLC gradient grade) and tetrachloroethylene (synthesis grade) were purchased from Scharlab (Barcelona, Spain). Triethylamine (analysis quality) and phenylisothiocyanate (PITC) of reagent grade were supplied by Panreac (Barcelona, Spain); 1,1-diphenyl-2-picrylhydrazyl (DPPH), butylated hydroxyanisole (BHA), butylated hydroxytoluene (BHT), and anthrone (reagent grade) were supplied by Sigma-Aldrich (St. Louis, MO, USA). Formic acid (synthesis grade) and amino acids (aspartic acid, glutamic acid, histidine, arginine, proline, valine, lysine, methionine, isoleucine, and phenylalanine) were provided by Merck (Darmstadt, Germany). Phenolic compounds were supplied as follows: gallic acid, protocatechuic acid, (−) epicatechin, ferulic acid, *p*-coumaric acid, vanillic acid, syringic acid, and (+) catechin by Sigma–Aldrich Chemie (Steinheim, Germany), and rutin and gentisic acid by Merck (Darmstadt, Germany). Ultrapure water was obtained from a Milli-Q system from Millipore (Bedford, MA, USA).

4.2. Algal Material

Three microalgae clonal strains of the Phylum Chlorophyta were selected and provided by the Culture Collection at the Spanish Bank of Algae (located in Taliarte; east coast of Gran Canaria): *Chloromonas. cf. reticulata* (BEA0990B; Order Chlamydomonadales), *Pseudopediastrum boryanum* (BEA0190B; Order Sphaeropleales), and *Chloroidium saccharophilum* (BEA0031B; Order Watanabeales). Strains were isolated from samples bioprospected in Gran Canaria (Canary Islands), except *C. saccharophilum*, which was collected in the Chihuahua region (Mexico). *C. reticulata* was obtained from a terrestrial habitat in the central countryside area of the island, and *P. boryanum* was collected from a moist rock curtain in the northwest. Cultures were scaled up to 2 L Erlenmeyer flasks under controlled conditions (temperature: 23 ± 2 °C; irradiance: <100 µmol photons m^{-2} s^{-1}; and photoperiod 16:8 light:dark (L:D)) in BG-11 culture medium (pH adjusted to 7.4), and continuous aeration was supplied with CO_2 pulses at a rate of 1 min per hour. Growth curves were followed, and samples were harvested at the final exponential growth phase. After centrifugation at 8000 rpm for 15 min, biomass samples were freeze dried (6.5 L Labconco, Kansas City, MO, USA) and kept in sealed vials in darkness before analysis.

4.3. Radical Scavenging Activity (RSA) Measurements

Freeze-dried microalgae biomass (25 mg) was mixed with methanol (1.5 mL) using a vortex (Vortex Ika Genius 3) for 20 min. The mixture was heated at 40 °C for 10 min and sonicated for 10 min (this step was performed twice). Then, the algal material was centrifuged for 10 min at 9000 rpm in a microcentrifuge (Thermo Scientific, Heraeus fresco 17) and removed by filtration. The filtrate was evaporated to dryness in a rotary vacuum evaporator, and the residue was dissolved in methanol (100 µL).

The RSA of samples was evaluated using the DPPH free radical assay described by Bondet et al. [69] with some modifications. Briefly, 25 µL of the samples and standards BHA and BHT (0.2 g L^{-1}) was mixed with 975 µL of DPPH solution (0.1 mM). The obtained mixture was vortexed and incubated for 20 min at room temperature in darkness. The neutralization of DPPH radical leads a decrease in absorbance monitored against a methanol blank at 515 nm using a Shimadzu 1800 UV-Vis spectrophotometer. The radical inhibition percentage was calculated by application of the equation

$$\text{RSA} = 100 \times (1 - \text{Abs in the presence of sample/Abs in the absence of sample}) \quad (1)$$

Measurements were taken in triplicate, and the results were averaged.

4.4. Phenolic Compounds Determination

The extraction of phenolic compounds was carried out as described in the above section (RSA measurements) by mixing 200 mg of biomass with 10 mL of methanol. Once the filtrate was evaporated, the dry residue was suspended in 5 mL of acidified water (pH 1.5). Then, solid phase extraction (SPE) was used following the procedure of Dvořáková et al. [70] with modifications. The cartridges (Chromabon Easy, Macherey-Nagel, 500 mg, particle size 93 µm) were conditioned by successive elution with 2 mL of water, 6 mL of methanol, and 2 mL of water. The suspension was passed through the cartridge, and the retained phenolics were eluted with acetone (6 mL) and evaporated to dryness in a rotary vacuum evaporator. Finally, the residue was dissolved in 200 µL of mobile phase, filtered through a 0.20 µm nylon syringe, and transferred to a vial. Three replicates were used for the quantification.

Chromatographic analysis was performed with a Jasco LC-4000 HPLC instrument equipped with a quaternary pump (PU-4180), an autosampler (AS-4150), photodiode array detector (MD-4015), and an LC-Net interface II. Data acquisition was carried out with ChromNav software. The phenolics were separated with a Varian C18 column (250 mm × 4.6 mm, 5 µm) and a guard column maintained at 30 °C. The gradient elution was performed using water with 0.1% formic acid as mobile phase A and methanol as mobile phase B with the following elution programme for eluent A: 0–5 min, 80% isocratic; 5–30 min, linear gradient from 80% to 40%. Finally, column was washed and reconditioned. Simultaneous monitoring was set at 270 nm (gallic acid, protocatechuic acid, catechin, vanillic acid, rutin, epicatechin, and syringic acid) and 324 nm (gentisic acid, coumaric acid, and ferulic acid) for quantification (Figure 3). Five different concentrations of each compound in the range of 1 to 50 mg L^{-1} were injected in triplicate. The presence of polyphenols in the extracts was confirmed by comparison of their retention times and overlaying of UV spectra with those of individual standard compounds.

4.5. Amino Acid Composition Determination

Extractions of amino acids were performed according to Machado et al. [47] with modifications. A total of 10 mg of freeze-dried biomass was mixed with 5 mL of deionized water for 2 min. The mixture was heated in a water bath at 70 °C for 30 min and was sonicated in an ultrasonic bath (Selecta, Spain) for 5 min followed by centrifugation for 10 min at room temperature (3000 rpm). The supernatant was collected and stored at −20 °C until analysis. Samples were examined under a microscope (Olympus BX40 model) to check complete cell wall lysis.

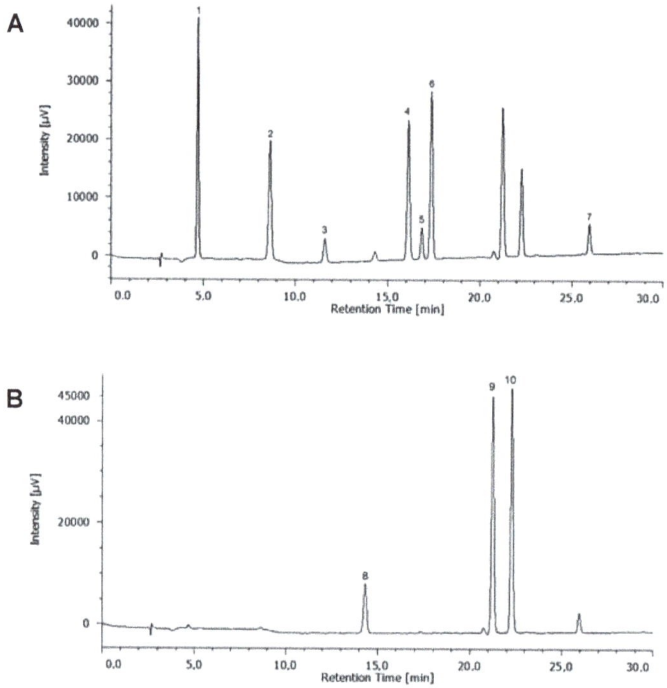

Figure 3. HPLC chromatograms of standard polyphenols: (**A**) 270 nm: 1, gallic acid; 2, protocatechuic acid; 3, catechin; 4, vanillic acid; 5, epicatechin; 6, syringic acid; 7, rutin. (**B**) 324 nm: 8, gentisic acid; 9, coumaric acid; 10, ferulic acid.

Protein hydrolysis for the total amino acid extraction was carried out as follows: HCl (6M, 2 mL) was added to 100 mg of freeze-dried microalgae in test tubes which were flushed under a N_2 stream and heated in an oven at 110 °C for 24 h. Then, the mixture was neutralised by adding NaOH (6M), and deionized water was added up to 5 mL. The extracts were stored at −20 °C until analysis.

The amino acid derivatization procedure of Shi et al. [71] was modified by adding 5 mL of sample solution to 2.5 mL of PITC (1M in acetonitrile) and 2.5 mL of triethylamine (1M in acetonitrile). The resulting solution was stirred for 1 h at room temperature. Subsequently, 5 mL of tetrachloroethylene was added, the mixture was vigorously shaken, and the upper layer was collected. This step was performed twice, and the final solution was filtered through a 0.22 µm nylon syringe filter. Three replicates were used for the analysis.

Chromatographic analysis of six essential amino acids (histidine, methionine, valine, lysine, isoleucine, and phenylalanine) and four non-essential (arginine, proline and glutamic, and aspartic acids) was carried out with a Jasco LC-4000 HPLC instrument, as described above. The amino acid derivatives were separated with a Phenomenex C18 column (250 mm × 4.6 mm, 5 µm) and a Phenomenex guard column maintained at 30 °C. The gradient elution was performed using water with 0.1% formic acid as mobile phase A and methanol as mobile phase B. The elution programme applied for eluent A was 0 min, 75%; 30 min, 40%; 40 min, 40%. Finally, column was washed and reconditioned. The flow rate was 1 mL min^{-1}, and the injection volume was 10 µL. The calibration curves were made by plotting the integrated peak areas of the samples versus the concentration. Five different concentrations of each amino acid in the range of 1 to 40 mg L^{-1} were injected in triplicate. The presence of amino acids was confirmed by comparing their retention times with those of standard compounds (Figure 4).

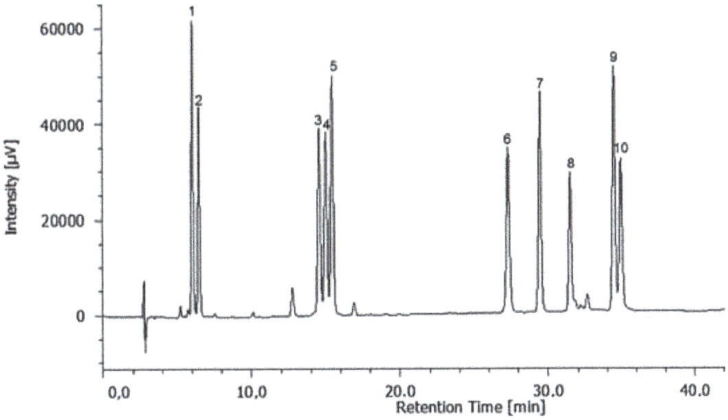

Figure 4. HPLC chromatogram of standard amino acids at 254 nm: 1, histidine; 2, arginine; 3, glutamic acid; 4, aspartic acid; 5, proline; 6, methionine; 7, valine; 8, lysine; 9, isoleucine; 10, phenylalanine.

4.6. Carbohydrates Quantification

Freeze-dried algal material was subjected to different pretreatments: (i) acid hydrolysis according to Templenton et al. [62], where the algal biomass (25 mg) was mixed with 250 µL of sulfuric acid (72 wt %) in a water bath at 30 °C for 1 h and then was heated at 121 °C in sulfuric acid (4 wt %) in an autoclave (Micro 8, JP Selecta SA); (ii) aqueous extraction was carried out by stirring the biomass (25 mg) with 2.5 mL of ultrapure water during 1 h at room temperature according to Jansen [72]. Both extracts were centrifuged (3500 rpm for 10 min) and recovered by filtration.

Carbohydrate contents were determined using the colorimetric method described by Brooks et al. [73] with modifications. Anthrone reagent was prepared fresh daily by dissolving anthrone (200 mg) in 72% sulfuric acid (100 mL). This reagent (2 mL) was mixed with 1 mL of each sample (microalgae extracts and standard solutions), vortexed for 30 s, and heated for 10 min at 100 °C in a water bath. The test tubes were cooled in an ice bath for 5 min, and the absorbance was recorded at 505 nm on a Shimadzu UV-1800 spectrophotometer. A standard calibration curve was prepared with solutions of glucose in the range of concentrations from 20 to 200 µg mL^{-1}. The results were expressed as grams of glucose equivalent (percentage of dry biomass). Three replicates were used for the determination of carbohydrate concentrations.

5. Conclusions

Two new simple, sensitive, accurate, and reproducible RP-HPLC methods were developed for detecting and quantifying 10 amino acids and 10 phenolic compounds in three novel selected microalgae strains. The antioxidant activities of extracts derived from *Chloromonas cf. reticulata*, *Psudopediastrum boryanum*, and *Chloroidium saccharophilum* determined in this study, as well as their amino acid, phenolic, and carbohydrate contents, confirm the potential possibilities of these microalgae species to be considered as novel source of bioactives for food, feed, and biomedical applications. Further research is needed to determine the impact of the growth conditions, illumination cycles during cell culture, their ability to accumulate metabolites, and, therefore, their specific potential.

Author Contributions: Conceptualization, A.R. and M.R.; Data curation, P.S.-D.; Formal analysis, P.S.-D., A.R. and M.R.; Funding acquisition, J.L.G.-P.; Investigation, P.S.-D., A.R., M.R. and J.L.G.-P.; Methodology, A.R., M.R. and J.L.G.-P.; Resources: J.L.G.-P.; Software, P.S.-D. and A.R.; Supervision, A.R.; Validation, P.S.-D. and A.R.; Visualization, P.S.-D., M.R. and J.L.G.-P.; Writing—original draft, M.R.; Writing—review and editing, P.S.-D., A.R., M.R. and J.L.G.-P. All authors have read and agreed to the published version of the manuscript.

Funding: The participation of Paula Santiago was funded through a PhD scholarship from the Universidad de Las Palmas de Gran Canaria (PIFULPGC-2019) to join the PhD Program in Oceanography and Global Change (DOYCAG), promoted by the Institute of Oceanography and Global Change (IOCAG). J.L. Gomez Pinchetti thanks the financial support of the European Territorial Cooperation Program PCT-MAC 2014-2020 through the Project REBECA-CCT (MAC/1.1.B/269) and the H2020-INFRADEV-03-2019 Project IS_MIRRI21 (No. 871129).

Data Availability Statement: The data used to support the findings of this study are included in the article.

Acknowledgments: The authors thank the Plataforma Oceánica de Canarias—Consorcio público—ICTS (plocan.eu) for allowing us to use the chromatographic equipment.

Conflicts of Interest: The authors declare no conflict of interest.

References

1. UN General Assembly. *Transforming Our World: The 2030 Agenda for Sustainable Development*; United Nations: New York, NY, USA, 2015.
2. FAO. *The Future of Food and Agriculture-Trends and Challenges*; FAO: Rome, Italy, 2017.
3. Ördög, V.; Stirk, W.A.; Lenobel, R.; Bancířová, M.; Strnad, M.; Van Staden, J.; Szigeti, J.; Németh, L. Screening Microalgae for Some Potentially Useful Agricultural and Pharmaceutical Secondary Metabolites. *J. Appl. Phycol.* **2004**, *16*, 309–314. [CrossRef]
4. Nethravathy, M.U.; Mehar, J.G.; Mudliar, S.N.; Shekh, A.Y. Recent Advances in Microalgal Bioactives for Food, Feed, and Healthcare Products: Commercial Potential, Market Space, and Sustainability. *Compr. Rev. Food Sci. Food Saf.* **2019**, *18*, 1882–1897. [CrossRef]
5. Martínez Andrade, K.A.; Lauritano, C.; Romano, G.; Ianora, A. Marine Microalgae with Anti-Cancer Properties. *Mar. Drugs* **2018**, *16*, 165. [CrossRef] [PubMed]
6. Bauer, S.; Jin, W.; Zhang, F.; Linhardt, R.J. The Application of Seaweed Polysaccharides and Their Derived Products with Potential for the Treatment of Alzheimer's Disease. *Mar. Drugs* **2021**, *19*, 89. [CrossRef]
7. de Jesus Raposo, M.F.; de Morais, A.M.M.B.; de Morais, R.M.S.C. Emergent Sources of Prebiotics: Seaweeds and Microalgae. *Mar. Drugs* **2016**, *14*, 27. [CrossRef]
8. Amna Kashif, S.; Hwang, Y.J.; Park, J.K. Potent Biomedical Applications of Isolated Polysaccharides from Marine Microalgae *Tetraselmis* Species. *Bioprocess. Biosyst. Eng.* **2018**, *41*, 1611–1620. [CrossRef] [PubMed]
9. Wu, G. Amino Acids: Metabolism: Functions, and Nutrition. *Amino Acids* **2009**, *37*, 1–17. [CrossRef] [PubMed]
10. Koyande, A.K.; Chew, K.W.; Rambabu, K.; Tao, Y.; Chu, D.-T.; Show, P.-L. Microalgae: A Potential Alternative to Health Supplementation for Humans. *Food Sci. Hum. Wellness* **2019**, *8*, 16–24. [CrossRef]
11. Safi, C.; Charton, M.; Ursu, A.V.; Laroche, C.; Zebib, B.; Pontalier, P.Y.; Vaca-Garcia, C. Release of Hydro-Soluble Microalgal Proteins Using Mechanical and Chemical Treatments. *Algal Res.* **2014**, *3*, 55–60. [CrossRef]
12. Kazir, M.; Abuhassira, Y.; Robin, A.; Nahor, O.; Luo, J.; Israel, A.; Golberg, A.; Livney, Y.D. Extraction of Proteins from Two Marine Macroalgae, *Ulva* sp. and *Gracilaria* sp., for Food Application, and Evaluating Digestibility, Amino Acid Composition and Antioxidant Properties of the Protein Concentrates. *Food Hydrocoll.* **2019**, *87*, 194–203. [CrossRef]
13. Vieira, E.F.; Soares, C.; Machado, S.; Correia, M.; Ramalhosa, M.J.; Oliva-Teles, M.T.; Paula Carvalho, A.; Domingues, V.F.; Antunes, F.; Oliveira, T.A.C.; et al. Seaweeds from the Portuguese Coast as a Source of Proteinaceous Material: Total and Free Amino Acid Composition Profile. *Food Chem.* **2018**, *269*, 264–275. [CrossRef] [PubMed]
14. Vitali, A. Proline-Rich Peptides: Multifunctional Bioactive Molecules as New Potential Therapeutic Drugs. *Curr. Protein Pept. Sci.* **2015**, *16*, 147–162. [CrossRef]
15. Zou, C.; Brewer, M.; Cao, X.; Zang, R.; Lin, J.; Deng, Y.; Li, C. Antitumor Activity of 4-(N-Hydroxyphenyl)Retinamide Conjugated with Poly(l-Glutamic Acid) against Ovarian Cancer Xenografts. *Gynecol. Oncol.* **2007**, *107*, 441–449. [CrossRef] [PubMed]
16. You, X.-Y.; Xu, Y.; Huang, Z.-B.; Li, Y.-P. Nonvolatile Taste Compounds of Jiangluobo (A Traditional Chinese Fermented Food). *J. Food Qual.* **2010**, *33*, 477–489. [CrossRef]
17. Trevino, S.R.; Scholtz, J.M.; Pace, C.N. Amino Acid Contribution to Protein Solubility: Asp, Glu, and Ser Contribute More Favorably than the Other Hydrophilic Amino Acids in RNase Sa. *J. Mol. Biol.* **2007**, *366*, 449–460. [CrossRef] [PubMed]
18. Wu, G.; Bazer, F.W.; Burghardt, R.C.; Johnson, G.A.; Kim, S.W.; Knabe, D.A.; Li, P.; Li, X.; McKnight, J.R.; Satterfield, M.C.; et al. Proline and Hydroxyproline Metabolism: Implications for Animal and Human Nutrition. *Amino Acids* **2011**, *40*, 1053–1063. [CrossRef]
19. Sansone, C.; Brunet, C. Promises and Challenges of Microalgal Antioxidant Production. *Antioxidants* **2019**, *8*, 199. [CrossRef]
20. Rico, M.; López, A.; Santana-Casiano, J.M.; González, A.G.; González-Dávila, M. Variability of the Phenolic Profile in the Diatom *Phaeodactylum tricornutum* Growing under Copper and Iron Stress. *Limnol. Oceanogr.* **2013**, *58*, 144–152. [CrossRef]
21. Coulombier, N.; Jauffrais, T.; Lebouvier, N. Antioxidant Compounds from Microalgae: A Review. *Mar. Drugs* **2021**, *19*, 549. [CrossRef]

22. Nova, P.; Martins, A.P.; Teixeira, C.; Abreu, H.; Silva, J.G.; Silva, A.M.; Freitas, A.C.; Gomes, A.M. Foods with Microalgae and Seaweeds Fostering Consumers Health: A Review on Scientific and Market Innovations. *J. Appl. Phycol.* **2020**, *32*, 1789–1802. [CrossRef]
23. Lafarga, T.; Acién-Fernández, F.G.; Castellari, M.; Villaró, S.; Bobo, G.; Aguiló-Aguayo, I. Effect of Microalgae Incorporation on the Physicochemical, Nutritional, and Sensorial Properties of an Innovative Broccoli Soup. *LWT* **2019**, *111*, 167–174. [CrossRef]
24. Žugčić, T.; Abdelkebir, R.; Barba, F.J.; Rezek-Jambrak, A.; Gálvez, F.; Zamuz, S.; Granato, D.; Lorenzo, J.M. Effects of Pulses and Microalgal Proteins on Quality Traits of Beef Patties. *J. Food Sci. Technol.* **2018**, *55*, 4544–4553. [CrossRef] [PubMed]
25. Fu, Y.; Chen, T.; Chen, S.H.Y.; Liu, B.; Sun, P.; Sun, H.; Chen, F. The Potentials and Challenges of Using Microalgae as an Ingredient to Produce Meat Analogues. *Trends Food Sci. Technol.* **2021**, *112*, 188–200. [CrossRef]
26. Fernandes, T.; Cordeiro, N. Microalgae as Sustainable Biofactories to Produce High-Value Lipids: Biodiversity, Exploitation, and Biotechnological Applications. *Mar. Drugs* **2021**, *19*, 573. [CrossRef] [PubMed]
27. Augustyniak, A.; Bartosz, G.; Čipak, A.; Duburs, G.; Horáková, L.; Łuczaj, W.; Majekova, M.; Odysseos, A.D.; Rackova, L.; Skrzydlewska, E.; et al. Natural and Synthetic Antioxidants: An Updated Overview. *Free Radic. Res.* **2010**, *44*, 1216–1262. [CrossRef]
28. Malavasi, V.; Soru, S.; Cao, G. Extremophile Microalgae: The Potential for Biotechnological Application. *J. Phycol.* **2020**, *56*, 559–573. [CrossRef] [PubMed]
29. Kröger, M.; Klemm, M.; Nelles, M. Extraction Behavior of Different Conditioned, *S. Rubescens*. *Energies* **2019**, *12*, 1336. [CrossRef]
30. De Farias Neves, F.; Demarco, M.; Tribuzi, G. Drying and Quality of Microalgal Powders for Human Alimentation. In *Microalgae: From Physiology to Application*; InTech Open: London, UK, 2019; ISBN 978-1-83880-035-2.
31. Pop, A.; Kiss, B.; Loghin, F. Endocrine Disrupting Effects of Butylated Hydroxyanisole (BHA–E320). *Clujul Med.* **2013**, *86*, 16–20. [PubMed]
32. Saito, M.; Sakagami, H.; Fujisawa, S. Cytotoxicity and Apoptosis Induction by Butylated Hydroxyanisole (BHA) and Butylated Hydroxytoluene (BHT). *Anticancer Res.* **2003**, *23*, 4693–4701. [PubMed]
33. Jerez-Martel, I.; García-Poza, S.; Rodríguez-Martel, G.; Rico, M.; Afonso-Olivares, C.; Gómez-Pinchetti, J.L. Phenolic Profile and Antioxidant Activity of Crude Extracts from Microalgae and Cyanobacteria Strains. *J. Food Qual.* **2017**, *2017*, 2924508. [CrossRef]
34. Quideau, S.; Deffieux, D.; Douat-Casassus, C.; Pouységu, L. Plant Polyphenols: Chemical Properties, Biological Activities, and Synthesis. *Angew. Chem. Int. Ed.* **2011**, *50*, 586–621. [CrossRef] [PubMed]
35. Corrêa da Silva, M.G.; Pires Ferreira, S.; Dora, C.L.; Hort, M.A.; Giroldo, D.; Prates, D.F.; Radmann, E.M.; Bemvenuti, R.H.; Costa, J.A.V.; Badiale-Furlong, E.; et al. Phenolic Compounds and Antioxidant Capacity of *Pediastrum boryanum* (Chlorococcales) Biomass. *Int. J. Environ. Health Res.* **2020**, 1–13. [CrossRef]
36. Blagojević, D.; Babić, O.; Rašeta, M.; Šibul, F.; Janjušević, L.; Simeunović, J. Antioxidant Activity and Phenolic Profile in Filamentous Cyanobacteria: The Impact of Nitrogen. *J. Appl. Phycol.* **2018**, *30*, 2337–2346. [CrossRef]
37. Jacobo-Velázquez, D.A.; Cisneros-Zevallos, L. Correlations of Antioxidant Activity against Phenolic Content Revisited: A New Approach in Data Analysis for Food and Medicinal Plants. *J. Food Sci.* **2009**, *74*, R107–R113. [CrossRef] [PubMed]
38. Klejdus, B.; Kopecký, J.; Benešová, L.; Vacek, J. Solid-Phase/Supercritical-Fluid Extraction for Liquid Chromatography of Phenolic Compounds in Freshwater Microalgae and Selected Cyanobacterial Species. *J. Chromatogr. A* **2009**, *1216*, 763–771. [CrossRef]
39. Onofrejová, L.; Vašíčková, J.; Klejdus, B.; Stratil, P.; Mišurcová, L.; Kráčmar, S.; Kopecký, J.; Vacek, J. Bioactive Phenols in Algae: The Application of Pressurized-Liquid and Solid-Phase Extraction Techniques. *J. Pharm. Biomed. Anal.* **2010**, *51*, 464–470. [CrossRef]
40. Hwang, H.-S.; Winkler-Moser, J.K.; Liu, S.X. Study on Antioxidant Activity of Amino Acids at Frying Temperatures and Their Interaction with Rosemary Extract, Green Tea Extract, and Ascorbic Acid. *J. Food Sci.* **2019**, *84*, 3614–3623. [CrossRef]
41. Marquis, V.; Smirnova, E.; Poirier, L.; Zumsteg, J.; Schweizer, F.; Reymond, P.; Heitz, T. Stress- and Pathway-Specific Impacts of Impaired Jasmonoyl-Isoleucine (JA-Ile) Catabolism on Defense Signalling and Biotic Stress Resistance. *Plant Cell Environ.* **2020**, *43*, 1558–1570. [CrossRef]
42. Rubino, J.T.; Franz, K.J. Coordination Chemistry of Copper Proteins: How Nature Handles a Toxic Cargo for Essential Function. *J. Inorg. Biochem.* **2012**, *107*, 129–143. [CrossRef]
43. Dietzen, D. Amino Acids, Peptides, and Proteins. In *Principles and Applications of Molecular Diagnostics*; Elsevier: Amsterdam, The Netherlands, 2018; pp. 345–380. ISBN 9780128160619.
44. Vendruscolo, R.G.; Facchi, M.M.X.; Maroneze, M.M.; Fagundes, M.B.; Cichoski, A.J.; Zepka, L.Q.; Barin, J.S.; Jacob-Lopes, E.; Wagner, R. Polar and Non-Polar Intracellular Compounds from Microalgae: Methods of Simultaneous Extraction, Gas Chromatography Determination and Comparative Analysis. *Food Res. Int.* **2018**, *109*, 204–212. [CrossRef]
45. Kolmakova, A.A.; Kolmakov, V.I. Amino Acid Composition of Green Microalgae and Diatoms, Cyanobacteria, and Zooplankton (Review). *Ecol. Physiol. Biochem. Hydrobionts* **2019**, *12*, 452–461. [CrossRef]
46. Peinado, I.; Girón, J.; Koutsidis, G.; Ames, J.M. Chemical Composition, Antioxidant Activity and Sensory Evaluation of Five Different Species of Brown Edible Seaweeds. *Food Res. Int.* **2014**, *66*, 36–44. [CrossRef]
47. Machado, M.; Machado, S.; Pimentel, F.B.; Freitas, V.; Alves, R.C.; Oliveira, M.B.P.P. Amino Acid Profile and Protein Quality Assessment of Macroalgae Produced in an Integrated Multi-Trophic Aquaculture System. *Foods* **2020**, *9*, 1382. [CrossRef]
48. Paliwal, C.; Mitra, M.; Bhayani, K.; Bharadwaj, S.V.V.; Ghosh, T.; Dubey, S.; Mishra, S. Abiotic Stresses as Tools for Metabolites in Microalgae. *Bioresour. Technol.* **2017**, *244*, 1216–1226. [CrossRef]

49. Granum, E.; Kirkvold, S.; Myklestad, S.M. Cellular and Extracellular Production of Carbohydrates and Amino Acids by the Marine Diatom *Skeletonema costatum*: Diel Variations and Effects of N Depletion. *Mar. Ecol. Prog. Ser.* **2002**, *242*, 83–94. [CrossRef]
50. Martin-Jézéquel, V.; Sournia, A.; Birrien, J.-L. A Daily Study of the Diatom Spring Bloom at Roscoff (France) in 1985. III. Free Amino Acids Composition Studied by HPLC Analysis. *J. Plankton Res.* **1992**, *14*, 409–421. [CrossRef]
51. Fleurence, J. Seaweed Proteins: Biochemical, Nutritional Aspects and Potential Uses. *Trends Food Sci. Technol.* **1999**, *10*, 25–28. [CrossRef]
52. Cobos, M.; Pérez, S.; Braga, J.; Vargas-Arana, G.; Flores, L.; Paredes, J.D.; Maddox, J.D.; Marapara, J.L.; Castro, J.C. Nutritional Evaluation and Human Health-Promoting Potential of Compounds Biosynthesized by Native Microalgae from the Peruvian Amazon. *World J. Microbiol. Biotechnol.* **2020**, *36*, 121. [CrossRef] [PubMed]
53. Biancarosa, I.; Espe, M.; Bruckner, C.G.; Heesch, S.; Liland, N.; Waagbø, R.; Torstensen, B.; Lock, E.J. Amino Acid Composition, Protein Content, and Nitrogen-to-Protein Conversion Factors of 21 Seaweed Species from Norwegian Waters. *J. Appl. Phycol.* **2017**, *29*, 1001–1009. [CrossRef]
54. Lourenço, S.O.; Barbarino, E.; Lavín, P.L.; Lanfer Marquez, U.M.; Aidar, E. Distribution of Intracellular Nitrogen in Marine Microalgae: Calculation of New Nitrogen-to-Protein Conversion Factors. *Eur. J. Phycol.* **2004**, *39*, 17–32. [CrossRef]
55. Sui, Y.; Muys, M.; Vermeir, P.; D'Adamo, S.; Vlaeminck, S.E. Light Regime and Growth Phase Affect the Microalgal Production of Protein Quantity and Quality with *Dunaliella salina*. *Bioresour. Technol.* **2019**, *275*, 145–152. [CrossRef]
56. Araya, M.; García, S.; Rengel, J.; Pizarro, S.; Álvarez, G. Determination of Free and Protein Amino Acid Content in Microalgae by HPLC-DAD with Pre-Column Derivatization and Pressure Hydrolysis. *Mar. Chem.* **2021**, *234*, 103999. [CrossRef]
57. Cheirsilp, B.; Torpee, S. Enhanced Growth and Lipid Production of Microalgae under Mixotrophic Culture Condition: Effect of Light Intensity, Glucose Concentration and Fed-Batch Cultivation. *Bioresour. Technol.* **2012**, *110*, 510–516. [CrossRef]
58. Wong, Y. Effects of Light Intensity, Illumination Cycles on Microalgae *Haematococcus pluvialis* for Production of Astaxanthin. *J. Mar. Biol. Aquac.* **2016**, *2*, 1–6. [CrossRef]
59. Seyfabadi, J.; Ramezanpour, Z.; Khoeyi, Z.A. Protein, Fatty Acid, and Pigment Content of Chlorella Vulgaris under Different Light Regimes. *J. Appl. Phycol.* **2011**, *23*, 721–726. [CrossRef]
60. Gorissen, S.H.M.; Crombag, J.J.R.; Senden, J.M.G.; Waterval, W.A.H.; Bierau, J.; Verdijk, L.B.; van Loon, L.J.C. Protein Content and Amino Acid Composition of Commercially Available Plant-Based Protein Isolates. *Amino Acids* **2018**, *50*, 1685–1695. [CrossRef]
61. Schulze, C.; Strehle, A.; Merdivan, S.; Mundt, S. Carbohydrates in Microalgae: Comparative Determination by TLC, LC-MS without Derivatization, and the Photometric Thymol-Sulfuric Acid Method. *Algal Res.* **2017**, *25*, 372–380. [CrossRef]
62. Templeton, D.W.; Quinn, M.; Van Wychen, S.; Hyman, D.; Laurens, L.M.L. Separation and Quantification of Microalgal Carbohydrates. *J. Chromatogr. A* **2012**, *1270*, 225–234. [CrossRef] [PubMed]
63. Visca, A.; Di Caprio, F.; Spinelli, R.; Altimari, P.; Cicci, A.; Iaquaniello, G.; Toro, L.; Pagnanelli, F. Microalgae Cultivation for Lipids and Carbohydrates Production. *Chem. Eng. Trans.* **2017**, *57*, 127–132. [CrossRef]
64. Stirk, W.A.; Bálint, P.; Vambe, M.; Lovász, C.; Molnár, Z.; van Staden, J.; Ördög, V. Effect of Cell Disruption Methods on the Extraction of Bioactive Metabolites from Microalgal Biomass. *J. Biotechnol.* **2020**, *307*, 35–43. [CrossRef] [PubMed]
65. Manivannan, K.; Thirumaran, G.; Devi, G.K.; Hemalatha, A.; Anantharaman, P. Biochemical Composition of Seaweeds from Mandapam Coastal Regions along Southeast Coast of India. *Am. Eurasian J. Bot.* **2008**, *1*, 32–37.
66. Suarez Garcia, E.; Suarez Ruiz, C.A.; Tilaye, T.; Eppink, M.H.M.; Wijffels, R.H.; van den Berg, C. Fractionation of Proteins and Carbohydrates from Crude Microalgae Extracts Using an Ionic Liquid Based-Aqueous Two Phase System. *Sep. Purif. Technol.* **2018**, *204*, 56–65. [CrossRef]
67. Jutur, P.P.; Nesamma, A.A.; Shaikh, K.M. Algae-Derived Marine Oligosaccharides and Their Biological Applications. *Front. Mar. Sci.* **2016**, *3*, 83. [CrossRef]
68. Costa, J.A.V.; Lucas, B.F.; Alvarenga, A.G.P.; Moreira, J.B.; de Morais, M.G. Microalgae Polysaccharides: An Overview of Production, Characterization, and Potential Applications. *Polysaccharides* **2021**, *2*, 759–772. [CrossRef]
69. Bondet, V.; Brand-Williams, W.; Berset, C. Kinetics and Mechanisms of Antioxidant Activity Using the DPPH·Free Radical Method. *LWT Food Sci. Technol.* **1997**, *30*, 609–615. [CrossRef]
70. Dvořáková, M.; Hulín, P.; Karabín, M.; Dostálek, P. Determination of Polyphenols in Beer by an Effective Method Based on Solid-Phase Extraction and High Performance Liquid Chromatography with Diode-Array Detection. *Czech J. Food Sci.* **2007**, *25*, 182–188. [CrossRef]
71. Shi, Z.; Li, H.; Li, Z.; Hu, J.; Zhang, H. Pre-Column Derivatization RP-HPLC Determination of Amino Acids in Asparagi Radix before and after Heating Process. *IERI Procedia* **2013**, *5*, 351–356. [CrossRef]
72. Jansen, E. Simple Determination of Sugars in Cigarettes. *J. Anal. Bioanal. Tech.* **2014**, *5*, 5–7. [CrossRef]
73. Brooks, J.R.; Griffin, V.K.; Kattan, M.W. A Modified Method for Total Carbohydrate Analysis of Glucose Syrups, Maltodextrins, and Other Starch Hydrolysis Products. *Cereal Chem.* **1986**, *63*, 465–466. [CrossRef]

marine drugs

MDPI

Article

Comparative Study Highlights the Potential of Spectral Deconvolution for Fucoxanthin Screening in Live *Phaeodactylum tricornutum* Cultures

Sean Macdonald Miller [1,*], Raffaela M. Abbriano [1], Anna Segecova [1,2], Andrei Herdean [1], Peter J. Ralph [1] and Mathieu Pernice [1]

[1] Climate Change Cluster (C3), Faculty of Science, University of Technology Sydney, Sydney, NSW 2007, Australia; raffaela.abbrianoburke@uts.edu.au (R.M.A.); segecova.a@czechglobe.cz (A.S.); andrei.herdean@uts.edu.au (A.H.); peter.ralph@uts.edu.au (P.J.R.); mathieu.pernice@uts.edu.au (M.P.)

[2] Global Change Research Centre, Academy of Science of the Czech Republic, 26101 Drasov, Czech Republic

* Correspondence: sean.a.macdonaldmiller@student.uts.edu.au

Abstract: Microalgal biotechnology shows considerable promise as a sustainable contributor to a broad range of industrial avenues. The field is however limited by processing methods that have commonly hindered the progress of high throughput screening, and consequently development of improved microalgal strains. We tested various microplate reader and flow cytometer methods for monitoring the commercially relevant pigment fucoxanthin in the marine diatom *Phaeodactylum tricornutum*. Based on accuracy and flexibility, we chose one described previously to adapt to live culture samples using a microplate reader and achieved a high correlation to HPLC (R^2 = 0.849), effectively removing the need for solvent extraction. This was achieved by using new absorbance spectra inputs, reducing the detectable pigment library and changing pathlength values for the spectral deconvolution method in microplate reader format. Adaptation to 384-well microplates and removal of the need to equalize cultures by density further increased the screening rate. This work is of primary interest to projects requiring detection of biological pigments, and could theoretically be extended to other organisms and pigments of interest, improving the viability of microalgae biotechnology as a contributor to sustainable industry.

Keywords: fucoxanthin; microalgae; green consumption; food consumption

Citation: Macdonald Miller, S.; Abbriano, R.M.; Segecova, A.; Herdean, A.; Ralph, P.J.; Pernice, M. Comparative Study Highlights the Potential of Spectral Deconvolution for Fucoxanthin Screening in Live *Phaeodactylum tricornutum* Cultures. *Mar. Drugs* **2022**, *20*, 19. https://doi.org/10.3390/md20010019

Academic Editor: Carlos Almeida

Received: 8 December 2021
Accepted: 21 December 2021
Published: 23 December 2021

Publisher's Note: MDPI stays neutral with regard to jurisdictional claims in published maps and institutional affiliations.

1. Introduction

Fucoxanthin is the most abundant marine carotenoid pigment, accounting for more than 10% of the total carotenoids produced naturally [1]. Fucoxanthin is of commercial interest for its antioxidant, anti-inflammatory, antibacterial, and anti-obesity characteristics [2–5], as well as having potential in inhibiting cancer cell growth [6]. Due to these beneficial properties, fucoxanthin is an established nutraceutical currently sourced from seaweed [7,8]. However, the marine diatom *Phaeodactylum tricornutum* (Bacillariophyta) exhibits up to 100 times higher fucoxanthin content (mg g^{-1} DW) than seaweed [5,9], which makes it an attractive alternative for commercial purposes. *P. tricornutum* has also been used extensively across a wide area of research and has full genome data [10]. This species has demonstrated genetic manipulability [11–13], which further supports its utility a biofactory template for fucoxanthin. Fucoxanthin in *P. tricornutum* can reach up to 59.2 mg g^{-1} under optimal growth conditions [14], and it has been improved by 69.3% (mg g^{-1} DW) in chemically-induced mutants [15] and by 45% per cell by introduction of the PSY (phytoene synthase) gene [16] compared to the wild-type strain.

Strategies that generate large mutant or transformant libraries (hundreds to thousands of clones) necessitate high-throughput methods to screen novel strains for pigments because the current benchmark of pigment detection is high-performance liquid chromatography (HPLC), which, while being accurate, is very expensive and time-consuming. Several approaches to screening fucoxanthin in *P. tricornutum* were investigated here and compared in terms of the accuracy of their prediction as compared to HPLC, as well as the effort invested to produce that result. Because fucoxanthin and chlorophyll *a* are associated with light harvesting, chlorophyll screening methods were included as fucoxanthin proxies alongside direct fucoxanthin quantifiers, and their assessment included both microplate reader and flow cytometer formats. Flow cytometry provides a means of analyzing samples at the resolution of individual cells, which negates effects from culture density and can be attached to fluorescence-activated cell sorting (FACS), while the microplate reader format provides an inexpensive and non-invasive way of analyzing many samples in a short period of time, therefore, both systems were included.

Three microplate reader methods and three flow cytometry methods were chosen for this study. First, two estimates for chlorophyll and fucoxanthin autofluorescence were measured using flow cytometry to investigate whether various excitation/emission gates could accurately predict average fucoxanthin content through analysis of single cells. A third flow cytometry method using Nile Red dye was included to investigate if dye fluorescence can improve the detection of fucoxanthin based on its relationship to lipids [17] and correlation to total carotenoids [15]. One approach for estimating chlorophyll in ethanol was tested as a proxy for measuring fucoxanthin in microplate format from Ritchie [18]. The second microplate reader method investigated in this study is theoretically similar to the equations from Ritchie [18], except that it aims to quantify fucoxanthin directly by using equations based on specific wavelength absorbance of pigment extracts related to fucoxanthin, rather than using chlorophyll as a proxy, found in Wang et al. [19]. The final microplate reader method is from Thrane et al. [20], developed from earlier work [21,22], and uses spectral deconvolution of absorbance spectra obtained from samples extracted using an organic solvent like ethanol. All samples were compared during exponential growth phase, as this is the easiest point to detect pigment content differences without compounding effects from media changes and cell senescence, and enables the screening methods with the best resolution to be highlighted accordingly.

This study aims to reconcile accuracy with practical considerations of screening methods, and to further optimize a promising method for high-throughput screening of fucoxanthin in *P. tricornutum*. Six published methods using two common pieces of analytical equipment were used to assess their correlation to HPLC, one being chosen for further optimisation based on this correlation as well as practical considerations like ease of use compared to cost and time investment. This work contributes to the field of microalgal biotechnology by improving the rate at which novel strains of *P. tricornutum* can be selected based on improved fucoxanthin content.

2. Results and Discussion

2.1. Culture Characteristics

The high light without nitrate (HL-N) treatment cell growth was statistically significantly higher ($p = 0.0005$) than both other HL treatments (~9.5×10^6 compared to an average 6.5×10^6 cells mL^{-1}) on the final day (Figure 1a). While the final pellet weights for all HL treatments were similar, the mean singlet forward scatter (FSC-A) of HL-N cultures was 50% less than the forward scatter of nitrogen-present HL treatments ($p < 0.0001$). Decreased cell size and volume due to nitrogen limitation has been observed previously [23] and explains the much higher cell density in HL-N treatments. Importantly, this indicates that *P. tricornutum* has impressive light stress tolerance mechanisms despite low or no nutrient availability, considering the analogous biomass productivity of HL treatments, which averages significantly higher ($p = 0.0206$) at up to two times that of the lowest LL biomass productivities.

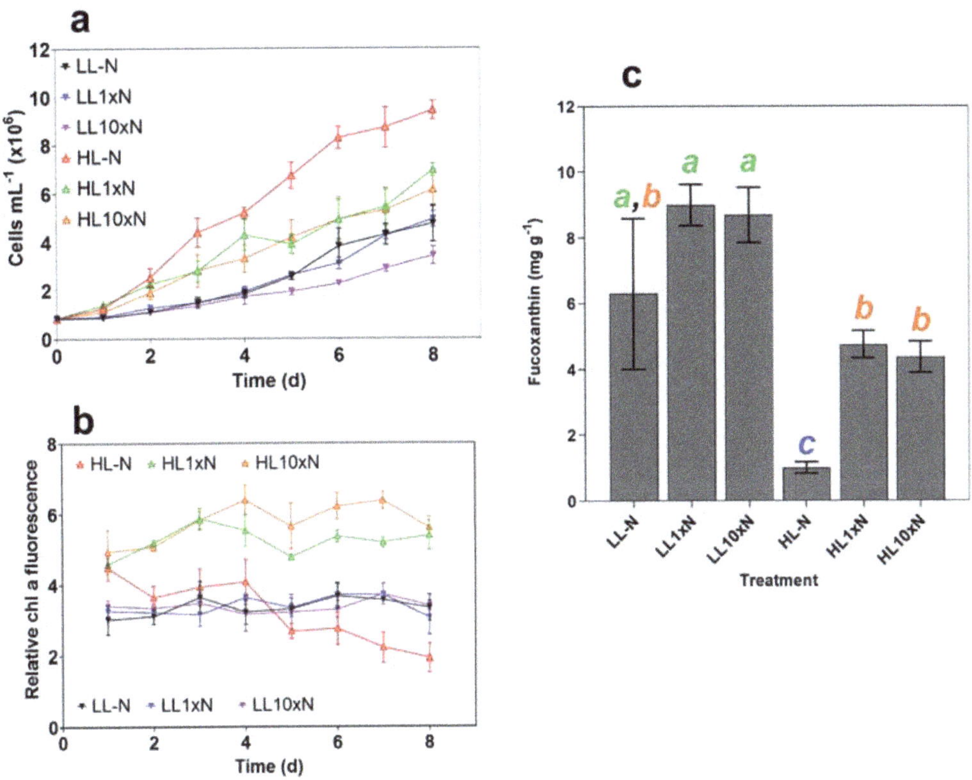

Figure 1. Measured culture characteristics over the 8-day experimental period. (**a**) Cell density in millions mL^{-1}; (**b**) Relative chlorophyll *a* fluorescence (chlorophyll *a* fluorescence/culture absorbance at 750 nm), which was used to determine optimal experiment termination time point; and (**c**) fucoxanthin content of freeze-dried samples measured using HPLC ($mg\ g^{-1}$). Treatment abbreviations are as follows: nitrate-free ASW media (-N), standard ($1 \times N$) nitrate media or media with $10\times$ nitrate ($10 \times N$), and either 10 (LL) or 200 (HL) μmol photons $m^{-2}\ s^{-1}$ of white light. Statistical significance was calculated using one-way ANOVA ($p < 0.05$) with letters denoting non-significant groupings. Error bars denote standard deviation ($n = 3$).

Pigment content was measured by relative chlorophyll *a*, and the experiment was terminated on day 8 when relative chlorophyll *a* was significantly different ($p < 0.0001$, ~1.9 for HL-N compared to ~3.3 for all LL treatments and ~5.5 for HL treatments with nitrate supplementation, Figure 1b). LL cultures displayed very similar chlorophyll *a* autofluorescence over the experimental period, while there was a clear distinction between nitrogen replete and deplete conditions under HL.

As expected, fucoxanthin content was a function of irradiance, specifically, its role as a light-harvesting pigment was demonstrated by its negative correlation with photon availability. Nitrogen limitation decreased overall fucoxanthin content and productivity under 200 μmol photons $m^{-2}\ s^{-1}$, while simply increasing the variability under 10 μmol photons $m^{-2}\ s^{-1}$ (Figure 1c). These treatments provided a sufficient range of fucoxanthin contents to assess high-throughput screen viability.

2.2. High-Throughput Screen Analysis

Data for live culture chlorophyll *a* autofluorescence on microplate reader were not shown because there was no correlation to HPLC ($R^2 < 0.002$). The microplate reader

Nile Red fluorescence data displayed a negative correlation, which can be attributed to disturbance by media noise and other resolution issues, and these were therefore not considered reliable methods.

Autofluorescence measurements using flow cytometry displayed the highest correlation to HPLC data with R^2 = 0.9335 and R^2 = 0.9485 for chlorophyll *a* (B690-50, Figure 2a) and fucoxanthin (Y710-50, Figure 2b), respectively. These were also higher than both Nile Red and the equations from Ritchie [18] and Wang et al. [19], which require either dyeing or extraction with an organic solvent. While chlorophyll *a* autofluorescence was included to estimate the B690-50 filter reliability at estimating fucoxanthin, using alternatives is likely to give more accurate predictions, as with the Y710-50 filter. Excitation at 561 nm with emission at 710 nm (Y710-50) has generally been avoided in the literature because there is little or no absorption of individual *P. tricornutum* pigments at these wavelengths. However, Premvhardan et al. [24] showed a shift in fucoxanthin absorbance when bound in Fucoxanthin-Chlorophyll Protein (FCP) to wider than 561 nm. Because the flow cytometry methods are measuring live cells rather than pigments in extracts, it is appropriate to assess fucoxanthin absorbance characteristics when bound to FCP. Also, an extra emission peak at 710 nm was observed by Fan et al. (2021) in *P. tricornutum*, who confirmed a relationship between emission intensity at 710 nm and fucoxanthin content measured using HPLC [25]. Using these wavelengths has the likely benefit of being able to measure fucoxanthin content by its relationship to cellular FCP rather than by chlorophyll content as a proxy, like when measuring extracts using equations from Ritchie et al. (2008) [18]. Despite the overall fluorescence intensity of Y710 being about 25% that of the B690 channel (data not shown), it appears sufficient as not only an accurate indicator, but more reliable in predicting fucoxanthin content than B690-50.

Measuring autofluorescence with flow cytometry is not only accurate for measuring fucoxanthin, it does not require extraction, dilution or dyeing, and this greatly improves screening time. Relevant to many projects that generate large mutant or transformant libraries, FACS can also be attached to this method to further improve screening and selection times, as also found in Gao et al. [17] and Fan et al. [25].

The use of Nile Red to correlate polar lipids with fucoxanthin has been performed previously using live cultures of *Tisochrysis lutea* with an R^2 value of 0.88 [17], there is a correlation between Nile Red and total carotenoids in *P. tricornutum* [15]. A relationship to HPLC was found herein (R^2 = 0.69) for *P. tricornutum* using the average single cell fluorescence of singlets after staining with Nile Red (Figure 2c). While a correlation was found, using different channels resulted in higher R^2 values without the need to dye first.

Microplate reader formats displayed overall lower correlation to HPLC, with the maximum found using the method from Ritchie [18] (R^2 = 0.9021, Figure 2d).

In our initial analysis, the method from Wang et al. [19] displayed reduced correlation to HPLC at R^2 = 0.0292 when the dilution of 1×10^6 cells mL^{-1} was used. Verification cultures in the original source use a cell density between 20–100 $\times 10^6$ cells mL^{-1}, which is likely to have contributed to the sensitivity and therefore accuracy reported therein ($R^2 \geq 0.946$). To improve the resolution of fucoxanthin predictions using the method from Wang et al. [19], experimental samples were analyzed again at higher cell concentration and an R^2 value of 0.8593 was found (Figure 2e). Despite this method displaying a slightly lower correlation than in the original source (R^2 = 0.8593 compared to $R^2 \geq 0.946$, see [19]), users performing equipment calibration can expect even better results and should keep in mind the changes to resolution when using low cell densities.

Despite displaying a relatively high correlation of R^2 = 0.7236 (Figure 2f), the spectral deconvolution method from Thrane et al. [20] did not require solvent extraction. This method also had the greatest potential for calibration and refinement, owing to its use of a much larger dataset and multiple programmable input scripts using RStudio version 1.3.1093 (RStudio, Boston, MA, USA). Therefore, the method was chosen for further optimisation.

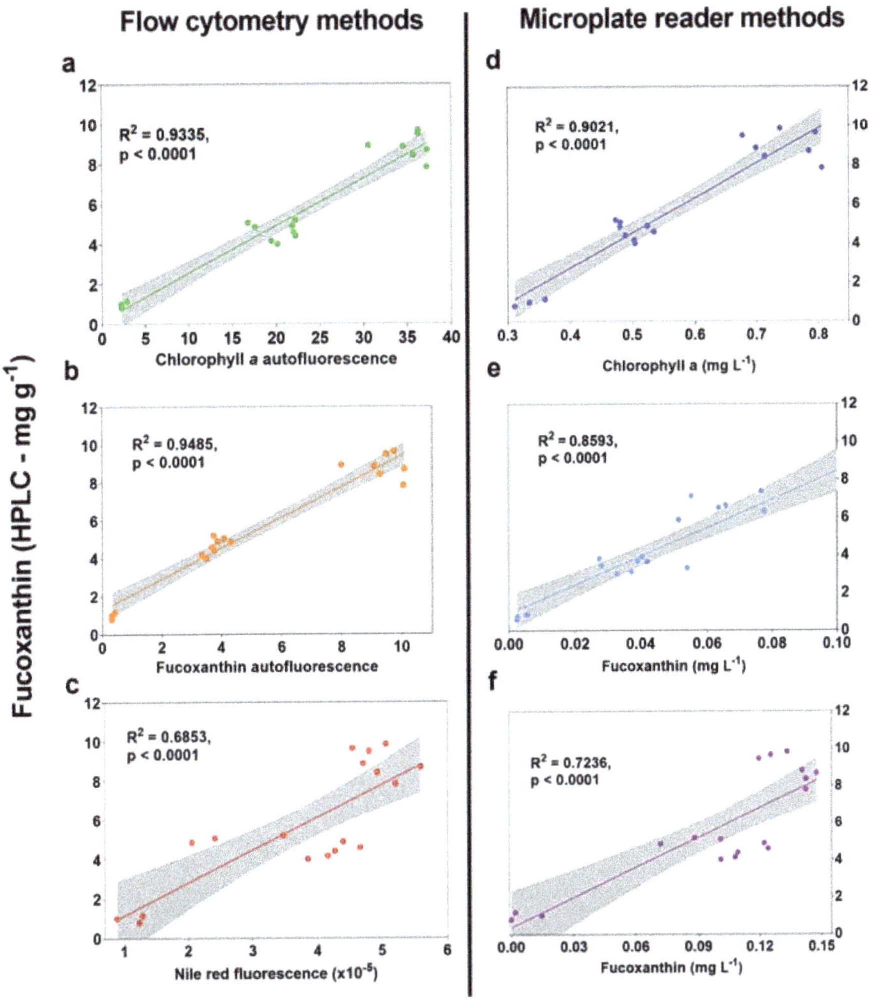

Figure 2. High-throughput screen results correlated to fucoxanthin (mg g^{-1}) measured using HPLC. (**a**) Mean single cell chlorophyll *a* autofluorescence measured on flow cytometer using blue excitation wavelength of 488 nm with 690/50 nm optical filter, (**b**) mean single cell fucoxanthin autofluorescence measured on flow cytometer using yellow excitation laser at 561 nm with 710/50 optical filter, (**c**) mean single cell fluorescence measured on flow cytometer using blue excitation laser at 488 nm with 610/20 optical filter after dyeing with Nile Red, (**d**) chlorophyll *a* content (mg L^{-1}) using equations for ethanol extracts from Ritchie (2008) on a microplate reader, (**e**) fucoxanthin content (mg L^{-1}) for concentrated ethanol extracts from Wang et al. (2018) on a microplate reader, and (**f**) spectral deconvolution method from Thrane et al. (2015) using raw culture absorbance spectra on microplate reader. Units on x axes for d, e, and f are simply what the sources for each use to determine fucoxanthin content, while samples for all three were extracted herein using an equal weight of biomass, effectively making x-axis units the weight of fucoxanthin per unit weight of biomass (like HPLC).

2.3. Optimisation of Spectral Deconvolution from Thrane et al. (2015)

The R scripts contained within Thrane et al. [20] were first tested using absorbance spectra obtained from uniformly-dilute ethanol-extracted experimental samples. It was noticed that predictions for fucoxanthin were either extremely low or zero. To allow the method to attribute lower wavelength spectral regions to the presence of fucoxanthin, pigments that are not present in *P. tricornutum*, or present in only minute amounts, were removed from the R script "gaussian.peak.parameters.txt" as well as from line 32 of the "pigments.function.R" script. It was found that removing all but the chlorophylls *a*, *c*1 and *c*2, and fucoxanthin from the method ensured detection of fucoxanthin in every sample. After removal of 'unnecessary' pigments and the reassigning of pathlength in the 'Sediments.R' file to z = 0.625 for microplate reader adaptation, the correlation between the script and the content of fucoxanthin detected using HPLC in mg g^{-1} was $R^2 = 0.801$.

The spectral qualities of a pigment are altered when measured in live cell thylakoid membranes compared to when measured as free-floating molecules in an organic solvent like ethanol. The cumulative effects of the silica frustule of *P. tricornutum*, the lipid layers associated with the thylakoid, effects from saltwater medium, and interference from other cellular components, are expected to have substantial effects on fucoxanthin spectral characteristics, so the ethanol-fucoxanthin coefficients used in the original source were replaced. Firstly, the *P. tricornutum* absorbance spectrum from cultures under LL+N (maximum fucoxanthin) were averaged and the spectrum between 400–550 nm was used to estimate new fucoxanthin Gaussian peaks using the "chl.b.fit.R" file from the original source. These new coefficients for peak height ("a"), wavelength ("xp") and peak halfwidth ("s") for live cultures were then used instead of the original coefficients for ethanol and used to compare the spectral deconvolution (mg L^{-1}) to fucoxanthin as measured using HPLC (mg g^{-1}), which resulted in an R^2 value of 0.849 (Figure 3a).

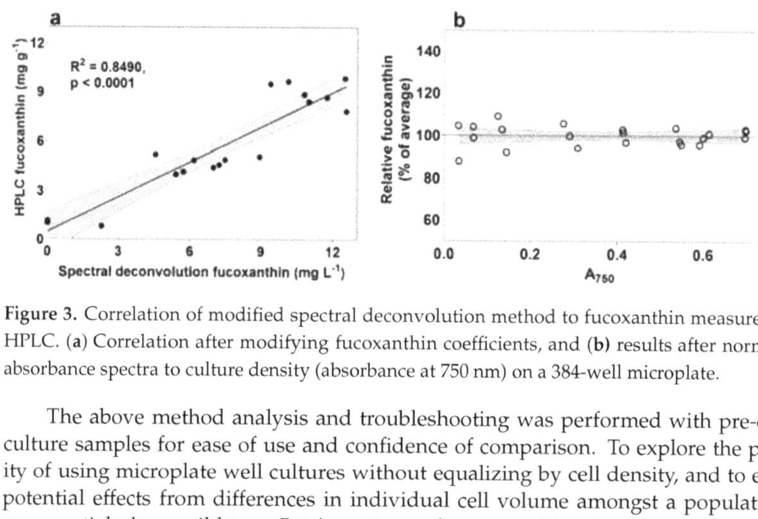

Figure 3. Correlation of modified spectral deconvolution method to fucoxanthin measured using HPLC. (**a**) Correlation after modifying fucoxanthin coefficients, and (**b**) results after normalizing absorbance spectra to culture density (absorbance at 750 nm) on a 384-well microplate.

The above method analysis and troubleshooting was performed with pre-diluted culture samples for ease of use and confidence of comparison. To explore the possibility of using microplate well cultures without equalizing by cell density, and to exclude potential effects from differences in individual cell volume amongst a population, an exponential-phase wild type *P. tricornutum* culture was diluted to various densities between approximately 0.5 and 20 × 10^6 cells mL^{-1} (A_{750} between 0.03 and 0.7) and their predicted fucoxanthin values compared on a 384-well microplate using spectral deconvolution of individual wavelengths normalized by the absorbance at 750 nm (Figure 3b), using the following equation:

$$\text{SpeDec Fx (mg L}^{-1}) = \text{spectra } [400 - 700 \text{ nm}]/(\text{Sample } A_{750} - \text{ASW } A_{750})$$

The choice to move to 384-well microplates instead of 96-well microplates was made to further accelerate the screening rate. Fucoxanthin predictions across replicates were within 8% of each other when only including samples at optical density of 0.4 or above. The success of using the absorbance at 750 nm to normalize the results of the spectral deconvolution method relies on each wavelength of a culture spectrum (400–700 nm) being divided by the culture A750 after blanking to culture media absorbance at the same wavelength. This method allows multiple wells to be measured without prior dilution, removing the need to account for screen bias towards denser cultures, but does not account for increased fucoxanthin due to culture shading effects.

The benefit of using this method in comparison to flow cytometry is that the latter is expensive to purchase and run, requires skilled users, and most importantly cannot screen using microplate format without a high risk of cross-well contamination. The modified spectral deconvolution method from Thrane et al. [20] allows users to screen libraries of hundreds to thousands of individuals cost-effectively without removing microplate lids or culture volume to preserve sterility, which also enables temporal evaluation across growth curves and different growth conditions. Method benefits are outlined in Table 1.

Table 1. Practical considerations of tested high-throughput screens for fucoxanthin in *P. tricornutum*. Culture contact refers to removing vessel lids, pipetting, or transferring into measurement vessels.

Fucoxanthin Screening Method	Method Letter	Correlation to HPLC (R^2)	No Dyeing Required	No Extraction Required	No Skilled Equipment Operators Required	No Culture Contact Required
Flow cytometry/FACS	A/B	0.949	✔	✔		
Nile Red	C	0.685		✔	✔	
Ritchie (2008) [18]	D	0.902	✔		✔	
Wang et al. (2018) [19]	E	0.859	✔		✔	
Thrane et al. (2015) [20]	F	0.849	✔	✔	✔	✔

3. Materials and Methods

3.1. Stock Culturing

Axenic *Phaeodactylum tricornutum* (CCAP 1055/1) stock cultures were obtained from the C3 culture collection and grown in Artificial Sea Water (ASW) medium (Darley & Volcani, 1969) under fluorescent light (200 μmol photons m^{-2} s^{-1}) with a 24:0 light cycle in shaking Erlenmeyer flasks (95 rpm) kept at 21 °C. Treatment flasks were inoculated at 8.5×10^5 cells mL^{-1} from cultures at exponential stage.

3.2. Experimental Design

Treatment flasks consisted of 250 mL conical flasks containing 100 mL of respective media. The highest fucoxanthin content was achieved with low light availability and high nitrate availability [14], so combinations of these two factors were used to maximise the expected culture fucoxanthin content range. Cells were inoculated into either nitrate-free ASW media (-N), standard (1 × N) nitrate media, or media with 10× nitrate (10 × N) from KNO3 and placed under either 10 (LL) or 200 (HL) μmol photons m^{-2} s^{-1} (Figure S1). To diminish variances in media composition introduced by pipetting culture volumes directly into experimental flasks, 30 mL culture aliquots were centrifuged and the pellets were then pipetted into experimental flasks (~2.5×10^7 cells total). Three × 30 mL stock cultures were also centrifuged, washed once with ultrapure water, flash-frozen in LN2, lyophilised, and weighed to estimate a starting value for biomass in each experimental flask. The treatments were left until there was a detectable difference observed in the relative chlorophyll *a* fluorescence across treatments.

3.3. Sampling

Each day, 200 μL from each flask was transferred into a clear-bottom, black 96-well microplate (Corning Inc., Corning, New York, NY, USA), and a plate reader (Tecan Infinite M1000 Pro) was used to measure absorbance at 750 nm as well as chlorophyll *a* fluorescence using an excitation wavelength of 440 nm and emission wavelength of 680 nm. Cells were counted daily using the Cytoflex LX flow cytometer (Beckman Coulter, Brea, CA, USA) by separating singlets via gating within an XY plot of forward scatter (FSC) and side scatter

(SSC). A second plot comparing forward scatter by cell area (FSC-A) to chlorophyll *a* fluorescence (excitation with blue laser at 488 nm with 690-50A optical filter) was constructed to separate singlets into live cells and dead cells/debris.

At the conclusion of the experimental period, flask culture volumes were measured, centrifuged, and washed once with ultrapure water to reduce dissolved salts. They were then flash-frozen in liquid nitrogen and lyophilized before being kept at −80 °C for measurement using HPLC. The remaining flask media was kept at −30 °C for nitrate analysis.

Culture aliquots were diluted to 1×10^6 cells mL^{-1}, and pigment extraction of both diluted and undiluted culture aliquots was performed by centrifugation and replacement of supernatant with equal volume of absolute ethanol kept at −30 °C. Extraction was furthered using 3 × 3-s pulses of an ultrasonic homogenizer (Qsonica Q125) at maximum amplitude with centrifugation in between pulses to confirm blanching of pellets and thus complete extraction of cellular pigments. Extracts and raw culture, both diluted and undiluted, were used to attain data for the various screening methods by pipetting 200 µL into clear-bottom, black 96-well microplates and measuring on microplate reader and by flow cytometry.

3.4. Chlorophyll and Fucoxanthin Autofluorescence Using Flow Cytometry (Method A and B)

Daily plots of chlorophyll *a* fluorescence (excitation with blue laser at 488 nm with 690/50 optical filter) for assisting in cell counts using flow cytometry were also utilized on the day of harvest to retrieve single-cell chlorophyll fluorescence data. A secondary plot for detecting fucoxanthin was also created on this day to assess whether a different excitation/emission arrangement would be more reliable for detecting fucoxanthin than that used for detecting chlorophyll (488/690). Firstly, HPLC data was compared to data for all available flow cytometry channel arrangements, and the channel with highest correlation to fucoxanthin was chosen to be displayed. This was a yellow excitation laser at 561 nm with 710/50 nm optical filter. The mean single-cell fluorescence for both channels was used to compare to fucoxanthin as measured using HPLC. To explore whether chlorophyll *a* fluorescence measured using microplate reader could be sufficient to predict fucoxanthin content, daily chlorophyll *a* measurements from Section 2.3. were also compared to fucoxanthin measured using HPLC.

3.5. Nile Red Fluorescence Using Flow Cytometry (Method C)

Nile Red fluorescence using both microplate reader and flow cytometry was included to test the sensitivity of this dye in both formats. One milliliter of each diluted culture sample was dyed with 0.75 µg Nile Red for ~10 min, and fluorescence was measured using microplate reader Excitation/Emission wavelengths of 484/583 nm, while the cytometry equivalent was measured by retrieving the mean single cell fluorescence using blue excitation laser at 488 nm with 610/20 optical filter.

3.6. Ritchie (2008) Using Microplate Reader (Method D)

The equations for chlorophyll for Ritchie [18] were applied to extracts to test whether they could reliably predict fucoxanthin in ethanol:

$$\text{Chl } a = -0.9394 \times A_{632} - 4.2774 \times A_{649} + 13.3914 \times A_{665}$$

3.7. Wang et al. (2018) Using Microplate Reader (Method E)

A similar method to the equations above, yet for direct quantification of fucoxanthin in *P. tricornutum*, was also developed, albeit without the need for removal of cell debris after extraction [19]. The equation from the original source is as follows (Cfuc′ = Fucoxanthin concentration in mg L^{-1}):

$$\text{Cfuc}' = 6.39 \times A_{445} - 5.18 \times A_{663} + 0.312 \times A_{750} - 5.27$$

3.8. Thrane et al. (2015) Using Microplate Reader (Method F)

The direct quantification method using visible light absorbance spectra developed by Thrane et al. [20], while requiring a more complex data analysis, was included to test the potential of 'gauss-peak' fitting originally established by Küpper, Spiller and Küpper [21]. The method is described thoroughly in Thrane et al. [20]. In short, individual pigment spectra are described by combinations of Gaussian peaks (termed the Gauss-peak spectra method) and are then modelled alongside background noise using non-negative least squares. The method utilizes R software to input coefficients for peak height, wavelength, and peak halfwidth to characterize pigments, and then compares this to sample absorbance spectra to estimate pigment content in mg L^{-1} using absorbance and pigment molar extinction coefficients. Thrane et al. [20] aimed to quantify multi-species pigment samples from lakes, whereas this work aims to adapt the method to a single species in a laboratory setting. The original source includes R scripts for readers to easily perform the method. To test if the method was further applicable to cultures without extraction, absorbance spectra obtained using diluted culture aliquots on a microplate reader were used instead of sample extract spectra.

3.9. HPLC for Pigment Detection

Freeze-dried pellet samples were weighed and re-suspended in a solvent ratio of between 2–3 mg of sample (DW) per mL of ethanol and sonicated with an ultrasonic homogenizer (Qsonica Q125) at 100% amplitude for 3 × 3-s pulses before being stored at −20 °C overnight once blanching of pellets was confirmed using a centrifuge. They were then filtered using 0.2 μm PTFE 13 mm syringe filters and stored in −80 °C until analysis. High Performance Liquid Chromatography (HPLC) was conducted using an Agilent Technologies 1290 Infinity, equipped with a binary pump with integrated vacuum degasser, thermostatted column compartment modules, Infinity 1290 auto-sampler, and PDA detector. Column separation was performed using a 4.6 mm × 150 mm Zorbax Eclipse XDB-C8 reverse-phase column (Agilent Technologies, Santa Clara, CA, USA) and guard column using a gradient of TBAA (tert-Butyl acetoacetate): Methanol mix (30:70) (solvent A) and Methanol (Solvent B) as follows: 0–22 min, from 5 to 95% B; 22–29 min, 95% B; 29–31 min, 5% B; and 31–40 min, column equilibration with 5% B. Column temperature was maintained at 55 °C. A complete pigment profile from 270 to 700 nm was recorded using PDA detector with 3.4 nm bandwidth. Example chromatograms can be found in supplementary files.

3.10. Additional Measurements

Media nitrate content was analyzed using an automated photometric analyzer (Gallery™ Discrete Analyzer, Thermo Fisher Scientific) by establishing standard calibration curves for nitrite and Total Oxidised Nitrogen (TON) from 0–20 mg L^{-1}, after which samples from 10 × nitrate flasks were diluted to 1/20th of their original concentration to fit within this range. Spiked recoveries were performed using three samples with dissimilar concentrations, and these resulted in an average recovery of 100 ± 7.5%.

3.11. Statistical Analysis

One-way ANOVA with a confidence level of 0.05 was performed followed by Tukey's post-hoc to determine significance among samples, using GraphPad Prism version 9.0.2 for Windows (GraphPad Software, San Diego, CA, USA).

4. Conclusions

An increasing screening rate is a critical factor for high-throughput projects regarding microalgal libraries. In this aspect, the spectral deconvolution method from Thrane et al. [20] holds great potential because samples can be analyzed with temporal and multiple-condition evaluation of individuals whilst maintaining culture sterility. This method removes the cost of most expendables and reduces the time spent extracting samples, which also reduces the likelihood of variability introduced through solvent evaporation or pipetting. Most notably, once using modified input scripts, the spectral deconvolution method from Thrane et al. [20] provides researchers with accurate measurement of fucoxanthin in live cultures of *P. tricornutum*, which, to our knowledge, has never been achieved, and significantly improves the screening rate of novel strains for biotechnological purposes. While this work is limited to fucoxanthin in *P. tricornutum*, future projects should look at adapting spectral deconvolution to other microalgal species as well as other pigments of interest.

Supplementary Materials: The following are available online at https://www.mdpi.com/article/10.3390/md20010019/s1, Figure S1: Representative HPLC chromatograms of Phaeodactylum tricornutum extract from each treatment group. Treatment abbreviations are as follows: nitrate-free ASW media (-N), standard (1 × N) nitrate media, or media with 10× nitrate (10 × N) and either 10 (LL) or 200 (HL) μmol photons m^{-2} s^{-1}. Peaks numbers are represented as follows: 1: Chlorophyll c; 2: Fucoxanthin; 3: Diadinoxanthin; 4: Chlorophyll a; 5: β-carotene.

Author Contributions: Conceptualization, S.M.M., R.M.A. and M.P.; Data curation, S.M.M. and A.S.; Formal analysis, S.M.M.; Investigation, S.M.M.; Methodology, S.M.M., R.M.A., A.H. and M.P.; Project administration, P.J.R. and M.P.; Resources, P.J.R. and M.P.; Software, S.M.M. and A.S.; Supervision, R.M.A., A.H., P.J.R. and M.P.; Visualization, S.M.M. and A.S.; Writing—original draft, S.M.M.; Writing—review & editing, S.M.M., R.M.A., A.S., A.H., P.J.R. and M.P. All authors have read and agreed to the published version of the manuscript.

Funding: This work was supported by a UTS Doctoral Scholarship to S.M.

Institutional Review Board Statement: Not applicable.

Informed Consent Statement: Not applicable.

Data Availability Statement: Not applicable.

Acknowledgments: Not applicable.

Conflicts of Interest: The authors declare no conflict of interest.

References

1. Matsuno, T. Aquatic animal carotenoids. *Fish. Sci.* **2001**, *67*, 771–783. [CrossRef]
2. Bae, M.; Kim, M.; Park, Y.K.; Lee, J.Y. Health benefits of fucoxanthin in the prevention of chronic diseases. *Biochim. Biophys. Acta Mol. Cell Biol. Lipids* **2020**, *1865*, 158618. [CrossRef] [PubMed]
3. Sangeetha, R.K.; Bhaskar, N.; Baskaran, V. Comparative effects of beta-carotene and fucoxanthin on retinol deficiency induced oxidative stress in rats. *Mol. Cell Biochem.* **2009**, *331*, 59–67. [CrossRef]
4. Karpiński, T.M.; Adamczak, A. Fucoxanthin—An antibacterial carotenoid. *Antioxidants* **2019**, *8*, 239. [CrossRef]
5. Koo, S.Y.; Hwang, J.H.; Yang, S.H.; Um, J.I.; Hong, K.W.; Kang, K.; Pan, C.H.; Hwang, K.T.; Kim, S.M. Anti-Obesity Effect of Standardized Extract of Microalga *Phaeodactylum tricornutum* Containing Fucoxanthin. *Mar. Drugs* **2019**, *17*, 311. [CrossRef] [PubMed]
6. Satomi, Y. Antitumor and cancer-preventative function of fucoxanthin: A marine carotenoid. *Anticancer Res.* **2017**, *37*, 1557–1562. [CrossRef]
7. Miyashita, K.; Nishikawa, S.; Beppu, F.; Tsukui, T.; Abe, M.; Hosokawa, M. The allenic carotenoid fucoxanthin, a novel marine nutraceutical from brown seaweeds. *J. Sci. Food Agric.* **2011**, *91*, 1166–1174. [CrossRef] [PubMed]
8. Miyashita, K.; Beppu, F.; Hosokawa, M.; Liu, X.; Wang, S. Nutraceutical characteristics of the brown seaweed carotenoid fucoxanthin. *Arch. Biochem. Biophys.* **2020**, *686*, 108364. [CrossRef]
9. Guo, B.; Liu, B.; Yang, B.; Sun, P.; Lu, X.; Liu, J.; Chen, F. Screening of Diatom Strains and Characterization of *Cyclotella cryptica* as A Potential Fucoxanthin Producer. *Mar. Drugs* **2016**, *14*, 125. [CrossRef] [PubMed]
10. Bowler, C.; Allen, A.E.; Badger, J.H.; Grimwood, J.; Jabbari, K.; Kuo, A.; Maheswari, U.; Martens, C.; Maumus, F.; Otillar, R.P.; et al. The *Phaeodactylum* genome reveals the evolutionary history of diatom genomes. *Nature* **2008**, *456*, 239–244. [CrossRef]
11. Fabris, M.; George, J.; Kuzhiumparambil, U.; Lawson, C.A.; Jaramillo-Madrid, A.C.; Abbriano, R.M.; Vickers, C.E.; Ralph, P. Extrachromosomal Genetic Engineering of the Marine Diatom *Phaeodactylum tricornutum* Enables the Heterologous Production of Monoterpenoids. *ACS Synth. Biol.* **2020**, *9*, 598–612. [CrossRef]
12. De Riso, V.; Raniello, R.; Maumus, F.; Rogato, A.; Bowler, C.; Falciatore, A. Gene silencing in the marine diatom *Phaeodactylum tricornutum*. *Nucleic Acids Res.* **2009**, *37*, e96. [CrossRef] [PubMed]

13. Yi, Z.; Xu, M.; Magnusdottir, M.; Zhang, Y.; Brynjolfsson, S.; Fu, W. Photo-oxidative stress-driven mutagenesis and adaptive evolution on the marine diatom *Phaeodactylum tricornutum* for enhanced carotenoid accumulation. *Mar. Drugs* **2015**, *13*, 6138–6151. [CrossRef] [PubMed]
14. McClure, D.D.; Luiz, A.; Gerber, B.; Barton, G.W.; Kavanagh, J.M. An investigation into the effect of culture conditions on fucoxanthin production using the marine microalgae *Phaeodactylum tricornutum*. *Algal Res.* **2018**, *29*, 41–48. [CrossRef]
15. Yi, Z.; Su, Y.; Xu, M.; Bergmann, A.; Ingthorsson, S.; Rolfsson, O.; Salehi-Ashtiani, K.; Brynjolfsson, S.; Fu, W. Chemical Mutagenesis and Fluorescence-Based High-Throughput Screening for Enhanced Accumulation of Carotenoids in a Model Marine Diatom *Phaeodactylum tricornutum*. *Mar. Drugs* **2018**, *16*, 272. [CrossRef]
16. Kadono, T.; Kira, N.; Suzuki, K.; Iwata, O.; Ohama, T.; Okada, S.; Nishimura, T.; Akakabe, M.; Tsuda, M.; Adachi, M. Effect of an introduced phytoene synthase gene expression on carotenoid biosynthesis in the marine diatom *Phaeodactylum tricornutum*. *Mar. Drugs* **2015**, *13*, 5334–5357. [CrossRef]
17. Gao, F.; Teles, I.; Ferrer-Ledo, N.; Wijffels, R.H.; Barbosa, M.J. Production and high throughput quantification of fucoxanthin and lipids in *Tisochrysis lutea* using single-cell fluorescence. *Bioresour. Technol.* **2020**, *318*, 124104. [CrossRef]
18. Ritchie, R.J. Universal chlorophyll equations for estimating chlorophylls a, b, c, and d and total chlorophylls in natural assemblages of photosynthetic organisms using acetone, methanol, or ethanol solvents. *Photosynthetica* **2008**, *46*, 115–126. [CrossRef]
19. Wang, L.-J.; Fan, Y.; Parsons, R.L.; Hu, G.-R.; Zhang, P.-Y.; Li, F.-L. A rapid method for the determination of fucoxanthin in diatom. *Mar. Drugs* **2018**, *16*, 33. [CrossRef] [PubMed]
20. Thrane, J.-E.; Kyle, M.; Striebel, M.; Haande, S.; Grung, M.; Rohrlack, T.; Andersen, T. Spectrophotometric analysis of pigments: A critical assessment of a high-throughput method for analysis of algal pigment mixtures by spectral deconvolution. *PLoS ONE* **2015**, *10*, e0137645. [CrossRef] [PubMed]
21. Küpper, H.; Spiller, M.; Küpper, F.C. Photometric method for the quantification of chlorophylls and their derivatives in complex mixtures: Fitting with Gauss-peak spectra. *Anal. Biochem.* **2000**, *286*, 247–256. [CrossRef] [PubMed]
22. Küpper, H.; Seibert, S.; Parameswaran, A. Fast, sensitive, and inexpensive alternative to analytical pigment HPLC: Quantification of chlorophylls and carotenoids in crude extracts by fitting with Gauss peak spectra. *Anal. Chem.* **2007**, *79*, 7611–7627. [CrossRef] [PubMed]
23. Li, W.; Gao, K.; Beardall, J. Interactive effects of ocean acidification and nitrogen-limitation on the diatom *Phaeodactylum tricornutum*. *PLoS ONE* **2012**, *7*, e51590. [CrossRef] [PubMed]
24. Premvardhan, L.; Sandberg, D.J.; Fey, H.; Birge, R.R.; Büchel, C.; van Grondelle, R. The charge-transfer properties of the S2 state of fucoxanthin in solution and in fucoxanthin chlorophyll-a/c2 protein (FCP) based on stark spectroscopy and molecular-orbital theory. *J. Phys. Chem. B* **2008**, *112*, 11838–11853. [CrossRef]
25. Fan, Y.; Ding, X.-T.; Wang, L.-J.; Jiang, E.-Y.; Van, P.N.; Li, F.-L. Rapid Sorting of Fucoxanthin-Producing *Phaeodactylum tricornutum* Mutants by Flow Cytometry. *Mar. Drugs* **2021**, *19*, 228. [CrossRef] [PubMed]

marine drugs

MDPI

Article

Transcriptomic and Proteomic Characterizations of the Molecular Response to Blue Light and Salicylic Acid in *Haematococcus pluvialis*

Xiaodong Wang [1,†], Chunxiao Meng [1,†], Hao Zhang [2], Wei Xing [1], Kai Cao [1], Bingkui Zhu [1], Chengsong Zhang [1], Fengjie Sun [3,*] and Zhengquan Gao [2,*]

1 School of Life Sciences and Medicine, Shandong University of Technology, Zibo 255049, China; 15110702040@stumail.sdut.edu.cn (X.W.); mengchunxiao@sdut.edu.cn (C.M.); 20410010842@stumail.sdut.edu.cn (W.X.); 21410010879@stumail.sdut.edu.cn (K.C.); 21410010901@stumail.sdut.edu.cn (B.Z.); 21410010898@stumail.sdut.edu.cn (C.Z.)

2 School of Pharmacy, Binzhou Medical University, Yantai 264003, China; hao073711@bzmc.edu.cn

3 School of Science and Technology, Georgia Gwinnett College, 1000 University Center Lane, Lawrenceville, GA 30043, USA

* Correspondence: fsun@ggc.edu (F.S.); zq7723@vip.163.com (Z.G.)

† These authors contributed equally to this work.

Abstract: *Haematococcus pluvialis* accumulates a large amount of astaxanthin under various stresses, e.g., blue light and salicylic acid (SA). However, the metabolic response of *H. pluvialis* to blue light and SA is still unclear. We investigate the effects of blue light and SA on the metabolic response in *H. pluvialis* using both transcriptomic and proteomic sequencing analyses. The largest numbers of differentially expressed proteins (DEPs; 324) and differentially expressed genes (DEGs; 13,555) were identified on day 2 and day 7 of the treatment with blue light irradiation (150 μmol photons $m^{-2}s^{-1}$), respectively. With the addition of SA (2.5 mg/L), a total of 63 DEPs and 11,638 DEGs were revealed on day 2 and day 7, respectively. We further analyzed the molecular response in five metabolic pathways related to astaxanthin synthesis, including the astaxanthin synthesis pathway, the fatty acid synthesis pathway, the heme synthesis pathway, the reactive oxygen species (ROS) clearance pathway, and the cell wall biosynthesis pathway. Results show that blue light causes a significant down-regulation of the expression of key genes involved in astaxanthin synthesis and significantly increases the expression of heme oxygenase, which shows decreased expression by the treatment with SA. Our study provides novel insights into the production of astaxanthin by *H. pluvialis* treated with blue light and SA.

Keywords: transcriptomics; proteomics; blue light; astaxanthin; fatty acid; heme; reactive oxygen species; cell wall; salicylic acid

Citation: Wang, X.; Meng, C.; Zhang, H.; Xing, W.; Cao, K.; Zhu, B.; Zhang, C.; Sun, F.; Gao, Z. Transcriptomic and Proteomic Characterizations of the Molecular Response to Blue Light and Salicylic Acid in *Haematococcus pluvialis*. *Mar. Drugs* **2022**, *20*, 1. https://doi.org/10.3390/md20010001

Academic Editor: Carlos Almeida

Received: 5 December 2021
Accepted: 17 December 2021
Published: 21 December 2021

1. Introduction

Astaxanthin (3,3′-dihydroxy-β-carotene-4,4′-dione) is a type of red-orange carotenoid with strong biological antioxidant capacity [1] and important applications in human health and in the nutraceutical, cosmetics, food, and feed industries [2]. During the process of astaxanthin synthesis, isopentenyl diphosphate (IPP) reacts with dimethylallyl diphosphate (DMAPP) to form geranyl pyrophosphate (GPP), which joins another molecule of IPP to form farnesyl pyrophosphate (FPP). FPP joins IPP to create geranylgeranyl pyrophosphate (GGPP). Two molecules of GGPP are connected to generate phytoene, which undergoes multiple dehydrogenation reactions to produce lycopene. Then, lycopene undergoes two reactions to produce β-carotene and ultimately to generate astaxanthin [3]. Compared with other species of microalgae, *Haematococcus pluvialis* has shown the highest accumulation of astaxanthin—up to 4% of dry cell weight (DCW) [4]. In the production of astaxanthin, *H. pluvialis* goes through two developmental stages, including the dividing stage (morphologically green, producing macrozooids or zoospores and microzooids) and the astaxanthin accumulating stage (morphologically red, producing palmella and hematocysts or aplanospores). In the astaxanthin accumulating stage, the green dividing cells of *H. pluvialis* are transformed into palmella, which further develop into red aplanospores, while cells of red aplanospores continue to produce astaxanthin [5]. As a type of secondary metabolite, astaxanthin is mainly accumulated once the cell divisions in the microalgae largely terminate [6]. When the cells of *H. pluvialis* stop dividing, both the chloroplast size and the chlorophyll content are reduced. Astaxanthin is generally deposited in cytoplasmic lipid droplets (CLD) of *H. pluvialis* [7]. The accumulation of astaxanthin could be caused by various types of environmental stresses, including high light intensities, low nitrogen, salt stress, and high temperature [8].

Due to its less heat dissipation with long life-expectancy and small size, the light-emitting diode (LED) has become an important factor in the microalgal industry [9]. Blue, green, and red lights are important for the growth and development of microalgae [10]. For example, blue light increases the cell size of algae, lipid accumulation, and growing period in *Chlamydomonas reinhardtii* [11,12], while green light of high intensities could significantly promote photosynthesis [13], and red light increases growth rates in *Dunaliella salina* and *Spirulina* [14]; meanwhile, a red LED at a relatively low intensity enhances cell growth in *H. pluvialis* [15]. Studies have shown that different light qualities increase the production of astaxanthin, biomass, and fatty acid content in *H. pluvialis* [16,17], while high-intensity blue light increases the accumulation of astaxanthin in *H. pluvialis* [18]. Carotenoids in microalgae primarily function as the accessory pigments in light-harvesting antenna by absorbing blue light that is inadequately absorbed by chlorophyll, thus improving light absorption. Furthermore, astaxanthin has been shown to help prevent damage to the photosynthesis system in algae [7].

Studies have shown that addition of chemical regulators enhances the yield of microalgal biomass and biological products [19]. As a type of important plant hormone and signal molecule, salicylic acid (SA) shows critical functions in plant growth, immunity, and development, which have been extensively investigated [20]. Specifically, SA is involved in the molecular regulation of metabolic response to both abiotic stresses, i.e., heat, cold, drought, heavy metal, salt stresses [21–23], and biotic stresses, e.g., a plant's defense against pathogens [24]. Our previous studies on SA inductions at the transcriptional level identified a group of plant-specific transcription factor families (e.g., MYB, AP2/ERF, WRKY, and HSF) in *H. pluvialis* in response to the treatment with SA [25,26]. Furthermore, studies have shown that SA regulates the astaxanthin synthesis of *H. pluvialis* by scavenging free radicals [27]. Under stressed conditions, a large amount of triacylglycerol (TAG) is also accumulated in *H. pluvialis* [28]. Astaxanthin mostly exists in the form of esterification, while an astaxanthin monoester joins an astaxanthin diester to form liposomes in the cytoplasm [29]. Moreover, under stresses, most of the de novo fatty acids are assembled into TAGs, which are esterified into cytosolic liposomes, along with astaxanthin esters through an unknown mechanism [30]. Therefore, it is important and beneficial to further explore the synthesis of fatty acids in order to identify novel experimental strategies to improve the accumulation of astaxanthin in microalgae.

As a key biological technology, RNA-seq is an important tool which is applied to investigate differential gene expression in various model organisms. For example, RNA-seq technology provides de novo gene sequence and annotation analyses [31]. With the complete genome of *H. pluvialis* recently sequenced [32], the molecular mechanism regulating the accumulation of astaxanthin in *H. pluvialis* could be further elucidated based on RNA-seq analysis. Previous studies based on transcriptome sequencing have investigated the molecular metabolism of *H. pluvialis* in different environments, e.g., strong light [33], different light qualities [10], salt ions [3,34], CO_2 [35], and trisodium citrate [36]. Furthermore, studies showed that the accumulation of astaxanthin in *H. pluvialis* was affected by pyruvate metabolism [37]. To date, transcriptome sequencing has become an essential method for studying the molecular metabolism of *H. pluvialis*. Furthermore, proteomic analysis has been commonly used to characterize the functions of various types of proteins in the development of plants and microalgae. Due to the large amount of protein sequence information available and the application of mass spectrometry in sensitive rapid protein identification, proteomic sequencing provides a novel approach for the analysis of complex functions in plants and microalgae.

In this study, we first investigated the response of *H. pluvialis*, also known as *H. lacustris* [38], to blue light, and further explored the synergistic effects of both blue light and SA stresses on the molecular response of *H. pluvialis*. We then identified and analyzed the differentially expressed genes (DEGs) of *H. pluvialis* under blue and white lights to explore the response and adaptation of *H. pluvialis* to the change of light wavelength. RNA-seq analyses were performed based on samples treated with white light (day 0) and blue light with or without the addition of SA (day 2 and day 7). Our results indicated that combined analysis of transcriptome and proteome sequencing was efficient in clarifying the metabolic mechanisms of astaxanthin production by *H. pluvialis*. Our study clearly demonstrated that the combined analysis of RNA-seq and proteomic sequencing comprehensively revealed the variations in the molecular mechanisms of *H. pluvialis* in response to different wavelengths of light and SA, further providing essential insights into the metabolic regulation involved in the synthesis of astaxanthin at molecular level. We noted that our study was unique, based on the following: firstly, we applied the combination of both the transcriptomic and proteomic analyses in our investigation of the molecular response of *H. pluvialis* to both blue light and SA. Furthermore, we applied a high-intensity blue light, whereas previous studies generally utilized either a low-intensity blue light or a white light of high intensity with SA to treat *H. pluvialis*, or different species, e.g., *Ulva prolifera* [23,39,40]. Moreover, we reported, for the first time, the differentially expressed genes (DEGs) and proteins (DEPs) involved in the heme pathway at the genomic level.

2. Results

2.1. Transcriptome Analysis

A total of 999,477,676 raw reads generated from RNA-seq were further processed with the adapters removed, yielding a total of 872,291,032 clean reads of 149.92 Gb of sequence data, including 24.02 Gb, 68.40 Gb, and 57.50 Gb derived from samples on day 0, day 2, and day 7, respectively (Table S1). Based on the de novo assembly, a total of 109,443 unigenes with the N50 value of 1193 bp were annotated based on the clean reads of the three groups of samples, ranging from 201 bp to 13,626 bp in size. The size distributions of these unigenes revealed that a total of 35,250 unigenes ranged between 200 and 300 bp, followed by a total of 16,338 unigenes ranging from 300 to 400 bp (Figure S2). In total, 21,504 (~19.65%) unigenes were annotated based on the SwissProt database, 27,965 (~25.55%) unigenes were identified in the non-redundant (nr) protein database, 26,248 (~23.98%) unigenes displayed significant similarities to known proteins in the Pfam database, and totals of 19,648 (~17.95), 24,798 (~22.66%), and 31,590 (~28.86%) unigenes were annotated in the Kyoto Encyclopedia of Genes and Genomes (KEGG), the Gene Ontology database (GO), and the EggNOG database, respectively. The top two species with the highest number of annotated unigenes were *Chlamydomonas eustigma* (4558 unigenes of ~16.3%) and *C. reinhardtii* (2782 unigenes of ~9.95%) (Figure S1). There were 24,798 unigenes annotated to 11,212 GO terms revealed in three categories of the GO database, i.e., molecular function, cellular component, and biological process. The top two GO terms in these three categories were the biological process and the oxidation–reduction process of the molecular function, cytoplasm, and nucleus of the cellular component category, and ATP binding and protein binding of the biological process category. A total of 19,648 unigenes were enriched in 6 and 19 KEGG metabolic pathways at level 1 and level 2, respectively. The results of enrichment analysis based on the KEGG database revealed that the top two categories of metabolic pathways with the largest numbers of unigenes annotated were the metabolism category (4440 unigenes) in the KEGG pathway at level 1 and the translation category (1717 unigenes) in the KEGG pathway at level 2 (Table S2).

2.2. Identification and Enrichment Analysis of DEGs and DEPs

As the treatment time increased from day 0 to day 7, the numbers of DEGs increased dramatically in the samples of *H. pluvialis*. The number of DEGs in the pairwise comparison of B257 vs. B07 was significantly higher than that of B252 vs. B02, suggesting that SA showed a weak effect on the microalgae in the early stage under the blue light culture and an enhanced effect in the late stage of the blue light treatment (Figure 1A). The most and the least numbers of DEGs of 29,295 and 76 were revealed in the pairwise comparisons of B07 vs. N and B252 vs. B02, respectively. As the treatment time of blue light irradiation increased, both the numbers of up-regulated and down-regulated genes increased significantly. In the groups treated with SA, there were only 76 DEGs, compared with the blue light group on day 2 of the treatment with blue light irradiation. However, on day 7 of the treatment with blue light irradiation, the numbers of up-regulated and down-regulated genes of the groups added with SA increased to 2187 and 9496, respectively (Figure 1A). The most and the least numbers of DEPs of 1385 and 52 were revealed in the pairwise comparisons of B07 vs. N and B257 vs. B07, respectively (Figure 1B).

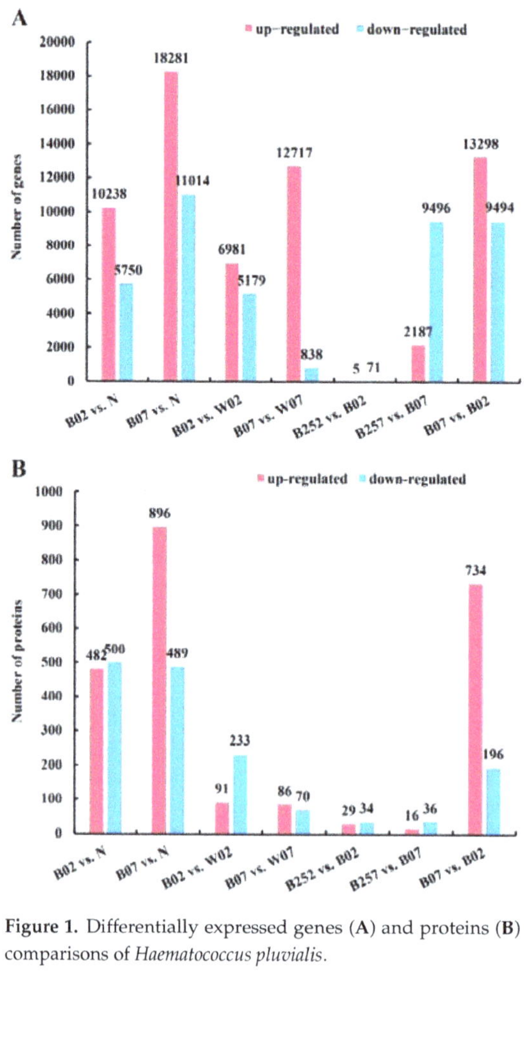

Figure 1. Differentially expressed genes (**A**) and proteins (**B**) identified among different pairwise comparisons of *Haematococcus pluvialis*.

To further investigate the effects of blue light and SA on *H. pluvialis*, we selected unigenes with significant changes in transcripts and proteins to perform the enrichment analyses based on both GO (Figure S3) and KEGG (Figure S4) databases (Table S3).

The results of the GO enrichment analysis showed that, on the second day of the treatment with blue light irradiation, most of the unigenes were annotated to carbon metabolism, e.g., glycolytic process, gluconeogenesis, and tricarboxylic acid cycle (Figure S3A). Notably, most genes, in response to cadmium ion, showed significant changes in their expressions. The results of the KEGG enrichment analysis showed that on the second day of the treatment with blue light irradiation, most of the unigenes were mainly enriched in metabolic pathways of energy metabolism, e.g., carbohydrate metabolism, metabolism of terpenoids and polyketides, and lipid metabolism (Figure S4A). On day 7 of the treatment with blue light irradiation, the unigenes showing significant changes in expression were mainly annotated to the GO terms related to proteins, e.g., translation, protein renaturation, and protein ubiquitin (Figure S3B), while the results of KEGG enrichment showed that the proportion of metabolic pathways of protein folding, sorting, and degradation was increased (Figure S4B). On the day 2 of the treatment with SA, no significantly expressed transcripts and proteins were observed. On day 7, with the addition of SA, most unigenes were annotated to GO terms of regulation of metabolic process, such as reactive oxygen species metabolic process, regulation of RNA metabolic process, and positive regulation by organization of apoptotic process (Figure S3C), while the results of KEGG enrichment analysis revealed that the significantly enriched metabolic pathways were related to infectious diseases (Figure S4C). These results were consistent with those reported previously, showing that SA, as one of the important signal molecules, played a vital role in disease resistance in plants [24].

2.3. DEGs and DEPs Involved in the Astaxanthin Metabolic Pathway

In *H. pluvialis*, isopentenyl diphosphate (IPP) is synthesized in the methyl-d-erythritol 4-phosphate (MEP) pathway [41] with a total of seven genes involved, including *dxr*, *dxs*, *ispD*, *ispE*, *ispF*, *ispG*, and *ispH* (Figure 2). In the early stage of exposure to blue light, the expression levels of *ispD* gene encoding 2-C-methyl-D-erythritol 4-phosphate cytidylyltransferase and *ispE* gene encoding 4-diphosphocytidyl-2-C-methyl-D-erythritol kinase were significantly lower than those treated with white light. Results showed that blue light showed little effect on the expression levels of *dxs* encoding 1-deoxy-D-xylulose-5-phosphate synthase, *dxr* encoding 1-deoxy-D-xylulose-5-phosphate reductoisomerase, *ispF* encoding 2-C-methyl-D-erythritol 2,4-cyclodiphosphate synthase, *ispG* encoding 1-hydroxy-2-methyl-2-(E)-butenyl 4-diphosphate synthase, and *ispH* encoding 4-hydroxy-3-methylbut-2-enyl diphosphate reductase (Figure 2).

The astaxanthin synthesis is initiated with the precursor of astaxanthin, i.e., β-carotene. The IPP produced in the MEP pathway is catalyzed by IPP isomerase (IPI) to generate DMAPP, which joins three IPP molecules to produce GGPP, catalyzed by farnesyl diphosphate synthase (FPPS) and GGPP synthase (GGPS). Subsequently, two GGPP molecules are catalyzed by phytoene synthase (PSY) to produce phytoene, which is used to generate ξ-carotene catalyzed by phytoene desaturase I (PDS). Then, ξ-carotene is used to generate lycopene, catalyzed by ξ-Carotene desaturase II (ZDS). Finally, lycopene cyclase (LCYB) catalyzes the transformation of lycopene into β-carotene, which is catalyzed by both beta-carotene hydroxylase (CRTR-B) and beta-carotene ketolase (BKT) to generate astaxanthin (Figure 3). Results of the transcriptome analysis showed that blue light exposure significantly up-regulated the expression of *ipi*, with its expression level escalated as the exposure time increased. However, on day 7 of the treatment with SA, the gene expression level of *ipi* was significantly decreased. On day 2 after exposure to blue light, compared with white light, the expression levels of *fpps*, *pds*, *psy*, and *crtR-B* (encoding β-carotene hydroxylase) were significantly decreased. On day 7 after the exposure to blue light, the expression level of *psy* was significantly increased, while the expression of *crtR-B* was significantly decreased. The addition of SA resulted in a significant decrease in the expression of *psy* on day 7 (Figure 3). On day 2 after the blue light exposure, the expression level of *crtR-B* was significantly down-regulated compared with that of *H. pluvialis*, treated with white light. However, the gene expression of *crtR-B* was increased as the exposure time increased. Blue light exposure showed no significant effect on the expression of gene *bkt* encoding β-carotene ketolase. However, with the increase in exposure time, the expression of *bkt* was significantly decreased.

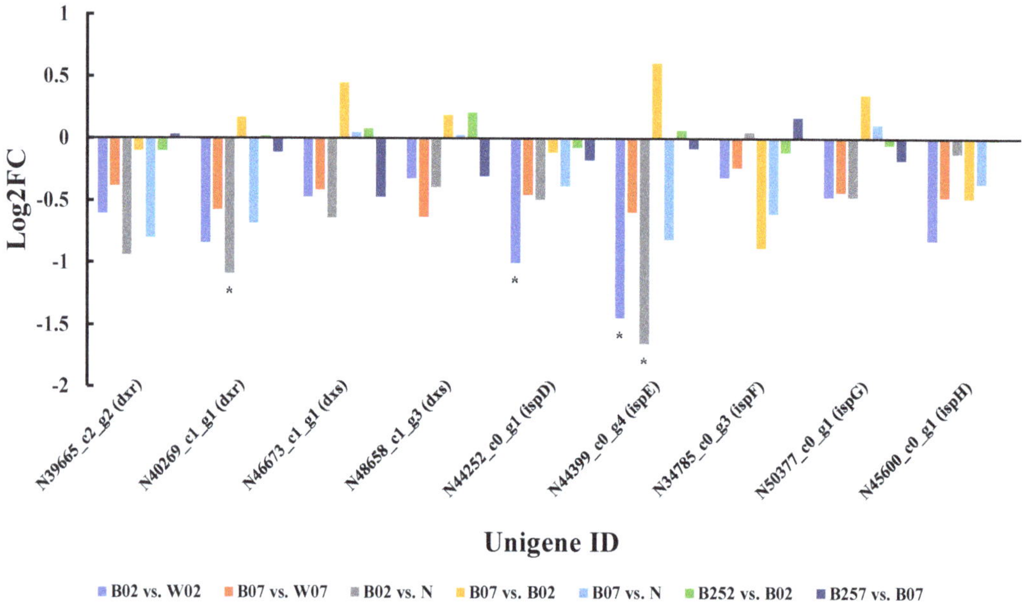

Figure 2. Expression profiles of the seven genes (i.e., *dxr*, *dxs*, *ispD*, *ispE*, *ispF*, *ispG*, and *ispH*) involved in MEP pathway based on transcriptomic analysis of pairwise comparisons in *Haematococcus pluvialis*. *—indicates significant down-regulation ($p < 0.05$ and Log2FC ≤ -1).

2.4. DEGs and DEPs Associated with the Lipid Metabolic Pathway

When *H. pluvialis* is cultured under stressed environments, astaxanthin accumulation is usually accompanied by the synthesis of a large amount of fatty acids. TAGs are usually synthesized de novo from fatty acids and combined with astaxanthin to form liposomes [29]. As the precursor of fatty acid synthesis, NADPH is derived from the pentose phosphate pathway (PPP) with the involvements of both glucose-6-phosphate dehydrogenase (G6PD) and 6-phosphogluconate dehydrogenase (6PGD) [42].

Our results showed that two and three unigenes were identified encoding 6PGD and G6PD, respectively (Figure 4A). Transcriptome analysis showed that on the second day of the treatment with blue light irradiation, the expression of *6pgd* was up-regulated, while the two transcripts of *g6pd* showed opposite expression patterns with one significantly up-regulated and the other significantly down-regulated. On day 7, the expressions of both 6PGD and G6PD were significantly up-regulated (Figure 4B). Compared with white light, no significant changes of both 6PGD and G6PD were observed under blue light irradiation on day 7. The addition of SA caused the down-regulation of the expression of both *6pgd* and *g6pd* on day 7. However, no significant changes caused by SA were observed at the protein level (Figure 4B).

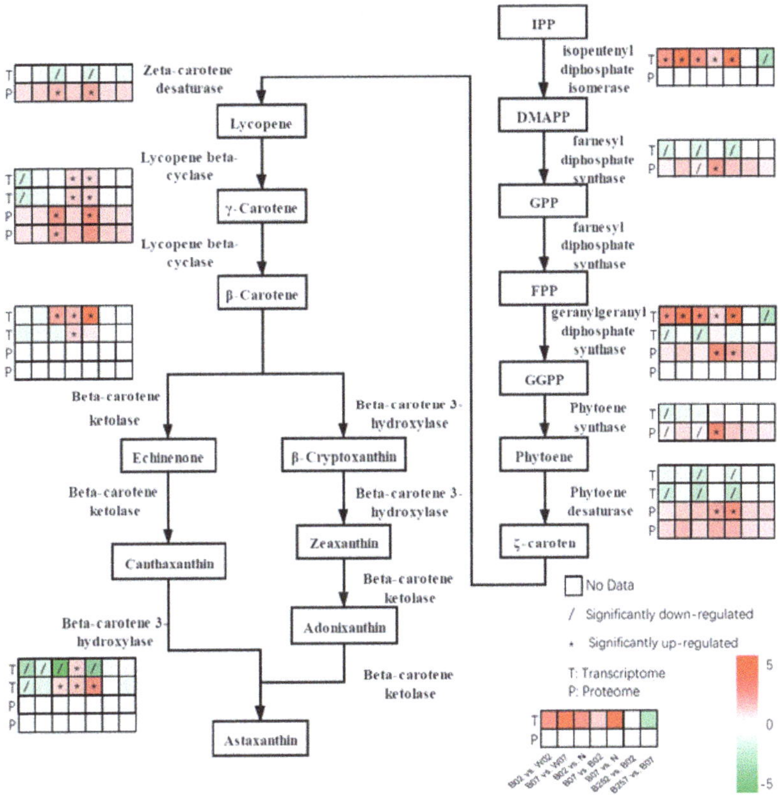

Figure 3. Expressions profiles of genes and proteins involved in astaxanthin synthesis pathway revealed in the pairwise comparisons in *Haematococcus pluvialis* based on transcriptomic and proteomic analyses.

In *H. pluvialis*, the fat acid synthesis is initiated with acetoyl-COA, which is catalyzed by acetyl-CoA carboxylase (ACC1) to synthesize malonyl-COA. The malonyl-COA reacts with acyl carrier protein (ACP) catalyzed by malonyl-CoA:ACP transacylase (FabD) to produce malonyl-[ACP] (Figure 5). After multiple reactions catalyzed by enzymes of beta-ketoacyl-[acyl-carrier-protein] synthase II (FabF), 3-oxoacyl-[acyl-carrier-protein] reductase (MabA), 3-hydroxyacyl-[acyl-carrier-protein] dehydratase (FabZ), and FabI, the C18:0-[ACP] is finally formed. Then, C18:0-[ACP] is catalyzed by fatty-acid synthase (FASN) to generate stearic acid [33].

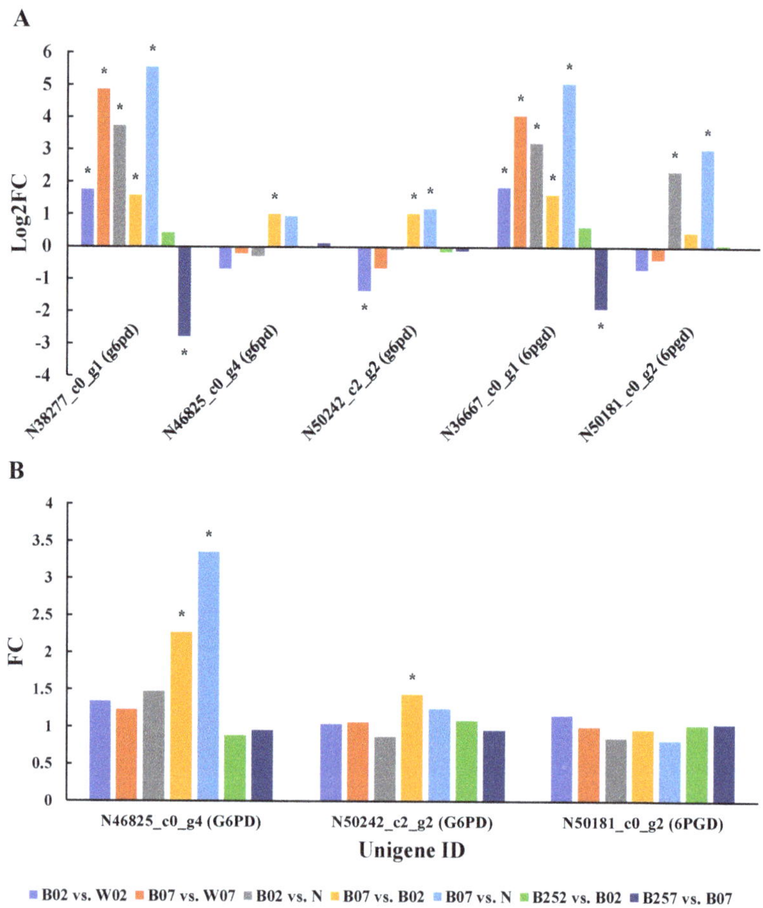

Figure 4. Expression profiles of genes *6pgd* and *g6pd* (**A**) and proteins 6PGD and G6PD (**B**) involved in the NADPH metabolism revealed in the pairwise comparisons of *Haematococcus pluvialis* based on transcriptomic and proteomic analyses, respectively. *—indicates significant up-regulation ($p < 0.05$ and Log2FC ≥ 1; $p < 0.05$ and FC ≥ 1.5) or significant down-regulation ($p < 0.05$ and Log2FC ≤ -1).

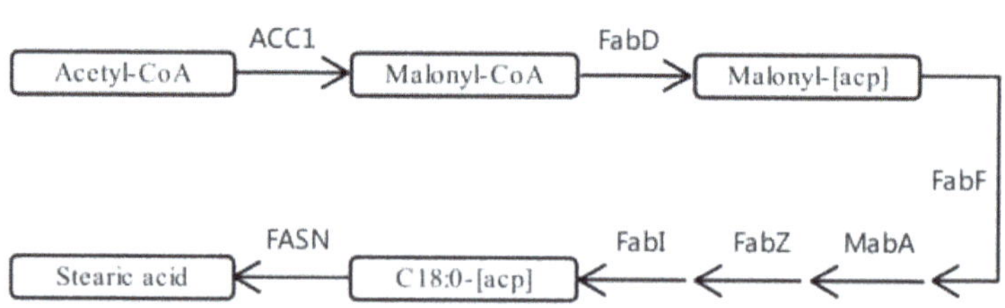

Figure 5. Fatty acid synthesis pathway in *Haematococcus pluvialis.*

Compared with white light, on the second day of the treatment with blue light irradiation, five of the seven genes (i.e., *acc1*, *fabF*, *fabI*, *fabZ*, and *fasn*) related to fatty acid synthesis were significantly down-regulated, while the other two genes (i.e., *fabD* and *mabA*) showed transcripts either significantly up-regulated or down-regulated. On day 7 of the treatment with blue light irradiation, both *fabD* and *mabA* were significantly up-regulated. No significant changes were revealed in the early stage (day 2) of the treatment with SA, whereas both *fabD* and *mabA* were significantly down-regulated on day 7 (Figure 6). At the protein level, ACC1 and FabD were significantly down-regulated on the second day of the treatment with blue light irradiation, while only MabA was significantly up-regulated on day 7 (Figure 7). No significant changes were observed on these proteins with the addition of SA in *H. pluvialis*.

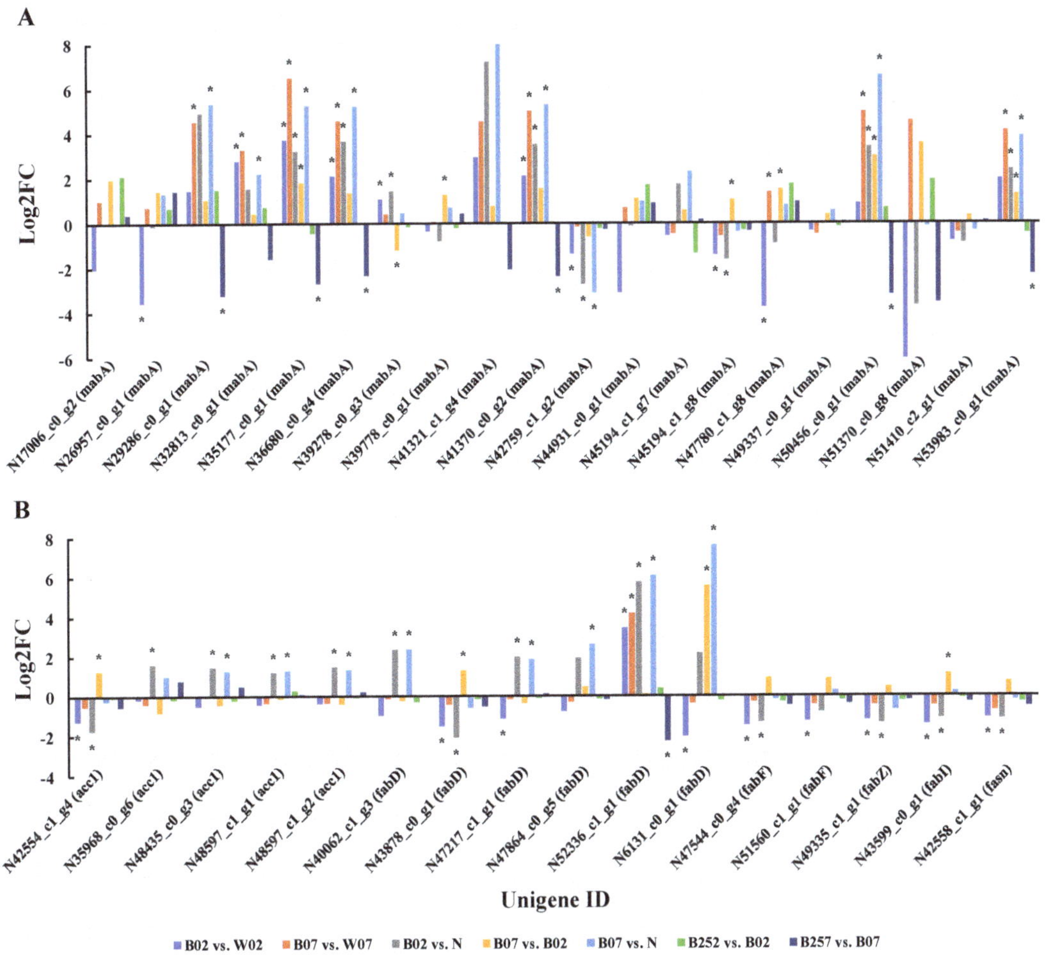

Figure 6. Expression profiles of genes *mabA* (**A**) and *acc1*, *fabD*, *fabF*, *fabI*, *fabZ*, and *fasn* (**B**) involved in the lipid metabolic pathway revealed in the pairwise comparisons of *Haematococcus pluvialis* based on transcriptomic analysis. *—indicates significant up-regulation ($p < 0.05$ and Log2FC ≥ 1) or significant down-regulation ($p < 0.05$ and Log2FC ≤ -1).

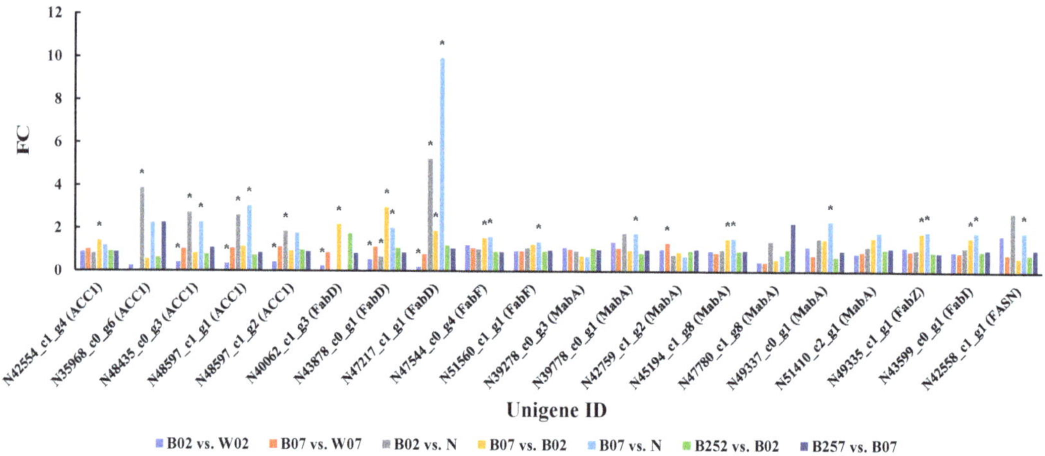

Figure 7. Expression profiles of proteins (i.e., ACC1, FabD, FabF, FabI, FabZ, FASN, and MabA) involved in the lipid metabolic pathway revealed in the pairwise comparisons of *Haematococcus pluvialis* based on proteomic analysis. *—indicates significant up-regulation ($p < 0.05$ and FC ≥ 1.5) or significant down-regulation ($p < 0.05$ and FC ≤ 0.67).

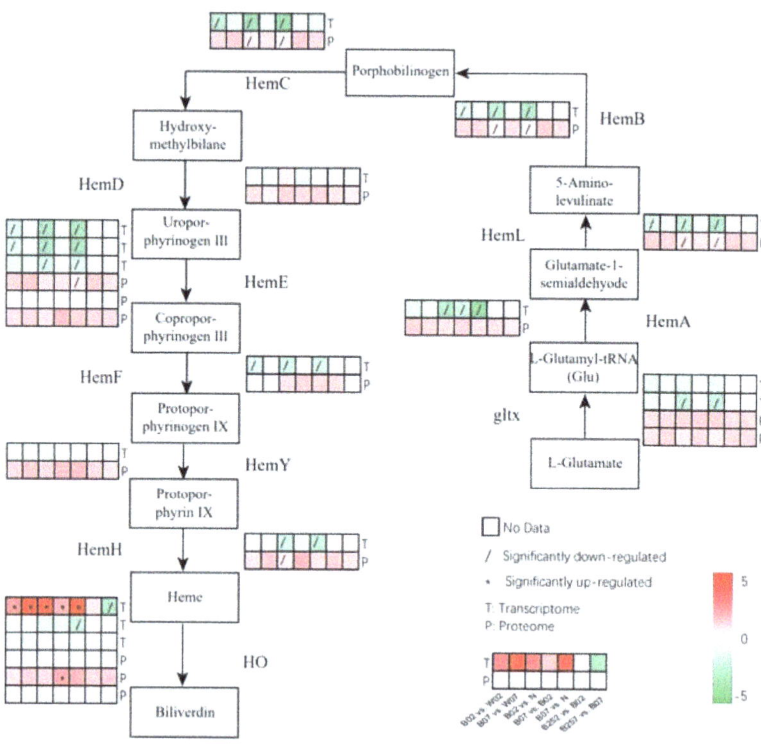

Figure 8. Expressions profiles of genes and proteins involved in the heme metabolic pathway revealed in the pairwise comparisons of *Haematococcus pluvialis* based on transcriptomic and proteomic analyses.

2.5. DEGs and DEPs Associated with the Generation of ROS

Studies have shown that *H. pluvialis* scavenged harmful ROS in cells by up-regulating genes involved in the removal of ROS [10]. Although the ROS is not beneficial for cell survival, it plays an important role in the production of astaxanthin [43]. Besides the generation of a large amount of astaxanthin, cells of *H. pluvialis* also prevent oxidative stress based on the involvements of a group of ROS-scavenging enzymes, including ascorbate peroxidase (APX), catalase (CAT), glutathione reductase (GR), peroxiredoxin (PRDX), monodehydroascorbate reductase (MDAR), ferritin, superoxide dismutase (SOD), glutathione peroxidase (GPX), glutaredoxin (GRX), and monothiol glutaredoxin (MGRX) [29,44]. Our results showed that on day 7, the expression levels of *sod*, *mdar*, *grx*, and *gr* in *H. pluvialis* treated with blue light were significantly higher than those treated with white light (i.e., pairwise comparisons of B02 vs. W02 and B07 vs. W07), the expression levels of *mgrx*, *cat*, and *gpx* under blue light were significantly higher than those in white light, whereas the expression of *apx* on day 2 after the exposure to blue light was significantly lower than that under white light (Figure 9). The expression level of gene encoding ferritin on day 7 of the treatment with blue light irradiation was significantly lower than that of white light. However, the expression patterns of *cat*, *ferritin*, *gpx*, *mgrx*, and *prdx* varied on day 2 after the treatment with blue light irradiation. On day 7 with SA added, the expression levels of *cat*, *gr*, *mdar*, *prdx*, and *sod* were significantly decreased. The expression levels of genes encoding ferritin and sod were decreased significantly on day 2 after SA was added. However, the expressions of related homologous genes were down-regulated. With the addition of SA, the expression levels of most ROS scavenging genes were decreased significantly and the down-regulated on day 7 after the treatment. Furthermore, the expression profiles of these genes were also changed with the treatment time of SA (Figure 9).

A total of nine proteins related to ROS clearance pathway were detected based on the proteomic sequencing (Figure 10). On day 2 after the treatment with blue light irradiation, the expression levels of GPX, MDAR, and PRDX increased, whereas the expression levels of CAT, SOD, and PRDX decreased (Figure 10). On day 7 of blue light irradiation, the expression levels of ferritin and GPX were slightly increased, and the expression of PRDX was significantly increased. However, no significant effect of the addition of SA was observed on the expression of related proteins in the ROS clearance pathway. With the addition of SA, the genes related to ROS clearance pathway were significantly down-regulated at the transcriptional level, whereas no significant changes were observed at the proteomic level.

2.6. DEGs and DEPs Associated with the Cell Wall Biosynthesis

When *H. pluvialis* is cultured in stressed environments, the cell walls of the microalgae are thickened to enhance the capability of resisting adverse factors [45]. Callose, cellulose, and pectins are fundamental structural components of the cell wall and are synthesized with the involvement of many enzymes including callose synthase (CalS), endoglucanase (BcsZ), and cellulose synthase (BcsA) [44]. In our study, a total of 6 unigenes encoding CalS, 1 unigene encoding BcsA, and 15 unigenes encoding BcsZ were identified. The expression levels of genes encoding CalS and BcsZ were decreased significantly on day 2 of the treatment with blue light irradiation compared with white light (Figure 11A). On day 7 of blue light irradiation, the expression levels of *Cals*, *bcsZ*, and *bcsA* were significantly higher than those under white light. The addition of SA did not significantly affect the expression levels of genes encoding CalS and BcsZ but significantly down-regulated the expression of the genes encoding CalS on day 7 of the treatment. The proteomic data showed that the expression of BcsZ was decreased significantly on day 2 after the treatment with blue light irradiation compared with white light. These results were consistent with those derived from the transcriptomic data. Compared with day 0, the expression of BcsZ was significantly increased on both days 2 and 7 of the treatment with blue light irradiation, indicating that the expression of BcsZ was less affected by blue light than by white light

(Figure 11A). These results were consistent with those derived from the proteomic data (Figure 11B).

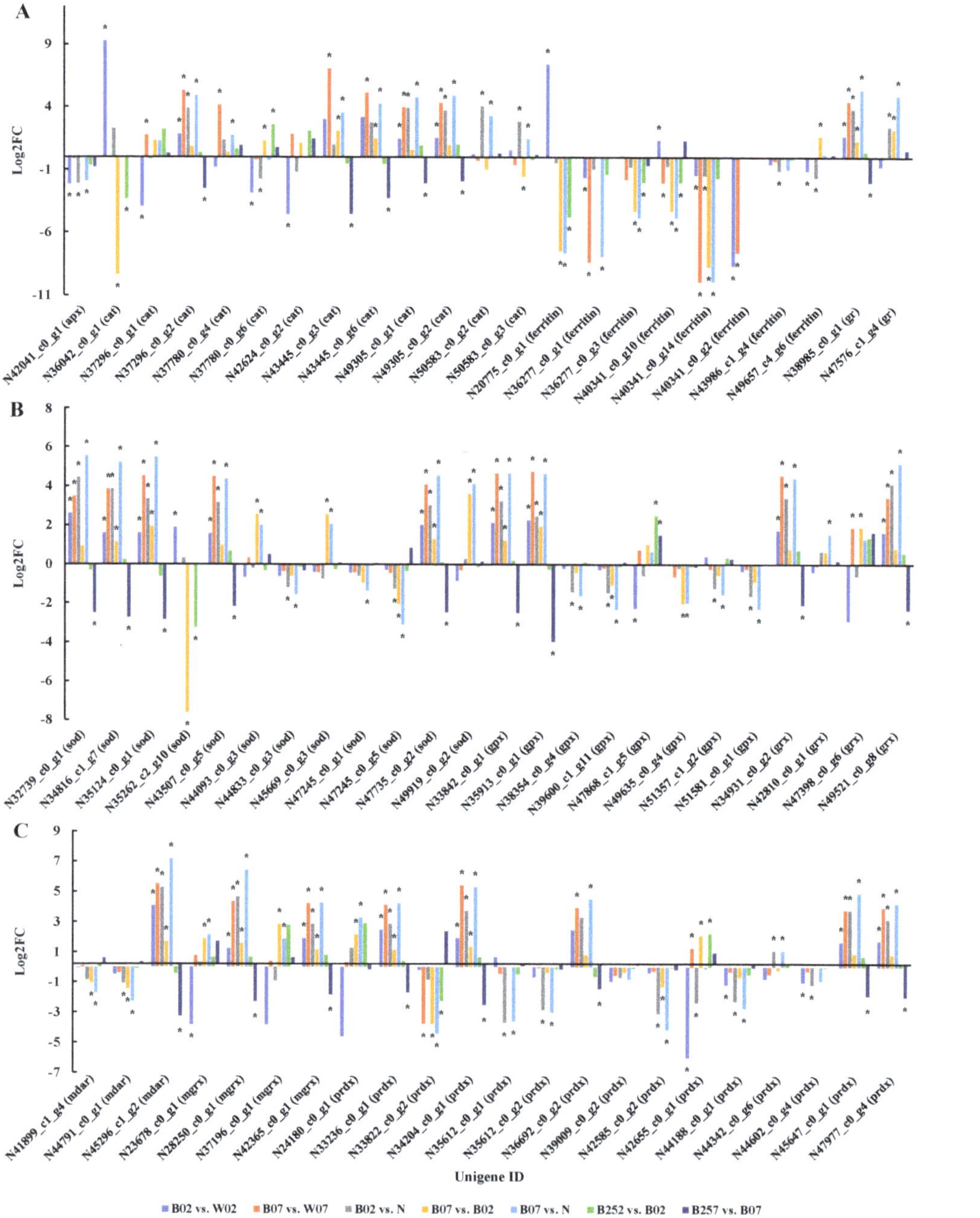

Figure 9. Expression profiles of genes—*apx*, *cat*, *ferritin*, and *gr* (**A**); *sod*, *gpx*, and *grx* (**B**); and *mdar*, *mgrx*, and *prdx* (**C**)—involved in the generation of ROS revealed in the pairwise comparisons of *Haematococcus pluvialis* based on transcriptomic analysis. *—indicates significant up-regulation ($p < 0.05$ and Log2FC ≥ 1) or significant down-regulation ($p < 0.05$ and Log2FC ≤ -1).

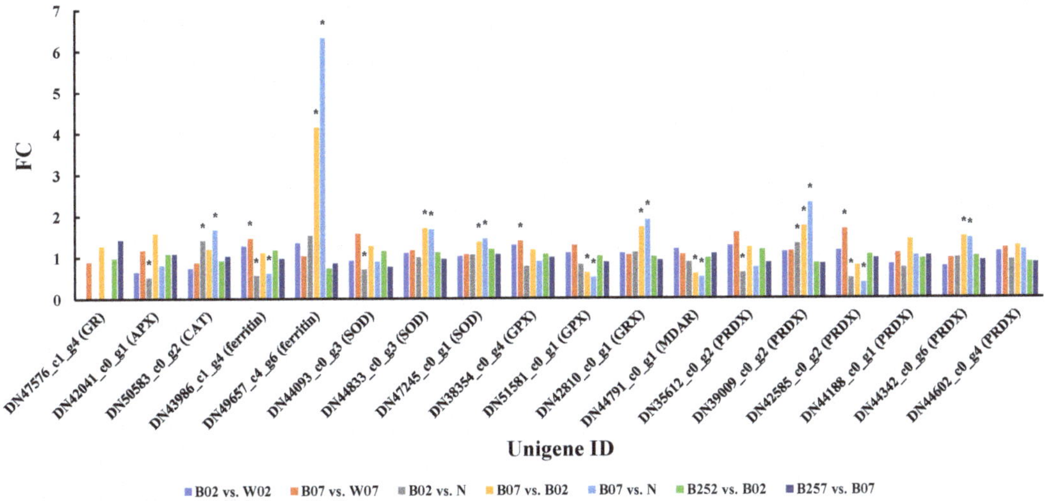

Figure 10. Expression profiles of proteins (i.e., APX, CAT, ferritin, GPX, GR, GRX, MDAR, PRDX, and SOD) involved in the generation of ROS revealed in the pairwise comparisons of *Haematococcus pluvialis* based on proteomic analysis. *—indicates significant up-regulation ($p < 0.05$ and FC ≥ 1.5) or significant down-regulation ($p < 0.05$ and FC ≤ 0.67).

2.7. DEGs Associated with the Blue Light Receptors

Plants utilize photoreceptors to regulate growth and development under light of different wavelengths. Both cryptochrome (CRY) and phototropin (PHOT) are blue light receptors, with conserved signal transduction in different plants. Members of eukaryotic CRYs include plant CRYs (pCRY), animal CRYs (aCRY), and *Drosophila*, *Arabidopsis*, *Synechocystis*, and Human (DASH) CRYs (dashCRY) [46,47]. Four CRYs (i.e., one pCRY, one aCRY, and two dashCRYs) were identified in *C. reinhardtii* [48] and seven CRYs (i.e., one pCRY, one aCRY, and five dashCRYs) were found in *Volvox carteri* [49]. The gene encoding pCRY contains multiple copies in higher plants but a single copy in eukaryotic green algae. To investigate the molecular mechanism regulating the astaxanthin accumulation in response to the blue LED, the CRYs were chosen for further transcriptomic analysis. In our study, seven CRYs were identified in *H. pluvialis* (Table 1). As the exposure time of blue light increased, the expression levels of genes encoding dashCRYs and CRYs were significantly decreased on day 2 and day 7 compared with those of day 0. Two genes encoding PHOT showed different expression patterns, with one showing no significant change and the other significantly down-regulated. The expression of one gene encoding dashCRY was significantly decreased from day 2 to day 7. On day 7 of the treatment with blue light exposure, the expression levels of genes encoding all blue light receptors were decreased compared with those of white light exposure. No significant changes in the expression of the genes encoding dashCRYs were observed with the addition of SA.

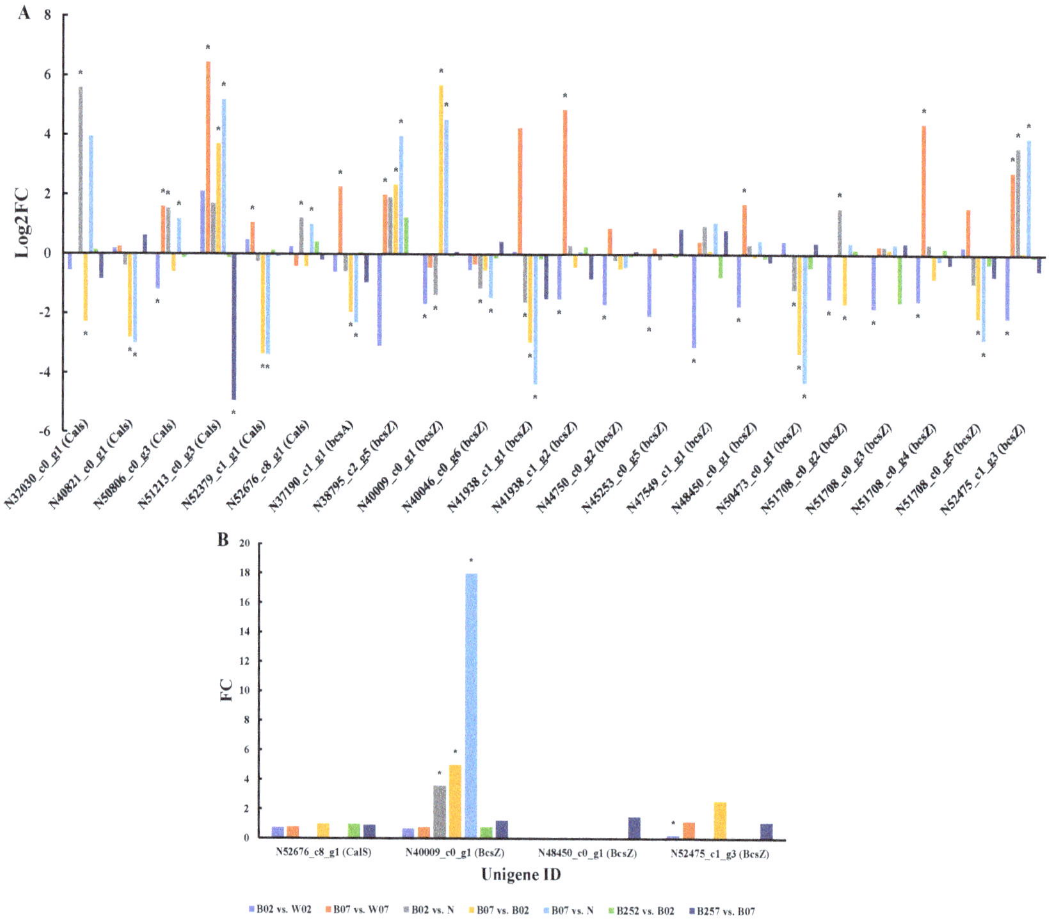

Figure 11. Expression profiles of genes *bcsA*, *bcsZ*, and *Cals* (**A**) and proteins BcsZ and CalS (**B**) involved in the cell wall biosynthesis revealed in the pairwise comparisons of *Haematococcus pluvialis* based on transcriptomic and proteomic analyses, respectively. *—indicates significant up-regulation ($p < 0.05$ and Log2FC ≥ 1; $p < 0.05$ and FC ≥ 1.5) or significant down-regulation ($p < 0.05$ and Log2FC ≤ -1; $p < 0.05$ and FC ≤ 0.67).

Table 1. Expression of genes encoding three classes of blue light receptors identified in different pairwise comparisons of *Haematococcus pluvialis*.

Class	Unigene ID	B02 vs. W02	B07 vs. W07	B02 vs. N	B07 vs. B02	B07 vs. N	B252 vs. B02	B257 vs. B07
dashCRY	N40695_c3_g2	−0.47	−0.12	−0.54	−0.39	−0.70	−0.17	0.38
dashCRY	N43038_c0_g10	−0.74	−0.49	−1.48	−0.09	−1.34	0.48	0.45
dashCRY	N49636_c1_g1	0.35	−0.33	−4.15	−1.54	−5.46	0.03	1.01
PHOT	N44442_c1_g2	−0.15	−0.03	0.32	−0.48	0.07	0.30	−0.12
PHOT	N50230_c0_g1	−0.32	−0.34	−0.93	−0.68	−1.38	0.14	0.02
CRY	N44179_c1_g4	0.00	−0.41	−1.40	−0.65	−1.83	0.16	−0.03
CRY	N50367_c0_g2	0.21	−0.11	−1.58	−0.97	−2.33	−0.20	0.06

2.8. Identification of Transcription Factors

Transcription factors (TFs) are important for the growth, development, and reproduction of organisms [50]. Understanding the metabolic response of TFs to the treatments with blue light and SA helps further investigate the molecular mechanism regulating both the astaxanthin and lipid synthesis. Plants improve their tolerance to a variety of stresses with the involvement of TFs, which in turn are affected by a variety of molecular factors [51]. In our study, a total of 54 TF families were identified in *H. pluvialis* based on the transcripts assembled de novo using iTAK. The top 10 TF families included C2H2 with 94 unigenes, TRAF (90 unigenes), SNF2 (75 unigenes), C3H (67 unigenes), GNAT (59 unigenes), SET (54 unigenes), MYB (48 unigenes), zn-clus (46 unigenes), MYB-related (44 unigenes), and HMG (42 unigenes). Results of the comparative transcriptomic analysis of the control and the experimental groups of *H. pluvialis* treated with blue light irradiation revealed a total of 503 putative genes assigned in 47 families encoding TFs in response to blue light exposure (Table S4). Compared with white light, the expression levels of genes encoding two TF families (i.e., SNF2 and zn-clus) were significantly increased on day 2 of the treatment with blue light irradiation.

Varied compositions of the dominant TFs were also revealed under different treatments of *H. pluvialis* (Table 2). For example, there was a lack of GNAT and MYB families of TFs on day 2 and day 7 after the treatment with blue light, respectively. On day 7 with the addition of SA, the TFs of the jumonji family were not identified. Studies have shown that these TFs play important roles in various biological processes. For example, zinc finger proteins (e.g., C2H2) regulate various cellular processes [52], TRAF induces the activation of several kinase cascades—eventually leading to the activation of signal transduction pathways and regulation of various cellular processes [53], and SNF2 regulates the structure of chromatin and participates in biological processes, such as cell differentiation and immune response [54]. As the potential regulators in the astaxanthin synthesis of *H. pluvialis* under blue light, the functions of these families of TFs identified in our study warrant further investigations based on combined genetic, molecular, and biochemical approaches.

Table 2. Top 10 transcription factor (TF) families with number of genes significantly expressed in different pairwise comparisons of *Haematococcus pluvialis*. Symbol "–" indicates the TFs not included in the pairwise comparisons.

Transcription Factor	B02 vs. W02	B07 vs. W07	B257 vs. B07
C2H2	45	30	44
C3H	20	20	16
GNAT	–	13	13
HMG	17	12	16
Jumonji	13	10	–
MYB	21	–	16
MYB-related	18	12	21
SET	13	17	16
SNF2	21	41	38
TRAF	38	57	50
zn-clus	31	43	40

3. Discussion

3.1. Synthesis of Astaxanthin and Fatty Acids under Blue Light Irradiation

FPPS is a key enzyme in isoprenoid biosynthesis [55], providing the precursors (i.e., sesquiterpene) for carotenoids. Our results showed that the expression of *fpps* was significantly regulated on the second day of blue light irradiation. Then, the flow of precursors of astaxanthin synthesis was reduced. Both PSY and PDS are the rate-limiting enzymes in the carotenoid biosynthesis pathway of *H. pluvialis* [56,57]. Furthermore, both BKT and CRTR-B are involved in the synthesis of astaxanthin in *H. pluvialis* [58]. Therefore, these four enzymes are essential for the synthesis of astaxanthin in *H. pluvialis*. Previous studies showed that, compared with white light of 150 μmol photons $m^{-2}s^{-1}$, blue light of

70 μmol photons $m^{-2}s^{-1}$ significantly increased the expression of *psy*, *lcy*, *bkt*, and *crtR-B* in *H. pluvialis* [9]. Our results showed that blue light reduced the expression of genes involved in astaxanthin synthesis in early stage of blue light irradiation, probably due to the different light intensities, suggesting that high-intensity blue light was not conducive for astaxanthin synthesis. It was expected that the carotenogenesis was correlated with the patterns of gene and protein expressions revealed in the current study. Future complementary studies are necessary to investigate the contents of astaxanthin and other important metabolites produced in *H. pluvialis*, i.e., carotenoids, chlorophylls, amino acids (i.e., glutamate), and lipids. Moreover, both the qRT-PCR and genetic knockout experiments are required to verify the expression patterns and functions of genes and proteins involved in astaxanthin biosynthesis pathway with the goals to explore the molecular mechanisms underlying the production of astaxanthin in *H. pluvialis* treated with high-intensity blue light and SA.

Previous studies showed that high light irradiation enhanced the synthesis of fatty acids in *H. pluvialis*, and the gene expression was significantly up-regulated with the increase in treatment time of high light irradiation [3,29]. However, our results showed that on the second day of the treatment with blue light irradiation, both gene and protein expressions involved in the production of acetyl-CoA and NADPH were significantly down-regulated, and the genes and proteins directly involved in the fatty acid synthesis were also significantly down-regulated, ultimately leading to a decrease in fatty acid synthesis. However, no significant changes were observed in these genes on day 7 of the treatment with blue light irradiation. Compared with day 7, the shorter treatment time (day 2) of blue light irradiation showed a greater effect on the production of astaxanthin of *H. pluvialis*, specifically resulting in the significant down-regulation of genes involved in astaxanthin synthesis (Figure 3). These results were consistent with those of the changes in gene and protein expressions involved in the biosynthesis of fatty acids (Figures 10 and 11). Because astaxanthin is esterified with fatty acids, the content of fatty acids is regulated with the change of astaxanthin content [59].

ROS burst in *H. pluvialis* was observed at the beginning of the treatment of light irradiation [44]. It is important for the cells of *H. pluvial is* to remove a large amount of ROS, which is also required in the production of astaxanthin. However, our results showed that many key genes involved in astaxanthin synthesis were significantly down-regulated on the second day of treatment with blue light irradiation, and genes in the fatty acid synthesis pathway were also significantly down-regulated, indicating that in the early stage of irradiation (day 0), the effect of blue light on astaxanthin production by *H. pluvialis* is less than that of white light. In the later stage of irradiation (day 7), blue light increased the astaxanthin production compared with white light. In the early stage of treatment (day 2), SA showed no significant effect on astaxanthin and fatty acid synthesis, while in the later stage (day 7), gene expression was significantly down regulated, probably because SA could regulate a variety of plant metabolic processes and the production of a variety of secondary metabolites to protect plants from abiotic stresses [22]. SA may also activate other defense pathways. For example, studies have shown that high light intensity and SA treatment for two days causes significant down-regulation of genes involved in astaxanthin and fatty acid synthesis [33], suggesting that *H. pluvialis* activates other metabolic processes to resist excessive ROS to adapt to the stressed environment. These results are consistent with our findings, showing that the gene expression was down-regulated on the 2nd day of blue light irradiation and on the 7th day of SA treatment.

3.2. Scavenging of Reactive Oxygen Species under Blue Light Irradiation

In general, organisms synthesize 5-ALA in one of the two ways, i.e., the C4 pathway and the C5 pathway. The C4 pathway mainly exists in animals and fungi, while the C5 pathway is utilized by plants, algae, and bacteria [60]. The 5-ALA is the precursor of chlorophyll, heme, and vitamin B12 biosynthesis pathways, with chlorophyll and heme synthesis pathways sharing several intermediate products. Our results showed that expression levels were decreased in genes (i.e., *hemL*, *hemB*, *hemC*, *hemD*, and *hemF*) involved in the shared portions between chlorophyll and heme synthesis pathways, whereas the expression of *ho* was significantly increased under blue light conditions (Figure 8). As the cofactor of catalase and many types of peroxidases, the increase in heme content removes excessive ROS produced in cells. Although the oxidation effect of the free heme damages cells under oxidative stress, cells develop a variety of protection mechanisms [61]. For example, studies have shown that heme oxygenase is involved in the antioxidant mechanism by catalyzing the degradation of heme to produce biliverdin, which shows antioxidant effect. The other two end products derived from biliverdin, i.e., CO and Fe/FTH, have also shown cytoprotective effect [62]. Compared with white light, the expression of heme oxygenase in blue light treatment group was significantly increased, leading to the increase in heme degradation products and the production of antioxidants. However, the expression of *ho* was decreased significantly on day 7 after the treatment with SA (Figure 8). As an important signal molecule, SA regulates a variety of oxidative stress responses to improve the efficiency of antioxidant system in plants. In *H. pluvialis*, a large amount of astaxanthin is generated to reduce the damage caused by the detrimental environment, complementing, or ultimately replacing, the antioxidant system of the microalgae. Therefore, *H. pluvialis* protects itself from the adverse effects of stress by accumulating more astaxanthin [29]. Our previous studies demonstrated that the addition of SA increased the astaxanthin content in *H. pluvialis* [4], suggesting that astaxanthin complements the antioxidant effect of heme degradation products and reduces the expression of *ho*. Similarly, in the blue light treatment groups, the expression of genes involved in astaxanthin synthesis (i.e., *psy*, *pds*, *lcyb*, and *crtr-b*) was also significantly down-regulated (Figure 6), whereas the expression of heme oxygenase was significantly up-regulated (Figure 8), suggesting that heme degradation products help maintain the intracellular homeostasis in *H. pluvialis*.

Previous studies showed that scavenging ROS was essential for the survival of *H. pluvialis* under high light intensity [63]. Our results showed that the expression levels of most genes related to ROS in *H. pluvialis* treated with blue light were significantly higher than those in white light, indicating that the damage of blue light to *H. pluvialis* was higher than that of white light (Figure 9). Previous studies showed that APX, CAT, GRX, GPX, and PRDX removed hydrogen peroxide [64], while SOD scavenged superoxide to form hydrogen peroxide [65]. Studies have shown that the antioxidant systems in different species of plants respond differently to the treatment with SA. For example, the activities of ascorbic APX and SOD were enhanced, and the activities of CAT were reduced in corn, whereas the activities of antioxidant enzymes CAT, peroxidase (POX), SOD, and glutathione reductase were decreased in rice [66]. Moreover, the level of H_2O_2 and the activities of antioxidant enzymes CAT, POX, and SOD were enhanced [66]. Our results showed that the treatment with SA led to the significant down-regulation of many genes involved in the ROS clearance pathway (Figure 9), suggesting the overall adjustment of resistance to stress in *H. pluvialis*. Furthermore, our previous studies showed that the addition of SA increased the content of astaxanthin in *H. pluvialis* [4]. These results suggested that SA reduced the expression of ROS scavenging genes, resulting in the production of a large amount of astaxanthin to maintain the homeostasis in cells of *H. pluvialis*.

3.3. Cell Wall Synthesis Inhibited in the Early Stage of the Treatment with Blue light Irradiation

Studies have shown that plants develop resistance to light stress of high intensity by regulating the composition of cell wall [67], while the light intensity and quality can affect the structure of cell wall [68]. For example, light regulates the sugar content of cell wall in peas [69]. Furthermore, studies have revealed that blue light showed a rapid inhibitory effect on the cell wall synthesis, while red light showed a long-term inhibitory effect in the pea seedlings [70]. Compared with white light, genes involved in cell wall synthesis were significantly down-regulated on day 2 but then significantly up-regulated on day 7 of the treatment with blue light irradiation, indicating that blue light showed less effect on *H. pluvialis* than that of white light in the early stage of blue light irradiation (day 2) but greater effect of long-term irradiation (day 7) on *H. pluvialis* than that of white light.

These results were consistent with those derived from the proteomic data (Figure 11B). Furthermore, these results also suggested that the blue light inhibited the cell wall synthesis of *H. pluvialis* in the early stage of the treatment with blue light irradiation.

3.4. Signal Transduction of Blue Light Receptors

PHOT is activated by blue light to participate in physiological processes such as chloroplast movement and plant phototropism [71]. PHOT contains both the serine/threonine kinase binding domains at the C-terminal and two light, oxygen, and voltage (LOV) domains at the N-terminal [72]. Exposure of LOV to blue light causes cysteine residues in the protein to covalently bind to the isopyrazine ring to generate conduction signal [73]. In *Arabidopsis*, PHOT regulates the transcripts in response to high-intensity blue light, though PHOT is not largely involved in transcriptional regulation [74,75]. However, PHOT causes significant changes in expression of genes involved in chlorophyll and carotenoid biosynthesis in *C. reinhardtii*, whereas the low-intensity blue light leads to a significant increase in the expression of *pds* [71]. Furthermore, the expression of *pds* is inhibited as the expression of *phot* is inhibited [71]. In our study, the expression of *phot* was decreased, but not significantly, under blue light, compared with white light. In the chlorophyll synthesis pathway (Figure 8), most genes were significantly down-regulated on the second day but not on day 7 of the treatment with blue light irradiation. In the astaxanthin synthesis pathway (Figure 6), most genes were significantly down-regulated on day 2 of the treatment with blue light irradiation, while no significant changes were observed on day 7.

The C-terminal domain of CRY is responsible for providing the interaction sites for DNA or downstream signal proteins, while the function of the N-terminal domain is to combine the second chromophore as an optical collection antenna [76]. Based on knockdown mutants, studies have shown that CRY regulated the transcriptional levels of various genes under blue light, caused by its oxidized fad chromophore. Under the blue light irradiation, the C-terminal domain interacts directly with downstream signal proteins, e.g., ubiquitin ligase. The photolyase homologous region (PHR) of cryptochrome can form homodimers, which are also essential for signal transduction [77].

The cryptochrome (CPH1) protein of *C. reinhardtii* undergoes photoinduced degradation under white, red, and blue lights [48]. Similar results were also observed in our study, showing that the blue light irradiation of high intensity significantly reduced the expression of *cry* in *H. pluvialis*. Furthermore, CPH1 also functions as a transcription regulator to regulate gene expression and induce the expression of both chlorophylls a and b binding protein and *hemL* [48]. Our results showed that the expression of *hemL* was significantly down-regulated on the second day of the treatment with blue light irradiation, while no significant down-regulation was observed on day 7, and *cry* was not significantly down-regulated under high-intensity blue light.

Plants respond to light through photoreceptors. Our study showed that the expression of blue light photoreceptors (i.e., PHOT and CRY) were not significantly down-regulated under blue and white light irradiations of high intensity, suggesting that in *H. pluvialis*, the metabolic pathways are not affected by the content of photoreceptors but by the signal transduction of photoreceptors. For example, ROS at high concentrations causes damage in cells but functions as signal molecules at low concentrations. Activated CRY can produce trace ROS as signal molecules in response to light. Furthermore, the high content of CRY causes the increased production of ROS in response to light in *Arabidopsis* [78]. Our results showed that, compared with normal light intensity, high-intensity blue light significantly inhibited the expression of CRYs, while the expression of PHOTs was decreased on day 7 of the treatment of high-intensity irradiation. Studies have revealed significant up-regulation of CRY and PHOT in *Dunaliella salina* treated with blue light irradiation of 50 μmol photons $m^{-2}s^{-1}$, whereas no significant changes were observed with the treatment with blue light with an increased intensity of 150 μmol photons $m^{-2}s^{-1}$ [79]. This is probably because that low-intensity blue light was sufficient enough to activate the photoreceptors [80]. It is speculated that, after the light intensity increases to a certain level, the effect of light quality on the function of blue light receptors is reduced.

4. Materials and Methods

4.1. Algal Materials

Haematococcus pluvialis strain 712 (FACHB-712) obtained from the Freshwater Algae Culture Collection at the Institute of Hydrobiology (FACHB) was cultured in 5000 mL beakers, containing BBM medium without aeration, under a light intensity of 25 μmol photons $m^{-2}s^{-1}$ with a photoperiod of 12 h/12 h light/dark at 22 °C. When the cell growth reached the logarithmic phase (10^5 cells mL^{-1}), the culture was evenly divided into 9 aliquots each of 1000 mL. Samples were divided into 3 groups, including the control group grown under constant white light of 150 μmol photons $m^{-2}s^{-1}$ at 22 °C (W0 group), the group treated with constant blue light of 150 μmol photons $m^{-2}s^{-1}$ at 22 °C (B0 group), and the group pre-treated with SA (2.5 mg/L) under constant blue light of 150 μmol photons $m^{-2}s^{-1}$ at 22 °C (B25 group). The selection of the concentration of SA (2.5 mg/L) was based on the previous study [81] and our pre-experiments, which showed that the production of astaxanthin was increased with reduced concentration of SA. Each group contained three replicates.

4.2. Transcriptome Sequencing

The three groups of microalgal samples were collected on day 0 (under white light) and after treatment on both 48 h (day 2) and 168 h (day 7) of blue light and SA. The sample collected on day 0 was named N. Other samples were named by group names and treatment times (day). Specifically, B02 and B07 represented samples treated with blue light irradiation for 2 and 7 days without SA, respectively. W02 and W07 represented samples treated with white light irradiation for 2 and 7 days without SA, respectively. B252 and B257 represented samples treated with blue light irradiation for 2 and 7 days with the addition of SA (2.5 mg/mL), respectively. Total RNA was extracted from *H. pluvialis* (10 mL) using Trizol Reagent (Invitrogen, Carlsbad, CA, USA). The algal sample (30~50 mg) was first ground in liquid nitrogen (less than 100 mg), then 1000 μL Trizol Reagent was added, and grinding continued until homogenized. The homogenate was transferred to a small tube and set still for 10 min before being centrifuged for 5 min at 12,000 g and 4 °C. The supernatant was collected and set still for 5 min, then added with 200 μL of chloroform, shaken vigorously for 15 sec, and kept for 3 min at room temperature. Then, the sample was centrifuged for 15 min at 12,000 g and 4 °C. The water phase was transferred, twice the volume of isopropanol was added, it was vortexed briefly, and kept at −20 °C overnight. Then, the sample was centrifuged for 15 min at 12,000 g and 4 °C. The supernatant was removed, and the sample was added to 1 mL ethanol (70%) and centrifuged for 5 min at 7500 g and 4 °C. The supernatant was removed, and the ethanol was removed from the sample, which was set to dry for 7 min at room temperature. Then, 50 μL of water was added to the centrifuge tube to dissolve the sample, which was then vortexed to mix well. The RNA integrity was assessed using 1% formaldehyde denaturing agarose gel electrophoresis. Then, mRNA was purified from total RNA by deploy-t with magnetic beads. The RNA was processed by mRNASeq sample preparation kit (Illumina, San Diego, CA, USA) to establish the sequencing system. Transcriptome sequencing was performed using Illumina Hiseq 4000 (LC Sciences, Houston, TX, USA).

4.3. De Novo Assembly and Unigene Annotation

Cutadapt [82] and in-house Perl scripts at Hangzhou Lianchuan Biotechnology Co. Ltd. (LC Sciences, Houston, TX, USA), were used to remove adaptor sequences and low-quality bases. The sequence quality was evaluated by FastQC (https://www.bioinformatics.babraham.ac.uk/projects/fastqc/, accessed on 16 March 2020). The de novo assembly of transcriptome from *H. pluvialis* was performed using Trinity2.4.0 [83]. All assembled unigenes were aligned against the non-redundant (nr) protein (http://www.ncbi.nlm.

nih.gov/, accessed on 16 March 2020), Gene Ontology (GO) (http://www.geneontology.org/, accessed on 16 March 2020), SwissProt (http://www.expasy.ch/sprot/, accessed on 16 March 2020), Kyoto Encyclopedia of Genes and Genomes (KEGG) (http://www.genome.jp/kegg/, accessed on 16 March 2020), and EggNOG (http://eggnogdb.embl.de/, accessed on 16 March 2020) databases using DIAMOND [84] with a threshold of E value < 0.00001. Transcription factors were identified using iTAK [85]. The selection of de novo assembly over genome-based analysis was based on two lines of evidence. First, the results of several biosynthetic pathways based on the genome-based analysis showed variations in the expression patterns of genes from those based on the de novo assembly. Therefore, we further investigated the cause of these inconsistent results by conducting the qRT-PCR analysis of genes involved in these biosynthetic pathways. Our results of qRT-PCR confirmed the expression patterns of genes based on the de novo assembly, whereas the results based on the genome-based assembly were largely not confirmed by qRT-PCR analysis. Second, our transcriptomic sequencing results based on the alignment program HISAT2 (https://daehwankimlab.github.io/hisat2/, accessed on 16 March 2020) revealed the overall alignment rate of only ~50% with the reference genome of *H. pluvialis* deposited at the NCBI database (GenBank accession number GCA_011766145.1 of BioProject PRJDB8952).

4.4. Differentially Expressed Unigenes Analysis

Salmon [86] was used to assess expression levels of unigenes by calculating the Transcripts Per Million (TPM) [87]. The differentially expressed unigenes were identified based on log2 (fold change) ≥ 1 or log2 (fold change) ≤ -1 with statistical significance ($p < 0.05$) evaluated by R package edgeR [88].

4.5. Proteome Sampling, Sequencing, and Bioinformatics

The microalgal cells of *H. pluvialis* were added with SDT lysate, homogenized, and broken, and boiled in hot water at 100 °C for 10 min. The samples were centrifuged for 15 min at 14,000 g with the supernatant collected. BCA method was used for protein quantification [89]. Filter aided proteome preparation (FASP) was used for enzymatic hydrolysis of proteins [90]. The peptides were desalted by C18 cartridge. After freeze-drying, 0.1% formic acid solution was added for re-dissolution, and the peptides were quantified. The peptide mixture was graded by Agilent 1260 infinity II HPLC system (Agilent Technologies, Santa Clara, CA, USA).

The proteomic sequencing was performed by Hangzhou Lianchuan Biotechnology Co., Ltd. (LC Sciences, Houston, TX, USA). The original mass spectrometry data were combined and analyzed by Spectronaut Pulsar X (version 12, Biognosys AG, Schlieren, Switzerland). TransDecoder (https://github.com/TransDecoder/TransDecoder/, accessed on 16 March 2020) was used to predict the transcript of unigenes to obtain the protein sequences and to establish the spectrum database by Spectronaut Pulsar X (version 12, Biognosys AG, Schlieren, Switzerland). Trypsin enzymolysis was set to allow two missing sites, with the following searching parameters included: (1) fixed modification: carbamidomethyl (c); (2) variable modification: oxidation (m) oxidation; and (3) N-terminal acetylation of acetyl (protein N-term) protein. A total of 1 μg of peptides from each sample was added with iRT peptides to mix and then to inject for separation using Nano-LC. Data were analyzed by online electrospray tandem mass spectrometry. The complete liquid–mass tandem system contained both the liquid system based on the Easy nLC system (Thermo Fisher Scientific, Waltham, MA, USA) and the mass spectrometry system using the Orbitrap Exploris 480 (Thermo Fisher Scientific, Waltham, MA, USA). Buffer solution A solution was 0.1% formic acid aqueous solution and B solution was 0.1% formic acid acetonitrile aqueous solution (80% acetonitrile). The sample was separated by a non-linearly increasing gradient in an analytical column (Nano Technology Column, 18 cm C18 column, with 1.9 μm C18 Resin, Catalog Number 26350-3) at a flow rate of 300 nL/min: 0–1 min, 2% B to 8% B; 1–99 min, 8% B to 27% B; 99–113 min, 27% B to 35% B; 113–117 min, 35% B to 100% B; and 117–120 min, 100% B. The electrospray voltage was set to 2056 V.

Raw data of DIA were processed and analyzed by Spectronaut 14.6 (Biognosys AG, Schlieren, Switzerland) with default settings with the Retention time prediction type set to dynamic iRT. Data extraction was determined by Spectronaut Pulsar X based on the extensive mass calibration. Spectronaut 14.6 was used to determine the ideal extraction window dynamically depending on iRT calibration and gradient stability. Q-value (FDR) cutoff on precursor and protein level was applied at 1%. Decoy generation was set to mutated which was similar to scrambled but only applied a random number of AA position swamps (min = 2, max = length/2). All selected precursors passing the filters were used for quantification. MS2 interference was used to remove most interfering fragment ions except for the 3 least interfering ones. The average top 3 filtered peptides which passed the 1% Q-value cutoff were used to calculate the major group quantities. Differentially expressed proteins were filtered based on Student's *t*-test ($p < 0.05$ and fold change > 1.5).

GO annotation based on unigenes was performed using Blast2GO [91]. KEGG enrichment analysis was performed based on the KEGG automatic annotation server (KAAS) [92].

4.6. Statistical Analysis

All experiments were performed in triplicates to ensure reproducibility. Data were presented as mean ± standard deviation (SD). The paired-samples t-test was performed using ANOVA with the statistically significant difference set at $p \leq 0.05$.

5. Conclusions

In this study, *H. pluvialis* was exposed to high-intensity blue light and SA, resulting in alterations in gene and protein expressions from those derived from exposure to white light of high intensity. Using transcriptomic and proteomic sequencing, we further analyzed the molecular responses in astaxanthin synthesis, fatty acid synthesis, heme synthesis, ROS scavenging pathway, and cell wall synthesis of *H. pluvialis* treated with blue light and SA. On day 2 of the treatment with blue light irradiation, genes showing significant changes in expression were mainly annotated in the pathways of carbon metabolism, while on day 7, the genes of significant regulation were mainly enriched in the pathways related to proteins. The addition of SA led to the enrichment of genes in disease-related pathways. Because low-intensity blue light caused the up-regulation of genes involved in astaxanthin synthesis, while high-intensity blue light led to the down-regulation of genes related to astaxanthin synthesis, indicating that the promotion of astaxanthin synthesis by blue light is probably related to light intensity. Therefore, it is proposed that, in order to improve the production of astaxanthin in *H. pluvialis*, high-intensity white light should be utilized in *H. pluvialis* cultures because the application of high-intensity blue light and SA decreased the expression of genes involved in the astaxanthin biosynthetic pathway, suggesting that the application of high-intensity blue light and SA was not beneficial for maximizing the astaxanthin production. In the cell wall synthesis pathway, the expression of proteins and genes was significantly decreased on day 2 of the treatment with blue light irradiation, whereas the gene expression was significantly increased on day 7. In *H. pluvialis*, the high-intensity blue light led to the significant up-regulation of ROS scavenging genes. Furthermore, blue light also reduced the expression of genes related to fatty acid synthesis. Our results showed that SA significantly down-regulated the ROS scavenging genes and genes involved in heme degradation pathway, suggesting that SA reduced the expression of enzymes related to ROS scavenging, thus promoting the production of astaxanthin by *H. pluvialis*. Moreover, different families of TFs were identified in *H. pluvialis*, induced with different treatment times of blue light irradiation. Our study provides novel insights into the molecular response of *H. pluvialis* to blue light and SA treatments based on both transcriptomic and proteomic analyses.

Supplementary Materials: The following are available online at https://www.mdpi.com/article/10.3390/md20010001/s1, Table S1: Summary of RNA-seq data in *Haematococcus pluvialis* based on three groups of samples on day 0, day 2, and day 7 after the treatments of white light, blue light, and salicylic acid (SA); Table S2: KEGG enrichment and GO annotation of unigenes identified in *Haematococcus pluvialis;* Table S3: Transcripts and proteins derived from significantly expressed unigenes in *Haematococcus pluvialis;* Table S4: Classification of differentially expressed genes annotated to transcription factors in *Haematococcus pluvialis*, Figure S1: Species distribution of annotated unigenes of *Haematococcus pluvialis*, Figure S2: The length distribution of unigenes of *Haematococcus pluvialis*, Figure S3: Enriched GO terms in three categories (i.e., biological process, cellular component, and molecular function) based on the unigenes with significant changes revealed in the pairwise comparisons of B02 vs. W02 (**A**), B07 vs. W07 (**B**), and B257 vs. B07 (**C**) of *Haematococcus pluvialis*, Figure S4: Enriched KEGG metabolic pathways based on the unigenes with significant changes revealed in the pairwise comparisons of B02 vs. W02 (**A**), B07 vs. W07 (**B**), and B257 vs. B07 (**C**) in *Haematococcus pluvialis*.

Author Contributions: Conceptualization, C.M. and Z.G.; methodology, X.W., C.M., F.S., Z.G. and H.Z.; formal analysis, X.W., W.X., F.S., H.Z., K.C. and B.Z.; resources, C.M. and Z.G.; data curation, X.W., C.M., F.S. and Z.G.; writing—original draft preparation, X.W., C.M., Z.G. and H.Z.; writing—review and editing, X.W., C.M., X.W., F.S., K.C., B.Z., Z.G., H.Z. and C.Z.; project administration, C.M., F.S. and Z.G.; funding acquisition, Z.G. and C.M. All authors have read and agreed to the published version of the manuscript.

Funding: This study was supported by the National Natural Science Foundation of China (31972815 and 42176124), the Natural Science Foundation of Shandong Province (ZR2019ZD17, ZR2020ZD23, and ZR2021MC051), the Scientific Research Fund of Binzhou Medical University (BY2021KYQD25 and BY2021KYQD28), and the Open Fund of Shandong Provincial Key Laboratory of Plant Stress (SPKLPS202001).

Institutional Review Board Statement: Not applicable.

Data Availability Statement: Raw Illumina sequencing data in this study have been deposited in the Sequence Read Archive (SRA) at the NCBI (http://www.ncbi.nlm.nih.gov/sra accessed on 16 March 2020) under accession number PRJNA766843. The mass spectrometry proteomics data have been deposited to the ProteomeXchange Consortium (http://proteomecentral.proteomexchange.org accessed on 16 March 2020) via the iProX (https://www.iprox.org/ accessed on 16 March 2020) partner repository with the dataset identifier PXD029109 or IPX0003602000.

Conflicts of Interest: The authors declare no conflict of interest.

References

1. Hussein, G.; Sankawa, U.; Goto, H.; Matsumoto, K.; Watanabe, H. Astaxanthin, a carotenoid with potential in human health and nutrition. *J. Nat. Prod.* **2006**, *69*, 443–449. [CrossRef] [PubMed]
2. Guerin, M.; Huntley, M.E.; Olaizola, M. Haematococcus astaxanthin: Applications for human health and nutrition. *Trends Biotechnol.* **2003**, *21*, 210–216. [CrossRef]
3. He, B.; Hou, L.; Dong, M.; Shi, J.; Huang, X.; Ding, Y.; Cong, X.; Zhang, F.; Zhang, X.; Zang, X. Transcriptome analysis in *Haematococcus pluvialis*: Astaxanthin induction by high light with acetate and Fe^{2+}. *Int. J. Mol. Sci.* **2018**, *19*, 175. [CrossRef]
4. Gao, Z.; Meng, C.; Zhang, X.; Xu, D.; Miao, X.; Wang, Y.; Yang, L.; Lv, H.; Chen, L.; Ye, N. Induction of salicylic acid (SA) on transcriptional expression of eight carotenoid genes and astaxanthin accumulation in *Haematococcus pluvialis*. *Enzyme Microb. Technol.* **2012**, *51*, 225–230. [CrossRef]
5. Jung, S.M.; Jeon, J.Y.; Park, T.H.; Yoon, J.H.; Lee, K.S.; Shin, H.W. Enhancing production-extraction and antioxidant activity of astaxanthin from *Haematococcus pluvialis*. *J. Environ. Biol.* **2019**, *40*, 924–931. [CrossRef]
6. Domínguez, A.; Pereira, S.; Otero, A. Does *Haematococcus pluvialis* need to sleep? *Algal Res.* **2019**, *44*, 101722. [CrossRef]
7. Pick, U.; Zarka, A.; Boussiba, S.; Davidi, L. A hypothesis about the origin of carotenoid lipid droplets in the green algae *Dunaliella* and *Haematococcus*. *Planta* **2019**, *249*, 31–47. [CrossRef]
8. Boussiba, S. Carotenogenesis in the green alga *Haematococcus pluvialis*: Cellular physiology and stress response. *Physiol. Plant.* **2000**, *108*, 111–117. [CrossRef]
9. Ma, R.; Thomas-Hall, S.R.; Chua, E.T.; Alsenani, F.; Eltanahy, E.; Netzel, M.E.; Netzel, G.; Lu, Y.; Schenk, P.M. Gene expression profiling of astaxanthin and fatty acid pathways in *Haematococcus pluvialis* in response to different LED lighting conditions. *Bioresour. Technol.* **2018**, *250*, 591–602. [CrossRef]

10. Lee, C.; Ahn, J.W.; Kim, J.B.; Kim, J.Y.; Choi, Y.E. Comparative transcriptome analysis of *Haematococcus pluvialis* on astaxanthin biosynthesis in response to irradiation with red or blue LED wavelength. *World J. Microbiol. Biotechnol.* **2018**, *34*, 96. [CrossRef]
11. Duarte, J.H.; Costa, J.A.V. Blue light emitting diodes (LEDs) as an energy source in *Chlorella fusca* and *Synechococcus nidulans* cultures. *Bioresour. Technol.* **2018**, *247*, 1242–1245. [CrossRef] [PubMed]
12. Oldenhof, H.; Zachleder, V.; Van Den Ende, H. Blue- and red-light regulation of the cell cycle in *Chlamydomonas reinhardtii* (Chlorophyta). *Eur. J. Phycol.* **2006**, *41*, 313–320. [CrossRef]
13. Mohsenpour, S.F.; Willoughby, N. Luminescent photobioreactor design for improved algal growth and photosynthetic pigment production through spectral conversion of light. *Bioresour. Technol.* **2013**, *142*, 147–153. [CrossRef] [PubMed]
14. Han, S.I.; Kim, S.; Lee, C.; Choi, Y.E. Blue-Red LED wavelength shifting strategy for enhancing beta-carotene production from halotolerant microalga, *Dunaliella salina*. *J. Microbiol.* **2019**, *57*, 101–106. [CrossRef] [PubMed]
15. Katsuda, T.; Lababpour, A.; Shimahara, K.; Katoh, S. Astaxanthin production by *Haematococcus pluvialis* under illumination with LEDs. *Enzyme Microb. Technol.* **2004**, *35*, 81–86. [CrossRef]
16. Pang, N.; Fu, X.; Fernandez, J.S.M.; Chen, S. Multilevel heuristic LED regime for stimulating lipid and bioproducts biosynthesis in *Haematococcus pluvialis* under mixotrophic conditions. *Bioresour. Technol.* **2019**, *288*, 121525. [CrossRef]
17. Park, E.K.; Lee, C.G. Astaxanthin production by *Haematococcus pluvialis* under various light intensities and wavelengths. *J. Microbiol. Biotechnol.* **2001**, *11*, 1024–1030.
18. Lababpour, A.; Hada, K.; Shimahara, K.; Katsuda, T.; Katoh, S. Effects of nutrient supply methods and illumination with blue light emitting diodes (LEDs) on astaxanthin production by *Haematococcus pluvialis*. *J. Biosci. Bioeng.* **2004**, *98*, 452–456. [CrossRef]
19. Paliwal, C.; Jutur, P.P. Dynamic allocation of carbon flux triggered by task-specific chemicals is an effective non-gene disruptive strategy for sustainable and cost-effective algal biorefineries. *Chem. Eng. J.* **2021**, *418*, 129413. [CrossRef]
20. Klessig, D.F.; Choi, H.W.; Dempsey, D.M.A. Systemic acquired resistance and salicylic acid: Past, present, and future. *Mol. Plant-Microbe Interact.* **2018**, *31*, 871–888. [CrossRef]
21. Miura, K.; Tada, Y. Regulation of water, salinity, and cold stress responses by salicylic acid. *Front. Plant Sci.* **2014**, *5*, 4. [CrossRef]
22. Khan, M.I.R.; Fatma, M.; Per, T.S.; Anjum, N.A.; Khan, N.A. Salicylic acid-induced abiotic stress tolerance and underlying mechanisms in plants. *Front. Plant Sci.* **2015**, *6*, 462. [CrossRef]
23. Fan, M.; Sun, X.; Xu, N.; Liao, Z.; Li, Y.; Wang, J.; Fan, Y.; Cui, D.; Li, P.; Miao, Z. Integration of deep transcriptome and proteome analyses of salicylic acid regulation high temperature stress in *Ulva prolifera*. *Sci. Rep.* **2017**, *7*, 11052. [CrossRef]
24. Corina Vlot, A.; Dempsey, D.M.A.; Klessig, D.F. Salicylic acid, a multifaceted hormone to combat disease. *Annu. Rev. Phytopathol.* **2009**, *47*, 177–206. [CrossRef]
25. Gao, Z.; Li, Y.; Wu, G.; Li, G.; Sun, H.; Deng, S.; Shen, Y.; Chen, G.; Zhang, R.; Meng, C.; et al. Transcriptome analysis in *Haematococcus pluvialis*: Astaxanthin induction by salicylic acid (SA) and jasmonic acid (JA). *PLoS ONE* **2015**, *10*, e0140609. [CrossRef]
26. Gao, Z.; Miao, X.; Zhang, X.; Wu, G.; Guo, Y.; Wang, M.; Li, B.; Li, X.; Gao, Y.; Hu, S.; et al. Comparative fatty acid transcriptomic test and iTRAQ-based proteomic analysis in *Haematococcus pluvialis* upon salicylic acid (SA) and jasmonic acid (JA) inductions. *Algal Res.* **2016**, *17*, 277–284. [CrossRef]
27. Raman, V.; Ravi, S. Effect of salicylic acid and methyl jasmonate on antioxidant systems of *Haematococcus pluvialis*. *Acta Physiol. Plant.* **2011**, *33*, 1043–1049. [CrossRef]
28. Zhekisheva, M.; Boussiba, S.; Khozin-Goldberg, I.; Zarka, A.; Cohen, Z. Accumulation of Triacylglycerols in *Haematococcus Pluvialis* is Correlated with that of Astaxanthin Esters. *J. Phycol.* **2002**, *38*, 40–41. [CrossRef]
29. Gwak, Y.; Hwang, Y.S.; Wang, B.; Kim, M.; Jeong, J.; Lee, C.G.; Hu, Q.; Han, D.; Jin, E. Comparative analyses of lipidomes and transcriptomes reveal a concerted action of multiple defensive systems against photooxidative stress in *Haematococcus pluvialis*. *J. Exp. Bot.* **2014**, *65*, 4317–4334. [CrossRef]
30. Chen, G.; Wang, B.; Han, D.; Sommerfeld, M.; Lu, Y.; Chen, F.; Hu, Q. Molecular mechanisms of the coordination between astaxanthin and fatty acid biosynthesis in *Haematococcus pluvialis* (Chlorophyceae). *Plant J.* **2015**, *81*, 95–107. [CrossRef]
31. Morozova, T.V.; Anholt, R.R.H.; Mackay, T.F.C. Phenotypic and transcriptional response to selection for alcohol sensitivity in *Drosophila melanogaster*. *Genome Biol.* **2007**, *8*, R231. [CrossRef]
32. Luo, Q.; Bian, C.; Tao, M.; Huang, Y.; Zheng, Y.; Lv, Y.; Li, J.; Wang, C.; You, X.; Jia, B.; et al. Genome and Transcriptome Sequencing of the Astaxanthin-Producing Green Microalga, *Haematococcus pluvialis*. *Genome Biol. Evol.* **2019**, *11*, 166–173. [CrossRef] [PubMed]
33. Hu, Q.; Huang, D.; Li, A.; Hu, Z.; Gao, Z.; Yang, Y.; Wang, C. Transcriptome-based analysis of the effects of salicylic acid and high light on lipid and astaxanthin accumulation in *Haematococcus pluvialis*. *Biotechnol. Biofuels* **2021**, *14*, 82. [CrossRef]
34. Wang, X.; Miao, X.; Chen, G.; Cui, Y.; Sun, F.; Fan, J.; Gao, Z.; Meng, C. Identification of microRNAs involved in astaxanthin accumulation responding to high light and high sodium acetate (NaAC) stresses in *Haematococcus pluvialis*. *Algal Res.* **2021**, *54*, 102179. [CrossRef]
35. Li, K.; Cheng, J.; Lu, H.; Yang, W.; Zhou, J.; Cen, K. Transcriptome-based analysis on carbon metabolism of *Haematococcus pluvialis* mutant under 15% CO_2. *Bioresour. Technol.* **2017**, *233*, 313–321. [CrossRef] [PubMed]
36. Du, F.; Hu, C.; Sun, X.; Zhang, L.; Xu, N. Transcriptome analysis reveals the promoting effect of trisodium citrate on astaxanthin accumulation in *Haematococcus pluvialis* under high light condition. *Aquaculture* **2021**, *543*, 736978. [CrossRef]
37. Fang, L.; Zhang, J.; Fei, Z.; Wan, M. Astaxanthin accumulation difference between non-motile cells and akinetes of *Haematococcus pluvialis* was affected by pyruvate metabolism. *Bioresour. Bioprocess.* **2020**, *7*, 5. [CrossRef]

38. Nakada, T.; Ota, S. What is the correct name for the type of *Haematococcus* Flot. (Volvocales, Chlorophyceae)? *Taxon* **2016**, *65*, 343–348. [CrossRef]
39. Cui, J.; Yu, C.; Zhong, D.B.; Zhao, Y.; Yu, X. Melatonin and calcium act synergistically to enhance the coproduction of astaxanthin and lipids in *Haematococcus pluvialis* under nitrogen deficiency and high light conditions. *Bioresour. Technol.* **2020**, *305*, 123069. [CrossRef]
40. Pereira, S.; Otero, A. *Haematococcus pluvialis* bioprocess optimization: Effect of light quality, temperature and irradiance on growth, pigment content and photosynthetic response. *Algal Res.* **2020**, *51*, 102027. [CrossRef]
41. Liang, C.; Zhang, W.; Zhang, X.; Fan, X.; Xu, D.; Ye, N.; Su, Z.; Yu, J.; Yang, Q. Isolation and expression analyses of methyl-d-erythritol 4-phosphate (MEP) pathway genes from *Haematococcus pluvialis*. *J. Appl. Phycol.* **2016**, *28*, 209–218. [CrossRef]
42. Li, W.; Wu, H.; Li, M.; San, K.Y. Effect of NADPH availability on free fatty acid production in *Escherichia coli*. *Biotechnol. Bioeng.* **2018**, *115*, 444–452. [CrossRef]
43. Zhang, Z.; Sun, D.; Mao, X.; Liu, J.; Chen, F. The crosstalk between astaxanthin, fatty acids and reactive oxygen species in heterotrophic *Chlorella zofingiensis*. *Algal Res.* **2016**, *19*, 178–183. [CrossRef]
44. Hu, C.; Cui, D.; Sun, X.; Shi, J.; Song, L.; Li, Y.; Xu, N. Transcriptomic analysis unveils survival strategies of autotrophic *Haematococcus pluvialis* against high light stress. *Aquaculture* **2019**, *513*, 734430. [CrossRef]
45. Shah, M.M.R.; Liang, Y.; Cheng, J.J.; Daroch, M. Astaxanthin-producing green microalga *Haematococcus pluvialis*: From single cell to high value commercial products. *Front. Plant Sci.* **2016**, *7*, 531. [CrossRef] [PubMed]
46. Heijde, M.; Zabulon, G.; Corellou, F.; Ishikawa, T.; Brazard, J.; Usman, A.; Sanchez, F.; Plaza, P.; Martin, M.; Falciatore, A.; et al. Characterization of two members of the cryptochrome/photolyase family from *Ostreococcus tauri* provides insights into the origin and evolution of cryptochromes. *Plant Cell Environ.* **2010**, *33*, 1614–1626. [CrossRef] [PubMed]
47. Petroutsos, D. Chlamydomonas Photoreceptors: Cellular Functions and Impact on Physiology. In *Chlamydomonas: Biotechnology and Biomedicine*, 1st ed.; Hippler, D., Ed.; Springer: Cham, Switzerland, 2017; Volume 31, pp. 1–19.
48. Reisdorph, N.A.; Small, G.D. The CPH1 gene of *Chlamydomonas reinhardtii* encodes two forms of cryptochrome whose levels are controlled by light-induced proteolysis. *Plant Physiol.* **2004**, *134*, 1546–1554. [CrossRef]
49. Böhm, M.; Kreimer, G. Orient in the World with a Single Eye: The Green Algal Eyespot and Phototaxis. *Prog. Bot.* **2020**, *82*, 259–304.
50. Udvardi, M.K.; Kakar, K.; Wandrey, M.; Montanari, O.; Murray, J.; Andriankaja, A.; Zhang, J.Y.; Benedito, V.; Hofer, J.M.I.; Chueng, F.; et al. Legume transcription factors: Global regulators of plant development and response to the environment. *Plant Physiol.* **2007**, *144*, 538–549. [CrossRef]
51. Joshi, R.; Wani, S.H.; Singh, B.; Bohra, A.; Dar, Z.A.; Lone, A.A.; Pareek, A.; Singla-Pareek, S.L. Transcription factors and plants response to drought stress: Current understanding and future directions. *Front. Plant Sci.* **2016**, *7*, 1029. [CrossRef] [PubMed]
52. Han, G.; Lu, C.; Guo, J.; Qiao, Z.; Sui, N.; Qiu, N.; Wang, B. C2H2 Zinc Finger Proteins: Master Regulators of Abiotic Stress Responses in Plants. *Front. Plant Sci.* **2020**, *11*, 115. [CrossRef]
53. Park, H.H. Structure of TRAF Family: Current Understanding of Receptor Recognition. *Front. Immunol.* **2018**, *9*, 1999. [CrossRef] [PubMed]
54. Li, M.; Xia, X.; Tian, Y.; Jia, Q.; Liu, X.; Lu, Y.; Li, M.; Li, X.; Chen, Z. Mechanism of DNA translocation underlying chromatin remodelling by Snf2. *Nature* **2019**, *567*, 409–413. [CrossRef] [PubMed]
55. Szkopińska, A.; Płochocka, D. Farnesyl diphosphate synthase; regulation of product specificity. *Acta Biochim. Pol.* **2005**, *52*, 45–55. [CrossRef]
56. Gong, M.; Bassi, A. Carotenoids from microalgae: A review of recent developments. *Biotechnol. Adv.* **2016**, *34*, 1396–1412. [CrossRef]
57. Steinbrenner, J.; Sandmann, G. Transformation of the green alga *Haematococcus pluvialis* with a phytoene desaturase for accelerated astaxanthin biosynthesis. *Appl. Environ. Microbiol.* **2006**, *72*, 7477–7484. [CrossRef]
58. Meng, C.X.; Teng, C.Y.; Jiang, P.; Qin, S.; Tseng, C.K. Cloning and characterization of β-carotene ketolase gene promoter in *Haematococcus pluvialis*. *Acta Biochim. Biophys. Sin.* **2005**, *37*, 270–275. [CrossRef] [PubMed]
59. Hu, C.; Cui, D.; Sun, X.; Shi, J.; Xu, N. Primary metabolism is associated with the astaxanthin biosynthesis in the green algae *Haematococcus pluvialis* under light stress. *Algal Res.* **2020**, *46*, 101768. [CrossRef]
60. Kang, Z.; Wang, Y.; Gu, P.; Wang, Q.; Qi, Q. Engineering *Escherichia coli* for efficient production of 5-aminolevulinic acid from glucose. *Metab. Eng.* **2011**, *13*, 492–498. [CrossRef]
61. Gozzelino, R.; Jeney, V.; Soares, M.P. Mechanisms of cell protection by heme Oxygenase-1. *Annu. Rev. Pharmacol. Toxicol.* **2010**, *50*, 323–354. [CrossRef]
62. Kikuchi, G.; Yoshida, T.; Noguchi, M. Heme oxygenase and heme degradation. *Biochem. Biophys. Res. Commun.* **2005**, *338*, 558–567. [CrossRef] [PubMed]
63. Han, D.; Wang, J.; Sommerfeld, M.; Hu, Q. Susceptibility and protective mechanisms of motile and non motile cells of *Haematococcus pluvialis* (chlorophyceae) to photooxidative stress. *J. Phycol.* **2012**, *48*, 693–705. [CrossRef] [PubMed]
64. Anjum, N.A.; Sharma, P.; Gill, S.S.; Hasanuzzaman, M.; Khan, E.A.; Kachhap, K.; Mohamed, A.A.; Thangavel, P.; Devi, G.D.; Vasudhevan, P.; et al. Catalase and ascorbate peroxidase—Representative H_2O_2-detoxifying heme enzymes in plants. *Environ. Sci. Pollut. Res.* **2016**, *23*, 19002–19029. [CrossRef]
65. Prasad, A.K.; Mishra, P.C. Mechanism of Action of Sulforaphane as a Superoxide Radical Anion and Hydrogen Peroxide Scavenger by Double Hydrogen Transfer: A Model for Iron Superoxide Dismutase. *J. Phys. Chem. B* **2015**, *119*, 7825–7836. [CrossRef]

66. Hayat, Q.; Hayat, S.; Irfan, M.; Ahmad, A. Effect of exogenous salicylic acid under changing environment: A review. *Environ. Exp. Bot.* **2010**, *68*, 14–25. [CrossRef]
67. Le Gall, H.; Philippe, F.; Domon, J.M.; Gillet, F.; Pelloux, J.; Rayon, C. Cell wall metabolism in response to abiotic stress. *Plants* **2015**, *4*, 112–166. [CrossRef]
68. Sasidharan, R.; Chinnappa, C.C.; Staal, M.; Elzenga, J.T.M.; Yokoyama, R.; Nishitani, K.; Voesenek, L.A.C.J.; Pierik, R. Light quality-mediated petiole elongation in arabidopsis during shade avoidance involves cell wall modification by xyloglucan endotransglucosylase/hydrolases. *Plant Physiol.* **2010**, *154*, 978–990. [CrossRef] [PubMed]
69. Masuda, Y.; Kamisaka, S.; Yanagisawa, H.; Suzuki, Y. Effect of Light on Growth and Metabolic Activities in Pea Seedlings I. Changes in Cell Wall Polysaccharides during Growth in the Dark and in the Light. *Biochem. Physiol. Pflanz.* **1981**, *176*, 23–34. [CrossRef]
70. Kigel, J.; Cosgrove, D.J. Photoinhibition of stem elongation by blue and red light: Effects on hydraulic and cell wall properties. *Plant Physiol.* **1991**, *95*, 1049–1056. [CrossRef]
71. Im, C.S.; Eberhard, S.; Huang, K.; Beck, C.F.; Grossman, A.R. Phototropin involvement in the expression of genes encoding chlorophyll and carotenoid biosynthesis enzymes and LHC apoproteins in *Chlamydomonas reinhardtii*. *Plant J.* **2006**, *48*, 1–16. [CrossRef]
72. Fedorov, R.; Schlichting, I.; Hartmann, E.; Domratcheva, T.; Fuhrmann, M.; Hegemann, P. Crystal structures and molecular mechanism of a light-induced signaling switch: The Phot-LOV1 domain from *Chlamydomonas reinhardtii*. *Biophys. J.* **2003**, *84*, 2474–2482. [CrossRef]
73. Freddolino, P.L.; Gardner, K.H.; Schulten, K. Signaling mechanisms of LOV domains: New insights from molecular dynamics studies. *Photochem. Photobiol. Sci.* **2013**, *12*, 1158–1170. [CrossRef] [PubMed]
74. Folta, K.M.; Kaufman, L.S. Phototropin 1 is required for high-fluence blue-light-mediated mRNA destabilization. *Plant Mol. Biol.* **2003**, *51*, 609–618. [CrossRef]
75. Ohgishi, M.; Saji, K.; Okada, K.; Sakai, T. Functional analysis of each blule light receptor, cry1, cry2, phot1, and phot2, by using combinatorial multiple mutants in *Arabidopsis*. *Proc. Natl. Acad. Sci. USA* **2004**, *101*, 2223–2228. [CrossRef] [PubMed]
76. Franz, S.; Ignatz, E.; Wenzel, S.; Zielosko, H.; Ngurah Putu, E.P.G.; Maestre-Reyna, M.; Tsai, M.D.; Yamamoto, J.; Mittag, M.; Essen, L.O. Structure of the bifunctional cryptochrome aCRY from *Chlamydomonas reinhardtii*. *Nucleic Acids Res.* **2018**, *46*, 8010–8022. [CrossRef]
77. Leopold, A.V.; Chernov, K.G.; Verkhusha, V.V. Optogenetically controlled protein kinases for regulation of cellular signaling. *Chem. Soc. Rev.* **2018**, *47*, 2454–2484. [CrossRef] [PubMed]
78. El-Esawi, M.; Arthaut, L.D.; Jourdan, N.; d'Harlingue, A.; Link, J.; Martino, C.F.; Ahmad, M. Blue-light induced biosynthesis of ROS contributes to the signaling mechanism of *Arabidopsis* cryptochrome. *Sci. Rep.* **2017**, *7*, 13875. [CrossRef]
79. Li, Y.; Cai, X.; Gu, W.; Wang, G. Transcriptome analysis of carotenoid biosynthesis in *Dunaliella salina* under red and blue light. *J. Oceanol. Limnol.* **2020**, *38*, 177–185. [CrossRef]
80. Matiiv, A.B.; Chekunova, E.M. Aureochromes—Blue Light Receptors. *Biochemistry* **2018**, *83*, 662–673. [CrossRef]
81. Gao, Z.; Meng, C.; Diao, X. The effect of extraneous salicylic acid on astaxanthin accumulation of alga *Haematoccus pluvialis*. *Fish. Sci.* **2007**, *7*, 377–380.
82. Martin, M. Cutadapt removes adapter sequences from high-throughput sequencing reads. *EMBnet J.* **2011**, *17*, 10–12. [CrossRef]
83. Grabherr, M.G.; Haas, B.J.; Yassour, M.; Levin, J.Z.; Thompson, D.A.; Amit, I.; Adiconis, X.; Fan, L.; Raychowdhury, R.; Zeng, Q.; et al. Full-length transcriptome assembly from RNA-Seq data without a reference genome. *Nat. Biotechnol.* **2011**, *29*, 644–652. [CrossRef]
84. Buchfink, B.; Xie, C.; Huson, D.H. Fast and sensitive protein alignment using DIAMOND. *Nat. Methods* **2014**, *12*, 59–60. [CrossRef] [PubMed]
85. Zheng, Y.; Jiao, C.; Sun, H.; Rosli, H.G.; Pombo, M.A.; Zhang, P.; Banf, M.; Dai, X.; Martin, G.B.; Giovannoni, J.J.; et al. iTAK: A Program for Genome-wide Prediction and Classification of Plant Transcription Factors, Transcriptional, Regulators, and Protein Kinases. *Mol. Plant* **2016**, *9*, 1667–1670. [CrossRef]
86. Patro, R.; Duggal, G.; Love, M.I.; Irizarry, R.A.; Kingsford, C. Salmon provides fast and bias-aware quantification of transcript expression. *Nat. Methods* **2017**, *14*, 417–419. [CrossRef] [PubMed]
87. Mortazavi, A.; Williams, B.A.; McCue, K.; Schaeffer, L.; Wold, B. Mapping and quantifying mammalian transcriptomes by RNA-Seq. *Nat. Methods* **2008**, *5*, 621–628. [CrossRef] [PubMed]
88. Robinson, M.D.; McCarthy, D.J.; Smyth, G.K. edgeR: A Bioconductor package for differential expression analysis of digital gene expression data. *Bioinformatics* **2009**, *26*, 139–140. [CrossRef] [PubMed]
89. Johnson, M. Protein Quantitation. *Mater. Methods* **2012**, *2*, 115. [CrossRef]
90. Wiśniewski, J.R.; Zougman, A.; Nagaraj, N.; Mann, M. Universal sample preparation method for proteome analysis. *Nat. Methods* **2009**, *6*, 359–362. [CrossRef]
91. Götz, S.; García-Gómez, J.M.; Terol, J.; Williams, T.D.; Nagaraj, S.H.; Nueda, M.J.; Robles, M.; Talón, M.; Dopazo, J.; Conesa, A. High-throughput functional annotation and data mining with the Blast2GO suite. *Nucleic Acids Res.* **2008**, *36*, 3420–3435. [CrossRef]
92. Moriya, Y.; Itoh, M.; Okuda, S.; Yoshizawa, A.C.; Kanehisa, M. KAAS: An automatic genome annotation and pathway reconstruction server. *Nucleic Acids Res.* **2007**, *35*, 182–185. [CrossRef] [PubMed]

marine drugs

Article

Immune Status and Hepatic Antioxidant Capacity of Gilthead Seabream *Sparus aurata* Juveniles Fed Yeast and Microalga Derived β-glucans

Bruno Reis [1,2,3,4,*], Ana Teresa Gonçalves [1,5], Paulo Santos [2,3], Manuel Sardinha [1,†], Luís E. C. Conceição [1], Renata Serradeiro [6], Jaume Pérez-Sánchez [7], Josep Calduch-Giner [7], Ulrike Schmid-Staiger [8], Konstantin Frick [9], Jorge Dias [1] and Benjamín Costas [2,3,*]

1 SPAROS Lda., Área Empresarial de Marim, Lote C, 8700-221 Olhão, Portugal; AnaGoncalves@sparos.pt (A.T.G.); manuelsardinha@sparos.pt (M.S.); LuisConceicao@sparos.pt (L.E.C.C.); jorgedias@sparos.pt (J.D.)
2 Centro Interdisciplinar de Investigação Marinha e Ambiental (CIIMAR), Universidade do Porto, Terminal de Cruzeiros de Leixões, Av. General Norton de Matos s/n, 4450-208 Matosinhos, Portugal; paulo.santos@ciimar.up.pt
3 Instituto de Ciências Biomédicas Abel Salazar (ICBAS-UP), Universidade do Porto, R. Jorge de Viterbo Ferreira 228, 4050-313 Porto, Portugal
4 Sorgal S.A., Estrada Nacional 109, Lugar da Pardala, 3880-728 São João de Ovar, Portugal
5 GreenCoLab—Associação Oceano Verde, Campus de Gambelas, Universidade do Algarve, 8005-139 Faro, Portugal
6 Riasearch, Rua do Farol, 131, Torrão do Lameiro, 3880-394 Ovar, Portugal; renataserradeiro@riasearch.pt
7 Nutrigenomics and Fish Growth Endocrinology Group, Institute of Aquaculture Torre de la Sal, IATS-CSIC, 12595 Castellón, Spain; jaime.perez.sanchez@csic.es (J.P.-S.); calduch@iats.csic.es (J.C.-G.)
8 Fraunhofer Institute for Interfacial Engineering and Biotechnology IGB, Innovation Field Algae Biotechnology—Development, Nobelstrasse 12, 70569 Stuttgart, Germany; ulrike.schmid-staiger@igb.fraunhofer.de
9 Institute of Interfacial Process Engineering and Plasma Technology, University of Stuttgart, Pfaffenwaldring 31, 70569 Stuttgart, Germany; konstantin.frick@igb.fraunhofer.de
* Correspondence: breis@ciimar.up.pt (B.R.); bcostas@ciimar.up.pt (B.C.); Tel.: +351-223-401-840 (B.R.); +351-223-401-838 (B.C.)
† Current affiliation: Biomar AS, Global RD Department, Havnegata 9, 7010 Trondheim, Norway.

Citation: Reis, B.; Gonçalves, A.T.; Santos, P.; Sardinha, M.; Conceição, L.E.C.; Serradeiro, R.; Pérez-Sánchez, J.; Calduch-Giner, J.; Schmid-Staiger, U.; Frick, K.; et al. Immune Status and Hepatic Antioxidant Capacity of Gilthead Seabream *Sparus aurata* Juveniles Fed Yeast and Microalga Derived β-glucans. *Mar. Drugs* **2021**, *19*, 653. https://doi.org/10.3390/md19120653

Academic Editor: Carlos Almeida

Received: 25 October 2021
Accepted: 17 November 2021
Published: 23 November 2021

Abstract: This work aimed to evaluate the effects of dietary supplementation with β-glucans extracted from yeast (*Saccharomyces cerevisiae*) and microalga (*Phaeodactylum tricornutum*) on gene expression, oxidative stress biomarkers and plasma immune parameters in gilthead seabream (*Sparus aurata*) juveniles. A practical commercial diet was used as the control (CTRL), and three others based on CTRL were further supplemented with different β-glucan extracts. One was derived from *S. cerevisiae* (diet MG) and two different extracts of 21% and 37% *P. tricornutum*-derived β-glucans (defined as Phaeo21 and Phaeo37), to give a final 0.06% β-glucan dietary concentration. Quadruplicate groups of 95 gilthead seabream (initial body weight: 4.1 ± 0.1 g) were fed to satiation three times a day for 8 weeks in a pulse-feeding regimen, with experimental diets intercalated with the CTRL dietary treatment every 2 weeks. After 8 weeks of feeding, all groups showed equal growth performance and no changes were found in plasma innate immune status. Nonetheless, fish groups fed β-glucans supplemented diets showed an improved anti-oxidant status compared to those fed CTRL at both sampling points (i.e., 2 and 8 weeks). The intestinal gene expression analysis highlighted the immunomodulatory role of Phaeo37 diet after 8 weeks, inducing an immune tolerance effect in gilthead seabream intestine, and a general down-regulation of immune-related gene expression. In conclusion, the results suggest that the dietary pulse administration of a *P. tricornutum* 37% enriched-β-glucans extract might be used as a counter-measure in a context of gut inflammation, due to its immune-tolerant and anti-oxidative effects.

Keywords: *Saccharomyces cerevisiae*; *Phaeodactylum tricornutum*; *Sparus aurata*; β-glucans; pulse feeding; immune tolerance

1. Introduction

Aquaculture is the fastest growing food-sector related industry, and as practices become more intensive, the risk of disease outbreaks increases accordingly [1,2]. In fact, animal health-related issues are nowadays the major constraint for aquaculture expansion and sustainability [3]. To date, one of the main strategies to cope with disease outbreaks in aquaculture has been the use of antibiotics. Although this issue has been mitigated in recent years with more restrictive legislation and regulations, antibiotics are still routinely used, leading to the emergence of new antibiotic-resistant bacteria [1]. In addition to vaccination, an alternative strategy to the use of antibiotics is the adoption of prophylactic measures through nutrition, such as the incorporation of immunostimulants and prebiotics in feeds to enhance fish disease resistance and general health [2,4,5].

Marine microalgae are a rich source of bioactive compounds [6] that are drawing increasing attention considering their use in different applications including functional feeds [7,8]. *Phaeodactylum tricornutum* is a marine diatom, unicellular brown microalgae rich in several health beneficial compounds such as β-glucans (BGs) [9–11]. BGs can be naturally found as cell wall components in bacteria, yeast, fungi, plants, micro- and macro-algae, and due to their promising biological activities, BGs have been extensively studied in vertebrates [2,12–14]. These polysaccharides can act as a prebiotic, enhancing the growth of commensal microbiota and by directly stimulating the innate immune system through interaction with specific cell receptors [4]. BGs bioactivity depends on their degree of branching, size and molecular structure [15]. However, those with higher biological activity show a common pattern: a repeating chain of (l-3)-linked β-D-glucopyranosyl units with randomly branched single β-D -glucopyranosyl units attached by l-6 or 1–4 linkages [2,15]. These repeating patterns, a feature shared with bacterial lipopolysaccharides (LPS), can be recognized by the host's cell pattern recognition receptors (PRR) and are termed pathogen-associated microbial patterns (PAMPs). Upon recognition. they can elicit an inflammatory response and activate the host's innate immune cells [12].

In mammals, dectin-1 is the best described BGs receptor, considered to be the most important for recognition and signal transduction. It is a C-type lectin receptor (CLR) which is predominantly expressed on cells from both the monocyte/macrophage and neutrophil lineages [12,16]. In a former study, European common carp (*Cyprinius carpio*) macrophages were activated with curdlan, a dectin-1-specific BG ligand in mammals, showing that immune modulatory effects in carp macrophages could be triggered by a member of the CLR family, although different from dectin-1 receptor [17]. In teleosts, the specific receptors involved in the recognition of BGs and consequent downstream signalling remain to be elucidated [14,18–20]. In contrast, the beneficial effects of BGs in fish innate immune response are well documented. Most of the studies focusing on fish showed that oral administration of BGs not only benefits innate immune response, such as the increase of phagocytic capacity, oxidative burst, lysozyme and complement activity [21–24], but also modulates immune gene expression in different organs [25–27]. However, the use of BG-rich microalgae cell extracts as feed supplements to modulate both the systemic and local immune response is still poorly explored. Therefore, the present study aimed to evaluate the effects of β-glucans extracted from microalga (*P. tricornutum*) and yeast (*Saccharomyces cerevisiae*), when applied as dietary supplements for juveniles of a valuable fish species for European aquaculture such as gilthead seabream (*Sparus aurata*).

2. Results

2.1. Growth Performance

Growth performance data are presented in Table 1. At the end of the trial (8 weeks), all fish showed similar final whole-body weight, regardless of dietary treatment. All groups showed similar feed conversion ratio (FCR) and relative growth rate (RGR) values (1.2 and 3.8% day^{-1}, respectively).

Table 1. Growth performance parameters in gilthead seabream juveniles after 8 weeks of feeding regimen. Data are the mean ± SEM (n = 4).

	8 Weeks			
DIETS	**CTRL**	**MG**	**Phaeo21**	**Phaeo37**
IBW	4.18 ± 0.04	4.12 ± 0.02	4.12 ± 0.05	4.15 ± 0.05
FBW	41.36 ± 0.81	42.48 ± 0.43	42.08 ± 0.53	41.93 ± 0.97
RGR	3.77 ± 0.03	3.83 ± 0.02	3.82 ± 0.03	3.8 ± 0.03
FCR	1.20 ± 0.01	1.19 ± 0.01	1.21 ± 0.02	1.19 ± 0.04

IBW: initial body weight (g); FBW: final body weight (g); RGR: relative growth rate (% average body weight/day) and FCR: feed conversion ratio.

2.2. Haematological Profile and Humoral Parameters

Peripheral cell dynamics were analysed at both sampling points (Table 2). The relative percentage of circulating lymphocytes increased in fish-fed Phaeo21 compared to CTRL and Phaeo37 groups after 2 weeks of feeding. In contrast, the same cell type showed decreased percentages in fish fed MG and Phaeo21 dietary treatments compared to those fed CTRL in the second (8 weeks) sampling. Furthermore, after 8 weeks, peripheral thrombocytes were higher in MG and Phaeo37 compared to CTRL. Monocytes and neutrophils numbers remained unaltered among dietary groups at both sampling points. Plasma humoral parameters (i.e., bactericidal activity, antiprotease activity and IgM) remained unchanged among dietary treatments at both 2 and 8 weeks (Table 3).

Table 2. Percentage values of peripheral blood leucocytes (thrombocytes, lymphocytes, monocytes and neutrophils) in gilthead seabream juveniles after 2 and 8 weeks of feeding regimen. Data are the mean ± SEM (n = 12).

	2 Weeks				8 Weeks			
DIETS	**CTRL**	**MG**	**Phaeo21**	**Phaeo37**	**CTRL**	**MG**	**Phaeo21**	**Phaeo37**
CELLS (%)								
Thrombocytes	65.2 ± 2.0	63.0 ± 2.2	60.3 ± 1.8	64.7 ± 1.6	71.2 ^{b} ± 2.6	78.5 ^{a} ± 1.4	76.4 ^{a,b} ± 2.5	81.0 ^{a} ± 0.8
Lymphocytes	24.1 ^{b} ± 1.5	28.3 ^{a,b} ± 1.9	30.7 ^{a} ± 1.4	24.8 ^{b} ± 1.3	18.4 ^{a} ± 2.5	13.0 ^{b} ± 1.0	13.1 ^{b} ± 2.6	13.4 ^{a,b} ± 0.8
Monocytes	5.3 ± 0.8	3.5 ± 0.7	4.4 ± 0.7	5.2 ± 0.4	3.4 ± 0.6	2.4 ± 0.7	3.1 ± 0.6	2.0 ± 0.5
Neutrophils	4.4 ± 1.0	4.1 ± 0.9	4.1 ± 0.7	4.8 ± 0.7	4.7 ± 0.4	5.6 ± 0.6	4.3 ± 0.7	3.6 ± 0.6

Different superscript letters indicate significant differences between diets ($p < 0.05$) within the same sampling point.

Table 3. Plasma immune parameters of gilthead seabream juveniles after 2 and 8 weeks feeding (antiprotease activity, bactericidal activity and immunoglobulin M). Data are the mean ± SEM (n = 12). Different letters indicate significant differences between dietary treatments ($p < 0.05$).

	2 Weeks				8 Weeks			
DIETS	**CTRL**	**MG**	**Phaeo21**	**Phaeo37**	**CTRL**	**MG**	**Phaeo21**	**Phaeo37**
Antiprotease act (%)	95.7 ± 0.6	96.0 ± 0.5	95.5 ± 0.7	95.4 ± 0.6	97.9 ± 0.2	98.1 ± 0.1	97.9 ± 0.2	98.2 ± 0.2
Bactericidal act (%)	45.0 ± 6.3	35.5 ± 4.0	40.3 ± 6.6	45.4 ± 4.4	53.3 ± 7.2	56.5 ± 5.8	57.4 ± 6.8	61.8 ± 4.8
IgM (OD 450 nm)	0.31 ± 0.03	0.32 ± 0.03	0.37 ± 0.05	0.29 ± 0.03	0.62 ± 0.05	0.53 ± 0.03	0.47 ± 0.05	0.55 ± 0.03

2.3. Oxidative Stress Biomarkers

Hepatic oxidative stress biomarkers showed significant differences at both sampling points (Figure 1). Lipid peroxidation decreased in seabream-fed MG and Phaeo37 diets compared to those fed CTRL at 2 weeks (early sampling), while the extent of lipid peroxidation was similar among groups after 8 weeks (final sampling). Catalase (CAT) and superoxide dismutase (SOD) activities were not affected by the dietary treatments at 2 weeks, however, CAT activity increased in Phaeo21 compared to Phaeo37 group at 8 weeks. SOD activity was enhanced in Phaeo37-fed fish compared to those fed CTRL and MG1000 dietary treatments. Total glutathione remained unaltered among dietary groups at both sampling points.

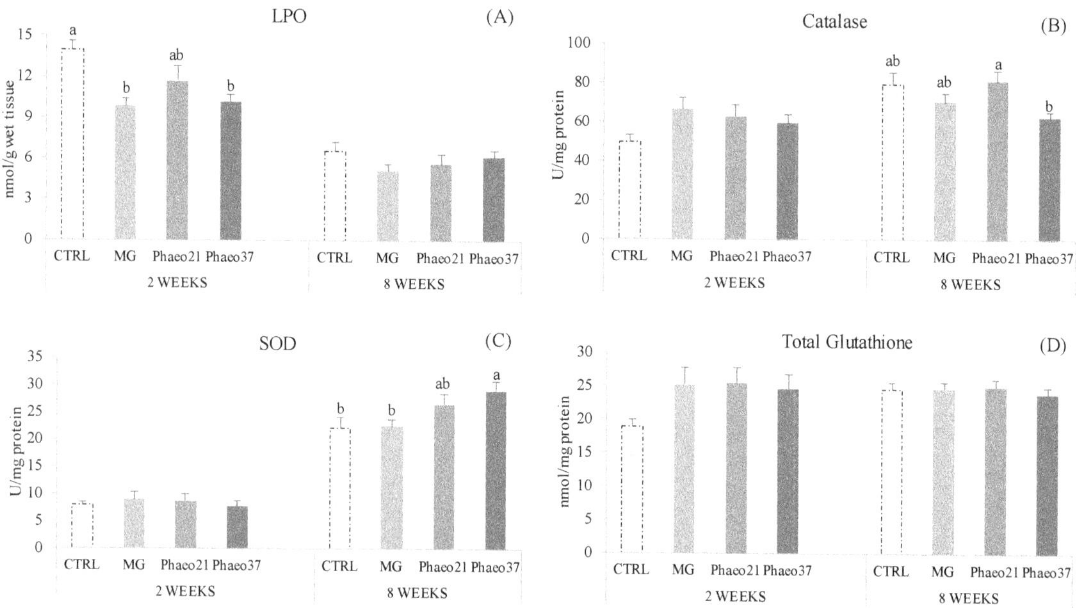

Figure 1. Liver oxidative stress biomarkers of gilthead seabream juveniles after 2 and 8 weeks feeding. Lipid peroxidation (LPO) (**A**); Catalase activity (CAT) (**B**); Superoxide dismutase activity (SOD) (**C**) and Total Glutathione (**D**). Data are the mean ± SEM (n = 12). Different lowercase letters indicate significant differences between dietary treatments ($p < 0.05$).

2.4. Multivariate Analysis from Physiological Parameters

An overall multivariate analysis combining raw data from haematological, humoral and hepatic oxidative stress biomarkers (using PCA-DA) was performed to discriminate the physiological effects caused by the experimental diets both at 2 and 8 weeks of feeding (Figure 2). The first two discriminant functions accounted for 95% of dataset variability at 2 weeks. Group discrimination was significant (Wilk's lambda = 0.3, $p = 0.01$) highlighting the differences between CTRL and BG groups ($p < 0.03$). This discrimination was loaded by lower lipid peroxidation, higher CAT and tGSH and to a lesser extent higher antiproteases activity in BG groups. At 8 weeks, groups were discriminated (Wilk's lambda = 0.2; $p < 0.001$) and the first two discriminant factors accounted for 84% of dataset variability. CTRL and MG dietary treatments were significantly discriminated from the Phaeo groups ($p < 0.04$) and this separation was loaded by a higher SOD activity and thrombocyte percentage in the last groups.

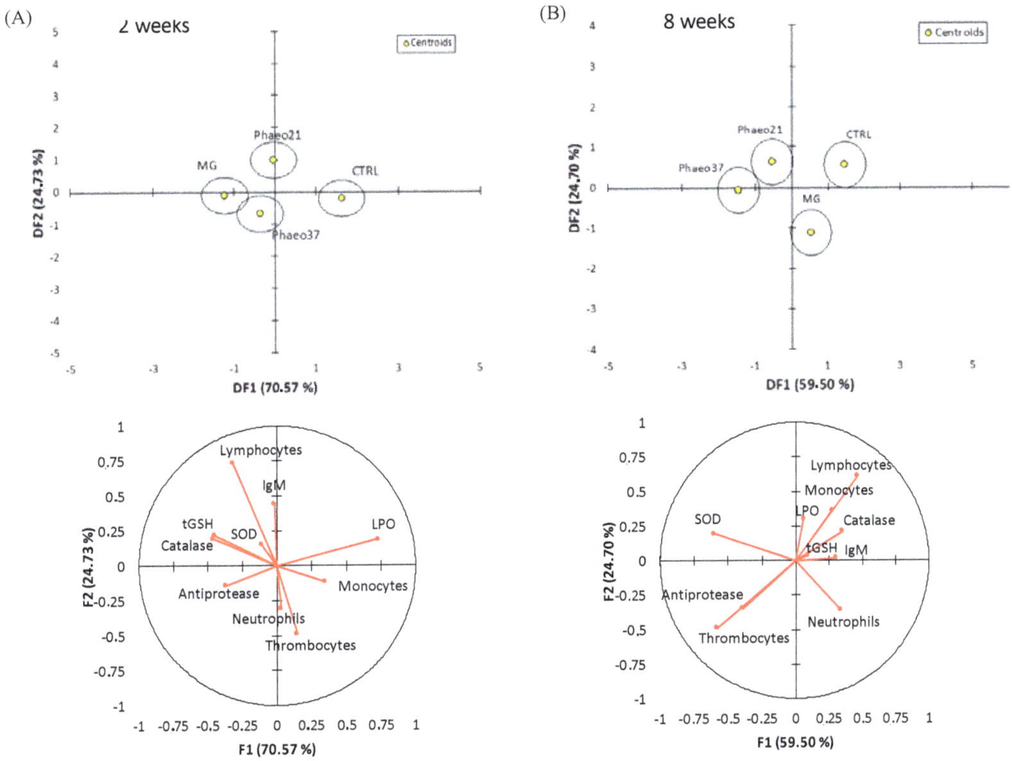

Figure 2. Discriminant analysis of experimental groups based on all physiological biomarkers analysed in the target tissues (yellow marker indicates the centroids of each group) and variables loads for DF1 and DF2 at 2 and 8 weeks. (**A**) 2 weeks; (**B**) 8 weeks.

2.5. Gene Expression Analysis

Different pathways represented in this gene array showed significant dietary effects in the proximal intestine. Proliferating cell nuclear antigen gene (*pcna*) was down-regulated in Phaeo37-fed fish at 2 weeks (Figure 3A). In contrast, different genes belonging to different molecular and cellular pathways were down-regulated in seabream-fed Phaeo37 after 8 weeks of feeding, in particular, fucolectin (*fcl*) (Figure 3D), as well as gap junction cx32.2 protein (*cx32.2*) (Figure 3B) genes. Moreover, the latter gene was down-regulated in fish fed MG as well as interleukin 10 (*il10*) (Figure 3C). Complete relative gene expression profile of the anterior intestine is provided as Supplementary Material (Tables S1 and S2). According to the Supplementary Tables, it was found that the mRNA expression levels of the other analysed genes showed no statistically significant differences among experimental groups.

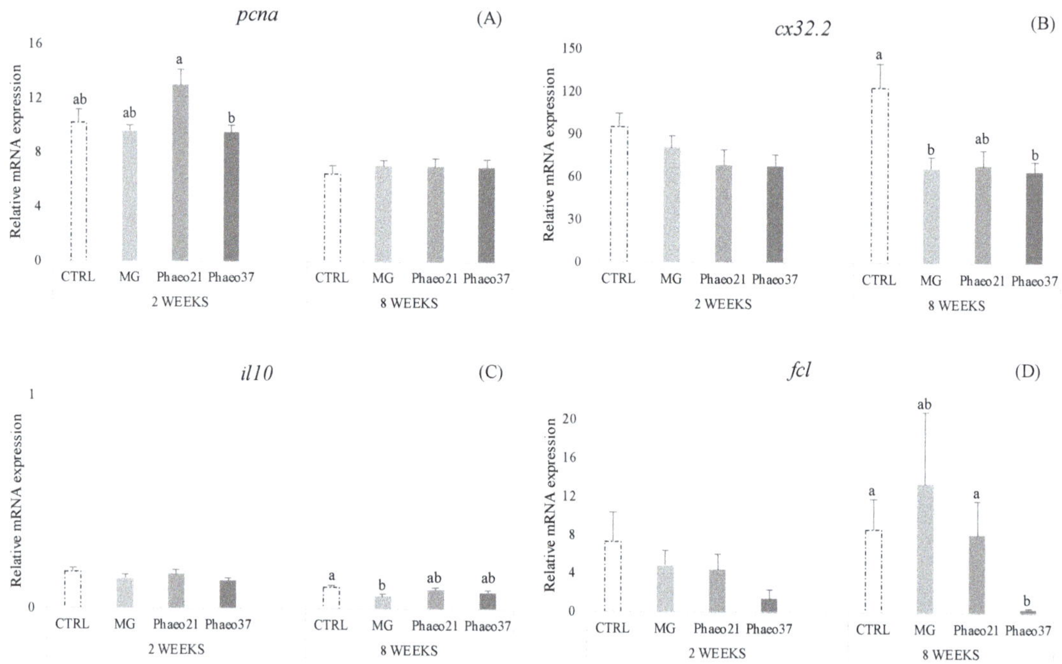

Figure 3. Relative mRNA expression of *pcna* (**A**), *cx32.2* (**B**), *il10* (**C**) and *fcl* (**D**) genes in the anterior intestinal tissue of gilthead seabream juveniles fed the experimental diets for 2 and 8 weeks. Data are the mean ± SEM (n = 9). All data values for each gene were in reference to the expression level of *cldn12* of CTRL fish with an arbitrary assigned value of 1. p values result from one-way ANOVA. Different lowercase letters indicate significant differences among dietary treatments ($p < 0.05$).

Further differences between fish fed the experimental diets in comparison to CTRL were highlighted by a clustering heatmap of gene expression after 8 weeks of feeding (Figure 4). This approach pointed to Phaeo37 as the experimental diet more apart from CTRL in the gene expression pattern. In order to get a clearer picture of the dietary effect on intestinal gene expression, an overall multivariate analysis combining raw data from the different genes (using PLS-DA) in CTRL and Phaeo 37 groups was performed. For the 2 weeks feeding period, the model was not able to show a clear separation between experimental groups (data not shown). However, at 8 weeks, the same approach showed that expression patterns can be summarized through two main components that explain 88.11% of total variance (Figure 5A). On the one hand, component 1 (63.59% of total variance, X-axis) appeared to be mostly related to diet effect, as it was able to clearly separate Phaeo37- and CTRL-fed fish. On the other hand, component 2 (19.52% of total variance, Y-axis) appeared to account for inner group variability. A total of 20 genes showed a VIP value > 1, highlighting their contribution to diet differences (Figure 5B).

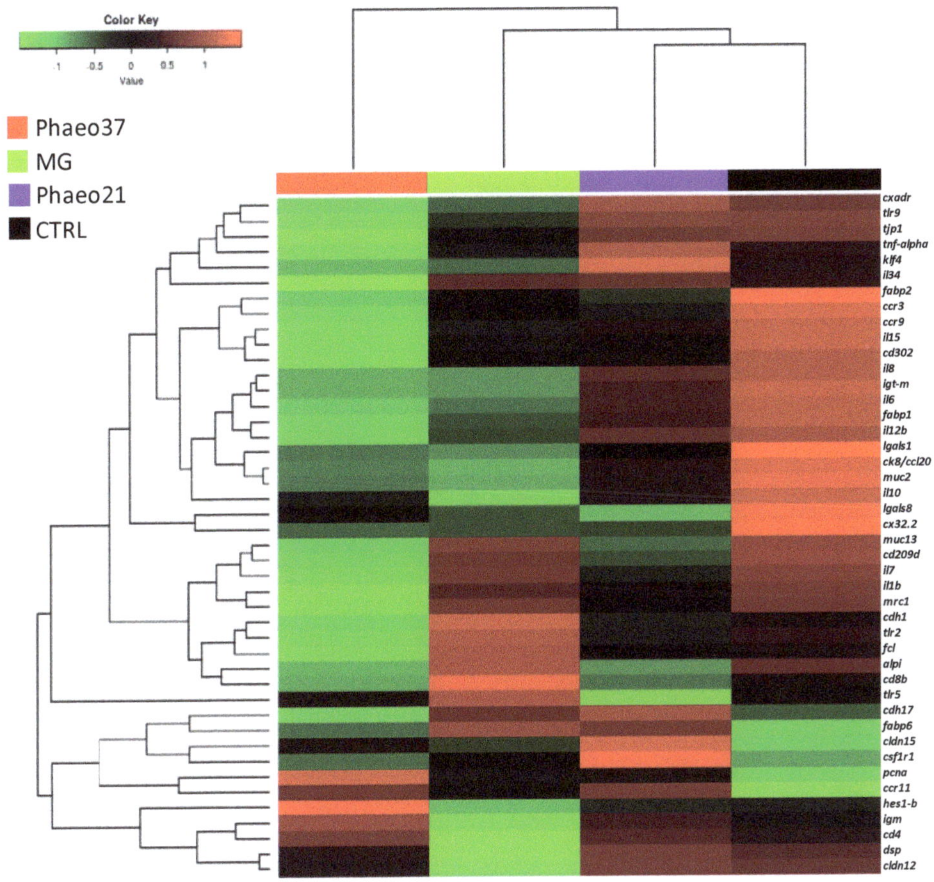

Figure 4. Heat map showing the normalized mRNA levels of selected genes in the anterior intestinal tissue of gilthead seabream juveniles after 8 weeks of feeding. Each block represents the mean mRNA level quantified by qPCR (n = 9).

(A)

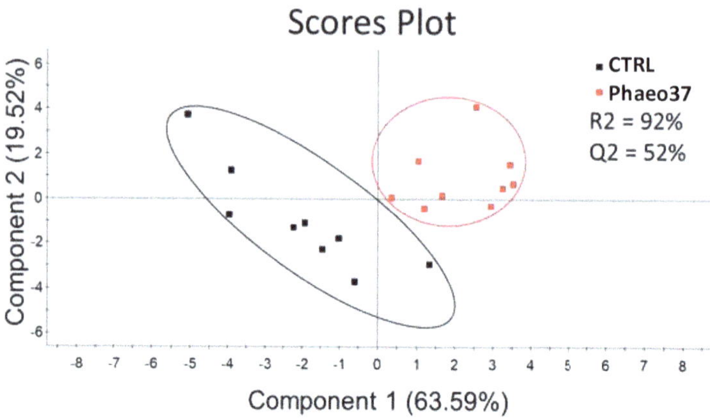

(B)

Gene	VIP[2] / FC
fcl	1.38 / -22.72
cd209d	1.27 / -4.14
il15	1.26 / -3.06
mrc1	1.29 / -2.53
cx32.2	1.71 / -1.91
il7	1.13 / -1.90
Igt-m	1.35 / -1.77
ccr3	1.22 / -1.76
tlr9	1.49 / -1.51
il8	1.04 / -1.46
fabp2	1.65 / -1.45
il12b	1.14 / -1.40
fabp1	1.10 / -1.34
ccr9	1.18 / -1.32
il34	1.20 / -1.32
lgals1	1.02 / -1.31
ck8/ccl20	1.20 / -1.27
tjp1	1.48 / -1.24
cxadr	1.13 / -1.19
muc13	1.02 / -1.11

Figure 5. (**A**) Partial least square discriminant analysis (PLS-DA) score plots of all gene expression biomarkers analysed in the proximal intestine of gilthead seabream juveniles along the two main components at 8 weeks. (**B**) Ordered list of markers by variable importance (VIP) in projection of PLS-DA model for group differentiation, as well as the fold-change (FC) in comparison to CTRL. Markers with VIP values > 1 after the first and second main components are represented.

3. Discussion

Intensive fish production creates stressful conditions that negatively affect immune function [28], increasing the risk of infection caused by opportunistic bacteria. Therefore, preventive strategies that can improve aquatic animal health and reduce the risk of disease outbreaks must be adopted, such as the use of prebiotics and immunostimulants in feed formulation [29–31]. Immunostimulant and prebiotic activities after β-(1,3; 1,6)-glucans administration are well-recognized; thus, these compounds have been suggested as potential nutraceuticals or vaccine adjuvants to enhance immune responses [4,32,33]. For that purpose, in the present study, gilthead seabream juveniles were fed microalgae (*P. tricornutum*) derived BGs in a 2-week cycle pulse-feeding regimen. This nutritional strategy was outlined, as care must be taken not to exhaust the fish immune system due to immunostimulant overexposure after long administration periods [34–37]. Moreover, intermittent administration seems a suitable approach as BGs apparently can induce long-lived effects in fish [38]. Several studies report increased innate immune parameters and pathogen resistance at least 2 weeks after BGs oral and intra-peritoneal (i.p.) administration [22,39,40].

Studies where BGs were orally administered to fish not only showed immunostimulatory effects but in some cases, improved growth performance [23,24,41,42]. Dawood et al. [43] showed that supplementing red seabream diets up to 0.1% (g/Kg feed) with a commercial BG product (85% purity) for 8 weeks improved final body weight and growth performance as well as lysozyme activity and higher tolerance against a low-salinity stress test when compared with a BG-free fed group. In the current study, β-glucan supplementation did not affect fish growth performance over the course of the trial (8 weeks), independently from its source. However, it did show immunomodulatory effects and improved oxidative stress status in accordance with the findings reported in other studies [23,44–47]. Dietary treatments appeared to modulate peripheral lymphocyte numbers. Results pointed to an immunostimulatory effect of diet Phaeo21 at 2 weeks feeding, with this particular BG extract apparently affecting the adaptive arm of the immune system with a rise in circulating lymphocytes. Previous works reported increased lymphocyte percentage in comparison to other leucocytes in pompano fish (*Trachinotus Ovatus*) fed 0.05% and 0.10% [45] and Persian sturgeon (*Acipenser persicus*) fed 0.2 and 0.3% yeast BGs [48] for 8 and 6 weeks, respectively. Nonetheless, in the current study, gilthead seabream fed Phaeo21 and MG dietary treatments showed a decrease in circulating lymphocytes percentage compared to those fed CTRL after 8 weeks. Kühlwein, et al. [49] reported no apparent effect on circulating lymphocytes when carp juveniles were fed 0.1, 1 and 2% yeast BGs continuously for 8 weeks.

On the other hand, non-specific humoral parameters (antiprotease, bactericidal activity and circulating IgM) were not affected by the supplementation with 0.06% BGs from *S. cerevisiae* or *P. tricornutum* throughout the experimental period. Accordingly, a study done in gilthead seabream fed a 0.1% supplemented feed with a macroalgae derived BG (laminarin) did not show changes in serum antiprotease activity and IgM levels after 4 weeks of feeding [23]. Yamamoto et al. [50] tested different levels, ranging from 0% to 0.8%, of microalgae (*Euglena gracilis*)-derived BGs in Nile tilapia both in vitro and in vivo. While exposing naïve head-kidney phagocytes directly to BGs facilitated the activation of immune cells increasing bactericidal activity against *Streptococcus iniae* and superoxide anion production, in vivo immune effects were found to be more moderate. Authors reported increased complement system activity but no effects on serum lysozyme and blood leukocytes respiratory burst. Still, previous studies in fish revealed a tendency of BGs oral administration to stimulate or modulate innate immune parameters [21,22,42,44,51,52].

Reactive oxygen species (ROS) are produced as a normal by-product of cellular metabolism but in excess, they can contribute to increased oxidative stress and cause cellular damage. Antioxidant enzymatic machinery is the principal cellular protective mechanism against oxidative stress in fish tissue [53]. BGs are reported to have antioxidant properties and modulate antioxidant enzymes activity as well as inhibiting lipid peroxidation in mammals [54]. In fish, BG injection increased SOD and CAT activities in the intestine [46] and blood erythrocytes [55], suggesting that BGs could improve anti-oxidative capacity. Accordingly, hepatic lipid peroxidation decreased in Phaeo37- and MG-fed animals at an early stage (2 weeks), while hepatic SOD showed a long-term (8 weeks) stimulation pattern, with the Phaeo37-fed group showing the highest activity. Zeng, et al. [46] reported a correlation between higher mRNA transcription of nuclear factor erythroid 2-related factor 2 (Nrf2) gene and increased SOD and CAT genes transcription, which translated in higher enzyme activity in fish injected with a 0.1% BG solution. Nrf proteins, under oxidative conditions translocate to the nucleus where they bind to the antioxidant response element (ARE) [56]. ARE is found in the promoters of several chemoprotective genes, including those involved in the response to oxidative stress [57]. Integrating all physiological responses into a multivariate analysis, dietary effects became clearer and differed between sampling points. At 2 weeks of feeding, all groups received BGs clustered together and were different from CTRL. Differences at this early stage pointed to a dietary effect mainly affecting the antioxidant defences and most prominently decreasing lipid peroxidation corroborating results from the one-way ANOVA. Previous studies reported higher antioxidant enzyme activity and lower lipid peroxidation when fish are previously treated with BGs through different administration routes (i.e., i.p. injection; oral route) [46,55,58]. After a toxicological insult, pre-treated fish with barley-derived BG were able to prevent intestinal Cu-induced lipid peroxidation [46]. Although BGs from the present study differ in origin and solubility, the observed early (2 weeks) beneficial effect was elicited by all BG-supplemented feeds, most likely due to the fact that both soluble and particulate BGs can act as exogenous ROS scavengers. Carballo et al. [59] reported that both a *P. tricornutum* chrysolaminarin-rich extract (soluble) and a yeast BG (particulate) show ROS scavenger activities. At a longer feeding period (8 weeks), the *P. tricornutum*-derived BG-groups showed higher SOD activity and clustered together independently from CTRL and MG. Furthermore, microalgae BG-treated groups also showed higher thrombocyte numbers. Although both Phaeo21 and 37 diets were supplemented with BG enriched extracts, other compounds such as *P. tricornutum* cell wall fragments might be present in the mixture and cannot be ruled out as immunomodulators. *P. tricornutum* cell wall is mainly composed of sulphated polysaccharides [60], which are known to interact with different toll-like receptors (TLRs). These compounds might act as antigens recognized by cell surface receptors activating different leucocyte types. In carp, peripheral thrombocytes constitutively express different TLR genes [61] and have been reported to have phagocytic activity and the ability to ingest particulate antigens possibly acting as an antigen presenting cell [62].

In the present study, at 2 weeks proliferating cell nuclear antigen (*pcna*) gene was down-regulated in the anterior gut of Phaeo37 fed fish. Former studies, report PCNA protein expression inhibition in mammalian cancer cells treated with different glucans including laminarin [63,64]. However, intestinal transcriptional changes were more significant at 8 weeks, where differences between CTRL and Phaeo37 gene profiles can be found. Furthermore, Phaeo37 and MG fed groups showed a down-regulation of different genes when compared to CTRL, namely *il10*, *cx32.2*, *fcl*. Hence, a multivariate analysis was performed (PLS-DA) allowing for a more comprehensive understanding of fish health status, whereas, at the same time identifying the most responsive gene biomarkers in fish intestine. VIP analysis with the first two components, highlighted that the top contributing genes for dietary differences in the gut were immune related (PRR- *fcl, cd209d, mrc1, tlr9, lgals1*; Interleukins- *il7, il8, il12β, il15, il34*; Immunoglobulin production- *igt-m*; chemokines and receptors- *ccr3, ccr9, ck8/cl20*). Phaeo37 dietary treatment caused a general down-regulation of gene transcription. Therefore, the effect of Phaeo37 supplemented diet was mostly immunomodulatory inducing a local anti-inflammatory state at molecular level, which as a consequence led to decreased immune cell activation in the gut. The down-regulation of intestinal immune-related genes can be understood as an immune tolerance effect that can be beneficial in an acute inflammation scenario, counterbalancing its negative and potentially dangerous effects. Falco et al. [25] also found an anti-inflammatory effect in common carp (*Cyprinius carpio L.*) intestine, with several inflammatory genes appearing down regulated when fish were fed a yeast BG supplemented diet for a 2-week period. Furthermore, even after a challenge (i.p. injection) with live bacteria (*Aeromonas salmonicida*), fish fed BGs showed decreased intestinal *il1b*, *il6* and *tnf-alpha* expression, while showing up-regulation in the head-kidney. In this particular case, it seems that BG may be preventing an acute response to infection in the gut, without compromising the systemic response. Additionally, the same down-regulation pattern (inflammatory genes) was seen in the spleen of rainbow trout after 37 days of feeding lentinan (soluble low molecular weight glucan)-supplemented diets [65]. These findings support the idea that BG can have localized specific effects depending on the target tissue.

Overall, some discrepancies can be observed among previous works and data gathered in the present study, which can be explained by different BG preparations. While some studies use crude BG extracts, others use purified compounds differing in molecular weight, branching and solubility. BGs solubility/insolubility seems to play a major role in ligand/receptor recognition and consequently immune cell activation [66,67]. In mammals, particulate BGs directly stimulate immune cell activation through a Dectin-1 recognition pathway, while soluble BG require complement-mediated opsonisation to activate a CR3-dependent pathway [66,68]. Still, in the present work, the particulate BG diet (MG) showed only mild effects mostly related with oxidative defenses after 2 weeks of feeding. In addition to solubility, molecular weight can play an important role in the biological effects of BGs. Different authors have found that in colitis-induced rat models, the dietary administration of low and high molecular weight oat BGs reduced the inflammatory response in colon and also ameliorated the local inflammation [69,70]. However, these authors found that low molecular mass BGs showed a significantly stronger anti-inflammatory effect, through the down-regulation of several pro-inflammatory cytokines and that the therapeutic effect is in evident relation with the molecular mass of the polymer. When comparing the different feeds used in the present study, Phaeo21 and 37 extracts show low molecular mass BGs (chrysolaminarin) [10] while, MG feed is supplemented with a high molecular weight BG (Baker´s yeast) [71,72]. Furthermore, *P. tricornutum* extract supplemented diets although having the same BG concentration, differ in purity, since Phaeo37 extract has a higher percentage of BGs compared to Phaeo21. Thus, the combination of low molecular mass BGs and higher extract purity might explain the higher overall immunomodulatory and oxidative protective effects of Phaeo37 dietary treatment.

In summary, novel feeds with increasingly higher percentages of terrestrial animal- and plant-derived ingredients have been shown to have anti-nutritional factors that often cause gut inflammation in fish, a condition that might lead to impaired nutrient absorption and the disruption of normal microbiota. The use of gut anti-inflammatory compounds can have special relevance nowadays in aquaculture, both as a prophylactic and therapeutic measure, as the industry decreases the use of FM, replacing it by the ingredients referred to above. In this regard, our results indicate that the dietary administration of a *P. tricornutum* 37% enriched-BG extract might be relevant in a context of extreme dietary formulation due to its anti-inflammatory and anti-oxidative effects.

4. Materials and Methods

4.1. *P. tricornutum* Extracts

Chrysolaminarin-rich biomass from *P. tricornutum* (SAG 1090-1b) grown under nitrogen-depleted conditions in flat panel airlift reactors was harvested and concentrated via centrifugation to 250–270 g L^{-1} (Clara 20, Alfa Laval, Lund, Sweden). Afterwards, the biomass was frozen at −20 °C. For further processing, the biomass was thawed and diluted to 100 g L^{-1} with deionized water. The cell disruption was performed according to Derwenskus, et al. [73] with a ball mill (PML-2, Bühler, Uzwil, Switzerland). Phaeo21 was freeze dried after cell disruption (VaCo 5, Zirbus, Bad Grund, Germany), while Phaeo37 was centrifuged and the supernatant freeze-dried (Avanti J-26 XP, Beckman Coulter, Brea, CA, USA).

4.2. Diet Composition

The trial comprised four isonitrogenous (63% crude protein) and isolipidic (17% crude fat) diets (Table 4). A high-quality, practical diet was used as control (CTRL) and 3 experimental diets based on CTRL were supplemented with either a commercial product derived from *S. cerevisiae* (diet MG) or different extracts of *P. tricornutum* (diets Phaeo21 and Phaeo37), to obtain a final concentration of 0.6 g β-glucans per Kg of feed (0.06%) in all supplemented diets. Diets were manufactured by SPAROS. All powder ingredients were mixed according to the target formulation in a double-helix mixer (model RM90, MAINCA, Barcelona, Spain) and ground (below 200 µm) in a micropulverizer hammer mill (model SH1, Hosokawa-Alpine, Augsburg, Germany). Subsequently, the oils were added to the mixtures, which were humidified with 20–25% water and agglomerated by a low-shear and low-temperature extrusion process (ITALPLAST, West Heidelberg, VIC, Australia). Extruded pellets (1.5 mm) were dried in a vibrating fluid bed dryer (model DR100, TGC Extrusion, Roullet-Saint-Estèphe, France). Diets were packed in sealed plastic buckets and shipped to the research site (Riasearch, Murtosa, Portugal) where they were stored at room temperature in a cool and aerated emplacement. Samples of each diet were taken for analytical characterization.

Table 4. Ingredients and proximate composition of experimental diets.

Ingredients %	CTRL	MG	Phaeo21	Phaeo37
Fishmeal [1]	20.00	20.00	20.00	20.00
Fish protein hydrolysate [2]	8.00	8.00	8.00	8.00
Squid meal [3]	21.00	21.00	21.00	21.00
Krill meal [4]	16.50	16.50	16.50	16.50
Wheat gluten [5]	11.50	11.50	11.50	11.50
Wheat meal [6]	0.29	0.19		0.13
Vitamin and mineral premix [7]	2.00	2.00	2.00	2.00
Lecithin [8]	4.30	4.30	4.30	4.30
Fish oil [9]	6.50	6.50	6.50	6.50
Binders, antioxidant and other additives [10]	9.91	9.91	9.91	9.91
Yeast beta-glucans [11]		0.10		
Algae beta-glucans Phaeo21 [12]			0.29	
Algae beta-glucans Phaeo37 [13]				0.16
Proximate composition				
Dry matter (DM) %	94.60	94.20	94.20	94.50
Ash, % DM	9.60	9.50	9.50	9.50
Crude protein, % DM	62.90	62.80	62.80	62.90
Crude fat, % DM	17.10	17.10	17.10	17.10
Gross energy (kJ g^{-1} DM)	22.90	22.90	22.90	22.90

[1] Super Prime: 66.3% CP, 11.5% CF, Pesquera Diamante, Peru; [2] CPSP 90: 82% CP 9% CF, Sopropêche, France; [3] Squid meal without guts: 83% CP, 4% CF, Sopropêche, France; [4] Krill meal: 61.1% CP, 17.4% CF, Aker Biomarine, Norway; [5] VITEN: 82% CP, 2.1% CF, Roquette, France; [6] Wheat meal: 10.2% CP; 1.2% CF, MOLISUR, Spain; [7] PREMIX Lda, Portugal: Vitamins (IU or mg/kg diet): DL-alpha tocopherol acetate, 200 mg; sodium menadione bisulphate, 50 mg; retinyl acetate, 40,000 IU; DL-cholecalciferol, 4000 IU; thiamin, 60 mg; riboflavin, 60 mg; pyridoxine, 40 mg; cyanocobalamin, 0.2 mg; nicotinic acid, 400 mg; folic acid, 30 mg; ascorbic acid, 1000 mg; inositol, 1000 mg; biotin, 6 mg; calcium panthotenate, 200 mg; choline chloride, 2000 mg, betaine, 1000 mg. Minerals (g or mg/kg diet): copper sulphate, 18 mg; ferric sulphate, 12 mg; potassium iodide, 1 mg; manganese oxide, 20 mg; sodium selenite, 0.02 mg; zinc sulphate, 27.5 mg; sodium chloride, 800 mg; excipient wheat middling's; [8] LECICO GmbH, Germany; [9] Sopropêche, France; [10] Confidential blend of constant binders and other additives; [11] Macrogard, 67.2% beta-glucans, Biorigin, Brazil; [12] Beta-glucan rich biomass of microalgae (*Phaeodactylum tricornutum* from SAG culture collection) with 21% beta-glucans; [13] Beta-glucan rich extract of microalgae (*Phaeodactylum tricornutum* from SAG culture collection) with 37% beta-glucans.

4.3. Fish Rearing Conditions and Feeding Trial

Fish were reared in a seawater recirculation system with aeration (mean dissolved oxygen above 6 mg L^{-1}) and water flow at 3 L min^{-1} (mean temperature 24.1 ± 0.6 °C; mean salinity 18.7 ± 0.1‰). Water quality parameters were monitored daily (mean dissolved oxygen 6.4 ± 1.0 mg L^{-1}; mean unionized ammonia levels 0.001 ± 0.002 mg L^{-1}). Diets were randomly assigned to quadruplicate groups of 95 gilthead seabream juveniles (initial body weight: 4.1 ± 0.1 g) that were fed to satiation three times a day for 8 weeks in a pulse-feeding regimen. Accordingly, in fish fed the different experimental diets, the CTRL diet was intercalated every 2 weeks, as shown in Figure 6.

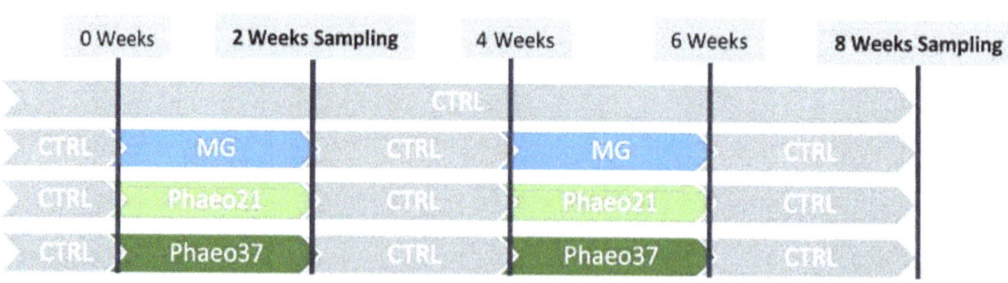

Figure 6. Schematic overview of the experimental design.

4.4. Sampling Procedures

Fish were individually weighed at the beginning and after 2 and 8 weeks of the feeding trial and feed consumption for each experimental replicate was registered daily. After 2 and 8 weeks, three fish per tank were euthanized with a 2-phenoxyethanol lethal dose (0.5 mL L^{-1}) [74], weighed and sampled for tissues (blood, head-kidney, liver and gut). Blood was collected from the caudal vein using heparinized syringes and centrifuged at 10,000× *g* during 10 min at 4 °C to obtain plasma samples. Plasma, head-kidney and liver samples were immediately frozen at −80 °C, and anterior intestine was preserved in RNA later until further analysis.

4.5. Haematological Procedures

Blood smears were prepared from peripheral blood, air dried and stained with Wright's stain (Haemacolor; Merck) after fixation for 1 min with formol–ethanol (10% formaldehyde in ethanol). Neutrophils were labelled through the detection of peroxidase activity revealed by the Antonow´s technique described by Afonso, et al. [75]. The slides were examined under oil immersion (1000×), and at least 200 leucocytes were counted and classified as thrombocytes, lymphocytes, monocytes and neutrophils. The relative percentage of each cell type was calculated.

4.6. Innate Humoral Parameters

Plasma bactericidal activity was determined following the method described by Machado, et al. [76] with some modifications. *Edwardsiella tarda* (*E. tarda*) strain ACC 53.1, gently provided by Prof. Alicia Toranzo (University of Santiago, Spain) was used in the protocol. Briefly, 20 µL of plasma were mixed with 20 µL of bacteria suspension (10^8 CFU mL^{-1}) in duplicate in a flat-bottom 96-well plate that was incubated for 2.5 h at 25 °C (positive control: 20 µL of TSB instead of plasma). Afterwards, 25 µL of 3-(4,5 dimethyl-2-yl)-2,5-diphenyl tetrazolium bromide (1 mg mL^{-1}; Sigma, St. Louis, MO, USA) was added to each well and incubated for 10 min at 25 °C to allow the formation of formazan precipitates. Plates were then centrifuged at 2000× *g* for 10 min, the supernatant was discarded and the precipitate was dissolved in 200 µL of dimethyl sulfoxide (Sigma, St. Louis, MO, USA). The absorbance was then measured at 560 nm. Bactericidal activity is expressed as a percentage calculated from the difference between surviving bacteria compared to the number of bacteria from positive controls (100%).

Anti-protease activity was determined as described by Ellis et al. [77] with some modifications. Briefly, 10 µL of plasma were incubated with the same volume of trypsin solution (5 mg mL^{-1} in $NaHCO_3$, 5 mg mL^{-1}, pH 8.3) for 10 min at 22 °C. After incubation, 100 µL of phosphate buffer (NaH_2PO_4, 13.9 mg mL^{-1}, pH 7.0) and 125 µL of azocasein solution (20 mg mL^{-1} in $NaHCO_3$, 5 mg mL^{-1}, pH 8.3) were added and incubated for 1 h at 22 °C. Finally, 250 µL of trichloroacetic acid were added to the reaction mixture and incubated for 30 min at 22 °C. The mixture was centrifuged at 10,000× *g* for 5 min at room temperature. Afterwards, 100 µL of the supernatant was transferred to a 96-well plate and mixed with 100 µL of NaOH (40 mg mL^{-1}). The OD was read at 450 nm in a Synergy HT microplate reader. Phosphate buffer instead of plasma and trypsin served as blank, whereas the reference sample was phosphate buffer instead of plasma. The sample inhibition percentage of trypsin activity was calculated as follows: 100 − ((sample absorbance/reference absorbance) × 100). All analyses were conducted in duplicate.

Plasma immunoglobulins (IgMs) were measured by an ELISA assay. Briefly, plasma samples were diluted (1:100) in Na_2CO_3 (50 mM, pH = 9.6). Diluted plasma samples (100 µL in duplicate) were incubated overnight (4 °C) in a 96 well plate, using Na_2CO_3 (100 µL) as a negative control. The samples (antigen) were then removed and 300 µL of blocking buffer (5% low fat milk in 0.1% Tween 20) was added to each well and incubated for 1 h at 22 °C. This mixture was then removed, followed by three consecutive washes with 300 µL of T-TBS (0.1% Tween 20). After properly cleaning and drying the wells, 100 µL of the anti-seabream primary IgM monoclonal antibody (1:200 dilution in blocking buffer;

Aquatic Diagnostics, UK) was added to each well and incubated for 1 h at 22 °C. The primary antibody was then removed by aspiration, with three consecutive washes being performed. Afterwards, the anti-mouse IgG-HRP, secondary antibody (1:1000 dilution in blocking buffer; SIGMA), was added and incubated for 1 h at 22 °C, then removed by aspiration. The wells were again washed three times and 100 µL of TMB substrate solution for ELISA (BioLegend #421101), was added to each well and incubated for 5 min. The reaction was stopped after 5 min by adding 100 µL of H_2SO_4 2 M and the optical density was read at 450 nm.

4.7. Analysis of Oxidative Stress Biomarkers

Liver samples were thawed and homogenized (1:10) in phosphate buffer 0.1 M (pH 7.4) using Precellys evolution tissue lyser homogenizer.

One aliquot of tissue homogenate was used to determine the extent of endogenous lipid peroxidation (LPO) by measuring thiobarbituric acid-reactive species (TBARS) as suggested by Bird and Draper [78]. To prevent artifactual lipid peroxidation, butylhydroxytoluene (BHT 0.2 mM) was added to the aliquot. Briefly, 1 mL of 100% trichloroacetic acid and 1 mL of 0.73% thiobarbituric acid solution (in Tris–HCl 60 mM pH 7.4 with DTPA 0.1 mM) were added to 0.2 mL of liver homogenate. After incubation at 100 °C for 60 min, the solution was centrifuged at 12,000× *g* for 5 min and LPO levels were determined at 535 nm.

The remaining tissue homogenate was centrifuged for 20 min at 10,000× *g* (4 °C) to obtain the post mitochondrial supernatant fraction (PMS). Total proteins in homogenates were measured by using Pierce™ BCA Protein Assay Kit, as described by the manufacturer.

Catalase (CAT) activity was determined in PMS by measuring substrate (H_2O_2) consumption at 240 nm according to Claiborne [79] adapted to microplate. Briefly, in a microplate well, 0.140 mL of phosphate buffer (0.05 M pH 7.0) and 0.150 mL H_2O_2 solution (30 mM in phosphate buffer 0.05 M pH 7.0) were added to 0.01 mL of liver PMS (0.7 mg ml^{-1} total protein). Enzymatic activity was determined in a microplate reader (BioTek Synergy HT) reading the optical density at 240 nm for 2 min every 15 sec interval.

Superoxide dismutase (SOD) activity was measured according to Flohé and Otting [80] adapted to microplate by Lima, et al. [81]. Briefly, in a microplate well, 0.2 mL of the reaction solution [1 part xantine solution 0.7 mM (in NaOH 1 mM) and 10 parts cytochrome c solution 0.03 mM (in phosphate buffer 50 mM pH 7.8 with 1 mM Na-EDTA)] was added to 0.05 mL of liver PMS (0.25 mg ml^{-1} total protein). Optical density was measured at 550 nm in a microplate reader (BioTek Synergy HT, Winooski, VT, USA) every 20-s interval for 3 min at 25° C.

Total glutathione (tGSH) content was determined with PMS fraction at 412 nm using a recycling reaction of reduced glutathione (GSH) with 5,5-dithiobis-(2-nitrobenzoic acid) (DTNB) in the presence of glutathione reductase (GR) excess [82,83]. TG content is calculated as the rate of TNB^{2-} formation with an extinction coefficient of DTNB chromophore formed, $\varepsilon = 14.1 \times 10^3\ M^{-1}cm^{-1}$.

4.8. Gene Expression

Total RNA isolation from target tissue (anterior intestine) was conducted with NZY Total RNA Isolation kit (NZYTech, Lisbon, Portugal) following the manufacturer's specifications. Reverse transcription (RT) of 500 ng total RNA was performed with random decamers using a High-Capacity cDNA Reverse Transcription Kit (Applied Biosystems, Foster City, CA, USA) according to the manufacturer's instructions. RT reactions were incubated for 10 min at 25 °C and 2 h at 37 °C. Negative control reactions were run without reverse transcriptase. Real-time quantitative PCR was carried out on a EpMotion 5070 Liquid Handling Robot (Eppendorf, Hamburg, Germany) using a 96-well PCR array layout with 44 genes designed for simultaneously profiling of anterior intestine (Table 5). Genes comprised in the array were selected for their involvement in gut integrity, health, immunity and signal transduction. Specific primer pair sequences are listed in Table S3.

Controls of general PCR performance were included on each array, and all the pipetting operations performed by means of the EpMotion 5070 Liquid Handling Robot (Eppendorf, Hamburg, Germany). Briefly, RT reactions were diluted to obtain the equivalent concentration of 660 pg of total input RNA which were used in a 25-μL volume for each PCR reaction. PCR wells contained a 2× SYBR Green Master Mix (Bio-Rad) and specific primers at a final concentration of 0.9 μM were used to obtain amplicons 50–150 bp in length. The program used for PCR amplification included an initial denaturation step at 95 °C for 3 min, followed by 40 cycles of denaturation for 15 s at 95 °C and annealing/extension for 60 s at 60 °C. The efficiency of PCR reactions was always higher than 90%, and negative controls without sample templates were routinely performed for each primer set. The specificity of reactions was verified by analysis of melting curves (ramping rates of 0.5 °C/10 s over a temperature range of 55–95 °C), and linearity of serial dilutions of RT reactions. Fluorescence data acquired during the PCR extension phase were normalized using the delta–delta Ct method [84]. Beta-actin (*actb*) was tested for gene expression stability using GeNorm software (M score = 0.13) and it was used as housekeeping gene in the normalization procedure. For comparing the mRNA expression level of a panel of genes in a given dietary treatment, all data values were in reference to the expression level of claudin 12 (*cldn12*) in CTRL fish, which was arbitrarily assigned a value of 1.

Table 5. PCR-array layout for gene expression profiling of anterior intestine in seabream.

Function	Gene	Symbol	GenBank
Epithelia integrity	proliferating cell nuclear antigen	*pcna*	KF857335
	transcription factor HES-1-B	*hes1-b*	KF857344
	krueppel-like factor 4	*klf4*	KF857346
	claudin-12	*cldn12*	KF861992
	claudin-15	*cldn15*	KF861993
	cadherin-1	*cdh1*	KF861995
	cadherin-17	*cdh17*	KF861996
	tight junction protein ZO-1	*tjp1*	KF861994
	desmoplakin	*dsp*	KF861999
	gap junction Cx32.2 protein	*cx32.2*	KF862000
	coxsackievirus and adenovirus receptor homolog	*cxadr*	KF861998
Nutrient transport	intestinal-type alkaline phosphatase	*alpi*	KF857309
	liver type fatty acid-binding protein	*fabp1*	KF857311
	intestinal fatty acid-binding protein	*fabp2*	KF857310
	ileal fatty acid-binding protein	*fabp6*	KF857312
Mucus production	mucin 2	*muc2*	JQ277710
	mucin 13	*muc13*	JQ277713
Interleukins	tumor necrosis factor-alpha	*tnf-alpha*	AJ413189
	interleukin 1 beta	*il1b*	AJ419178
	interleukin 6	*il6*	EU244588
	interleukin 7	*il7*	JX976618
	interleukin 8	*il8*	JX976619
	interleukin 10	*il10*	JX976621
	interleukin 12 subunit beta	*il12b*	JX976624
	interleukin 15	*il15*	JX976625
	interleukin 34	*il34*	JX976629
Cell markers	cluster differentiation 4	*cd4*	AM489485
	cluster differentiation 8 beta	*cd8b*	KX231275
	C-C chemokine receptor 3	*ccr3*	KF857317
	C-C chemokine receptor 9	*ccr9*	KF857318
	C-C chemokine receptor 11	*ccr11*	KF857319
	C-C chemokine ck8/C-C motif chemokine ligand 20	*ck8/ccl20*	GU181393
	macrophage colony-stimulating factor 1 receptor	*csf1r*	AM050293
Ig production	immunoglobulin M	*igm*	JQ811851
	immunoglobulin T membrane-bound form	*igt-m*	KX599201

Table 5. *Cont.*

Function	Gene	Symbol	GenBank
Pathogen associated microbial pattern (PAMP)	galectin 1	*lgals1*	KF862003
	galectin 8	*lgals8*	KF862004
	toll like receptor 2	*tlr2*	KF857323
	toll like receptor 5	*tlr5*	KF857324
	toll like receptor 9	*tlr9*	AY751797
	CD209 antigen-like protein D	*cd209d*	KF857327
	CD302 antigen	*cd302*	KF857328
	macrophage mannose receptor 1	*mrc1*	KF857326
	fucolectin	*fcl*	KF857331

4.9. Data Analysis

All results are expressed as mean ± standard error (mean ± SE). Residuals were tested for normality (Shapiro–Wilk´s test) and homogeneity of variance (Levene's test). When residuals did not meet the assumptions, data were transformed by a Box-Cox transformation. One-way ANOVAs were performed for all datasets, with "dietary treatment" as the fixed effect, followed by multiple comparison Tukey post-hoc tests. The factor "time" was not considered in the analysis since it is not a goal of the study to evaluate how time affects the measured variables, and groups were treated independently for 2 and 8 weeks.

In an attempt to discriminate and classify individuals by the existing groups, a multivariate canonical discriminant analysis was performed on the physiological dataset (obtained from blood, plasma and liver tissues analyses) to evaluate the linear combinations of the original variables that will best separate the groups (discriminant functions). Each discriminant function explains part of total variance of the dataset and is loaded by variables contributing the most for that variation. Discriminatory effectiveness was assessed by Wilk's λ test, and the distance between groups' centroids was measured by squared Mahalanobis distance, and Fisher's F statistic was applied to infer significance. All statistical analyses were performed using the computer package SPSS 26 for WINDOWS.

Gene expression results were evaluated with an unsupervised multivariate analysis by principal component analysis (PCA) as an unbiased statistical method to observe intrinsic trends in the dataset, using EZ-INFO® v3.0 (Umetrics, Sweden). To achieve the maximum separation among the groups, a supervised multivariate analysis by partial least-squares discriminant analysis (PLS-DA) was sequentially applied, using EZ-INFO® v3.0 (Umetrics, Umeå, Sweden). Potential differential genes were selected according to the Variable Importance in the Projection (VIP) values. Variables with VIP > 1 were considered to be influential for the separation of samples in PLS-DA analysis [85–87]. The level of significance used was $p \leq 0.05$ for all statistical tests. Heat map of transcript levels were produced with the R package gplots, using the average linkage method and Euclidean distance.

Supplementary Materials: The following are available online at https://www.mdpi.com/article/10.3390/md19120653/s1, Table S1. Relative gene expression profiling of anterior intestine in gilthead seabream juveniles fed experimental diets for 2 weeks; Table S2. Relative gene expression profiling of anterior intestine in gilthead seabream juveniles fed experimental diets for 8 weeks; Table S3: Primers for qPCR amplification in seabream.

Author Contributions: Conceptualization, B.R., M.S., L.E.C.C., R.S., J.D. and B.C.; Data curation, B.R., A.T.G., J.P.-S. and J.C.-G.; Formal analysis, B.R., A.T.G., P.S., J.P.-S. and J.C.-G.; Funding acquisition, L.E.C.C., U.S.-S., J.D. and B.C.; Investigation, B.R., P.S., R.S., U.S.-S. and K.F.; Project administration, J.D. and B.C.; Resources, U.S.-S. and K.F.; Supervision, L.E.C.C., J.D. and B.C.; Writing—original draft, B.R.; Writing—review and editing, A.T.G., M.S., L.E.C.C., R.S., J.P.-S., J.C.-G., U.S.-S., J.D. and B.C. All authors have read and agreed to the published version of the manuscript.

Funding: This work has received funding from the Bio Based Industries Joint Undertaking (BBI JU) under the European Union's Horizon 2020 research and innovation programme under grant agreement No. 745754 (project MAGNIFICENT). This output reflects the views only of the author(s), and the European Union and BBI JU cannot be held responsible for any use which may be made of the information contained therein. This work is also a result of the project ATLANTIDA (ref. NORTE-01-0145-FEDER-000040), supported by the Norte Portugal Regional Operational Programme (NORTE 2020), under the PORTUGAL 2020 Partnership Agreement and through the European Regional Development Fund (ERDF), and by national funds through FCT—Foundation for Science and Technology within the scope of UIDB/04423/2020 and UIDP/04423/2020. BR and BC were supported by FCT (PD/BDE/129262/2017 and IF/00197/2015, respectively).

Institutional Review Board Statement: The experiment was carried out in compliance with the Guidelines of the European Union Council (Directive 2010/63/EU) and Portuguese legislation for the use of laboratory animals at Riasearch facilities (Portugal).

Informed Consent Statement: Not applicable.

Data Availability Statement: Not applicable.

Conflicts of Interest: The authors declare that they have no known competing financial interests or personal relationships that could have appeared to influence the work reported in this paper.

References

1. Van Boeckel, T.P.; Brower, C.; Gilbert, M.; Grenfell, B.T.; Levin, S.A.; Robinson, T.P.; Teillant, A.; Laxminarayan, R. Global trends in antimicrobial use in food animals. *Proc. Natl. Acad. Sci. USA* **2015**, *112*, 5649–5654. [CrossRef] [PubMed]
2. Meena, D.K.; Das, P.; Kumar, S.; Mandal, S.C.; Prusty, A.K.; Singh, S.K.; Akhtar, M.S.; Behera, B.K.; Kumar, K.; Pal, A.K.; et al. Beta-glucan: An ideal immunostimulant in aquaculture (a review). *Fish Physiol. Biochem.* **2013**, *39*, 431–457. [CrossRef]
3. Adams, A. Progress, challenges and opportunities in fish vaccine development. *Fish Shellfish Immunol.* **2019**, *90*, 210–214. [CrossRef]
4. Song, S.K.; Beck, B.R.; Kim, D.; Park, J.; Kim, J.; Kim, H.D.; Ringø, E. Prebiotics as immunostimulants in aquaculture: A review. *Fish Shellfish Immunol.* **2014**, *40*, 40–48. [CrossRef] [PubMed]
5. Yilmaz, S.; Yilmaz, E.; Dawood, M.A.O.; Ringø, E.; Ahmadifar, E.; Abdel-Latif, H.M.R. Probiotics, prebiotics, and synbiotics used to control vibriosis in fish: A review. *Aquaculture* **2022**, *547*, 737514. [CrossRef]
6. Guedes, A.C.; Amaro, H.M.; Malcata, F.X. Microalgae as sources of high added-value compounds-a brief review of recent work. *Biotechnol. Prog.* **2011**, *27*, 597–613. [CrossRef] [PubMed]
7. Nazih, H.; Bard, J.-M. Chapter 10—Microalgae in Human Health: Interest as a Functional Food. In *Microalgae in Health and Disease Prevention*; Levine, I.A., Fleurence, J., Eds.; Academic Press: Cambridge, MA, USA, 2018; pp. 211–226.
8. Molino, A.; Iovine, A.; Casella, P.; Mehariya, S.; Chianese, S.; Cerbone, A.; Rimauro, J.; Musmarra, D. Microalgae Characterization for Consolidated and New Application in Human Food, Animal Feed and Nutraceuticals. *Int. J. Environ. Res. Public Health* **2018**, *15*, 2436. [CrossRef]
9. Kim, S.M.; Jung, Y.-J.; Kwon, O.-N.; Cha, K.H.; Um, B.-H.; Chung, D.; Pan, C.-H. A Potential Commercial Source of Fucoxanthin Extracted from the Microalga *Phaeodactylum tricornutum*. *Appl. Biochem. Biotechnol.* **2012**, *166*, 1843–1855. [CrossRef]
10. Gügi, B.; Le Costaouec, T.; Burel, C.; Lerouge, P.; Helbert, W.; Bardor, M. Diatom-Specific Oligosaccharide and Polysaccharide Structures Help to Unravel Biosynthetic Capabilities in Diatoms. *Mar. Drugs* **2015**, *13*, 5993–6018. [CrossRef]
11. Gilbert-López, B.; Barranco, A.; Herrero, M.; Cifuentes, A.; Ibáñez, E. Development of new green processes for the recovery of bioactives from *Phaeodactylum tricornutum*. *Food Res. Int.* **2017**, *99*, 1056–1065. [CrossRef]
12. Dalmo, R.A.; Bøgwald, J. ß-glucans as conductors of immune symphonies. *Fish Shellfish Immunol.* **2008**, *25*, 384–396. [CrossRef]
13. Legentil, L.; Paris, F.; Ballet, C.; Trouvelot, S.; Daire, X.; Vetvicka, V.; Ferrières, V. Molecular Interactions of β-(1→3)-Glucans with Their Receptors. *Molecules* **2015**, *20*, 9745–9766. [CrossRef] [PubMed]
14. Petit, J.; Bailey, E.C.; Wheeler, R.T.; de Oliveira, C.A.F.; Forlenza, M.; Wiegertjes, G.F. Studies Into β-Glucan Recognition in Fish Suggests a Key Role for the C-Type Lectin Pathway. *Front. Immunol.* **2019**, *10*, 280. [CrossRef]
15. Soltanian, S.; Stuyven, E.; Cox, E.; Sorgeloos, P.; Bossier, P. Beta-glucans as immunostimulant in vertebrates and invertebrates. *Crit. Rev. Microbiol.* **2009**, *35*, 109–138. [CrossRef] [PubMed]
16. Brown, G.D.; Herre, J.; Williams, D.L.; Willment, J.A.; Marshall, A.S.; Gordon, S. Dectin-1 mediates the biological effects of beta-glucans. *J. Exp. Med.* **2003**, *197*, 1119–1124. [CrossRef] [PubMed]
17. Pietretti, D.; Vera-Jimenez, N.I.; Hoole, D.; Wiegertjes, G.F. Oxidative burst and nitric oxide responses in carp macrophages induced by zymosan, MacroGard® and selective dectin-1 agonists suggest recognition by multiple pattern recognition receptors. *Fish Shellfish Immunol.* **2013**, *35*, 847–857. [CrossRef]
18. Ainsworth, A.J. A β-glucan inhibitable zymosan receptor on channel catfish neutrophils. *Vet. Immunol. Immunopathol.* **1994**, *41*, 141–152. [CrossRef]

19. Esteban, M.A.; Rodrıguez, A.; Meseguer, J. Glucan receptor but not mannose receptor is involved in the phagocytosis of *Saccharomyces cerevisiae* by seabream (*Sparus aurata* L.) blood leucocytes. *Fish Shellfish Immunol.* **2004**, *16*, 447–451. [CrossRef]
20. Kiron, V.; Kulkarni, A.; Dahle, D.; Vasanth, G.; Lokesh, J.; Elvebo, O. Recognition of purified beta 1,3/1,6 glucan and molecular signalling in the intestine of Atlantic salmon. *Dev. Comp. Immunol.* **2016**, *56*, 57–66. [CrossRef] [PubMed]
21. El-Boshy, M.E.; El-Ashram, A.M.; AbdelHamid, F.M.; Gadalla, H.A. Immunomodulatory effect of dietary *Saccharomyces cerevisiae*, β-glucan and laminaran in mercuric chloride treated Nile tilapia (*Oreochromis niloticus*) and experimentally infected with *Aeromonas hydrophila*. *Fish Shellfish Immunol.* **2010**, *28*, 802–808. [CrossRef]
22. Chang, C.S.; Huang, S.L.; Chen, S.; Chen, S.N. Innate immune responses and efficacy of using mushroom beta-glucan mixture (MBG) on orange-spotted grouper, *Epinephelus coioides*, aquaculture. *Fish Shellfish Immunol.* **2013**, *35*, 115–125. [CrossRef]
23. Guzmán-Villanueva, L.T.; Tovar-Ramírez, D.; Gisbert, E.; Cordero, H.; Guardiola, F.A.; Cuesta, A.; Meseguer, J.; Ascencio-Valle, F.; Esteban, M.A. Dietary administration of β-1,3/1,6-glucan and probiotic strain *Shewanella putrefaciens*, single or combined, on gilthead seabream growth, immune responses and gene expression. *Fish Shellfish Immunol.* **2014**, *39*, 34–41. [CrossRef] [PubMed]
24. Dawood, M.A.O.; Koshio, S.; Ishikawa, M.; Yokoyama, S. Interaction effects of dietary supplementation of heat-killed *Lactobacillus plantarum* and β-glucan on growth performance, digestibility and immune response of juvenile red sea bream, *Pagrus major*. *Fish Shellfish Immunol.* **2015**, *45*, 33–42. [CrossRef] [PubMed]
25. Falco, A.; Frost, P.; Miest, J.; Pionnier, N.; Irnazarow, I.; Hoole, D. Reduced inflammatory response to *Aeromonas salmonicida* infection in common carp (*Cyprinus carpio* L.) fed with β-glucan supplements. *Fish Shellfish Immunol.* **2012**, *32*, 1051–1057. [CrossRef]
26. Falco, A.; Miest, J.J.; Pionnier, N.; Pietretti, D.; Forlenza, M.; Wiegertjes, G.F.; Hoole, D. β-Glucan-supplemented diets increase poly(I:C)-induced gene expression of Mx, possibly via Tlr3-mediated recognition mechanism in common carp (*Cyprinus carpio*). *Fish Shellfish Immunol.* **2014**, *36*, 494–502. [CrossRef]
27. Dawood, M.A.O.; Metwally, A.E.-S.; El-Sharawy, M.E.; Atta, A.M.; Elbialy, Z.I.; Abdel-Latif, H.M.R.; Paray, B.A. The role of β-glucan in the growth, intestinal morphometry, and immune-related gene and heat shock protein expressions of Nile tilapia (*Oreochromis niloticus*) under different stocking densities. *Aquaculture* **2020**, *523*, 735205. [CrossRef]
28. Pickering, A.D. Growth and stress in fish production. *Aquaculture* **1993**, *111*, 51–63. [CrossRef]
29. Vaseeharan, B.; Thaya, R. Medicinal plant derivatives as immunostimulants: An alternative to chemotherapeutics and antibiotics in aquaculture. *Aquac. Int.* **2014**, *22*, 1079–1091. [CrossRef]
30. Zanuzzo, F.S.; Sabioni, R.E.; Montoya, L.N.F.; Favero, G.; Urbinati, E.C. Aloe vera enhances the innate immune response of pacu (*Piaractus mesopotamicus*) after transport stress and combined heat killed *Aeromonas hydrophila* infection. *Fish Shellfish Immunol.* **2017**, *65*, 198–205. [CrossRef]
31. El-Saadony, M.T.; Alagawany, M.; Patra, A.K.; Kar, I.; Tiwari, R.; Dawood, M.A.O.; Dhama, K.; Abdel-Latif, H.M.R. The functionality of probiotics in aquaculture: An overview. *Fish Shellfish Immunol.* **2021**, *117*, 36–52. [CrossRef] [PubMed]
32. Rørstad, G.; Aasjord, P.M.; Robertsen, B. Adjuvant effect of a yeast glucan in vaccines against furunculosis in Atlantic salmon (*Salmo salar* L.). *Fish Shellfish Immunol.* **1993**, *3*, 179–190. [CrossRef]
33. Pilarski, F.; Ferreira de Oliveira, C.A.; Darpossolo de Souza, F.P.B.; Zanuzzo, F.S. Different β-glucans improve the growth performance and bacterial resistance in Nile tilapia. *Fish Shellfish Immunol.* **2017**, *70*, 25–29. [CrossRef] [PubMed]
34. Sakai, M. Current research status of fish immunostimulants. *Aquaculture* **1999**, *172*, 63–92. [CrossRef]
35. Couso, N.; Castro, R.; Magariños, B.; Obach, A.; Lamas, J. Effect of oral administration of glucans on the resistance of gilthead seabream to pasteurellosis. *Aquaculture* **2003**, *219*, 99–109. [CrossRef]
36. Bricknell, I.; Dalmo, R.A. The use of immunostimulants in fish larval aquaculture. *Fish Shellfish Immunol.* **2005**, *19*, 457–472. [CrossRef]
37. Álvarez-Rodríguez, M.; Pereiro, P.; Reyes-López, F.E.; Tort, L.; Figueras, A.; Novoa, B. Analysis of the Long-Lived Responses Induced by Immunostimulants and Their Effects on a Viral Infection in Zebrafish (*Danio rerio*). *Front. Immunol.* **2018**, *9*, 1975. [CrossRef] [PubMed]
38. Petit, J.; Wiegertjes, G.F. Long-lived effects of administering β-glucans: Indications for trained immunity in fish. *Dev. Comp. Immunol.* **2016**, *64*, 93–102. [CrossRef] [PubMed]
39. Bagni, M.; Archetti, L.; Amadori, M.; Marino, G. Effect of long-term oral administration of an immunostimulant diet on innate immunity in sea bass (*Dicentrarchus labrax*). *J. Vet. Med. B Infect. Dis. Vet. Public Health* **2000**, *47*, 745–751. [CrossRef] [PubMed]
40. Paredes, M.; Gonzalez, K.; Figueroa, J.; Montiel-Eulefi, E. Immunomodulatory effect of prolactin on Atlantic salmon (*Salmo salar*) macrophage function. *Fish Physiol. Biochem.* **2013**, *39*, 1215–1221. [CrossRef] [PubMed]
41. Cook, M.T.; Hayball, P.J.; Hutchinson, W.; Nowak, B.F.; Hayball, J.D. Administration of a commercial immunostimulant preparation, EcoActiva™ as a feed supplement enhances macrophage respiratory burst and the growth rate of snapper (*Pagrus auratus*, Sparidae (Bloch and Schneider)) in winter. *Fish Shellfish Immunol.* **2003**, *14*, 333–345. [CrossRef]
42. Ai, Q.; Mai, K.; Zhang, L.; Tan, B.; Zhang, W.; Xu, W.; Li, H. Effects of dietary β-1, 3 glucan on innate immune response of large yellow croaker, *Pseudosciaena crocea*. *Fish Shellfish Immunol.* **2007**, *22*, 394–402. [CrossRef] [PubMed]
43. Dawood, M.A.O.; Koshio, S.; Ishikawa, M.; Yokoyama, S.; El Basuini, M.F.; Hossain, M.S.; Nhu, T.H.; Moss, A.S.; Dossou, S.; Wei, H. Dietary supplementation of β-glucan improves growth performance, the innate immune response and stress resistance of red sea bream, *Pagrus major*. *Aquac. Nutr.* **2017**, *23*, 148–159. [CrossRef]

44. Bagni, M.; Romano, N.; Finoia, M.G.; Abelli, L.; Scapigliati, G.; Tiscar, P.G.; Sarti, M.; Marino, G. Short- and long-term effects of a dietary yeast β-glucan (Macrogard) and alginic acid (Ergosan) preparation on immune response in sea bass (*Dicentrarchus labrax*). *Fish Shellfish Immunol.* **2005**, *18*, 311–325. [CrossRef] [PubMed]
45. Do Huu, H.; Sang, H.M.; Thanh Thuy, N.T. Dietary β-glucan improved growth performance, Vibrio counts, haematological parameters and stress resistance of pompano fish, *Trachinotus ovatus* Linnaeus, 1758. *Fish Shellfish Immunol.* **2016**, *54*, 402–410. [CrossRef]
46. Zeng, L.; Wang, Y.-H.; Ai, C.-X.; Zhang, J.-S. Differential effects of β-glucan on oxidative stress, inflammation and copper transport in two intestinal regions of large yellow croaker *Larimichthys crocea* under acute copper stress. *Ecotoxicol. Environ. Saf.* **2018**, *165*, 78–87. [CrossRef]
47. Carballo, C.; Pinto, P.I.S.; Mateus, A.P.; Berbel, C.; Guerreiro, C.C.; Martinez-Blanch, J.F.; Codoñer, F.M.; Mantecon, L.; Power, D.M.; Manchado, M. Yeast β-glucans and microalgal extracts modulate the immune response and gut microbiome in Senegalese sole (*Solea senegalensis*). *Fish Shellfish Immunol* **2019**, *92*, 31–39. [CrossRef]
48. Aramli, M.S.; Kamangar, B.; Nazari, R.M. Effects of dietary β-glucan on the growth and innate immune response of juvenile Persian sturgeon, *Acipenser persicus*. *Fish Shellfish Immunol.* **2015**, *47*, 606–610. [CrossRef] [PubMed]
49. Kühlwein, H.; Merrifield, D.L.; Rawling, M.D.; Foey, A.D.; Davies, S.J. Effects of dietary β-(1,3)(1,6)-D-glucan supplementation on growth performance, intestinal morphology and haemato-immunological profile of mirror carp (*Cyprinus carpio* L.). *J. Anim. Physiol. Anim. Nutr.* **2014**, *98*, 279–289. [CrossRef]
50. Yamamoto, F.Y.; Sutili, F.J.; Hume, M.; Gatlin, D.M., III. The effect of β-1,3-glucan derived from Euglena gracilis (Algamune™) on the innate immunological responses of Nile tilapia (*Oreochromis niloticus* L.). *J. Fish Dis.* **2018**, *41*, 1579–1588. [CrossRef]
51. Sahoo, P.K.; Mukherjee, S.C. Effect of dietary β-1,3 glucan on immune responses and disease resistance of healthy and aflatoxin B1-induced immunocompromised rohu (*Labeo rohita* Hamilton). *Fish Shellfish Immunol.* **2001**, *11*, 683–695. [CrossRef] [PubMed]
52. Pionnier, N.; Falco, A.; Miest, J.; Frost, P.; Irnazarow, I.; Shrive, A.; Hoole, D. Dietary β-glucan stimulate complement and C-reactive protein acute phase responses in common carp (*Cyprinus carpio*) during an *Aeromonas salmonicida* infection. *Fish Shellfish Immunol.* **2013**, *34*, 819–831. [CrossRef] [PubMed]
53. Mourente, G.; Díaz-Salvago, E.; Bell, J.G.; Tocher, D.R. Increased activities of hepatic antioxidant defence enzymes in juvenile gilthead sea bream (*Sparus aurata* L.) fed dietary oxidised oil: Attenuation by dietary vitamin E. *Aquaculture* **2002**, *214*, 343–361. [CrossRef]
54. Bobek, P.; Nosálová, V.; Cerná, S. Effect of pleuran (beta-glucan from *Pleurotus ostreatus*) in diet or drinking fluid on colitis in rats. *Food/Nahrung* **2001**, *45*, 360–363. [CrossRef]
55. Kim, Y.-S.; Ke, F.; Zhang, Q.-Y. Effect of β-glucan on activity of antioxidant enzymes and Mx gene expression in virus infected grass carp. *Fish Shellfish Immunol.* **2009**, *27*, 336–340. [CrossRef] [PubMed]
56. Sant, K.E.; Hansen, J.M.; Williams, L.M.; Tran, N.L.; Goldstone, J.V.; Stegeman, J.J.; Hahn, M.E.; Timme-Laragy, A. The role of Nrf1 and Nrf2 in the regulation of glutathione and redox dynamics in the developing zebrafish embryo. *Redox Biol.* **2017**, *13*, 207–218. [CrossRef] [PubMed]
57. Wasserman, W.W.; Fahl, W.E. Functional antioxidant responsive elements. *Proc. Natl. Acad. Sci. USA* **1997**, *94*, 5361–5366. [CrossRef]
58. Neamat-Allah, A.N.F.; Abd El Hakim, Y.; Mahmoud, E.A. Alleviating effects of β-glucan in *Oreochromis niloticus* on growth performance, immune reactions, antioxidant, transcriptomics disorders and resistance to *Aeromonas sobria* caused by atrazine. *Aquac. Res.* **2020**, *51*, 1801–1812. [CrossRef]
59. Carballo, C.; Chronopoulou, E.G.; Letsiou, S.; Maya, C.; Labrou, N.E.; Infante, C.; Power, D.M.; Manchado, M. Antioxidant capacity and immunomodulatory effects of a chrysolaminarin-enriched extract in Senegalese sole. *Fish Shellfish Immunol.* **2018**, *82*, 1–8. [CrossRef]
60. Le Costaouëc, T.; Unamunzaga, C.; Mantecon, L.; Helbert, W. New structural insights into the cell-wall polysaccharide of the diatom *Phaeodactylum tricornutum*. *Algal. Res.* **2017**, *26*, 172–179. [CrossRef]
61. Fink, I.R.; Ribeiro, C.M.S.; Forlenza, M.; Taverne-Thiele, A.; Rombout, J.H.W.M.; Savelkoul, H.F.J.; Wiegertjes, G.F. Immune-relevant thrombocytes of common carp undergo parasite-induced nitric oxide-mediated apoptosis. *Dev. Comp. Immunol.* **2015**, *50*, 146–154. [CrossRef]
62. Nagasawa, T.; Nakayasu, C.; Rieger, A.M.; Barreda, D.R.; Somamoto, T.; Nakao, M. Phagocytosis by Thrombocytes is a Conserved Innate Immune Mechanism in Lower Vertebrates. *Front. Immunol.* **2014**, *5*, 445. [CrossRef] [PubMed]
63. Akramiene, D.; Aleksandraviciene, C.; Grazeliene, G.; Zalinkevicius, R.; Suziedelis, K.; Didziapetriene, J.; Simonsen, U.; Stankevicius, E.; Kevelaitis, E. Potentiating effect of beta-glucans on photodynamic therapy of implanted cancer cells in mice. *Tohoku J. Exp. Med.* **2010**, *220*, 299–306. [CrossRef]
64. Lavi, I.; Nimri, L.; Levinson, D.; Peri, I.; Hadar, Y.; Schwartz, B. Glucans from the edible mushroom *Pleurotus pulmonarius* inhibit colitis-associated colon carcinogenesis in mice. *J. Gastroenterol.* **2012**, *47*, 504–518. [CrossRef]
65. Djordjevic, B.; Škugor, S.; Jørgensen, S.M.; Øverland, M.; Mydland, L.T.; Krasnov, A. Modulation of splenic immune responses to bacterial lipopolysaccharide in rainbow trout (*Oncorhynchus mykiss*) fed lentinan, a beta-glucan from mushroom *Lentinula edodes*. *Fish Shellfish Immunol.* **2009**, *26*, 201–209. [CrossRef]

66. Goodridge, H.S.; Reyes, C.N.; Becker, C.A.; Katsumoto, T.R.; Ma, J.; Wolf, A.J.; Bose, N.; Chan, A.S.; Magee, A.S.; Danielson, M.E.; et al. Activation of the innate immune receptor Dectin-1 upon formation of a 'phagocytic synapse'. *Nature* **2011**, *472*, 471–475. [CrossRef]
67. Qi, C.; Cai, Y.; Gunn, L.; Ding, C.; Li, B.; Kloecker, G.; Qian, K.; Vasilakos, J.; Saijo, S.; Iwakura, Y.; et al. Differential pathways regulating innate and adaptive antitumor immune responses by particulate and soluble yeast-derived β-glucans. *Blood* **2011**, *117*, 6825–6836. [CrossRef] [PubMed]
68. Bose, N.; Chan, A.S.; Guerrero, F.; Maristany, C.; Walsh, R.; Ertelt, K.; Jonas, A.; Gorden, K.; Dudney, C.; Wurst, L.; et al. Binding of Soluble Yeast β-Glucan to Human Neutrophils and Monocytes is Complement-Dependent. *Front. Immunol.* **2013**, *4*, 230. [CrossRef]
69. Kopiasz, Ł.; Dziendzikowska, K.; Gajewska, M.; Wilczak, J.; Harasym, J.; Żyła, E.; Kamola, D.; Oczkowski, M.; Królikowski, T.; Gromadzka-Ostrowska, J. Time-Dependent Indirect Antioxidative Effects of Oat Beta-Glucans on Peripheral Blood Parameters in the Animal Model of Colon Inflammation. *Antioxidants* **2020**, *9*, 375. [CrossRef] [PubMed]
70. Żyła, E.; Dziendzikowska, K.; Gajewska, M.; Wilczak, J.; Harasym, J.; Gromadzka-Ostrowska, J. Beneficial Effects of Oat Beta-Glucan Dietary Supplementation in Colitis Depend on Its Molecular Weight. *Molecules* **2019**, *24*, 3591. [CrossRef] [PubMed]
71. Sonck, E.; Stuyven, E.; Goddeeris, B.; Cox, E. The effect of β-glucans on porcine leukocytes. *Vet. Immunol. Immunopathol.* **2010**, *135*, 199–207. [CrossRef]
72. Russo, R.; Barsanti, L.; Evangelista, V.; Frassanito, A.M.; Longo, V.; Pucci, L.; Penno, G.; Gualtieri, P. *Euglena gracilis* paramylon activates human lymphocytes by upregulating pro-inflammatory factors. *Food Sci. Nutr.* **2017**, *5*, 205–214. [CrossRef]
73. Derwenskus, F.; Metz, F.; Gille, A.; Schmid-Staiger, U.; Briviba, K.; Schließmann, U.; Hirth, T. Pressurized extraction of unsaturated fatty acids and carotenoids from wet *Chlorella vulgaris* and *Phaeodactylum tricornutum* biomass using subcritical liquids. *GCB Bioenergy* **2019**, *11*, 335–344. [CrossRef]
74. Mylonas, C.C.; Cardinaletti, G.; Sigelaki, I.; Polzonetti-Magni, A. Comparative efficacy of clove oil and 2-phenoxyethanol as anesthetics in the aquaculture of European sea bass (*Dicentrarchus labrax*) and gilthead sea bream (*Sparus aurata*) at different temperatures. *Aquaculture* **2005**, *246*, 467–481. [CrossRef]
75. Afonso, A.; Ellis, A.E.; Silva, M.T. The leucocyte population of the unstimulated peritoneal cavity of rainbow trout (*Oncorhynchus mykiss*). *Fish Shellfish Immunol.* **1997**, *7*, 335–348. [CrossRef]
76. Machado, M.; Azeredo, R.; Díaz-Rosales, P.; Afonso, A.; Peres, H.; Oliva-Teles, A.; Costas, B. Dietary tryptophan and methionine as modulators of European seabass (*Dicentrarchus labrax*) immune status and inflammatory response. *Fish Shellfish Immunol.* **2015**, *42*, 353–362. [CrossRef]
77. Ellis, A.E.; Cavaco, A.; Petrie, A.; Lockhart, K.; Snow, M.; Collet, B. Histology, immunocytochemistry and qRT-PCR analysis of Atlantic salmon, *Salmo salar* L., post-smolts following infection with infectious pancreatic necrosis virus (IPNV). *J. Fish Dis.* **2010**, *33*, 803–818. [CrossRef] [PubMed]
78. Bird, R.P.; Draper, H.H. Comparative studies on different methods of malonaldehyde determination. *Methods Enzymol.* **1984**, *105*, 299–305. [PubMed]
79. Claiborne, A. Catalase activity. In *Handbook of Methods for Oxygen Radical Research*; Greenwald, R.A., Ed.; CRC Press Inc.: Boca Raton, FL, USA, 1984; pp. 283–284.
80. Flohé, L.; Otting, F. Superoxide dismutase assays. *Methods Enzymol.* **1984**, *105*, 93–104.
81. Lima, I.; Moreira, S.M.; Osten, J.R.-V.; Soares, A.M.V.M.; Guilhermino, L. Biochemical responses of the marine mussel *Mytilus galloprovincialis* to petrochemical environmental contamination along the North-western coast of Portugal. *Chemosphere* **2007**, *66*, 1230–1242. [CrossRef] [PubMed]
82. Tietze, F. Enzymic method for quantitative determination of nanogram amounts of total and oxidized glutathione: Applications to mammalian blood and other tissues. *Anal. Biochem.* **1969**, *27*, 502–522. [CrossRef]
83. Baker, M.A.; Cerniglia, G.J.; Zaman, A. Microtiter plate assay for the measurement of glutathione and glutathione disulfide in large numbers of biological samples. *Anal. Biochem.* **1990**, *190*, 360–365. [CrossRef]
84. Livak, K.J.; Schmittgen, T.D. Analysis of relative gene expression data using real-time quantitative PCR and the 2(-Delta Delta C(T)) Method. *Methods* **2001**, *25*, 402–408. [CrossRef]
85. Wold, S.; Sjöström, M.; Eriksson, L. PLS-regression: A basic tool of chemometrics. *Chemometr. Intell. Lab. Syst.* **2001**, *58*, 109–130. [CrossRef]
86. Li, H.; Ma, M.-L.; Luo, S.; Zhang, R.-M.; Han, P.; Hu, W. Metabolic responses to ethanol in *Saccharomyces cerevisiae* using a gas chromatography tandem mass spectrometry-based metabolomics approach. *Int. J. Biochem. Cell Biol.* **2012**, *44*, 1087–1096. [CrossRef]
87. Kieffer, D.A.; Piccolo, B.D.; Vaziri, N.D.; Liu, S.; Lau, W.L.; Khazaeli, M.; Nazertehrani, S.; Moore, M.E.; Marco, M.L.; Martin, R.J.; et al. Resistant starch alters gut microbiome and metabolomic profiles concurrent with amelioration of chronic kidney disease in rats. *Am. J. Physiol.-Renal Physiol.* **2016**, *310*, F857–F871. [CrossRef] [PubMed]

marine drugs

Article

Exogenous Antioxidants Improve the Accumulation of Saturated and Polyunsaturated Fatty Acids in *Schizochytrium* sp. PKU#Mn4

Sai Zhang [1,2,†], Xiaohong Chen [1,†], Biswarup Sen [1], Mohan Bai [1], Yaodong He [1,*] and Guangyi Wang [1,3,*]

1 Center for Marine Environmental Ecology, School of Environmental Science and Engineering, Tianjin University, Tianjin 300072, China; zhangsai@tju.edu.cn (S.Z.); xh_chen2011@tju.edu.cn (X.C.); bsen@tju.edu.cn (B.S.); bmh@zju.edu.cn (M.B.)
2 Polar Research Institute of China, Shanghai 200136, China
3 Key Laboratory of Systems Bioengineering (Ministry of Education), Tianjin University, Tianjin 300072, China
* Correspondence: yaodong.he@tju.edu.cn (Y.H.); gywang@tju.edu.cn (G.W.)
† The two authors contributed equally to this paper.

Abstract: Species of *Schizochytrium* are well known for their remarkable ability to produce lipids intracellularly. However, during their lipid accumulation, reactive oxygen species (ROS) are generated inevitably as byproducts, which if in excess results in lipid peroxidation. To alleviate such ROS-induced damage, seven different natural antioxidants (ascorbic acid, α-tocopherol, tea extract, melatonin, mannitol, sesamol, and butylated hydroxytoluene) were evaluated for their effects on the lipid accumulation in *Schizochytrium* sp. PKU#Mn4 using a fractional factorial design. Among the tested antioxidants, mannitol showed the best increment (44.98%) in total fatty acids concentration. However, the interaction effects of mannitol (1 g/L) and ascorbic acid (1 g/L) resulted in 2.26 ± 0.27 g/L and 1.45 ± 0.04 g/L of saturated and polyunsaturated fatty acids (SFA and PUFA), respectively, in batch fermentation. These concentrations were further increased to 7.68 ± 0.37 g/L (SFA) and 5.86 ± 0.03 g/L (PUFA) through fed-batch fermentation. Notably, the interaction effects yielded 103.7% and 49.6% increment in SFA and PUFA concentrations in batch fermentation. The possible mechanisms underlining those increments were an increased maximum growth rate of strain PKU#Mn4, alleviated ROS level, and the differential expression of lipid biosynthetic genes andupregulated catalase gene. This study provides an applicable strategy for improving the accumulation of SFA and PUFA in thraustochytrids by exogenous antioxidants and the underlying mechanisms.

Keywords: thraustochytrids; antioxidants; saturated fatty acids; polyunsaturated fatty acids; reactive oxygen species; transcriptomics

Citation: Zhang, S.; Chen, X.; Sen, B.; Bai, M.; He, Y.; Wang, G. Exogenous Antioxidants Improve the Accumulation of Saturated and Polyunsaturated Fatty Acids in *Schizochytrium* sp. PKU#Mn4. *Mar. Drugs* **2021**, *19*, 559. https://doi.org/10.3390/md19100559

Academic Editor: Carlos Almeida

Received: 3 September 2021
Accepted: 27 September 2021
Published: 30 September 2021

1. Introduction

Thraustochytrids are marine unicellular heterotrophic protists that can naturally accumulate lipids up to 55% of their dry biomass [1] and are promising cell factories for high-value polyunsaturated fatty acids (PUFA), saturated fatty acids (SFA), and terpenoids [2]. The thraustochytrid strains belonging to genera *Aurantiochytrium*, *Schizochytrium*, and *Thraustochytrium* are particularly known for their extraordinary ability to produce lipids during glucose or glycerol fermentation [3–6]. Over the past two decades, several strategies towards enhancing the production of high-value fatty acids in oleaginous thraustochytrids were reported [7–12]. The environmental stressors inducing cellular stress in thraustochytrids, such as nitrogen starvation [13,14] and low temperature [15,16], are traditionally employed in improving the biomass and fatty acids content. However, the major limitation of most enhancement strategies is that they also induce intracellular accumulation of reactive oxygen species (ROS), which are reported to inhibit growth and lower the fatty acids content [17].

Considerable efforts were made in the past few decades to optimize the culture conditions for the efficient production of fatty acids by oleaginous microalgae [18]. Particularly, strategies based on ROS regulation were recently proposed to avoid the undesirable cell death and peroxidation of fatty acids induced by excessive ROS [18,19]. For example, the manipulation of the enzymatic antioxidant system through the overexpression of superoxide dismutase has shown up to a 38.5% decline in ROS level along with 32.9% higher PUFA production in the engineered strain of *Schizochytrium* sp. [20]. Similarly, an exogenous superoxide reductase gene expression in cyanobacteria also suppressed ROS accumulation and lipid peroxidation [21]. However, public health concerns over genetically modified strains limit the application of such strategies in the nutraceutical industry.

Approaches that enhance the intracellular antioxidant capacity seem promising for overcoming the ROS-induced adverse effects. One such approach towards the manipulation of antioxidant capacity includes the application of antioxidant supplementation [22]. Exogenous antioxidants such as ascorbic acid, phenolic compounds, etc., are reported as an important component of the ROS scavenging system [23]. These compounds could alleviate ROS levels by acting as an electron donor or by suppressing ROS generation through the chelation of transition metal ions [24]. Previous studies have shown promising results of exogenous antioxidants on the growth and lipid production in microalgae. For instance, the addition of melatonin lowered the ROS levels in *Monoraphidium* sp. QLY-1 and increased lipid accumulation by 1.22-fold [25]. The addition of sesamol increased the cell growth and PUFA synthesis in *Crypthecodinium cohnii* [26]. With the exogenous addition of ascorbic acid (9 g/L), the yield of docosahexaenoic acid (DHA) was increased by 30.4% in *Schizochytrium* sp. [27]. Similarly, the addition of flaxseed oil [28] or propyl gallate [29] to the culture medium of *Schizochytrium* sp. significantly improved the DHA yield or lipid accumulation, respectively. However, most of these studies focused on the evaluation of singular antioxidants and not on the interaction effects of mixed antioxidants. Moreover, the mechanisms reinforcing the elevated accumulation of lipids in thraustochytrids as a result of exogenous antioxidants addition remain poorly understood.

In this study, seven different inexpensive, natural antioxidants were screened for their effects on the cell growth and lipid production capacity of *Schizochytrium* sp. PKU# Mn4. The optimal levels of the best antioxidants and their interaction effects were determined and the possible mechanisms underlining the effects of antioxidant supplementation were also investigated. This study provides several lines of evidence that suggest that the supplementation of exogenous antioxidants is an effective strategy towards improving the accumulation of SFA and PUFA in thraustochytrids.

2. Results

2.1. Screening Potential Antioxidants

The ANOVA models for dry cell weight (DCW) and total fatty acids (TFA) yield were found to be significant ($p < 0.05$) in the 2-level fractional factorial design (FFD) screening experiment (Tables 1 and 2). α-tocopherol and mannitol were significant model terms ($p < 0.05$) for DCW, while α-tocopherol, melatonin, and ascorbic acid interactions with mannitol, butylated hydroxytoluene (BHT), or sesamol were significant ($p < 0.05$) terms for TFA yield. The antioxidants, namely α-tocopherol, melatonin, and mannitol with significant effects on the DCW and TFA yield were selected for further concentration optimization experiments.

Table 1. Low- and high-level concentrations of individual antioxidants.

Factor	Antioxidant	Low Level (−1)	High Level (+1)
A	Ascorbic acid	9.0 g/L	13.5 g/L
B	α-tocopherol	0.50 g/L	0.75 g/L
C	Tea extract	0.50 g/L	0.75 g/L
D	Melatonin	0.25 mg/L	0.375 mg/L
E	Mannitol	1.0 g/L	1.5 g/L
F	Sesamol	70 mg/L	105 mg/L
G	BHT	2 mg/L	3 mg/L

Table 2. Significance of ANOVA model terms for the response variables DCW and TFA yield.

Source	*p*-Value	Source	*p*-Value
Response 1: DCW (g/L)		Response 2: TFA (mg/g DCW)	
Model	0.0447	Model	0.0180
A: ascorbic acid	0.3917	A: ascorbic acid	0.6802
B: α-tocopherol	0.0406	B: α-tocopherol	0.0203
C: tea extract	0.1094	D: melatonin	0.0206
D: melatonin	0.1397	E: mannitol	0.1194
E: mannitol	0.0141	F: sesamol	0.1643
F: sesamol	0.4890	G: BHT	0.6473
AE	0.1069	AE	0.0487
AF	0.1061	AF	0.0166
		AG	0.0047
		BD	0.0697

2.2. *Effects of Different Concentrations of Single Antioxidants*

The concentration optimization of α-tocopherol, melatonin, and mannitol revealed that these exogenous antioxidants show concentration-dependent effects. While these antioxidants showed little impact on DCW (Figure S1), their effects on the TFA accumulation were significant (Figure 1). The supplementation of α-tocopherol at 50 mg/L increased the TFA concentration by 45.85% (2.12 ± 0.10 g/L), and the PUFA and SFA concentrations by 70.85% (1.13 ± 0.07 g/L) and 26.21% (0.96 ± 0.04 g/L), respectively. With the optimal melatonin concentration (0.4 mg/L), the TFA concentration increased by 33.55% (1.94 ± 0.06 g/L), and the PUFA and SFA concentrations increased by 35.71% (0.90 ± 0.03 g/L) and 32.18% (1.0 ± 0.02 g/L), respectively. On the other hand, the optimal mannitol concentration (1.0 g/L) increased the TFA concentration by 44.98% (2.22 ± 0.19 g/L), and the PUFA and SFA concentrations by 56.31% (1.03 ± 0.11 g/L) and 28.96% (1.13 ± 0.08 g/L), respectively. Of these three tested antioxidants, the supplementation of mannitol (1.0 g/L) yielded the best TFA concentration.

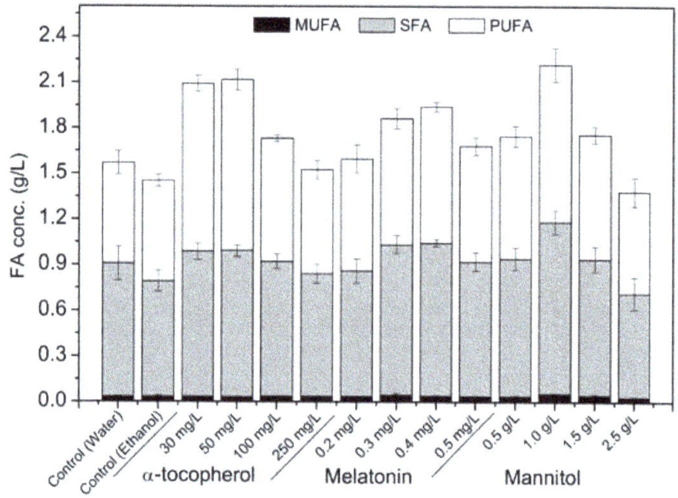

Figure 1. Effects of α-tocopherol, melatonin, and mannitol on the concentrations of different fatty acids accumulated by *Schizochytrium* PKU#Mn4 in batch culture. The data are expressed as mean ± SD of triplicate experiments.

2.3. Interaction Effects of Ascorbic Acid and Mannitol in Batch Culture

The DCW and TFA concentration declined with the increasing concentration of ascorbic acid in the mannitol and ascorbic acid supplement (Table 3). With the mannitol (1 g/L) and ascorbic acid (1 g/L) supplementation (MA), the DCW and TFA concentration increased by up to 54.1% (8.49 ± 0.18 g/L) and 78.2% (3.78 ± 0.26 g/L), respectively. The TFA produced with MA constituted 2.26 ± 0.27 g/L SFA, 1.45 ± 0.04 g/L PUFA, and 0.08 ± 0.01 g/L MUFA, which increased by 103.7%, 49.6%, and 82.0%, respectively. Furthermore, the addition of ascorbic acid at the concentration range of 1 to 6 g/L showed a notable rise in the SFA/PUFA ratio when compared with that without supplementation.

Table 3. Interaction effects of various combinations of antioxidants on the biomass and fatty acids production in batch culture of *Schizochytrium* PKU#Mn4.

Treatment	DCW (g/L)	MUFA (g/L)	SFA (g/L)	PUFA (g/L)	SFA/PUFA
Control	5.89 ± 0.23	0.03 ± 0.00	0.88 ± 0.11	0.66 ± 0.08	1.32
Mannitol (1 g/L)	5.51 ± 0.33	0.04 ± 0.01	1.11 ± 0.17	0.97 ± 0.13	1.14
Mannitol (1 g/L) + Ascorbic acid (1 g/L)	8.49 ± 0.18	0.08 ± 0.01	2.26 ± 0.27	1.45 ± 0.04	1.56
Mannitol (1 g/L) + Ascorbic acid (2 g/L)	7.96 ± 0.28	0.06 ± 0.01	2.26 ± 0.07	1.19 ± 0.09	1.89
Mannitol (1 g/L) + Ascorbic acid (3 g/L)	6.79 ± 0.12	0.04 ± 0.00	1.87 ± 0.16	1.01 ± 0.14	1.85
Mannitol (1 g/L) + Ascorbic acid (6 g/L)	4.67 ± 0.33	0.04 ± 0.00	1.31 ± 0.19	0.73 ± 0.07	1.80
Mannitol (1 g/L) + Ascorbic acid (9 g/L)	3.38 ± 0.00	0.02 ± 0.00	0.79 ± 0.06	0.59 ± 0.02	1.35
Mannitol (1 g/L) + Ascorbic acid (12 g/L)	1.84 ± 0.03	0.01 ± 0.00	0.45 ± 0.02	0.34 ± 0.03	1.32
Mannitol (1 g/L) + Ascorbic acid (15 g/L)	1.22 ± 0.02	0.01 ± 0.00	0.29 ± 0.02	0.28 ± 0.03	1.07

Note: The results are expressed as mean ± SD of triplicate experiments.

The time course of DCW, SFA, and PUFA concentrations upon MA supplementation exhibited sigmoid patterns (Figure 2), and the modified Gompertz model fitted these data well (Figure S2). The model predicted a maximum growth rate of 0.118 $g/L \cdot h^{-1}$ for the culture with the supplement, which was 15.7% higher than that of the culture without the supplement (Table 4). Similarly, the model also predicted the maximum accumulation rates of SFA and PUFA to be 0.047 g/L·h^{-1} and 0.056 g/L·h^{-1}, respectively, which were 123.8% and 86.7% higher than that of without supplementation. Both experimental (Figure 2) and predicted (Figure S2) data revealed a marked increase in DCW, SFA, and PUFA concentrations of the culture with supplementation compared with that of without supplementation from 36 h until the end of fermentation.

Table 4. Estimates of modified Gompertz model parameters after fitting experimental data.

Dependent Variable		*a* (g/L)	*R* (g/L·h^{-1})	λ (h)	Residual Standard Error
DCW	w/o MA	5.799 ***	0.102 ***	−1.791	0.2223
	w/MA	9.276 ***	0.118 ***	3.478	0.2045
SFA	w/o MA	0.977 ***	0.021 ***	1.763	0.0434
	w/MA	2.191 ***	0.047 ***	9.832 *	0.1017
PUFA	w/o MA	0.729 ***	0.030 ***	9.069 ***	0.0194
	w/MA	1.369 ***	0.056 **	13.723 **	0.0948

Note: 'w/o MA' and 'w/MA' stand for without and with mannitol (1 g/L) and ascorbic acid (1 g/L) supplementation; significance codes: *** 0.001, ** 0.01, * 0.05.

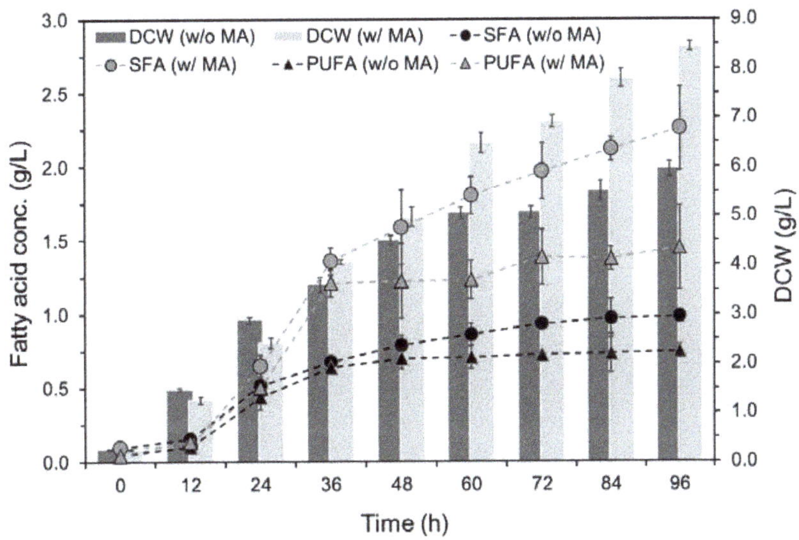

Figure 2. Time course of dry cell weight and concentrations of fatty acids accumulated by *Schizochytrium* PKU#Mn4 in batch culture supplemented with mannitol (1 g/L) and ascorbic acid (1 g/L) mixture. 'w/o MA' and 'w/MA' stand for without and with mannitol (1 g/L) and ascorbic acid (1 g/L) supplementation. The data are expressed as mean ± SD of triplicate experiments.

The time course of intracellular ROS level in the batch culture with and without MA supplementation exhibited somewhat different patterns (Figure 3). Without the supplement, the ROS level declined until the first 24 h but increased thereafter until the end of fermentation. Whereas, with the supplement, the ROS level decreased until 48 h and then increased gradually until the end of fermentation. Notably, the ROS levels were significantly ($p < 0.05$) lower in the culture with MA supplementation from 36 h until the end of fermentation.

2.4. Effects of Mannitol and Ascorbic Acid Supplemention on Fed-Batch Culture

In the fed-batch fermentation, the accumulation of the fatty acid with supplementation increased until the end of fermentation (i.e., day 8), while their accumulation without supplementation increased only until day 6 of fermentation (Figure 4). The maximum growth with or without MA was achieved at the same time (i.e., day 6); however, the growth with MA remained significantly higher from day 4 until the end of the fermentation. The maximum DCW achieved with MA in fed-batch fermentation was 25.03 ± 0.51 g/L, an increase of 19.71%. The accumulation patterns of SFA and PUFA were strongly ($p < 0.001$) correlated with the growth patterns during fermentation with and without MA. However, the accumulation of these fatty acids was generally better with MA. The MA supplementation yielded the maximum SFA and PUFA concentrations of 7.68 ± 0.37 g/L and 5.86 ± 0.03 g/L, which were 38.1% and 40.2% greater than that of the culture without MA supplementation.

2.5. Effects of Mannitol and Ascorbic Acid Supplementation at the Transcriptome Level

The transcriptome-level changes were analyzed for the cells harvested from the culture samples collected at 48 h of batch fermentation. A total of 33,813,639 (with MA group) and 21,621,092 (without MA group) clean reads were obtained, and over 94.42% and 93.42% of these reads were mapped to the reference genome. Further analyses identified 1210 differentially expressed genes (DEGs), of which 577 and 633 were upregulated and downregulated in the group with MA supplementation, respectively. The DEGs were further annotated against the GO database for functional information, which revealed that 39.79%, 36.81%, and 23.4% of the DEGs belonged to the biological process category, cellular component, and molecular function, respectively (Figure S3).

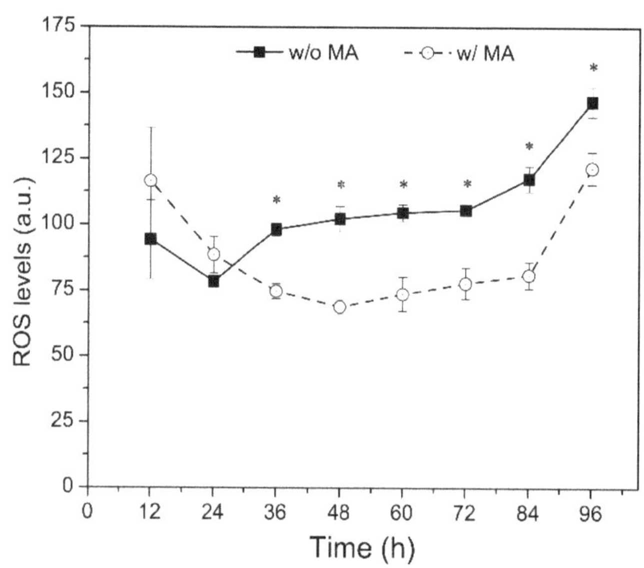

Figure 3. Time course of intracellular ROS levels in the batch culture of *Schizochytrium* PKU# Mn4 supplemented with mannitol (1 g/L) and ascorbic acid (1 g/L) mixture. * indicates the data have statistical significance at $p < 0.05$. 'w/o MA' and 'w/MA' stand for without and with mannitol (1 g/L) and ascorbic acid (1 g/L) supplementation. The data are expressed as mean ± SD of triplicate experiments.

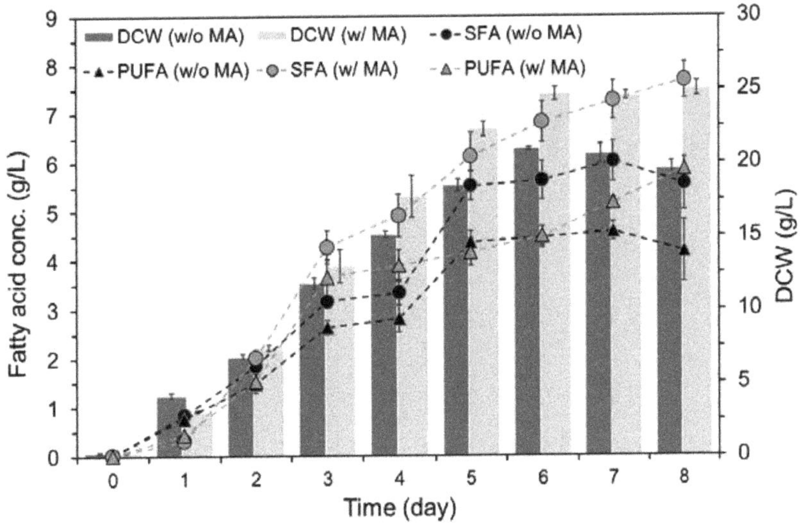

Figure 4. Time course of dry cell weight and concentrations of fatty acids accumulated by *Schizochytrium* PKU#Mn4 in fed-batch culture (5 L) supplemented with the antioxidant mixture (MA). 'w/o MA' and 'w/MA' stand for without and with mannitol (1 g/L) and ascorbic acid (1 g/L) supplementation. The data are expressed as mean ± SD of triplicate experiments.

Several DEGs and transcripts related to the generation of acetyl-CoA and malonyl-CoA were upregulated, including genes encoding glyceraldehyde-3-phosphate dehydrogenase (GAPDH), pyruvate dehydrogenase transcripts (PDHX1, PDHX2, and PDHB), malonate-semialdehyde dehydrogenase (ALDH6A1), and acetyl-CoA carboxylase (ACC) (Figure 5). Two transcripts of the ketoacyl-reductase (KR) gene involved in the biosynthesis of fatty acids from the pool of acetyl-CoA and malonyl-CoA were also upregulated. However, one transcript of the KR gene and two genes encoding ω-3 desaturase (FAD3) and polyketide synthase (PKS) were downregulated at the same time. Furthermore, two transcripts encoding diacylglycerol pyrophosphate phosphatase 1 ($DPP1_1$ and $DPP1_2$) and one transcript encoding glycerol-3-phosphate O-acyltransferase 1 (GPAT1) involved in the process of lipid biosynthesis were upregulated. While the gene encoding triacylglycerol lipase (LIP) was downregulated concurrently.

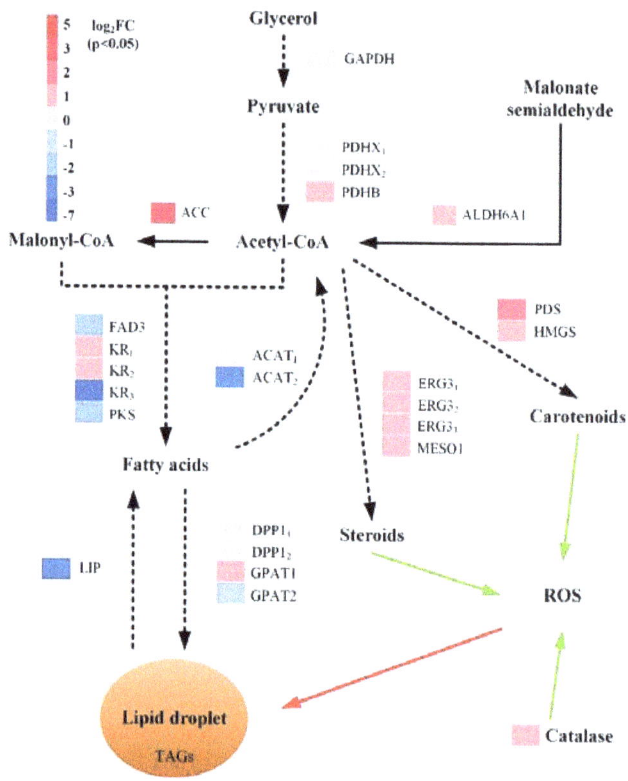

Figure 5. Differentially abundant genes and transcripts involved in lipid metabolism and ROS detoxification upon supplementation of mannitol (1 g/L) and ascorbic acid (1 g/L) mixture in batch fermentation. The solid and dotted black arrows indicate single-step and multiple-step pathways, respectively. The red solid arrow indicates the route of ROS-induced lipid peroxidation and the green solid arrow indicates the routes for reduction in ROS. GAPDH, glyceraldehyde-3-phosphate dehydrogenase; PDHX, pyruvate dehydrogenase complex; PDHB, pyruvate dehydrogenase E1 component subunit beta; ALDH6A1, malonate-semialdehyde dehydrogenase; ACC, acetyl-CoA carboxylase; FAD3, ω-3 desaturase; KR, ketoacyl-reductase; PKS, polyketide synthase; ACAT, acetyl-CoA acetyltransferase; DPP1, diacylglycerol pyrophosphate phosphatase 1; GPAT1, glycerol-3-phosphate O-acyltransferase 1; GPAT2, glycerol-3-phosphate O-acyltransferase 2; LIP, triacylglycerol lipase; ERG3, delta(7)-sterol 5(6)-desaturase; MESO1, methylsterol monooxygenase; PDS, phytoene desaturase; HMGS, hydroxymethylglutaryl-CoA synthase; CAT, catalase.

In addition to the genes and transcripts involved in the lipid biosynthetic pathway, certain genes and transcripts encoding enzymes that catalyze the biosynthesis of steroids and carotenoids were also differentially expressed. Those included three transcripts of gene encoding delta(7)-sterol 5(6)-desaturase ($ERG3_1$, $ERG3_2$, and $ERG3_3$), and genes encoding methylsterol monooxygenase 1 (MSMO1), phytoene desaturase (PDS), and hydroxymethylglutaryl-CoA synthase (HMGS), which were all upregulated. Furthermore, the gene encoding the enzyme catalase (CAT), which neutralizes hydrogen peroxide, was upregulated.

3. Discussion

3.1. Antioxidants Affect Growth and Fatty Acids Accumulation

The supplementation of antioxidants during fermentation were one of the most effective strategies to control the intracellular ROS levels and improve the production of fatty acids for several microbial strains [22]. In this study, the effects of seven different antioxidants and their interactions on the growth and fatty acid production were first evaluated and then the optimal concentrations of the potential antioxidants were determined. The most potent antioxidants, namely α-tocopherol, mannitol, and melatonin, at their optimal concentrations could significantly improve the production of fatty acids (Figure 1). However, each of these antioxidants yielded different effects on the accumulation of SFA and PUFA. For example, the supplementation of α-tocopherol and mannitol increased the accumulation of PUFA more than SFA, while the melatonin addition uniformly raised the accumulation of SFA and PUFA. Such differential effects of antioxidants can be useful while manipulating the selective fatty acid production for application in the biodiesel or nutraceuticals industry.

This study revealed that not all seven antioxidants are effective in improving the accumulation of the fatty acids in the strain PKU#Mn4. For instance, BHT was reported to enhance the lipid content of *Haematococcus pluvialis* LUGU and stimulate lipid accumulation in *Schizochytrium* sp. S31 [29,30]. However, in this study, the BHT supplementation did not show a promising effect on the lipid production of strain PKU#Mn4. Similarly, ascorbic acid, which was previously reported to improve the yield of certain PUFA in *Schizochytrium* sp. HX-308 by 30.44% [27], did show any significant effects on the accumulation of fatty acids in strain PKU#Mn4. Nevertheless, our results showed that melatonin, which was previously investigated for its application in improving fatty acid productivity in autotrophic microalgae [31–33], could markedly increase the accumulation of SFA and PUFA in thraustochytrids.

Based on the ANOVA results of the 2-level FFD experiment (Table 1), we evaluated the potential interaction effects of mannitol and ascorbic acid mixture by varying the concentration of ascorbic acid in the mixture. Our results suggested that the interaction between these two antioxidants could significantly promote the accumulation of SFA and PUFA under both batch and fed-batch fermentation conditions. Furthermore, it was realized that alleviation of ROS level during fermentation (Figure 3), and increased maximum growth rate of strain PKU#Mn4 (Table 4) were probable physiological mechanisms underlining the increased growth and accumulation of SFA and PUFA with MA supplement.

3.2. Exogenous Antioxidants Trigger Differential Expression of Lipid Biosynthetic Genes

The possible molecular mechanisms supporting the increased accumulation of fatty acids with MA supplementation were inferred by identifying the DEGs involved in lipid metabolism. Several DEGs were identified which catalyze reactions in the acetyl-CoA, fatty acids, lipids, steroids, and carotenoid biosynthetic pathways. One of those DEGs, encoding acetyl-CoA carboxylase, catalyzes the conversion of acetyl-CoA to malonyl-CoA—the rate-limiting step of fatty acid biosynthesis [34]. The upregulation of this gene and concomitant increase of fatty acids observed in the present study was in close agreement with the previous studies which have shown that increased expression of *acc* gene increases the precursor pool and promotes fatty acid biosynthesis [35–37]. Furthermore, the downregulation of the gene encoding enzyme LIP, which catalyzes the hydrolysis of triacylglycerols, suggested that antioxidant supplement may also suppress lipid degradation in strain PKU#Mn4. Overall, our study provides evidence that the MA supplementation can upregulate the build-up of the acetyl-CoA pool, and simultaneously suppress the oxidation of fatty acids, thereby increasing the accumulation of SFA and PUFA in thraustochytrids.

The importance of exogenous antioxidants supplementation in cellular ROS scavenging is well established [23]. However, it is not clear through which mechanisms the exogenous antioxidants alleviate cellular ROS levels. Previous studies have reported that exogenous melatonin alleviates the ROS level, suppresses lipid peroxidation, and elevates the activity of antioxidant enzymes in *Monoraphidium* sp. QLY-1 [25]. The addition of fulvic acid and EDTA during the cultivation of *Schizochytrium* sp. HX-308 showed a 44.7% and 81.9% increase in the activities of superoxide dismutase and catalase [19]. Along similar lines, our study revealed that the addition of exogenous antioxidants leads to the upregulation of catalase in strain PKU#Mn4, which is one of the key antioxidant enzymes that mediates intracellular ROS scavenging.

In conclusion, the interaction effects of 1 g/L each of mannitol and ascorbic acid could significantly augment growth, SFA, and PUFA accumulation, and also alleviate the ROS level during fermentation. More importantly, the findings of this study significantly advance the knowledge about the mechanisms that underpin the upscaled accumulation of lipids in thraustochytrids under the influence of exogenous antioxidants.

4. Materials and Methods

4.1. Microorganism and Batch Fermentation

Schizochytrium sp. PKU#Mn4, previously isolated from the Pearl River Delta region of China [38], was used in this study. The strain was maintained on 1% modified Vishniac's (MV) agar plates at room temperature and subcultured every 4 weeks [39].

Seed culture medium and conditions were used as described in our previous study [20]. A 5% (*v/v*) seed culture was transferred to a 100 mL shake flask containing 40 mL of fermentation medium and cultured at 28 °C with a constant orbital shaking speed of 170 rpm for 4 days. The fermentation medium contained 40 g/L glycerol, 2.5 g/L yeast extract, 0.25 g/L KH_2PO_4, and 33 g/L sea salt [40]. A 10× stock solution of each antioxidant was prepared by dissolving in water or ethanol (EtOH) based on its solubility and then filtered before addition into the medium. Ascorbic acid, mannitol, and tea extracts were dissolved in water, while α-tocopherol, melatonin, sesamol, and BHT were dissolved in ethanol. A control flask was prepared to contain an equal amount of water or ethanol.

4.2. Experimental Design and Statistical Analyses

Seven different natural antioxidants, namely ascorbic acid, α-tocopherol, tea extract, melatonin, mannitol, sesamol, and BHT were screened based on a two-level FFD (Table 1). According to the FFD, the total number of experimental runs was 16 (Table S1), each carried out in triplicate. The antioxidant compounds were added to the fermentation culture and not to the seed culture. The response variables, DCW (g/L), and TFA yield (mg/g DCW), for each run, were measured at 96 h of batch cultivation. Design-Expert software (version 10.0.7, Stat-Ease Co. Inc., Minneapolis, MN, USA) was used for the experimental design and regression analysis of the experimental data. A modified Gompertz model [41] was fitted to the experimental data to estimate the parameters a (maximum growth potential, g/L), R (maximum growth rate, g/L·h^{-1}), and λ (lag time, h) using R software (version 4.0.0, https://www.r-project.org, accessed on 8 May 2019).

4.3. Quantification of Dry Cell Weight and Fatty Acids

The cells in the culture broth were collected, lyophilized, and their dry weight was determined following the procedures described in our previous study [20]. Using the lyophilized cells (30 mg), a one-step transesterification process was performed to prepare the fatty acids methyl esters (FAMEs) [20], which were then resolved on a gas chromatography 7890B (Agilent Technologies, Santa Clara, CA, USA) system equipped with DB-23 capillary column (60 m × 0.25 μm × 0.32 μm) (Agilent Technologies, Santa Clara, CA, USA) and nitrogen as the carrier gas. The FAMEs were quantified by comparing the retention time of each peak with the one in 37 Component FAME Mix Standard (Supelco Co. Inc., Bellefonte, PA, USA). C16:0 was the major SFA constituent, C22:6 and C22:5 were the major PUFA constituents, and C14:1, C15:1, and C17:1 were the MUFA constituents.

4.4. Determination of Intracellular ROS

The intracellular ROS levels were measured with ROS Assay Kit (Beyotime Biotechnology, Shanghai, China) following the manufacturer's instructions. In brief, 1×10^6 to 2×10^7 cells were harvested by centrifugation at 4000 rpm for 10 min, and the pellet was then washed and re-suspended in 1 mL artificial seawater. One µL of 2′-7′-dichlorofluorescin diacetate (DCFH-DA) was added to the cell suspension and incubated at 37 °C for 20 min. After DCFH-DA treatment, the cell suspension was washed thrice with artificial seawater to remove the excess DCFH-DA. The fluorescence intensity was detected with a fluorescence spectrophotometer (F97 Pro, Lengguang Technology Co. Inc., Shanghai, China) at the excitation and emission wavelengths of 488 nm and 525 nm, respectively.

4.5. Fed-Batch Fermentation

Fed-batch fermentation was performed in a 5-L bioreactor (Shanghai Dong Ming Industrial Co. Ltd., Shanghai, China) at 28 °C for 8 days as described in our previous study [40]. In brief, 300 mL seed culture was prepared as described in Section 4.1 and inoculated into a 2 L fermentation medium with or without antioxidant mixture (1 g/L each of mannitol and ascorbic acid). The agitation rate of the bioreactor was altered to maintain the dissolved oxygen (DO) at 50% from day 0 to day 2 and at 10% for the rest of the fermentation period. On day 2 and day 3, a 100 mL feed medium (400 g/L glycerol and 25 g/L yeast extract) was added to the fermentation broth. An aliquot of the culture was harvested every 24 h to determine the DCW and fatty acids concentration following the methods described in Section 2.3.

4.6. Transcriptome Analysis

The changes in the transcriptome of strain PKU#Mn4 under the influence of antioxidants (1 g/L each of mannitol and ascorbic acid) supplementation in the batch culture were investigated by the RNA-seq method. Three parallel samples (10 mL) of culture broth at 48 h of fermentation, without (control group) or with (test group) antioxidants, were collected and the cells were harvested, frozen with liquid nitrogen, and stored at −80 °C. Total RNA extraction, its quality, integrity, and quantity checks were conducted by BioMarker Technologies (Beijing, China). For each sample, 1 µg RNA was used to generate the library with the NEBNext Ultra TM RNA Library Prep Kit for Illumina (New England BioLabs Inc., Ipswich, MA, USA). The clustering of the index-coded samples was carried out on a cBot Cluster Generation System using TruSeq PE Cluster Kit v4-cBot-HS (Illumina Inc., San Diego, CA, USA) according to the manufacturer's instructions. The library was sequenced on an Illumina platform and paired-end reads were generated. All analyses were based on clean data with high-quality reads after removing those that contained adapter, poly-N, or low-quality reads from the raw data. The clean reads were first mapped to the reference genome with HISAT2 [42] and then assembled with StringTie [43]. The gene expression levels were quantified by the FPKM method [44], and the differential expression analysis was carried out with DEGseq [45]. The genes or transcripts with a false discovery rate (FDR) less than 0.05 and the absolute value of $\log_2$ (fold change (FC)) > 1 were regarded as the differentially expressed genes (DEGs). RNA sequencing was performed by BioMarker Technologies (Beijing, China), and the sequencing data analysis was done by BMKCloud (www.biocloud.net, accessed on 8 May 2019).

Supplementary Materials: The following are available online at https://www.mdpi.com/article/10.3390/md19100559/s1, Figure S1: Effects of α-tocopherol, melatonin, and mannitol on the dry cell weight (DCW) of *Schizochytrium* PKU#Mn4 culture; Figure S2: Experimental data and prediction using modified Gompertz model; Figure S3: GO classification of differentially expressed genes; Table S1: Experiment design and response values for screening antioxidants.

Author Contributions: Conceptualization, S.Z., G.W. and B.S.; methodology, S.Z., X.C., Y.H. and M.B.; formal analysis, S.Z. and B.S.; investigation, S.Z., X.C. and M.B.; resources, X.C., Y.H. and M.B.; data curation, S.Z. and B.S.; visualization, S.Z. and B.S.; writing—original draft preparation, S.Z. and

B.S.; writing—review and editing, B.S. and G.W.; supervision, G.W. and Y.H.; project administration, X.C., Y.H. and M.B.; funding acquisition, G.W. All authors have read and agreed to the published version of the manuscript.

Funding: This research was funded by the National Natural Science Foundation of China (NSFC) (#32170063) granted to G.W.

Institutional Review Board Statement: Not applicable.

Informed Consent Statement: Not applicable.

Data Availability Statement: Not applicable.

Conflicts of Interest: The authors declare no conflict of interest. The funders had no role in the design of the study; in the collection, analyses, or interpretation of data; in the writing of the manuscript, or in the decision to publish the results.

References

1. Gupta, A.; Barrow, C.J.; Puri, M. Omega-3 biotechnology: Thraustochytrids as a novel source of omega-3 oils. *Biotechnol. Adv.* **2012**, *30*, 1733–1745. [CrossRef] [PubMed]
2. Du, F.; Wang, Y.Z.; Xu, Y.S.; Shi, T.Q.; Liu, W.Z.; Sun, X.M.; Huang, H. Biotechnological production of lipid and terpenoid from thraustochytrids. *Biotechnol. Adv.* **2021**, *48*, 107725. [CrossRef] [PubMed]
3. Chandrasekaran, K.; Roy, R.K.; Chadha, A. Docosahexaenoic acid production by a novel high yielding strain of Thraustochytrium sp. of Indian origin: Isolation and bioprocess optimization studies. *Algal Res.* **2018**, *32*, 93–100. [CrossRef]
4. Leyton, A.; Flores, L.; Shene, C.; Chisti, Y.; Larama, G.; Asenjo, J.A.; Armenta, R.E. Antarctic Thraustochytrids as Sources of Carotenoids and High-Value Fatty Acids. *Mar. Drugs* **2021**, *19*, 386. [CrossRef]
5. Zhang, A.; Xie, Y.; He, Y.; Wang, W.; Sen, B.; Wang, G. Bio-based squalene production by *Aurantiochytrium* sp. through optimization of culture conditions, and elucidation of the putative biosynthetic pathway genes. *Bioresour. Technol.* **2019**, *287*, 121415. [CrossRef]
6. Aasen, I.M.; Ertesvåg, H.; Heggeset, T.M.B.; Liu, B.; Brautaset, T.; Vadstein, O.; Ellingsen, T.E. Thraustochytrids as production organisms for docosahexaenoic acid (DHA), squalene, and carotenoids. *Appl. Microbiol. Biotechnol.* **2016**, *100*, 4309–4321. [CrossRef]
7. Guo, D.-S.; Ji, X.-J.; Ren, L.-J.; Li, G.-L.; Sun, X.-M.; Chen, K.-Q.; Gao, S.; Huang, H. Development of a scale-up strategy for fermentative production of docosahexaenoic acid by *Schizochytrium* sp. *Chem. Eng. Sci.* **2018**, *176*, 600–608. [CrossRef]
8. Huang, T.Y.; Lu, W.C.; Chu, I.M. A fermentation strategy for producing docosahexaenoic acid in Aurantiochytrium limacinum SR21 and increasing C22:6 proportions in total fatty acid. *Bioresour. Technol.* **2012**, *123*, 8–14. [CrossRef]
9. Li, J.; Liu, R.; Chang, G.; Li, X.; Chang, M.; Liu, Y.; Jin, Q.; Wang, X. A strategy for the highly efficient production of docosahexaenoic acid by Aurantiochytrium limacinum SR21 using glucose and glycerol as the mixed carbon sources. *Bioresour. Technol.* **2015**, *177*, 51–57. [CrossRef]
10. Ling, X.; Guo, J.; Liu, X.; Zhang, X.; Wang, N.; Lu, Y.; Ng, I.S. Impact of carbon and nitrogen feeding strategy on high production of biomass and docosahexaenoic acid (DHA) by *Schizochytrium* sp. LU310. *Bioresour. Technol.* **2015**, *184*, 139–147. [CrossRef]
11. Qu, L.; Ji, X.J.; Ren, L.J.; Nie, Z.K.; Feng, Y.; Wu, W.J.; Ouyang, P.K.; Huang, H. Enhancement of docosahexaenoic acid production by *Schizochytrium* sp. using a two-stage oxygen supply control strategy based on oxygen transfer coefficient. *Lett. Appl. Microbiol.* **2011**, *52*, 22–27. [CrossRef] [PubMed]
12. Ren, L.; Ji, X.; Huang, H.; Qu, L.; Feng, Y.; Tong, Q.; Ouyang, P. Development of a Stepwise Aeration Control Strategy for Efficient Docosahexaenoic Acid Production by *Schizochytrium* sp. *Appl. Microbiol. Biotechnol.* **2010**, *87*, 1649–1656. [CrossRef] [PubMed]
13. Heggeset, T.M.B.; Ertesvåg, H.; Liu, B.; Ellingsen, T.E.; Vadstein, O.; Aasen, I.M. Lipid and DHA-production in Aurantiochytrium sp.—Responses to nitrogen starvation and oxygen limitation revealed by analyses of production kinetics and global transcriptomes. *Sci. Rep.* **2019**, *9*, 19470. [CrossRef] [PubMed]
14. Jakobsen, A.N.; Aasen, I.M.; Josefsen, K.D.; Strøm, A.R. Accumulation of Docosahexaenoic Acid-Rich Lipid in Thraustochytrid *Aurantiochytrium* sp. strain T66: Effects of N and P Starvation and O_2 Limitation. *Appl. Microbiol. Biotechnol.* **2008**, *80*, 297–306. [CrossRef]
15. Chodchoey, K.; Verduyn, C. Growth, fatty acid profile in major lipid classes and lipid fluidity of Aurantiochytrium mangrovei Sk-02 as a function of growth temperature. *Braz. J. Microbiol.* **2012**, *43*, 187–200. [CrossRef]
16. Taoka, Y.; Nagano, N.; Okita, Y.; Izumida, H.; Sugimoto, S.; Hayashi, M. Influences of Culture Temperature on the Growth, Lipid Content and Fatty Acid Composition of Aurantiochytrium sp. Strain mh0186. *Mar. Biotechnol.* **2009**, *11*, 368–374. [CrossRef]
17. Shi, K.; Gao, Z.; Shi, T.-Q.; Song, P.; Ren, L.-J.; Huang, H.; Ji, X.-J. Reactive Oxygen Species-Mediated Cellular Stress Response and Lipid Accumulation in Oleaginous Microorganisms: The State of the Art and Future Perspectives. *Front. Microbiol.* **2017**, *8*, 793. [CrossRef]
18. Sajjadi, B.; Chen, W.-Y.; Raman, A.A.A.; Ibrahim, S. Microalgae lipid and biomass for biofuel production: A comprehensive review on lipid enhancement strategies and their effects on fatty acid composition. *Renew. Sustain. Energy Rev.* **2018**, *97*, 200–232. [CrossRef]

19. Sun, X.M.; Ren, L.J.; Ji, X.J.; Huang, H. Enhancing biomass and lipid accumulation in the microalgae *Schizochytrium* sp. by addition of fulvic acid and EDTA. *AMB Express* **2018**, *8*, 150. [CrossRef]
20. Zhang, S.; He, Y.; Sen, B.; Chen, X.; Xie, Y.; Keasling, J.D.; Wang, G. Alleviation of reactive oxygen species enhances PUFA accumulation in *Schizochytrium* sp. through regulating genes involved in lipid metabolism. *Metab. Eng. Commun.* **2018**, *6*, 39–48. [CrossRef]
21. Kitchener, R.L.; Grunden, A.M. Methods for enhancing cyanobacterial stress tolerance to enable improved production of biofuels and industrially relevant chemicals. *Appl. Microbiol. Biotechnol.* **2018**, *102*, 1617–1628. [CrossRef]
22. Zhang, S.; He, Y.; Sen, B.; Wang, G. Reactive oxygen species and their applications toward enhanced lipid accumulation in oleaginous microorganisms. *Bioresour. Technol.* **2020**, *307*, 123234. [CrossRef]
23. Santos-Sánchez, N.F.; Salas-Coronado, R.; Villanueva-Cañongo, C.; Hernández-Carlos, B. Antioxidant Compounds and Their Antioxidant Mechanism. In *Antioxidants*; Shalaby, E., Ed.; IntechOpen: London, UK, 2019.
24. Ahmad, P.; Jaleel, C.A.; Salem, M.A.; Nabi, G.; Sharma, S. Roles of enzymatic and nonenzymatic antioxidants in plants during abiotic stress. *Crit. Rev. Biotechnol.* **2010**, *30*, 161–175. [CrossRef]
25. Zhao, Y.; Li, D.; Xu, J.-W.; Zhao, P.; Li, T.; Ma, H.; Yu, X. Melatonin enhances lipid production in *Monoraphidium* sp. QLY-1 under nitrogen deficiency conditions via a multi-level mechanism. *Bioresour. Technol.* **2018**, *259*, 46–53. [CrossRef]
26. Liu, B.; Liu, J.; Sun, P.; Ma, X.; Jiang, Y.; Chen, F. Sesamol Enhances Cell Growth and the Biosynthesis and Accumulation of Docosahexaenoic Acid in the Microalga *Crypthecodinium cohnii*. *J. Agric. Food. Chem.* **2015**, *63*, 5640–5645. [CrossRef]
27. Ren, L.; Sun, X.; Ji, X.; Chen, S.; Guo, D.; Huang, H. Enhancement of Docosahexaenoic Acid Synthesis by Manipulation of Antioxidant Capacity and Prevention of Oxidative Damage in *Schizochytrium* sp. *Bioresour. Technol.* **2017**, *223*, 141–148. [CrossRef]
28. Gaffney, M.; O'Rourke, R.; Murphy, R. Manipulation of fatty acid and antioxidant profiles of the microalgae *Schizochytrium* sp. through flaxseed oil supplementation. *Algal Res.* **2014**, *6*, 195–200. [CrossRef]
29. Singh, D.; Mathur, A.S.; Tuli, D.K.; Puri, M.; Barrow, C.J. Propyl gallate and butylated hydroxytoluene influence the accumulation of saturated fatty acids, omega-3 fatty acid and carotenoids in thraustochytrids. *J. Funct. Foods* **2015**, *15*, 186–192. [CrossRef]
30. Zhao, Y.; Yue, C.; Ding, W.; Li, T.; Xu, J.-W.; Zhao, P.; Ma, H.; Yu, X. Butylated hydroxytoluene induces astaxanthin and lipid production in *Haematococcus pluvialis* under high-light and nitrogen-deficiency conditions. *Bioresour. Technol.* **2018**, *266*, 315–321. [CrossRef]
31. Na, C.; Junmu, X.; Yongjie, F.; Yongteng, Z.; Xuya, Y.; Jun-Wei, X.; Tao, L.; Peng, Z. Antioxidants enhance lipid productivity in *Heveochlorella* sp. Yu. *Algal Res.* **2021**, *55*, 102235.
32. Li, D.; Zhao, Y.; Ding, W.; Zhao, P.; Xu, J.W.; Li, T.; Ma, H.; Yu, X. A strategy for promoting lipid production in green microalgae *Monoraphidium* sp. QLY-1 by combined melatonin and photoinduction. *Bioresour. Technol.* **2017**, *235*, 104–112. [CrossRef] [PubMed]
33. Ding, W.; Zhao, P.; Peng, J.; Zhao, Y.; Xu, J.-W.; Li, T.; Reiter, R.J.; Ma, H.; Yu, X. Melatonin enhances astaxanthin accumulation in the green microalga Haematococcus pluvialis by mechanisms possibly related to abiotic stress tolerance. *Algal Res.* **2018**, *33*, 256–265. [CrossRef]
34. Davis, M.S.; Solbiati, J.; Cronan, J.E., Jr. Overproduction of acetyl-CoA carboxylase activity increases the rate of fatty acid biosynthesis in Escherichia coli. *J. Biol. Chem.* **2000**, *275*, 28593–28598. [CrossRef] [PubMed]
35. Liu, J.; Pei, G.; Diao, J.; Chen, Z.; Liu, L.; Chen, L.; Zhang, W. Screening and transcriptomic analysis of Crypthecodinium cohnii mutants with high growth and lipid content using the acetyl-CoA carboxylase inhibitor sethoxydim. *Appl. Microbiol. Biotechnol.* **2017**, *101*, 6179–6191. [CrossRef]
36. Fathy, W.; Essawy, E.; Tawfik, E.; Khedr, M.; Abdelhameed, M.S.; Hammouda, O.; Elsayed, K. Recombinant overexpression of the Escherichia coli acetyl-CoA carboxylase gene in *Synechocystis* sp. boosts lipid production. *J. Basic Microbiol.* **2021**, *61*, 330–338. [CrossRef]
37. Che, R.; Huang, L.; Xu, J.-W.; Zhao, P.; Li, T.; Ma, H.; Yu, X. Effect of fulvic acid induction on the physiology, metabolism, and lipid biosynthesis-related gene transcription of *Monoraphidium* sp. FXY-10. *Bioresour. Technol.* **2017**, *227*, 324–334. [CrossRef]
38. Liu, Y.; Singh, P.; Sun, Y.; Luan, S.; Wang, G. Culturable Diversity and Biochemical Features of Thraustochytrids from Coastal Waters of Southern China. *Appl. Microbiol. Biotechnol.* **2014**, *98*, 3241–3255. [CrossRef]
39. Wang, Q.; Ye, H.; Xie, Y.; He, Y.; Sen, B.; Wang, G. Culturable Diversity and Lipid Production Profile of Labyrinthulomycete Protists Isolated from Coastal Mangrove Habitats of China. *Mar. Drugs* **2019**, *17*, 268. [CrossRef]
40. Wang, Q.; Ye, H.; Sen, B.; Xie, Y.; He, Y.; Park, S.; Wang, G. Improved production of docosahexaenoic acid in batch fermentation by newly-isolated strains of *Schizochytrium* sp. and *Thraustochytriidae* sp. through bioprocess optimization. *Synth. Sys. Biotechnol.* **2018**, *3*, 121–129. [CrossRef]
41. Tjørve, K.M.C.; Tjørve, E. The use of Gompertz models in growth analyses, and new Gompertz-model approach: An addition to the Unified-Richards family. *PLoS ONE* **2017**, *12*, e0178691. [CrossRef]
42. Kim, D.; Langmead, B.; Salzberg, S.L. HISAT: A fast spliced aligner with low memory requirements. *Nat. Methods* **2015**, *12*, 357–360. [CrossRef] [PubMed]
43. Pertea, M.; Pertea, G.M.; Antonescu, C.M.; Chang, T.C.; Mendell, J.T.; Salzberg, S.L. StringTie enables improved reconstruction of a transcriptome from RNA-seq reads. *Nat. Biotechnol.* **2015**, *33*, 290–295. [CrossRef]
44. Mortazavi, A.; Williams, B.A.; McCue, K.; Schaeffer, L.; Wold, B. Mapping and quantifying mammalian transcriptomes by RNA-Seq. *Nat. Methods* **2008**, *5*, 621–628. [CrossRef] [PubMed]

45. Wang, L.; Feng, Z.; Wang, X.; Wang, X.; Zhang, X. DEGseq: An R package for identifying differentially expressed genes from RNA-seq data. *Bioinformatics* **2010**, *26*, 136–138. [CrossRef] [PubMed]

marine drugs

Article

Response to Static Magnetic Field-Induced Stress in *Scenedesmus obliquus* and *Nannochloropsis gaditana*

Génesis Serrano [1], Carol Miranda-Ostojic [1], Pablo Ferrada [2,*], Cristian Wulff-Zotelle [3], Alejandro Maureira [1], Edward Fuentealba [2], Karem Gallardo [4], Manuel Zapata [1] and Mariella Rivas [1,*]

1 Laboratorio de Biotecnología Ambiental Aplicada, Departamento de Biotecnología, Universidad de Antofagasta, Av. Angamos 601, Antofagasta 1270300, Chile; genesis.lsd1@gmail.com (G.S.); carol.miranda@ua.cl (C.M.-O.); alejandro.maureira.calderon@ua.cl (A.M.); manuel.zapata@uantof.cl (M.Z.)

2 Centro de Desarrollo Energético Antofagasta, Universidad de Antofagasta, Av. Angamos 601, Antofagasta 1270300, Chile; edward.fuentealba@uantof.cl

3 Laboratorio de Biología Celular, Molecular y Genética, Departamento Biomédico, Universidad de Antofagasta, Av. Angamos 601, Antofagasta 1270300, Chile; cristian.wulff@uantof.cl

4 Centro de Investigación Tecnológica de Agua en el Desierto (CEITSAZA), Universidad Católica del Norte, Av. Angamos 0610, Antofagasta 1270709, Chile; kgallardo@ucn.cl

* Correspondence: pablo.ferrada@uantof.cl (P.F.); mariella.rivas@uantof.cl (M.R.)

Abstract: Magnetic fields in biological systems is a promising research field; however, their application for microalgae has not been fully exploited. This work aims to measure the enzymatic activity and non-enzymatic activity of two microalgae species in terms of superoxide dismutase (SOD), catalase (CAT), and carotenoids, respectively, in response to static magnetic fields-induced stress. Two magnet configurations (north and south) and two exposure modes (continuous and pulse) were applied. Two microalgae species were considered, the *Scenedesmus obliquus* and *Nannochloropsis gaditana*. The SOD activity increased by up to 60% in *S. obliquus* under continuous exposure. This trend was also found for CAT in the continuous mode. Conversely, under the pulse mode, its response was hampered as the SOD and CAT were reduced. For *N. gaditana*, SOD increased by up to 62% with the south configuration under continuous exposure. In terms of CAT, there was a higher activity of up to 19%. Under the pulsed exposure, SOD activity was up to 115%. The CAT in this microalga was increased by up to 29%. For *N. gaditana*, a significant increase of over 40% in violaxanthin production was obtained compared to the control, when the microalgae were exposed to SMF as a pulse. Depending on the exposure mode and species, this methodology can be used to produce oxidative stress and obtain an inhibitory or enhanced response in addition to the significant increase in the production of antioxidant pigments.

Keywords: enzymatic activity; fluid dynamics; microalgae; oxidative stress; static magnetic fields; violaxanthin

Citation: Serrano, G.; Miranda-Ostojic, C.; Ferrada, P.; Wulff-Zotelle, C.; Maureira, A.; Fuentealba, E.; Gallardo, K.; Zapata, M.; Rivas, M. Response to Static Magnetic Field-Induced Stress in *Scenedesmus obliquus* and *Nannochloropsis gaditana*. *Mar. Drugs* **2021**, *19*, 527. https://doi.org/10.3390/md19090527

Academic Editor: Carlos Almeida

Received: 3 August 2021
Accepted: 13 September 2021
Published: 21 September 2021

1. Introduction

There are several cases where magnetism is present in living matter. For instance, the magnetotactic bacterium can align its body according to the Earth's magnetic field. This feature is possible due to chains of single-domain, biogenic magnetite (Fe_3O_4) able to sense geomagnetic fields [1]. Macroscopic organisms, such as birds, are known to orient themselves to magnetic field lines and to travel long distances [2]. In fact, all matter falls into one of these categories: diamagnetism, paramagnetism, ferromagnetism, ferrimagnetism or antiferrimagnetism [3]. Firstly, magnetic fields (MF) conceived as a perturbation, which can be used by organisms in a natural way, or MF applied to biological systems to promote changes at several levels, can be classified according to the broader and general perspective of electromagnetic (EM) fields. The detection/sensing or the response to MF can be associated with different mechanisms occurring at the tissue, cell, membrane, and molecular levels, including the medium where the organisms live [4]. To understand interaction mechanisms and effects of static magnetic fields (SMF) on cells, a categorization of EM fields is needed and provided in the following.

Electromagnetic fields (EM) can be classified according to four main parameters: time-dependence, distribution or homogeneousness, direction, and exposure time. The first parameter can be split into two branches, namely, time-constant (static EM fields) or time-varying (dynamic EM fields). The static EM fields are grouped as stationary fields not exhibiting mutual coupling between the electric and magnetic component.

Regarding the second parameter for classification, the distribution of the fields can be homogeneous when same magnitude occurs in all space or inhomogeneous when the magnitude varies with one or more than a spatial coordinate. The third parameter refers to the direction of the field. When implementing an array of electro or permanent magnets, depending on location and orientation, many patterns and intensities are possible [5]. The fourth parameter refers to how much time a specimen or biological system is exposed to the MF. Different processes within a biological system can occur at times, which can differ in orders of magnitude [6].

Parameters such as the magnet materials used, magnet dimensions, pole configuration, measurement of the field strength, frequency of application, duration and site of application, magnet support device, target tissue, and distance from magnet surface have been discussed by Colbert et al. [7]. These authors performed an analysis based on 56 articles focusing on the quality of reporting SMF and those parameters concluding that 61% of the studies failed to provide enough experimental details regarding SMF. Consequently, the possibility for other researchers to replicate protocols and give satisfactory explanations might be reduced. These findings highlight the importance of providing proper conditions for SMF exposure and characterization.

Regarding effects of SMF on different cell parameters, authors found that cell size, shape, orientation, and membrane roughness changed due to the exposure [6]. Researchers showed an increased number and size of holes in cell membrane [8]. Thus, the volume force due to magnetic gradient led to higher membrane permeability. The magnetic gradient just mentioned was examined by Zablotskii et al. [9] where the discussion was centered on the mechanism by which MFs act on tissues. It was argued that it is the high magnetic gradient and not the field itself which can cause several effects in cells and their functions. They showed that high MF gradient can be involved in the change of the ion-channel on/off probability, suppression of cell growth by magnetic pressure, magnetically induced cell division and cell reprograming, and forced migration of membrane receptor proteins. Fundamentals regarding how MF can alter the rate, yield, and product distribution of chemical reactions can be seen in terms of magnetokinetics [10]. A review on the effects of SMF on cellular systems provides a wide insight regarding the most influential factors and mechanisms [11]. Authors point out that the radical pair recombination and the diamagnetic anisotropy can cause changes and subsequent effects on the susceptibility of biomolecules, intracellular structural modifications, and changes in the enzymatic reactions. Some studies have used SMF to control algal growth [12], and changes in biochemical composition and photosynthesis have also been described [12]. In contrast, other studies propose their use as biostimulants for growth [13]. It has been shown that SMF promotes growth and oxygen production in *Scenedesmus obliquus* [14]. Moreover, an increase in growth (100%), stimulation of antioxidant defense in *Chlorella vulgaris* has been demonstrated [15], and production of exopolysaccharides [16].

The main cellular physiological processes generate reactive oxygen species (ROS). To classify ROS into categories, there are free radicals and non-radical species [17]. The radicals are understood as molecular species, which contain an unpaired electron in an atomic orbital [18]. Consequently, a free radical tends to remove an electron from a stable molecule to reach electrochemical stability. Accordingly, oxygen, hydrogen atom, and transition metals constitute free radicals [17]. Particularly, this unpaired electron makes these species paramagnetic and susceptible to magnetic fields. In the case where the field is present in biological systems, it may affect ROS dynamics. Particularly, in the photosynthetic cells as the case of microalgae, the main organelles generating ROS are mitochondria, chloroplasts, and to a lesser extent, peroxisomes, where they are formed as a consequence of oxidative energy metabolism and the intense flow of electrons that exists in the mitochondrial and thylakoidal membranes [19]. The ROS are highly reactive and can oxidize essential macromolecules. All cells possess an antioxidant defense system that neutralizes or metabolizes ROS. This system includes antioxidant enzymes (catalase, superoxide dismutase, ascorbate peroxidase, glutathione peroxidase, among others) and low molecular weight metabolites (glutathione, ascorbic acid, carotenes, phenols, flavonoids) [20]. If this system of defenses is overcome, the cellular oxide-reduction equilibrium shifts to a pro-oxidant situation and a state of oxidative stress is reached in which oxidative damage occurs to proteins, nucleic acids, and lipids [21]. In addition, various environmental factors and the entry of contaminants into the cells can intensify the production of these molecules; the most harmful being the superoxide anion ($O_2^{\bullet-}$), hydrogen peroxide (H_2O_2), and hydroxyl radical (OH^-) [22,23].

The non-enzymatic antioxidant system includes compounds having a low molecular weight, where the most important are the reduced glutathione, vitamin E, vitamin C, such as ascorbic acid, and vitamin A. In addition, there are flavonoids, phenolic acids, α-lipoic acid, uric acid, bilirubin, some sugars, and amino acids. In the defense mechanism, these compounds capture free radicals avoiding chain reactions. They can be associated with the membrane (vitamin E, carotenoids, phycobiliproteins), dispersed in the cytoplasm (mycosporine-like amino acids), or linked to water-soluble reducing compounds, such as GHS and vitamin C [24,25].

In the present work, we investigated the influence of SMF on the response of two microalgae, growing in different culture media. The model species used were the *S. obliquus* and the *Nannochloropsis gaditana*, as they have known structure, functionality, and rapid growth. Furthermore, these species exhibit biotechnological potential in the generation of antioxidant biomolecules. *S. obliquus* is dominant in freshwater bodies, lakes, and rivers [26], it is cultivated at industrial scale to produce bioactive molecules, and under stress conditions it has been shown to accumulate β-carotene, astaxanthin isomers, lutein, and canthaxanthin [27,28]. Moreover, *N. gaditana* is known to produce several compounds such as fatty acids, lipids, including pigments such as zeaxanthin, astaxanthin, and canthaxanthin [28,29]. The focus of this work is the analysis of the oxidative stress of the selected species to deepen the understanding of the effects of SMF and consequently the potential production of antioxidant pigments. Nevertheless, the pigment composition of algae varies from species to species and according to environmental conditions including different stress conditions such as temperature, salinity, and irradiance, which may induce carotenoid production in algal species. Therefore, analysis of a complete pigment spectrum was intended for both microalgae in the control condition and treatments. For this purpose, we used two magnet configurations and two exposure modes and characterized the changes in terms of enzymatic activity and non-enzymatic production. We then discussed the results taking into consideration the characterization related to a single exposure mode previously reported in Ferrada et al. [30], and thus providing a broader analysis.

2. Results

The physicochemical medium parameters such as temperature (T), pH, electrical conductivity (σ), salinity (S), and dissolved oxygen (DO) were characterized for *S. obliquus* and *N. gaditana* for the control condition and under the influence of SMF (see Table 1 for a summary). Regardless of the species and treatment used, i.e., control or using a magnet configuration and exposure mode, T and pH remained at similar values in the range of 21–22 °C and 8.6–8.7, respectively. However, the electrical conductivity, salinity, and dissolved oxygen concentration were different depending on the medium. *N. gaditana* grows in a medium, in which σ is ~46 times higher and S is ~60 times higher than that of *S. obliquus*. Conversely, DO for *S. obliquus* is 20% higher than the medium for *N. gaditana*. Qualitatively, these results mean that the changes induced by SMF were not observed at the macroscopic level, that is, in terms of physicochemical parameters. This highlights the need to look inside the cell and quantify the oxidative stress, which is exposed in Section 2.1. A visualization of the magnetic flux density $\boldsymbol{B}$ is shown below.

Table 1. Summary of the physicochemical medium parameters for *S. obliquus* and *N. gaditana*. Averages and standard deviations were computed as each magnet configuration and control were implemented in triplicate. NC: north-continuous; NP: north-pulse; SC: south-continuous; SP: south-pulse.

Species	Treatment	T (°C)	pH	σ (S/m)	S (PSU)	DO (%)
S. obliquus	Control for North	21.1 ± 2.3	8.7 ± 0.8	0.09 ± 0.01	0.47 ± 0.07	6.5 ± 0.9
	NC	21.0 ± 2.4	8.5 ± 0.8	0.1 ± 0.01	0.49 ± 0.05	6.8 ± 0.9
	NP	21.0 ± 2.5	9.0 ± 1.1	0.09 ± 0.01	0.46 ± 0.08	6.6 ± 1.1
	Control for South	22.8 ± 0.3	8.9 ± 0.5	0.10 ± 0.01	0.48 ± 0.05	5.7 ± 0.3
	SC	23.0 ± 0.5	8.6 ± 0.7	0.09 ± 0.03	0.42 ± 0.14	5.9 ± 0.6
	SP	23.0 ± 0.8	8.8 ± 0.8	0.09 ± 0.02	0.47 ± 0.11	6.0 ± 0.6
N. gaditana	Control for North	21.2 ± 1.4	8.4 ± 0.4	3.7 ± 0.4	23.7 ± 2.5	5.2 ± 0.3
	NC	21.3 ± 1.8	8.6 ± 0.2	4.4 ± 0.6	28.5 ± 4.6	5.4 ± 0.3
	NP	21.3 ± 1.5	8.6 ± 0.2	4.4 ± 0.6	28.9 ± 4.5	5.2 ± 0.4
	Control for South	21.1 ± 0.2	8.6 ± 0.2	4.2 ± 0.9	27.5 ± 6.3	5.0 ± 0.3
	SC	21.3 ± 0.1	8.6 ± 0.2	4.3 ± 0.6	28.2 ± 4.1	5.1 ± 0.2
	SP	21.2 ± 0.2	8.6 ± 0.2	5.0 ± 0.6	33.0 ± 4.4	4.9 ± 0.3

The magnetic flux density norm ($|\boldsymbol{B}|$) is visualized in Figure 1 corresponding to cases when all north or south poles are pointing towards the center of the flask. Figure 1A depicts circles to indicate where the field is specified (Figure 1B). Similarly, Figure 1C indicates where the field is computed (Figure 1D). The case of all north poles oriented to the center results in field lines which oppose each other and leads zero magnitude of $\boldsymbol{B}$ at the center of the flask at $z = 0$.

2.1. Response of Microalgae to SMF

S. obliquus and *N. gaditana* microalgae were exposed to SMF for 96 h and their growth was determined by optical density at 680 nm. The control condition corresponds to microalgae without exposure to SMF and in the presence of a SMF equal to 4500 G (referred as intensity at the side of the magnet, see Figure 1) with north and south arrangement in pulse or continuous form during 96 h. For *S. obliquus* after 48 h of growth, significant differences between the control and the exposed microalgae are evidenced, in the presence of SMF North-C and North-P a decrease was observed from 48 to 96 h (OD680 nm equal to 0.46 and 0.53, respectively) compared to the superior growth of the control condition with an OD680 nm equal to 0.95 (Figure 2A). On the contrary, in the presence of SMF South-C and P, less growth was observed in the microalgae exposed continuously at 48 h and 72 h (equal to OD680 nm of 0.52 and 0.71) (Figure 2C). For *N. gaditana*, clear effects of SMF on growth were observed from 48 h. Lower growth was observed in both treatments with north continuous and pulse exposure, which was more evident at 72 h or 96 h (Figure 2B). In response to exposure to south configuration, reduced growth is observed from 24 h of culture (Figure 2D).

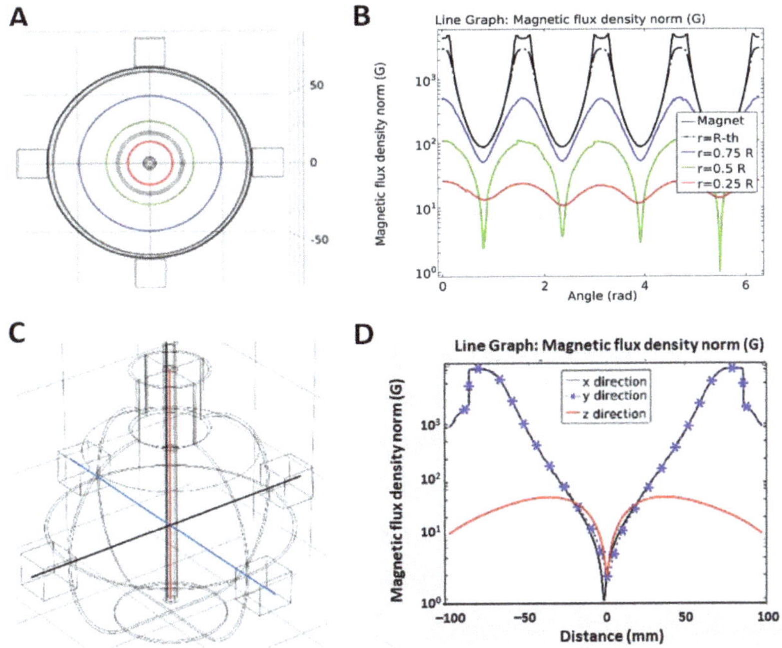

Figure 1. (**A**) Top view indicating along the curve where the magnetic flux density norm $|\boldsymbol{B}|$ is evaluated. (**B**) Plot of $|\boldsymbol{B}|$ where all north poles are oriented towards the center of the flask, evaluated along curves shown in (**A**). (**C**) Lateral view to specify $\boldsymbol{B}$ field along x, y, and z directions. (**D**) Plot of $|\boldsymbol{B}|$, evaluated along curves shown in (**C**).

Enzymatic activity was determined after 48 h of exposure to SMF. Figures 3 and 4 show the measured activity of two antioxidant enzymes, the superoxide dismutase (SOD) and catalase (CAT), respectively, whereas Figure 5 depicts metabolites of interest. It is observed that the SMF influences the enzymatic activity in the *S. obliquus* since the SOD levels under continuous exposure increased up to 1.27 mU mg^{-1} for north and 1.41 mU mg^{-1} for the south configuration, representing a 65% and 42% relative difference with respect to the control condition. Nevertheless, the behavior is apparently reverted to an inhibitory effect under the pulse exposure mode since SOD activity with respect to the control values reduced by 26.3% for the north and 50.5% for the south magnet configurations, corresponding to 0.55 and 0.44 mU mg^{-1} of enzymatic activity (Figure 3A,B). In the case of *N. gaditana* SOD increased up to 0.94 mU mg^{-1} with the north and 1.20 mU mg^{-1} with the south magnet configuration under the continuous exposure mode, representing 38% and 62% higher production compared to the control values. Under the pulse exposure mode, SOD activity was 78% and 116% above that of the control for the north and south magnet configuration, respectively (equivalent to 1.2 mU mg^{-1} and 1.6 mU mg^{-1}, respectively) (Figure 3C,D).

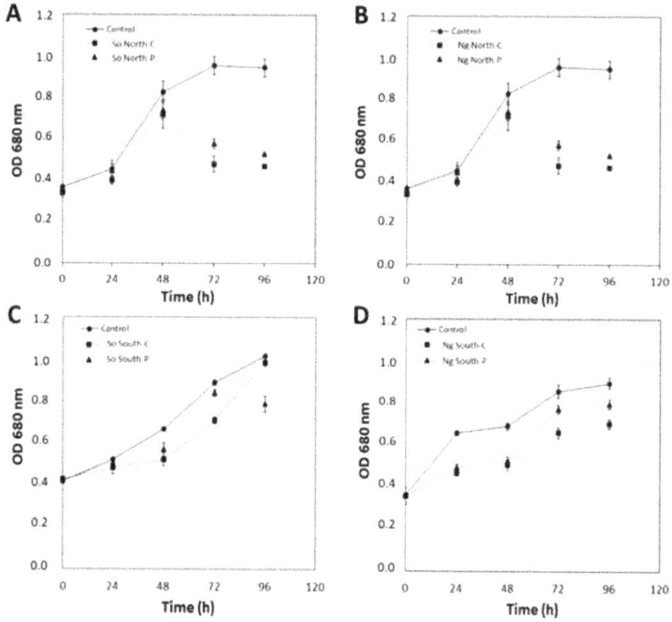

Figure 2. Growth kinetics of *S. obliquus* and *N. gaditana* in response to pulsed and continuous SMF exposure. North arrangement for *S. obliquus* (**A**) and for *N. gaditana* (**B**). South configuration for *S. obliquus* (**C**) and for *N. gaditana* (**D**). Data are presented as mean ± standard deviation analyzed from three replicates. Square corresponds to the control without exposure to SMF. So: *S. obliquus*; Ng: *N. gaditana*; North-C: north-continuous exposure; North-P: north-pulse-type exposure; South-C: south-continuous exposure; South-P: south-pulse-type exposure.

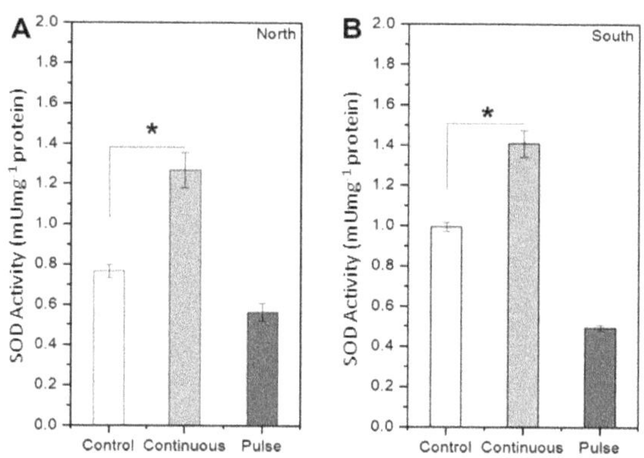

Figure 3. *Cont.*

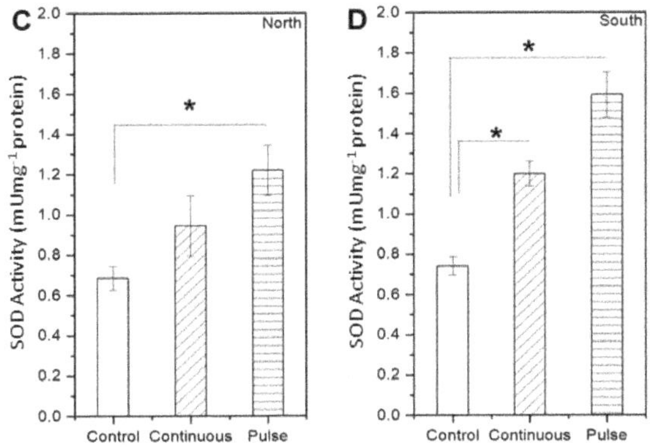

Figure 3. Measurements of the enzymatic activity in terms of SOD for the two microalgae unexposed or control and exposed to SMF with north and south configurations under continuous and pulse exposure modes. (**A**) North and (**B**) south magnet configuration for the *S. obliquus;* (**C**) north and (**D**) south magnet configuration for the *N. gaditana*. Data represent the average of three replicates and the standard deviation (±). Significant differences between the control and experimental treatments with univariate statistical models using ANOVA analysis with a confidence of 95% (p values < 0.05 are indicated by *).

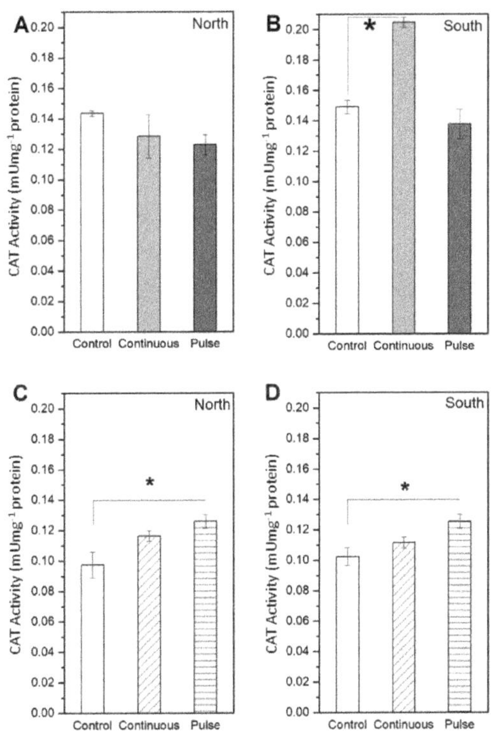

Figure 4. Measurements of the enzymatic activity in terms of CAT for the two microalgae unexposed or control and exposed to SMF with north and south configurations under continuous and pulse exposure

modes. (**A**) North and (**B**) south magnet configuration for the *S. obliquus*. (**C**) North and (**D**) south magnet configuration for the *N. gaditana*. Data represent the average of three replicates and the standard deviation (±). Significant differences between the control and experimental treatments with univariate statistical models using ANOVA analysis with a confidence of 95% (p values < 0.05 are indicated by *).

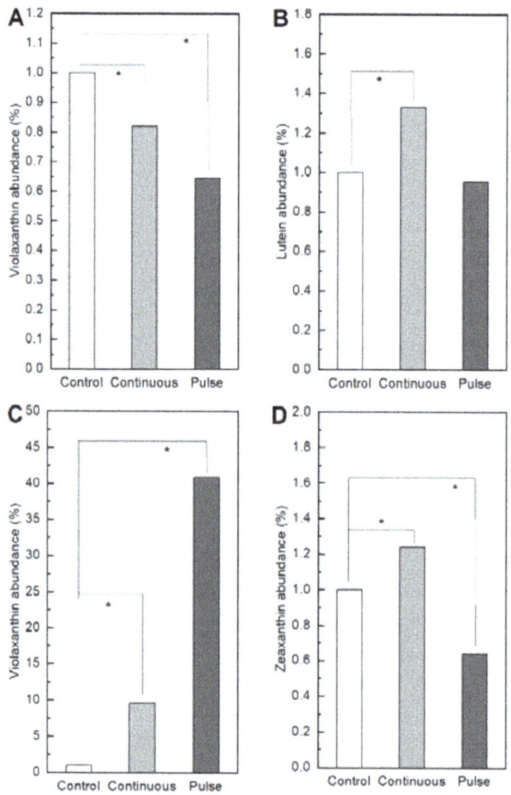

Figure 5. Quantification of metabolites of interest. (**A**,**B**) show the abundance of violaxanthin and lutein for *S. obliquus*; while (**C**,**D**) correspond to the abundance of violaxanthin and zeaxanthin for *N. gaditana*. Data represent the average of three replicates and the standard deviation (±). Significant differences between the control and experimental treatments with univariate statistical models using ANOVA analysis with a confidence of 95% (p values < 0.05 are indicated by *).

For *S. obliquus* the upward trend is also found for the CAT activity in the continuous mode combined with the south configuration, since there was a 37% increase in production compared to the control case (Figure 4B). The inhibition is also observed for the CAT activity under both exposure modes, since the reduction in CAT activity reached 14% for the north and 7.6% for the south magnet configuration (Figure 4A,B). On the other hand, for *N. gaditana* in terms of CAT, there was a higher activity expressed with a 19% and 9% higher production with respect to control values for the north south magnet configuration under continuous exposure mode, respectively (Figure 4C,D). For pulse exposure mode, CAT activity in this microalga reached 0.13 mU mg^{-1} for north and south configuration, meaning 29% and 23% with respect to their control conditions, respectively (Figure 4C,D).

We quantified metabolites of interest after 96 h of SMF exposure. Figure 5A,B indicates respectively that for *S. obliquus* there is a slight decrease in the production of violaxanthin in both exposure modes, a higher production of lutein under continuous exposure, but the same behavior as the control under pulsed exposure. In both cases the variation in the production of violaxanthin and lutein for the *S. obliquus* represents less than 1% with respect to the control. Conversely, Figure 5C reveals that for *N. gaditana*, violaxanthin increased by 10% under continuous SMF and 40% under pulsed exposure compared to the control. Figure 5D shows that the changes in zeaxanthin for *N. gaditana* with respect to the control are below 1%. These results highlight the effectiveness of the SMF on *N. gaditana*, especially for the pulse exposure mode.

3. Discussion

Magnetic fields are capable of inducing effects in many biological systems both in vivo and in vitro [12–17]. Some studies have been successful in stimulating the production of an antioxidant response to physiological stress induced by SMF or MF, as well as stimulating growth in microalgae such as *Chlorella vulgaris* [15,16] and *Dunaliella salina* [31]. However, the strength of the SMF may influence the effect on microalgae as described by Hunt et al. [32]. In contrast, in our study, only a significant increase in growth was observed for *S. obliquus* exposed to a north SMF in pulse form at 72 and 96 h. In all the other conditions analyzed, a decrease in growth or no clear effect on growth was determined, as was the case for *N. gaditana*. However, we note that in other studies the strength of SMF used is lower, for example, for *C. vulgaris* an increase in growth is described with exposure to a 1500 G SMF for 384 h, while high-intensity magnetic fields inhibit the growth of algal biomass [16].

The antioxidant enzymes SOD and CAT are characterized to present paramagnetic elements (Fe, Mn, and Cu) in their catalytic domain, and these could be strongly affected as consequence of the magnetic fields to which the algae were exposed during the experiments. From our study, it is possible to generalize that for *S. obliquus*, under the continuous exposure mode, there is a tendency for a higher production of SOD and CAT. Contrariwise, the pulse exposure mode led to an inhibitory effect on this microalga supported by a decrease in the SOD activity (20–50%) and CAT activity (10–20%). A possible explanation is that due to a saturation of ROS production inside the cell, it affects the metal ions present in the active site of the enzymes, preventing the substrate-enzyme assembly [33]. These results agree with those obtained by Corpas et al. [34], where the ROS production by lead in the *Arabidopsis thaliana* plant inhibits the catalase activity with respect to the control. A similar effect was observed by Chokshi et al. [35], where the salinity-induced oxidative stress in the *Acutodesmus dimorphus* microalgae alters the enzymatic activity in terms of SOD and CAT, as it was below the control values over time. Another point is that despite having a medium with a low electrical conductivity, interacting electrons in molecules are affected by magnetic fields, adding that the cell structure has a slightly rigid wall, allowing greater permeability to internal molecular changes [36]. This peculiarity in the functioning of enzymes was observed by Ferrada et al. [30], where an inhibition of enzymatic activity with respect to the control *in S. obliquus* exposed to SMF occurs, leading to state that the amount of ROS produced exceeds the threshold for the proper cells' antioxidant capacity.

In the same way, the results for the *N. gaditana* allows for stating that, regardless of the exposure mode and magnet configuration, there is always a significant increase in the SOD and CAT production with respect to the control condition. These results also agree with Ferrada et al. [30], where an increase in the enzymatic activity in terms of SOD is observed in *N. gaditana* when exposed to SMF. Furthermore, in *C. vulgaris*, [16] demonstrated a higher induction of CAT activity at 101.37 Umg^{-1} using a 400 G SMF. However, as they increase the SMF intensity up to 1500 G, CAT activity decreases in agreement with the results observed in *S. obliquus* (Figure 4) exposed to approximately 4500 G in our study. Considering the SOD and CAT results, it is inferred that SMF impacts the antioxidant response of algae, although the effects on different enzymes and algae are inconsistent. It has been described that high magnetic field strengths produce changes in enzyme conformation, which in turn affects cellular biochemical reactions [37]. Furthermore, SMFs affect the transition metal ions in some enzymes which become paramagnetic, producing an increase in enzyme activity [38].

Previous research has shown that the use of SMF can induce a non-enzymatic defense response [28,39] as carotenoids and in turn, these pigments are known for their industrial applications [39]. The aim of our study was to determine the effect of SMF on the composition of carotenoid pigments in both microalgae. As described by Ambati et al. [28], microalgae of the genus *Scenedesmus* spp. are known to produce β-carotene, lutein, canthaxanthin, astaxanthin, and fucoxanthin. However, the pigment compositions of algae vary from species to species and also according to environmental conditions including different stress conditions such as temperature, salinity, and irradiance which may induce carotenoid production in algal species. Therefore, in our study by U-HPLC analysis, the complete pigment spectrum was determined for both microalgae in the control condition and treatments. Subsequently, using Xcalibur software, a quantitative analysis and identification of the pigments according to the chromatograms was performed. The peaks that showed significant differences between the control and the treatments were detected, and this analysis is shown in Figure 5, only for the pigments that showed differences between the treatments with respect to the control. In *S. obliquus*, the activity in terms of metabolites such as violaxanthin and lutein shows changes below 1%. On the contrary, in *N. gaditana* a significant change was observed in terms of the abundance of violaxanthin, where the increase in production under continuous mode reached 10% and 40% under the pulse exposure mode. This outcome indicates that the production of ROS occurs as a consequence of the oxidative stress in cells caused by the exposure to SMF, coinciding with Guo et al. and Lozano et al. [40,41], where the ROS formation increases the enzymatic activity with respect to the control. In addition, in the future it will be interesting to use pulse SMF exposure as an alternative to increase violaxanthin production as other stress conditions such as temperature, salinity, and irradiance may induce carotenoid production in algal species [42].

4. Materials and Methods

4.1. Algal Strain and Growth Conditions

The microalgae species, the *S. obliquus* (Fitoplancton Marino S.L., Cadiz, Spain) and the *N. gaditana* CCMP27, were cultured in 1 L volumes. The bold basal medium (BBM) was used for the *S. obliquus* [43,44], which is prepared from fresh water. In the case of *N. gaditana.*, the medium was a modified f/2 [45] prepared from sea water. Both cultures were prepared in glass flasks of 1000 mL with constant air inflow through a tube (see Figure 6). Cultures were incubated for 4 days at 23 ± 1 °C under continuous light using cold-white fluorescent lights (see Figure 7) at 100 μE $m^{-2}s^{-1}$ (irradiances were measured with a calibrated quantum sensor and spectroradiometer, LI-250A, LI-COR1800, Li-COR, Biosciences, Lincoln, NE, USA). The initial concentrations of *S. obliquus* and *N. gaditana* in all groups were 1×10^{5} cells mL^{-1}. Every 24 h samples of each culture were collected by centrifugation (10,000× *g*, 10 min at 4 °C) and washed twice with distilled water. Cells were measured at optical density 680 nm (OD_{680}) in a Spectroquant Prove 300 spectrophotometer (Merck, Germany). The growth curve for both microalgae were obtained.

Figure 6. Photograph of the experimental setup for a single flask having the culture inside, surrounded by the permanent magnets, and under illumination.

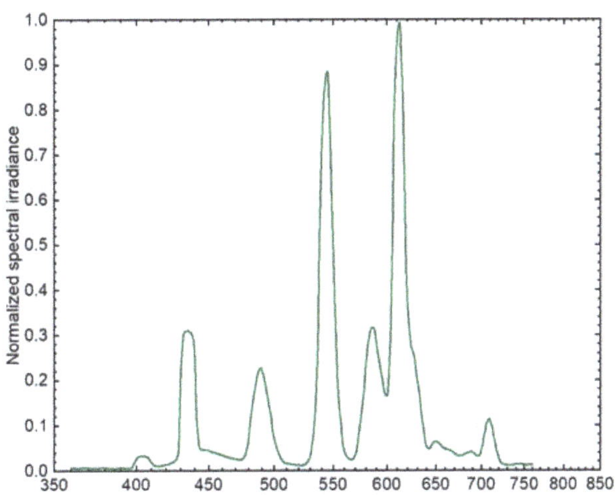

Figure 7. Spectrum of the light source. This spectrum is characterized by sharp peaks of intensities near the maximum magnitude at 545 nm and 613 nm, and several ones of intensities below 30% of the maximum at 435 nm, 489 nm, 587 nm, and 709 nm.

4.2. Experimental Setup

The setup consists of a flat bottom flask surrounded by an array of neodymium magnets as shown in Figure 8. The neodymium magnets ($Nd_2Fe_{14}B$ of grade N33) are coated with Ni/Cu/Ni and have dimensions of 20 × 20 × 20 mm^3. The rare earth (RE) content in the RE-Fe alloy of the magnet ranges between 30 and 35 wt%. According to manufacturing specifications, the corresponding remanent flux density B_r, coercivity values BH_c and jH_c, and maximum stored energy $(BH)_{max}$ are, respectively, 1.164 T, 11.43 kOe and 25.99 kOe, and 33.06 MGOe. The flask contains microalgae media, subject to illumination (see Figure 8). In addition, a thin tube entering from the top was used to introduce air near the bottom. The air inflow results in bubbles which in turn, due to buoyancy, lead to fluid circulation. Regarding the groups, there were 3 flasks for 2 different configurations and 2 exposure modes (continuous and pulse), and additionally, 3 flasks for a control case without SMF exposure (18 flasks for each species). The mode "continuous" means that the biological system was continuously exposed to the SMF during a period of 96 h. The mode "pulse" refers to an intermittent application of the SMF, meaning that the culture was exposed only 1 h per day for 96 h. The latter means that for each microalga, a control condition was considered without exposure to SMF, North-P, North-C, South-P, and South-C in triplicate. The implemented cases under study are enumerated in the following.

- Control (C): No magnetic field applied.
- South-continuous (SC): All south poles are oriented towards the center of the flask and continuous exposure to the magnetic flux density (***B***).
- South-pulse (SP): All south poles oriented towards the center of the flask and pulsed exposure to ***B***.
- North-continuous (NC): All north poles oriented towards the center of the flask and continuous exposure to ***B***.
- North-pulse (NP): All north poles oriented towards the center of the flask and pulsed exposure to ***B***.

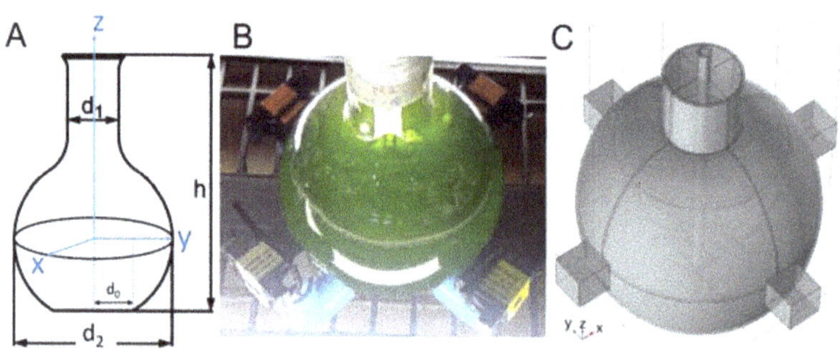

Figure 8. (**A**) Geometrical measures of the flask: $d_0 = 30$ mm, $d_1 = 42$ mm, $d_2 = 131$ mm, and $h = 190$ mm. (**B**) Glass flask surrounded by an array of permanent magnets at $z = 0$ and a thin tube to introduce air. (**C**) Geometry model in 3D. The neodymium magnets ($Nd_2Fe_{14}B$ of grade N33) are coated with Ni/Cu/Ni and have dimensions of $20 \times 20 \times 20$ mm^3. The rare earth (RE) content in the RE-Fe alloy of the magnet ranges between 30 and 35 wt%. According to manufacturing specifications, the corresponding remanent flux density B_r, coercivity values BH_c and jH_c, and maximum stored energy $(BH)_{max}$ are, respectively, 1.164 T, 11.43 kOe and 25.99 kOe, and 33.06 MGOe.

4.3. Measurement Procedures

4.3.1. Physicochemical Medium Parameters

Temperature, pH, conductivity (σ), salinity, pressure, and dissolved oxygen were characterized with an HI 98194 multiparameter probe from Hanna Instruments. Samples were taken from the control and treated groups of both media at the exponential growth phase. The measurements were taken every 24 h for a period of 96 h.

4.3.2. Magnetic Flux Density

The magnetic flux density norm was determined for every individual magnet used in all magnet configurations by using an AC/DC magnetic meter composed of a plate-like Hall sensor with a temperature compensation unit. The device is conceived for measurements of homogenous magnetic flux density at the level of the active part (magnet). The field direction of the measurement is uniaxial, sensing the main magnetic component perpendicular to the chip [46], $|\boldsymbol{B}|$, in gauss (G), and mT units with a resolution of up to 0.1 G in both DC and AC mode and an accuracy of $\pm 5\%$ at 23 ± 5 °C. For good measurements, [47] pointed out that the current density and the magnetic field must be as much orthogonal as possible and that the shape of the active region has to facilitate the measurement of the Hall voltage. Considering measurements of $|\boldsymbol{B}|$ for each configuration, the average values ranged from 4441 to 4547 G with a standard deviation from 0.8% to 3.6% in the worst case. The $|\boldsymbol{B}|$ values at the side of each magnet were used to visualize $\boldsymbol{B}$ in space, which leads to the oxidative stress.

The investigation of the $\boldsymbol{B}$ distribution was performed by means of a mutiphase turbulent bubbly flow coupled with Maxwell equations implemented in COMSOL Multiphysics v5.5. The required inputs such as the remanent magnetic flux density (B_r) of each magnet was obtained by using the measured magnetic flux density norm $|\boldsymbol{B}|$ at the side of magnets in Equation (1) and solved for B_r, where L, W, and D are the length, width, and depth of a rectangular magnet [48–50]. The computed $|\boldsymbol{B_r}|$ results in values in the range of 1.03 ± 0.02 T. Replacing densities and viscosities [51] at the temperature of the experiment (20 °C) as well as the measured conductivities [30] for both fresh and seawater gives a Reynolds number $Re > 10,000$ [52] and a magnetic Reynolds number $Re_m \ll 1$ [53]. Thus, the fluid is turbulent, and $\boldsymbol{B}$ is determined by boundary conditions, not the fluid flow. Thus, the visualization of $\boldsymbol{B}$ can be computed as that field in absence of fluid flow.

$$|\boldsymbol{B}(d)| = \frac{B_r}{\pi}\left\{tg^{-1}\left(\frac{L \cdot W}{2d\sqrt{4d^2 + L^2 + W^2}}\right) - tg^{-1}\left(\frac{L \cdot W}{2(D+d)\sqrt{4(D+d)^2 + L^2 + W^2}}\right)\right\} \quad (1)$$

4.3.3. Enzymatic Activity

The superoxide dismutase (SOD) and catalase (CAT) total enzymatic activities were measured, since both groups of enzymes are characterized to present paramagnetic elements (Fe, Mn, and Cu) in their catalytic domain, and therefore affected by the magnetic fields in which the algae were exposed during the experiment. The cell break was carried out in a mortar with liquid nitrogen and glass beads. The homogenization was performed with 2 mL of extraction buffer (K_2HPO_4-KH_2PO_4) KH_2PO_4, 50 mm, pH 7.0, containing Triton X-100 (0.01% v/v) [54]. In the following, the homogenate was centrifuged at 13,000$\times$ g at 4 °C for 15 min [55]. Proteins were quantified according to the Bradford method [56]. The extract was immediately used to quantify the antioxidant enzymes. The measurement of the SOD activity was measured for both microalgae based on the photochemical reduction of Nitro blue tetrazolium chloride (NBT) described by Donahue et al. [57] and modified by Cartes et al. [55]. For CAT activity, the evaluation was performed through spectrophotometry by the disappearance of hydrogen peroxide according to [58].

4.3.4. Pigments Determination by Ultra High-Performance Liquid Chromatography (UHPLC) Mass-to-Mass

The extracted total pigments were filtered at 0.45 µm and stored at −80 °C until use, according to the protocol described in [59]. The high-performance liquid chromatography (UHPLC) method was used for the quantitative identification of carotenoids as described by Fu et al. [59]. For the separation of carotenoids, a Dionex Ultimate 3000 chromatograph with a diode array detector connected to an Orbitrap Q Exactive Focus (Thermo Scientific, ThermoFisher Scientific, MA, USA) was used. The results were analyzed using Thermo Xcalibur Sequence Setup software (Thermo Scientific, ThermoFisher Scientific, Waltham, MA, USA).

4.3.5. Statistical Analysis

The data of the different treatments were analyzed with univariate statistical models using analysis of variances (ANOVA) with a confidence of 95%. These analyses were performed with the Minitab 17 software.

5. Conclusions

In this study we evaluate the response of two different microalgae to the presence of static magnetic fields (SMF) exposed in continuous and pulsed form. The SMFs used have north and south configurations. For both microalgae, *S. obliquus* and *N. gaditana*, a clear negative effect on growth was observed in the presence of the north configuration, both in pulse and continuous exposure, after 48 h of culture. Other effects were evidenced in the alteration of antioxidant enzyme activity in SOD and CAT enzymes. In *S. obliquus*, an over-regulation of SOD in the presence of SMF in continuous exposure and a down-regulation in pulse exposure to SMF were observed for both the north and south configurations. In contrast, for CAT only an increase in enzyme activity is observed only in continuous exposure to SMF. In the case of *N. gaditana*, an over-regulation of SOD and CAT enzyme activity is observed upon continuous and pulse exposure to SMF in both north and south configurations. The latter coincides with a significant increase in antioxidant pigments in *N. gaditana*, especially violaxanthin, which is part of the non-enzymatic defense system, but may have an interesting use as a bioactive molecule, which in future studies could be scaled up and optimized using SMF.

Author Contributions: Conceptualization, M.Z. and P.F.; methodology, C.M.-O., C.W.-Z., K.G. and G.S.; software, C.M.-O., K.G. and P.F.; validation, C.M.-O., M.Z., M.R. and P.F.; formal analysis, A.M., G.S., C.M.-O., P.F. and M.R.; investigation, C.M.-O., A.M. and G.S.; resources, M.Z., E.F., P.F. and C.W.-Z.; data curation, P.F. and C.M.-O.; writing—original draft preparation, P.F. and M.R.; writing—review and editing, P.F. and M.R.; funding acquisition, E.F., P.F., M.Z. and M.R. All authors have read and agreed to the published version of the manuscript.

Funding: This research was funded by ANID/FONDEF/IDEA, grant number ID15I10487, and in part by the Programa Semilleros de Investigación: "Caracterización de una línea base para el análisis y determinación de los components celulares", by the Fondo para el Desarrollo en investigación científica y/o tecnológica de actividades de titulación de pregrado 2018: "Producción de metabolitos de alto valor económico mediante exposición a un campo magnético estático tipo continuo y tipo pulso en las microalgas *Scenedesmus obliquus* y *Nannochloropsis gaditana*", and by Programa Asignación año 2021 de Asistentes de investigación de la Universidad de Antofagasta código ANT20992.

Institutional Review Board Statement: The study was conducted according to the guidelines of the Declaration of Helsinki, and approved by the Comité de Ética de Investigación Científica CEIC-UA in 2018, Folio no 178.

Informed Consent Statement: Not applicable.

Acknowledgments: The authors would like to thank the Universidad de Antofagasta for their support in performing this study.

Conflicts of Interest: The authors declare no conflict of interest.

References

1. Nadkarni, R.; Barkley, S.; Fradin, C. A comparison of methods to measure the magnetic moment of magnetotactic bacteria through analysis of their trajectories in external magnetic fields. *PLoS ONE* **2013**, *8*, e82064. [CrossRef]
2. Günther, A.; Einwich, A.; Sjulstok, E.; Feederle, R.; Bolte, P.; Koch, K.W.; Solov'yov, I.A.; Mouritsen, H. Double-cone localization and seasonal expression pattern suggest a role in magnetoreception for european robin cryptochrome 4. *Curr. Biol.* **2018**, *28*, 211–223.e4. [CrossRef] [PubMed]
3. Kirschvink, J.L.; Kobayashi-Kirschvink, A.; Diaz-Ricci, J.C.; Kirschvink, S.J. Magnetite in human tissues: A mechanism for the biological effects of weak ELF magnetic fields. *Bioelectromagnetics* **1992**, *13*, 101–113. [CrossRef] [PubMed]
4. Ghodbane, S.; Lahbib, A.; Sakly, M.; Abdelmelek, H. Bioeffects of static magnetic fields: Oxidative stress, genotoxic effects, and cancer studies. *BiomMed Res. Int.* **2013**, *2013*, 602987. [CrossRef] [PubMed]

5. Joshi, P.; Williams, P.S.; Moore, L.R.; Caralla, T.; Boehm, C.; Muschler, G.; Zborowski, M. Circular halbach array for fast magnetic separation of hyaluronan-expressing tissue progenitors. *Anal. Chem.* **2015**, *87*, 9908–9915. [CrossRef]
6. Zhang, X.; Yarema, K.; Xu, A. Impact of static magnetic field (SMFs) on cells. In *Biological Effects of Static Magnetic Fields*; Springer: Singapore, 2017; pp. 81–132. [CrossRef]
7. Colbert, A.P.; Wahbeh, H.; Harling, N.; Connelly, E.; Schiffke, H.C.; Forsten, C.; Gregory, W.L.; Markov, M.S.; Souder, J.J.; Elmer, P.; et al. Static magnetic field therapy: A critical review of treatment parameters. *Evid. Based Complement. Altern. Med.* **2009**, *6*, 133–139. [CrossRef]
8. Liu, Y.; Qi, H.; Sun, R.G.; Chen, W.F. An investigation into the combined effect of static magnetic fields and different anticancer drugs on K562 cell membranes. *Tumori* **2011**, *97*, 386–392. [CrossRef]
9. Zablotskii, V.; Polyakova, T.; Lunov, O.; Dejneka, A. How a high-gradient magnetic field could affect cell life. *Sci. Rep.* **2016**, *6*, 37407. [CrossRef]
10. Steiner, U.E.; Ulrich, T. Magnetic field effects in chemical kinetics and related phenomena. *Chem. Rev.* **1989**, *89*, 51–147. [CrossRef]
11. Albuquerque, W.W.C.; Costa, R.M.P.B.; de Salazar e Fernandes, T.; Porto, A.L.F. Evidences of the static magnetic field influence on cellular systems. *Prog. Biophys. Mol. Biol.* **2016**, *121*, 16–28. [CrossRef]
12. Small, D.P.; Hüner, N.P.A.; Wan, W. Effect of static magnetic fields on the growth photosynthesis and ultrastructure of *Chlorella kessleri* microalgae. *Bioelectromagnetics* **2012**, *33*, 298–308. [CrossRef] [PubMed]
13. Oliveira, M. Magnetic stimulation on the growth of the microalga *Nannochloropsis oculata*. In *Electronic Thesis and Dissertation Repository*; The University of Western Ontario: London, ON, Canada, 2017; p. 4597. Available online: http://ir.lib.uwo.ca/etd/4597 (accessed on 25 July 2021).
14. Tu, R.; Jin, W.; Xi, T.; Yang, Q.; Han, S.F.; Abomohra, A.E.F. Effect of static magnetic field on the oxygen production of *Scenedesmus obliquus* cultivated in municipal wastewater. *Water Res.* **2015**, *86*, 132–138. [CrossRef] [PubMed]
15. Wang, H.; Zeng, X.; Guo, S.; Li, Z. Effects of magnetic field on the antioxidant defense system of recirculation-cultured *Chlorella vulgaris*. *Bioelectromagnetics* **2008**, *29*, 39–46. [CrossRef] [PubMed]
16. Luo, X.; Zhang, H.; Li, Q.; Zhang, J. Effects of static magnetic field on *Chlorella vulgaris*: Growth and extracellular polysaccharide (EPS) production. *J. Appl. Phycol.* **2020**, *32*, 2819–2828. [CrossRef]
17. Phaniendra, A.; Jestadi, D.B.; Periyasamy, L. Free radicals: Properties, sources, targets, and their implication in various diseases. *Ind. J. Clin. Biochem.* **2015**, *30*, 11–26. [CrossRef]
18. Lobo, V.; Patil, A.; Phatak, A.; Chandra, N. Free radicals, antioxidants and functional foods: Impact on human health. *Pharmacogn. Rev.* **2010**, *4*, 118–126. [CrossRef]
19. Foyer, C.H. Reactive oxygen species, oxidative signaling and the regulation of photosynthesis. *Environ. Exp. Bot.* **2018**, *154*, 134–142. [CrossRef]
20. Gill, S.S.; Tuteja, N. Reactive oxygen species and antioxidant machinery in abiotic stress tolerance in crop plants. *Plant Physiol. Biochem.* **2010**, *48*, 909–930. [CrossRef] [PubMed]
21. Birben, E.; Sahiner, U.M.; Sackesen, C.; Erzurum, S.; Kalayci, O. Oxidative stress and antioxidant defense. *World Allergy Organ. J.* **2012**, *5*, 9–19. [CrossRef]
22. Ajayan, K.V.; Selvaraju, M. Heavy metal induced antioxidant defense system of green microalgae and its effective role in phycoremediation of tannery effluent. *Pak. J. Biol. Sci.* **2012**, *15*, 1056–1062. [CrossRef]
23. Valavanidis, A.; Vlahogianni, T.; Dassenakis, M.; Scoullos, M. Molecular biomarkers of oxidative stress in aquatic organisms in relation to toxic environmental pollutants. *Ecotoxicol. Environ. Saf.* **2006**, *64*, 178–189. [CrossRef] [PubMed]
24. Rezayian, M.; Niknam, V.; Ebrahimzadeh, H. Oxidative damage and antioxidative system in algae. *Toxicol. Rep.* **2019**, *6*, 1309–1313. [CrossRef]
25. He, L.; He, T.; Farrar, S.; Ji, L.; Liu, T.; Ma, X. Antioxidants maintain cellular redox homeostasis by elimination of reactive oxygen species. *Cell. Physiol. Biochem.* **2017**, *44*, 532–553. [CrossRef] [PubMed]
26. Mandotra, S.K.; Kumar, P.; Suseela, M.R.; Ramteke, P.W. Fresh water green microalga Scenedesmus abundans: A potential feedstock for high quality biodiesel production. *Bioresour. Technol.* **2014**, *156*, 42–47. [CrossRef] [PubMed]
27. Ho, S.H.; Chan, M.C.; Liu, C.C.; Chen, C.Y.; Lee, W.L.; Lee, D.J.; Chang, J.S. Enhancing lutein productivity of an indigenous microalga *Scenedesmus obliquus* FSP-3 using light-related strategies. *Bioresour. Technol.* **2014**, *152*, 275–282. [CrossRef] [PubMed]
28. Ambati, R.R.; Deepika, G.; Ravishankar, G.; Sarada, R.; Panduranga, N.B.; Lei, B.; Yuepeng, S. Industrial potential of carotenoid pigments from microalgae: Current trends and future prospects. *Crit. Rev. Food Sci. Nutr.* **2019**, *59*, 1880–1902. [CrossRef]
29. Solovchenko, A.; Lukyanov, A.; Solovchenko, O.; Didi-Cohen, S.; Boussiba, S.; Khozin-Goldberg, I. Interactive effects of salinity, high light, and nitrogen starvation on fatty acid and carotenoid profiles in *Nannochloropsis oceanica* CCALA 804. *Eur. J. Lipid Sci. Technol.* **2014**, *166*, 635–644. [CrossRef]
30. Ferrada, P.; Rodríguez, S.; Serrano, G.; Miranda-Ostojic, C.; Maureira, A.; Zapata, M. An analytical–experimental Approach to quantifying the effects of static magnetic fields for cell culture applications. *Appl. Sci.* **2020**, *10*, 531. [CrossRef]
31. Yamaoka, Y.; Takimura, O.; Fuse, H.; Kaminura, K. Effect of magnetism on growth of *Dunaliella salina*. *Res. Photosynth.* **1992**, *3*, 87–90.
32. Hunt, R.W.; Zavalin, A.; Bhatnagar, A.; Chinnasamy, S.; Das, K.C. Electromagnetic biostimulation of living cultures for biotechnology, biofuel and bioenergy applications. *Int. J. Mol. Sci.* **2009**, *10*, 4515–4558. [CrossRef] [PubMed]

33. Gebicka, L.; Krych-Madej, J. The role of catalases in the prevention/promotion of oxidative stress. *J. Inorg. Biochem.* **2019**, *197*, 110699. [CrossRef] [PubMed]
34. Corpas, F.J.; Barroso, J.B. Lead-induced stress, which triggers the production of nitric oxide (NO) and superoxide anion (O_2^-) in *Arabidopsis* peroxisomes, affects catalase activity. *Nitric Oxide Biol. Chem.* **2017**, *68*, 103–110. [CrossRef]
35. Chokshi, K.; Pancha, I.; Ghosh, A.; Mishra, S. Salinity induced oxidative stress alters the physiological responses and improves the biofuel potential of green microalgae Acutodesmus dimorphus. *Bioresour. Technol.* **2017**, *244*, 1376–1383. [CrossRef] [PubMed]
36. Bux, F.; Chisti, Y. *Algae Biotechnology: Products and Processes*; Springer International Publishing: Cham, Switzerland, 2016; ISBN 978-3-319-12334-9. [CrossRef]
37. Wang, H.Y.; Zeng, X.B.; Guo, S.Y. Growth of *Chlorella vulgaris* under different magnetic treatments. *Prog. Mod. Biomed.* **2006**, *6*, 106–108.
38. Katz, E.; Lioubashevski, O.; Willner, I. Magnetic field effects on bioelectrocatalytic reactions of surface-confined enzyme systems: enhanced performance of biofuel cells. *J. Am. Chem. Soc.* **2005**, *127*, 3979–3988. [CrossRef]
39. Lu, H.; Wang, X.; Hu, S.; Han, T.; He, S.; Zhang, G.; He, M.; Lin, X. Bioeffect of static magnetic field on photosynthetic bacteria: Evaluation of bioresources production and wastewater treatment efficiency. *Water Environ Res.* **2020**, *92*, 1131–1141. [CrossRef]
40. Guo, J.; Peng, J.; Lei, Y.; Kanerva, M.; Li, Q.; Song, J.; Guo, J.; Sun, H. Comparison of oxidative stress induced by clarithromycin in two freshwater microalgae *Raphidocelis subcapitata* and *Chlorella vulgaris*. *Aquat. Toxicol.* **2020**, *219*, 105376. [CrossRef] [PubMed]
41. Lozano, P.; Trombini, C.; Crespo, E.; Blasco, J.; Moreno-Garrido, I. ROI-scavenging enzyme activities as toxicity biomarkers in three species of marine microalgae exposed to model contaminants (copper, Irgarol and atrazine). *Ecotoxicol. Environ. Saf.* **2014**, *104*, 294–301. [CrossRef]
42. Minhas, A.K.; Hodgson, P.; Barrow, C.J.; Adholeya, A. A review on the assessment of stress conditions for simultaneous production of microalgal lipids and carotenoids. *Front. Microbiol.* **2016**, *7*, 546. [CrossRef]
43. Bold, H.C. The morphology of *Chlamydomonas chlamydogama*, sp. nov. *Bull. Torrey Bot. Club* **1949**, *76*, 101. [CrossRef]
44. Bischoff, H.W.; Bold; Harold, C. *Some Soil Algae from Enchanted Rock and Related Algal Species*; University of Texas: Austin, TX, USA, 1963.
45. Guillard, R.R.; Ryther, J.H. Studies of marine planktonic diatoms. I. Cyclotella nana Hustedt, and Detonula confervacea (cleve) Gran. *Can. J. Microbiol.* **1962**, *8*, 229–239. [CrossRef] [PubMed]
46. Bottura, L.; Henrichsen, K.N. *Field Measurements*; CAS–CERN Accelerator School: Superconductivity and Cryogenics for Accelerators and Detectors: Geneva, Switzerland, 2002; pp. 118–148. [CrossRef]
47. Popovic, R. Not-plate-like Hall magnetic sensors and their applications. *Sens. Actuators A Phys.* **2000**, *85*, 9–17. [CrossRef]
48. Yang, Z.J.; Johansen, T.H.; Bratsberg, H.; Helgesen, G.; Skjeltorp, A.T. Potential and force between a magnet and a bulk $Y_1Ba_2Cu_3O_7$-δsuperconductor studied by a mechanical pendulum. *Supercond. Sci. Technol.* **1990**, *3*, 591–597. [CrossRef]
49. Camacho, J.M.; Sosa, V. Alternative method to calculate the magnetic field of permanent magnets with azimuthal symmetry. *Rev. Mex. Fis. E* **2013**, *59*, 8–17.
50. Furlani, E.P.; Furlani, E.P. Permanent magnet applications. In *Perm. Magn. Electromech. Devices*; Elsevier, B.V.: Amsterdam, The Netherlands, 2001; pp. 207–333. ISBN 978-0-12-269951-1. [CrossRef]
51. ITTC Recommended Procedures. The Specialist Committee on Uncertainty Analysis. In Proceedings of 26th International Towing Tank Conference, Rio de Janeiro, Brazil, 28 August–3 September 2011; Fresh Water and Seawater Properties. Available online: http://ittc.info/media/5526/08.pdf (accessed on 25 July 2021).
52. Schlichting, H.; Gersten, K. *Boundary Layer Theory*, 9th ed.; Springer: Berlin/Heidelberg, Germany, 2018; ISBN 9783662570951.
53. Davidson, P.A. Kinematics of MHD: Advection and diffusion of a magnetic field. In *An Introduction to Magnetohydrodynamics*; Cambridge University Press: Cambridge, UK, 2010; pp. 102–116. ISBN 9780511626333.
54. Janknegt, P.J.; Rijstenbil, J.W.; van de Poll, W.H.; Gechev, T.S.; Buma, A.G.J. A comparison of quantitative and qualitative superoxide dismutase assays for application to low temperature microalgae. *J. Photochem. Photobiol. B Biol.* **2007**, *87*, 218–226. [CrossRef]
55. Cartes, P.; McManus, M.; Wulff-Zottele, C.; Leung, S.; Gutiérrez-Moraga, A.; de la Luz Mora, M. Differential superoxide dismutase expression in ryegrass cultivars in response to short term aluminium stress. *Plant Soil* **2012**, *350*, 353–363. [CrossRef]
56. Bradford, M.M. A rapid and sensitive method for the quantitation of microgram quantities of protein utilizing the principle of protein-dye binding. *Anal. Biochem.* **1976**, *72*, 248–254. [CrossRef]
57. Donahue, J.L.; Okpodu, C.M.; Cramer, C.L.; Grabau, E.A.; Alscher, R.G. Responses of antioxidants to paraquat in pea leaves (relationships to resistance). *Plant Physiol.* **1997**, *113*, 249–257. [CrossRef]
58. Pinhero, R.G.; Rao, M.V.; Paliyath, G.; Murr, D.P.; Fletcher, R.A. Changes in activities of antioxidant enzymes and their relationship to genetic and paclobutrazol-induced chilling tolerance of maize seedlings. *Plant Physiol.* **1997**, *114*, 695–704. [CrossRef]
59. Fu, W.; Magnúsdóttir, M.; Brynjólfson, S.; Palsson, B.; Paglia, G. UPLC-UV-MSE analysis for quantification and identification of major carotenoid and chlorophyll species in algae. *Anal. Bioanal. Chem.* **2012**, *404*, 3145–3154. [CrossRef]

marine drugs

Review

Humic Substances as Microalgal Biostimulants—Implications for Microalgal Biotechnology

Daria Gabriela Popa [1,2], Carmen Lupu [2], Diana Constantinescu-Aruxandei [2,*] and Florin Oancea [1,2,*]

[1] Faculty of Biotechnologies, University of Agronomic Sciences and Veterinary Medicine of Bucharest, Mărăști Blv, No. 59, Sector 1, 011464 Bucharest, Romania; daria.popa@icechim.ro

[2] Bioproducts Team, Bioresources Department, National Institute for Research & Development in Chemistry and Petrochemistry—ICECHIM, Splaiul Independenței No. 202, Sector 6, 060021 Bucharest, Romania; carmen.lupu@icechim.ro

* Correspondence: diana.constantinescu@icechim.ro (D.C.-A.); florin.oancea@icechim.ro (F.O.)

Abstract: Humic substances (HS) act as biostimulants for terrestrial photosynthetic organisms. Their effects on plants are related to specific HS features: pH and redox buffering activities, (pseudo) emulsifying and surfactant characteristics, capacity to bind metallic ions and to encapsulate labile hydrophobic molecules, ability to adsorb to the wall structures of cells. The specific properties of HS result from the complexity of their supramolecular structure. This structure is more dynamic in aqueous solutions/suspensions than in soil, which enhances the specific characteristics of HS. Therefore, HS effects on microalgae are more pronounced than on terrestrial plants. The reported HS effects on microalgae include increased ionic nutrient availability, improved protection against abiotic stress, including against various chemical pollutants and ionic species of potentially toxic elements, higher accumulation of value-added ingredients, and enhanced bio-flocculation. These HS effects are similar to those on terrestrial plants and could be considered microalgal biostimulant effects. Such biostimulant effects are underutilized in current microalgal biotechnology. This review presents knowledge related to interactions between microalgae and humic substances and analyzes the potential of HS to enhance the productivity and profitability of microalgal biotechnology.

Keywords: humic substances; microalgae cultivation; hormetic effects; increased nutrient availability; improved protection against abiotic stress; higher accumulation of bioactive ingredients; enhanced microalgal productivity

Citation: Popa, D.G.; Lupu, C.; Constantinescu-Aruxandei, D.; Oancea, F. Humic Substances as Microalgal Biostimulants—Implications for Microalgal Biotechnology. *Mar. Drugs* **2022**, *20*, 327. https://doi.org/10.3390/md20050327

Academic Editor: Carlos Almeida

Received: 25 April 2022
Accepted: 12 May 2022
Published: 16 May 2022

1. Introduction

Humic substances (HS) are a significant part of soil organic matter. HS are considered important for soil fertility due to their ability to retain water and nutrients, to improve soil cation exchange capacity (CEC), to increase nutrient availability and to generate aerated soil structure [1–3]. Soil scientists separate HS into humic acids (HA), soluble at alkaline pH and insoluble at acidic pH, and fulvic acids (FA), which are soluble both at alkaline and acidic pH [4,5].

HS are supramolecular structures resulting from the association of small molecules derived from slow degradation of biological material, especially plant residues, under specific conditions [6,7]. The supramolecular structure of HS comprises hydrophilic portions, including -OH and -COOH groups, and hydrophobic portions [6,8]. The hydrophobic portions are represented mainly by networks of polyaromatic hydrocarbons derived from lignin [9,10]. The following organic molecules are linked to this hydrophobic scaffold: organic acids, mainly derived from fermentative carbohydrate metabolism [11]; substances derived from protein metabolism, including amino acids [12–14] and polyamines [15]; aliphatic hydrocarbons, which result from waxes; and cross-linked fatty acids [16]. FA are more polar than HA, containing smaller amounts of hydrophobic fragments and larger quantities of organic (poly)acids than HA [17]. Humic acids have a more pronounced amphiphilic structure than FA [18]. Such structures promote aggregation in the supramolecular structures stabilized by hydrophobic interactions [19].

The genesis of humic substances in the soil is still under dispute [20]. The majority consider that HS result from polymerization and polycondensation of the components generated in soil by the decomposition of plant residues and soil microorganisms [21–23]. Such (medium) molecular weight structures are further aggregated into supramolecular structures, according to their polarity index and/or hydrophobic/hydrophilic ratio [24,25], in a reaction that is catalyzed by soil microorganisms [26–28] or soil inorganic materials [27,29]. Another opinion considers HS generation as being the result of the "selective preservation" and "progressive decomposition" of biological material in soils [1].

Despite the controversies related to their formation, the HS-specific structure determines the HS-specific features. The phenolic and carboxylic groups are responsible for the HS weak acid behavior and pH buffering [30]. The quinones–phenols switch is involved in redox buffering activity [25,31]. The coexistence of the hydrophobic and hydrophilic portions in HS supramolecular structure determines several properties of HS, such as the (pseudo)emulsifier effect and tendency to form micelles [32,33] and potential encapsulation of labile hydrophobic compounds in the hydrophobic pocket [34]. Such properties are related to the HS effects on biological systems. For example, the HS reactivity toward different (micro)biocenosis components is related to the ratio between hydrophilic and hydrophobic components [35,36].

HS are plant biostimulants, promoting nutrient uptake and nutrient use efficiency. Root dry weight increases by more than 20% after exogenous HS application [37]. A large body of evidence demonstrated the auxin-like activity of the humic and fulvic acids, including the rhizogenesis and induction of the proton-pump H^+-ATPase [38]. Activation of the secondary ion transport occurs after the change of radicular cell membrane potential [39]. An increase in ionic nutrient uptake following HS application was demonstrated for nitrate [40,41], phosphate [42], and nitrate and sulfate [43]. Ionic nutrient uptake is also stimulated due to enhanced nutrient availability resulting from the chelating effect of humic substances [44], combined with their redox properties [3]. HS also promote primary anabolic pathways, such as nitrate reduction [45] and carbon catabolism [46], leading to enhanced nutrient use efficiency.

Transcriptomic and metabolomic analyses reveal significant modifications of metabolic pathways in both dicotyledonous and monocotyledonous plants under HA treatments. The HA extracted from black peat in alkaline conditions and separated from fulvic acids by HCl precipitation activated the carbon (C)-, nitrogen (N)- and sulfur (S)- metabolic pathways in rapeseed seedlings (*Brassica napus* var. Capitol) [43]. In sugarcane, *Saccharum officinarum* cv. RB 96 7515, application of HA extracted from vermicompost activated the metabolic pathways related to plant stress response and cellular growth [47]. In corn, *Zea mays* cv. PAN 3Q-240, soil application of HS formulation determined several effects according to the nutrient status of the plant. In plants supplied with normal quantities of mineral nutrients, HS stimulated the metabolic pathways related to primary metabolisms—tricarboxylic acid (TCA) cycle and amino acid metabolism. In plants under mineral nutrient deficit, HS stimulated the secondary metabolism pathways related to stress response [48].

Induction of H^+-ATP-ase in root cells by HS elicits the internal cell signals, including an increased level of nitric oxide (NO) [49] and calcium ion influx, followed by the activation of the calcium-dependent protein-kinase (CDPK) [50].

Humic substances activate plant secondary metabolism by (bio)chemical priming [51]. The mechanism of chemical priming by HS in plant tissues and cells involves an increased level of reactive oxygen species (ROS) and modulation of polyamine metabolism [44]. Activation of secondary metabolism determines higher plant tolerance to abiotic stress and increased accumulation of compounds resulting from the plant secondary metabolism in the edible yield [52].

The complex activity of HS on terrestrial plants is related to the complexity of the HS chemical systems. Complexity in its "emergence" aspect is an essential characteristic of first-generation plant biostimulants [53]. The HS complex chemical system presents emergent features, i.e., unexpected properties resulting from component interaction. One illustrative example is the synergic interaction between hydrophobic and hydrophilic moieties in water-holding in arid soil, according to relative humidity. Another feature related to HS system complexity is the context-dependent characteristics, such as redox properties [54]. Due to this intrinsic complexity, combined methods are needed to characterize the HS used as plant biostimulants [55].

In contrast to terrestrial plants, studies related to HS effects on microalgae are relatively scarce. The initial driver of studying the HS effects on microalgae was an ecotoxicological one. Due to the growing interest in carbon sequestration by microalgae, HS has been proposed as a biostimulant for microalgae in recent years [56,57].

HS effects on microalgae are more pronounced than on terrestrial plants. In aqueous solutions/suspensions, HS generates a more dynamic supramolecular structure than in soil [58]. Due to the dynamic structure of HS in solution/suspension, the main HS features, such as pH and redox buffering, ionic nutrient complexation, encapsulation of hydrophobic molecules, and emulsifying characteristics are significantly enhanced [23]. The biological activities related to these HS characteristics are also enhanced. Therefore, the potential of HS to act as a biostimulant, including as support to enhance tolerance to various stresses, is higher in microalgae than in terrestrial photosynthetic organisms. At present, such potential is underutilized in microalgal biotechnology.

This paper reviews the interactions between HS and microalgae and analyzes the practical implications for microalgal biotechnology resulting from the specific effects of HS on microalgae.

2. Hormetic Effects of Humic Substances on Microalgae

As was already mentioned, the initial studies investigated the influence of HS leached from soil on phytoplankton. The positive effect of humic substances on phytoplankton growth was reported almost 50 years ago [59]. More recent studies demonstrated that higher doses of HS from various sources determined inhibition of microalgal growth. Such a dual response is typical of hormesis. Hormesis is a dose–response phenomenon characterized by low-dose stimulation and high-dose inhibition. Hormetic effects are typically graphed as an inverted U-shaped dose–response and a J-shaped dose–response, depending on the endpoint evaluated.

The microalgal response to humic substances is consistent at different cellular and biochemical levels. Various HA concentrations were used to treat *Scenedesmus capricornus* microalgae. At HA concentrations lower than 2.0 mg C L^{-1}, the growth of *S. capricornus* was promoted slightly, and above 2.0 mg C L^{-1}, it was inhibited [60]. The same study reported that an increase in polysaccharide content was observed at low HA concentrations (less than 2.0 mg C L^{-1}), and that with an increment in HA concentration, it decreased. At 10.0 mg C L^{-1} HA, the average polysaccharide concentration was only 12.02 mg L^{-1} compared with the 18.43 mg L^{-1} average for the control group.

When studying the impact of a range of concentrations of HA (0.001–0.007%) isolated from six soil types on *Chlorella vulgaris* microalgae strains, the results illustrated an adverse effect of HA for all preparations at HA concentrations above 0.003%. At concentrations higher than 0.003%, the photosynthesis rate in *C. vulgaris* cells decreased, and respiration increased abruptly [61].

HS influenced the photosynthetic performance and stress response of two green algae, *Raphidocelis subcapitata*, strain 61.81 and *Monoraphidium braunii* strain 2006. These eukaryotic microalgae were exposed to four different concentrations of HS—0.17, 0.42, 1.67, and 4.17 mM dissolved organic carbon (DOC). The results highlighted that the dry weight per cell ratio decreased with increasing HS concentration. In contrast, the exposure to lower concentrations of HS stimulated better growth of the phototrophs and increased the quantum efficiency of photosystem II [62]. In the same paper, the authors reported a different response by the prokaryotic microalgae (cyanobacteria) *Synechocystis* sp. (PCC 6803) and *Microcystis aeruginosa* (PCC 7806). The tested *Synechocystis* strain was less sensitive to HS. The quantum efficiency of photosystem II was not increased in *M. aeruginosa* PCC 7806. However, in this cyanobacterial strain, the chlorophyll *a* content increased at the highest HS concentration tested compared to control.

The difference in response to HS by eukaryotic and prokaryotic microalgae species was demonstrated in an initial paper of the Steinberg group. The prokaryotic strain *Chroococcus minutus* 276-4b and the eukaryotic strain *Desmodesmus communis* 41.71 were compared. The authors considered that prokaryotic cells that lack internal membrane-delimited organelles would generally be more sensitive to HS than eukaryotic cells, with a cellular organization with several internal membrane-delimited compartments. When exposing the two species to HS at concentrations of 0.3 and 1.5 mg L^{-1} DOC, the authors reported an increased the number of cell of *D. communis* under low HS concentrations and an inhibitory effect at the highest concentrations. The cyanobacterium showed reduced photosynthetic activity and reduced population growth across the entire concentration range of HS tested [63].

HS extracted from lignite (Biomin) with 61.2% C composition had the same effect on eukaryotic (*Scenedesmus acutus* Meyen Tomaselli 8, *C. vulgaris* C-3) as compared to prokaryotic (*Anabaena variabilis* 786, *Nostoc commune*) microalgae strains. The microalgae cultures showed an average biomass increase of 18 and 15% compared to control for 1 and 10 mg L^{-1} Biomin, respectively. Negative effects of treatment with concentrations higher than 1 g L^{-1} HS were recorded, such as very significant decreases in protein, carbohydrate, and chlorophyll contents [64].

Papers presenting such dual/hormetic responses of microalgae to HS are summarized in Table 1.

Table 1. Dual response of microalgae to humic substances according to the dose.

Humic Substances	Tested Microalgae	Concentration–Effect	Reference
Humic acid extracted from lignite	*Scenedesmus acutus* Meyen Tomaselli 8 *Chlorella vulgaris* C-3 *Anabaena variabilis* 786 *Nostoc commune*	Up to 10 mg L^{-1} enhances biomass accumulation 100 mg L^{-1} enhances protein accumulation 1 g L^{-1}—inhibition	Pouneva, 2005 [64]
Humic acids extracted from lake sediments	*Desmodesmus communis* 41.71 *Chroococcus minutus* 276-4b	0.3 mg L^{-1} increases the number of *D. communis* cells Inhibitory effects on *C. minutus*	Prokhotskaya and Steinberg, 2007 [63]
Humic acid extracted from lignite, Artificial humic acid	*Raphidocelis subcapitata* 61.81 *Monoraphidium braunii* 2006 *Synechocystis* PCC 6803 *Microcystis aeruginosa* PCC 7806	0.17 mM stimulates photosynthesis 4.7 mM reduces cell development and inhibits photosynthesis	Bährs et al., 2012 [62]
Humic acids extracted from soils	*Chlorella vulgaris* co. 157	0.01–0.03% activation >0.03% inhibition	Toropkina et al., 2017 [61]
Commercial (Suwannee River) humic acids	*Scenedesmus capricornus* FACHB-271 *Chlorella* spp. FACHB-271	0.05–0.1 mg L^{-1} stimulation 1.0 mg L^{-1} inhibition	Zheng et al., 2022 [60]

In the aquatic system, HS are a component of dissolved organic matter. HS molecules also represent a C and N source for microalgae at low concentrations. Additional beneficial effects of HS result from the overlapping effects of enhanced availability of nutrients and (bio)chemical priming (similar to terrestrial plants) generated by the higher physiological level of reactive oxygen and nitrogen species. At higher concentrations, the HS effects on microalgae are dominated by their interference with photosynthetic structures and the production of the pathophysiological level of reactive oxygen [65] and nitrogen species [66]. In Figure 1, the overlapping mechanisms of HS action in microalgae leading to stimulatory or inhibitory effects are detailed.

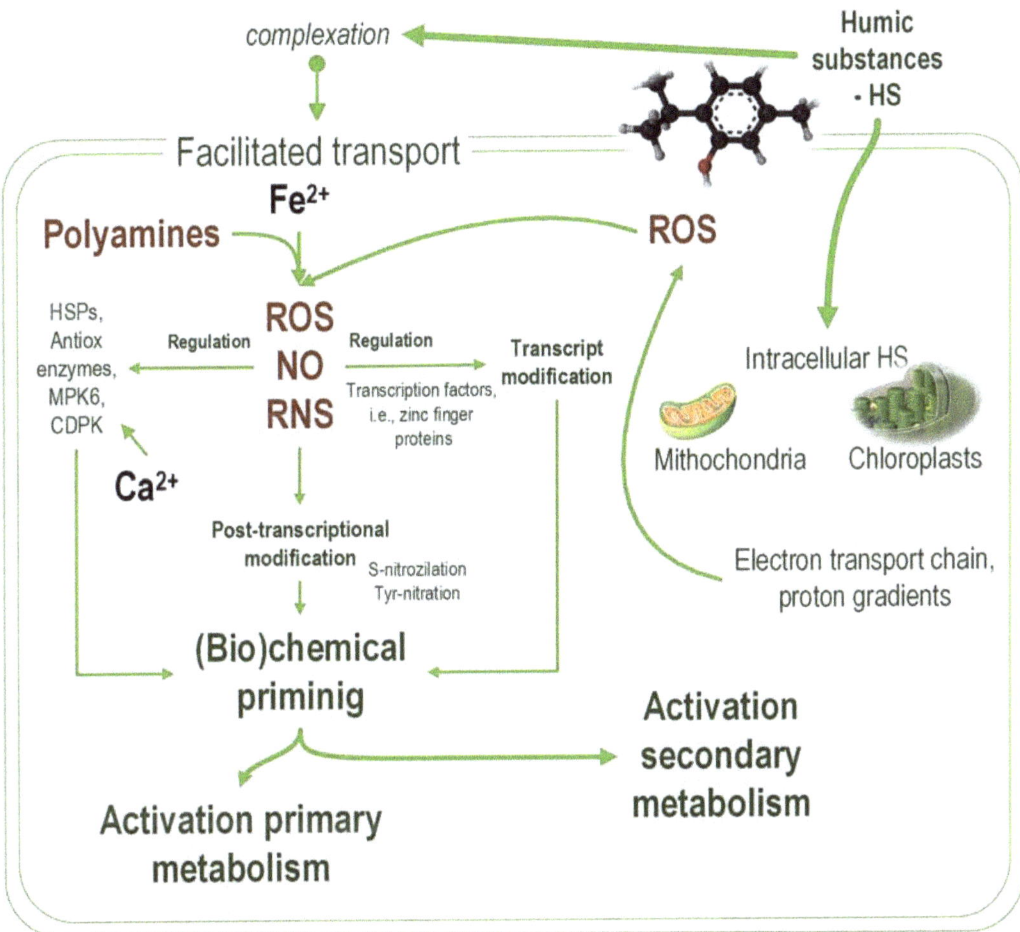

Figure 1. The mechanisms of humic substances (HS) effects on microalgae cells. HS increase membrane permeability for calcium and ferrous ions and diffuse through the plasmatic membrane. Ferrous ions promote the formation of reactive oxygen species (ROS) by redox reaction and activate the formation of nitric oxide (NO) from polyamines. Calcium ions activate specific protein kinases involved in cellular signaling. Intracellular HS interfere with the electron transport chain in chloroplasts and mitochondria, producing a higher level of reactive oxygen species. The simultaneous increase in NO and ROS levels causes an accumulation of reactive nitrogen species (RNS). The resulting nitrosative stress causes the development of physiological compensation mechanisms, which ultimately lead to the activation of primary and secondary metabolism. An increase in oxidative and nitrosative stress over the physiological thresholds damages cell function. HSPs—thermal shock proteins; MPK6—mitogenically activated protein kinases; CDPK—calcium-dependent protein kinases.

Modifying membrane permeability for various ionic species due to various HS effects, including complexation, generates elevated intracellular levels of redox-active ionic species, such as Fe^{2+}, and ionic species acting as secondary messengers, e.g., Ca^{2+} [67,68]. The ability of HS to complex ionic species, especially iron ionic species, is important for the HS effect on microalgae [69]. An increased level of redox-active iron determines higher levels of reactive oxygen species [70,71]. The hydrophobic–hydrophilic ratio of humic substances is important for the effect on microalgae and the complexation of iron ionic species. The hydrophobic components determine high growth rates of *R. subcapitata* (synonym used *Pseudokirchneriella subcapitata*). The hydrophilic components inhibit microalgae growth due to reduced iron ionic species bioavailability [72]. The permeability change in the membrane of *R. subcapitata* depends on pH. The passive diffusion of fluorescent tracers through the membranes of *R. subcapitata* (synonym used *Selenastrum capricornutum*) is higher at pH 5 than at pH 7 [67].

Another HS mechanism of action is related to chloroplasts and mitochondria as a source of ROS involved in controlling the cellular response to stress factors [73]. In terrestrial plants, it was demonstrated that chloroplasts have an essential role in plant defense against stresses [74,75]. In plants, chloroplasts also dominate the primary metabolic pathways from plants—i.e., C-, N- and S-assimilation [76]. Therefore, their activation should determine an enhanced primary metabolism and, consequently, enhanced nutrient uptake and utilization. Activation of plant defense by chloroplast is associated with activation of the secondary metabolism through retrograde signaling [77]. In microalgae, the chloroplast retrograde signaling mechanism is similar to that described in plants because initially, such mechanisms evolved in the algal ancestors of the terrestrial plants [78]. ROS production by chloroplasts through photosystem II is involved in the microalgae intracellular communication network [79].

HS interactions with microalgae cells modulate the function of photosystem II. Due to their amphiphilic structure, humic substances diffuse through the microalgae cell membrane [18]. The intracellular humic substances modify chloroplast inner and outer membrane permeability [60]. At low concentrations, HS support the plastoquinone function as a trans-membrane proton shuttle, increasing the efficiency of photosystem II [62]. At higher concentrations, HS cause plastid homeostasis imbalance due to the destructuration of the proton gradient across the inner chloroplast membrane [80]. This proton gradient is fundamental for converting light energy into chemical energy (ATP) and is essential for the efficient functioning of photosystem II [81].

The complex nature of humic substances determines various effects on microalgae. Huminfeed®,(Humintech, Grevenbroich, Germany) an extract from leonardite, a highly oxidized lignite with high C: CH_2 and C: H ratios, did not influence the growth or photosynthetic rate of the strain 61.81 *R. subcapitata* (synonym used *P. subcapitata*), strain 276-4d *Desmodesmus armatus*, or strain no. 2006 *Monoraphidium braunii* in a concentration of 2–20 mg C L^{-1}. However, the chemistry of photosystem II is modified under treatment with Huminfeed® (Humintech, Grevenbroich), as demonstrated by the modification of thermoluminescent light emission [82].

The humic lakes, developed primarily due to leaching of dissolved organic matter (DOM) from soil into continental water bodies, are natural aquatic ecosystems wherein the hormetic effects of humic substances influence the microalgae community structure [83]. DOM ranges from 0.5 to 100 mg C L^{-1} in the natural aquatic system [84]. For such concentrations, at the ecosystem level, the overlapping mechanisms of HS action on microalgae promote the development of species able to develop in multicellular structures, such as filamentous (Diatoms) or colonial (Cyanobacteria) structures [85], more resistant to oxidative stress than unicellular species [86].

3. Protective Effects of Humic Substances on Microalgae

Due to their specific properties, low concentrations of humic substances protect microalgae against the toxicity of potentially toxic elements and xenobiotics. The following main mechanisms are involved: (*i*) reduction of the bioavailability of potentially toxic elements and xenobiotics and modification of cellular uptake due to the formation of a protective coating of adsorbed HS-toxic complex on the microalgal cell wall; (*ii*) activation of photodegradation mechanism due to soluble HS–metal ion complexes; and (*iii*) higher tolerance to oxidative stress due to HS redox buffering activity and/or activation of secondary metabolism. These main mechanisms are presented in Figure 2.

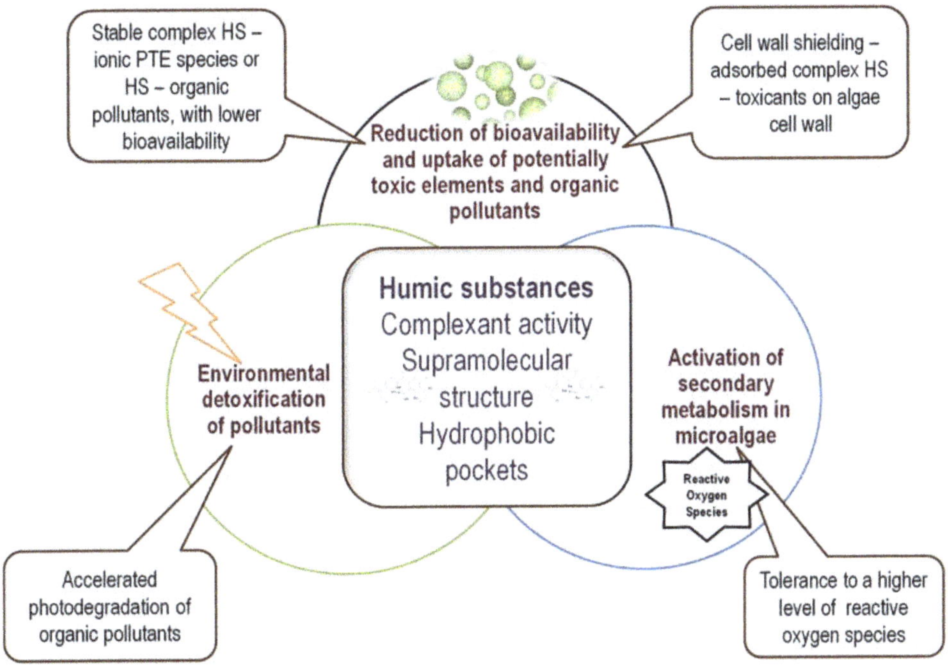

Figure 2. Illustration of the main mechanisms involved in microalgae protection by humic substances and the characteristics that are involved in these mechanisms.

Reduced cellular uptake due to the protective coating formed on the microalgae cell wall is the primary mechanism of action for protection against heavy metals and hydrophobic pollutants. The activation of photodegradation is usually induced by the HS–metal ion complex and is active on photosensitive pollutants. Increased tolerance to oxidative stress due to (bio)chemical priming mitigates the effects of compounds that induce reactive oxygen and nitrogen species at cellular levels. Illustrative examples of such mechanisms of protective HS effects on microalgae are presented in Table 2 and are further discussed.

Bioavailability is considered a key concept linking the changes in the concentrations of potentially toxic elements to their detrimental effects on biota [93]. Humic substances with low molecular weights (i.e., fulvic acids), soluble in water, increase the bioavailability of potentially toxic elements (PTE) in ionic forms and enhance the uptake of these ions by microalgae cells [94]. In the case of humic substances with supramolecular structures, i.e., humic acids, the bioavailability is decreased due to the adsorption of HA-ionic PTE on microalgae cell walls.

In the presence of a standard HA product (Suwannee River Humic Acid, SRHA), the Ag^{+} ion toxicity to microalgae *Chlamydomonas reinhardtii* and *R. subcapitata* (synonym used *P. subcapitata*) decreased, although the microalgae cells took up a higher amount of free silver ions. Most of the silver ions were bound to the cell walls and recovered in the cell debris fractions [95].

Table 2. Examples of protective effects of humic substances (HS) against the ionic forms of potentially toxic elements and xenobiotics/chemical pollutants.

Aquatic Pollutant	Tested Microalgae	Main Mechanism	Reference
Cd^{2+}, Zn^{2+}	*Raphidocelis subcapitata*	Supramolecular structure adsorbed in the cell wall surface, which complexes toxic ions	Koukal et al., 2003 [87]
Pb^{2+}	*Chlorella kesslerii*	HS–Pb^{2+} complexes are adsorbed on microalgae cell walls. HS photoalteration reduce the adsorption of HS–Pb^{2+} to microalgae surface	Spierings et al., 2011 [88]
Cu^{2+}	*Chlorella vulgaris*	HS addition reduces bioavailability of Cu^{2+} and decreases the secretion of exopolysaccharide matrix involved in Cu^{2+} toxicity	Shi et al., 2021 [89]
Microplastics	*Chlorella vulgaris*	HA decrease electrostatic interactions between polystyrene nanoplastics and microalgae and ameliorate cellular aggregation	Hanachi et al., 2022 [90]
Tetracycline	*Coelastrella* sp.	Reduction of oxidative stress damage (due to biochemical priming)	Tong et al., 2020 [91]
Graphene family materials (GFMs)	*Chlorella pyrenoidosa*	Reduction of absorption due to steric hindrance HA- GFMs Reduction of oxidative stress damage	Zhao et al., 2019 [92]

By increasing Cu ion addition to *C. vulgaris* culture, the microalgae registered evident growth inhibition and oxidative damage. The Cu-induced toxicity damage was alleviated when the culture was supplemented with HS, in an HS concentration-dependent manner [89]. Another study was performed to determine the influence of humic substances, applied at concentrations of 1 and 5 mg L^{-1}, on the toxicity of Zn and Cd ions. The concentrations of metal ions were 390 μg L^{-1} for Zn^{2+} and 200 μg L^{-1} for Cd^{2+}. The tested humic substances were Suwannee River fulvic acids and the humic acids extracted from peat and soil. HS and metal ions were tested in a microalgae photosynthesis inhibition assay using *R. subcapitata*. The effect was additionally studied by using a tangential flow ultrafiltration unit, which separated the colloidal HS from the dissolved HS. The humic acids significantly reduced the toxicity of metal ions. The results suggest that HS reduces Cd and Zn ions toxicity in two ways. The colloidal HS, i.e., the humic acids extracted from peat and soil, reduce the metal ions' bioavailability because the formed complexes are relatively stable. The same high molecular weight supramolecular structures could adsorb onto algal surfaces and shield the cells from free Cd and Zn ions [87].

The shielding coat effect of the colloidal stable HS complex with metal ions that forms on the algal cell wall and acts as an inhibitor of subsequent adsorption of metal ions was demonstrated also for *Scenedesmus quadricauda* exposed to 100 μM Cd, Ni, Pb, and Hg ions for 24 h. The microalgae were cultivated for 30 days before their biomass was exposed to metal ions. The HS were tested at three concentrations of 1, 5, and 10 mg L^{-1}. The authors pointed out that metals were mainly accumulated when no HS was added.

Another study investigated the effects of different irradiated HS against Pb^{2+} applied to freshwater microalga *Chlorella kesslerii*. The microalgae were exposed to 10^{-6} M Pb^{2+} applied as $Pb(NO_3)_2$ for 1 h in the presence of 10 mg C L^{-1} irradiated HS. Different commercial HS were studied: Suwannee River humic acid (SRHA), Suwannee River fulvic acid (SRFA), and Aldrich humic acid (AHA). The irradiation was performed with a solar simulator to mimic natural conditions. The experiment also intended to consider the irradiation factor and the effects of HS photoalteration on the bioavailability of toxic metals to microalgae.

The results indicated that photoalteration of the tested HS decreased the amount of HS adsorbed to the microalgae cell wall. The adsorption of HS to algae is dependent on the HS composition and seems to be significant for HS with high hydrophobicity [88].

Microscale algal growth inhibition (μ-AGI) was developed in a high-throughput bioassay. This method was used to test the influence of humic substances on the toxicity of heavy metals (Hg, Cu) and hydrophobic organic pollutants (HOPs, such as pesticides and polycyclic aromatic hydrocarbons). The bioassay was performed on *C. reinhardtii*, strain C-239: UTEX-90, mt+. According to the International Humic Substances Society, several humic acid (HA) preparations were made and compared with two commercial preparations. Initially, the range of concentrations of HA was from 1 to 40 mg L^{-1}. However, concentrations higher than 10 mg L^{-1} significantly inhibited microalgae growth. To mitigate the toxicity of heavy metals and HOPs, a concentration of 10 mg L^{-1} HA was used and proved to be effective. This μ-AGI method confirmed that the carboxylic acid content and molecular weight of HAs are essential for mitigating the toxic effects of Hg and Cu ions on *C. reinhardtii*. In the case of HOPs, mitigation of toxicity toward microalgae is directly related to the aromaticity and polarity of HAs [96].

The reduction in micro- and nano-sized plastic toxicity to microalgae by humic acids is also related to a mechanism involving a protective adsorbent coating that mitigates the adverse effects of micro/nano-plastic particles [90].

The humic acid–metal complexes are well-known for their role in environmental detoxification [97]. Detoxification of various aquatic pollutants reduces their toxicity on microalgae. One example is the photodegradation of glyphosate, a widely used herbicide, by the complex formed between HA–Fe^{3+}. Glyphosate is highly toxic for the algae, inducing oxidative effects on the lipophobic intracellular environment of microalgae [98].

The experiments on enhanced photodegradation of glyphosate by complex compounds formed between humic acids and Fe^{3+} (HA-Fe^{3+}) demonstrated that the optimum concentrations for the highest degradation rate were 20 mg L^{-1} HA and 0.5 mmol L^{-1} Fe^{3+} when the glyphosate concentration was 50 mg L^{-1} [99].

Humic acids reduce the toxicity of tetracycline toward microalgae. Tetracycline determined a dual response in the bioassay with *Coelastrella* sp. Tetracycline stimulated biomass and protein accumulation at concentrations lower than 2 mg L^{-1}. Tetracycline concentrations higher than 2 mg L^{-1} reduced microalgae growth by more than 50%. Sodium salts of humic acids, at concentrations of 2 mg L^{-1} and 5 mg L^{-1}, significantly reduced tetracycline toxicity. A high concentration of tetracycline induced high oxidative stress. Humic acid addition reduced oxidative damage and associated oxidative stress biomarkers, most probably due to (bio)chemical priming of secondary metabolism and the microalgae detoxification and defense systems [91].

Humic acids reduce the toxicity of graphene family materials (GFMs) toward microalgae, *Tetradesmus obliquus* (synonym used *Scenedesmus obliquus*) [100], and *Chlorella pyrenoidosa* [92]. The toxicity mechanism of GFMs is the membrane damage associated with oxidative stress [101]. Humic acids interact with GFMs and decrease GFM aggregation and adhesion to the microalgae cell membrane due to steric hindrance [92]. An additional protective effect is related to the reduced oxidative damage of GMFs in microalgae. Besides the (bio)chemical priming and activation of the tolerance against oxidative stress, another mechanism is related to the oscillating antioxidant–prooxidant characteristics of HS. The chinone–phenol moieties of HS have context-dependent redox characteristics [54]. In the "crowded" cellular environment with a high level of reactive oxygen species, HA exert antioxidant activity. The HA have a strong redox buffering activity due to the switch between oxidized quinone and reduced phenol [23,31].

The same effect of reducing the toxicity of graphene material, graphene oxide (GO), by humic acids due to the antioxidant activity was also reported for *C. vulgaris*, strain FACHB-8. GO also determines mutagenic effects (revealed by comet assay) due to the high level of induced reactive oxygen species. The concomitant HA applications reduced these mutagenic effects. At the same time, the nanoparticles formed by aggregation of the dissolved/suspended natural organic matter (NOM) acted as promotor of (geno) toxic GO effects [102].

4. Humic Substances as Microalgae Biostimulants

The scientific community has largely accepted the concept of plant biostimulants (PB) for more than a decade [103]. Plant biostimulants are a class of agrochemical inputs situated between fertilizers and plant protection products. The PB-specific effects are increased nutrient uptake and nutrient use efficiency, enhanced plant tolerance to abiotic stress, and improved edible crop quality [104]. Nowadays, plant biostimulants are classified as microbial and non-microbial plant biostimulants [105]. Non-microbial plant biostimulants are further classified into organic plant biostimulants and inorganic plant biostimulants [104]. Organic plant biostimulants are seaweed and botanical extracts [106–109], humic and fulvic acids [110,111], protein hydrolysates [112,113], chitosan [114,115], and other biopolymers [116–118]. Inorganic plant biostimulants are plant-beneficial elements, such as silicon [119,120] or selenium [121,122], which determine PB-specific biological effects when applied to cultivated plants.

The first generation of plant biostimulants was defined as a "formulated product of biological origin that improves plant productivity as a consequence of the novel or emergent properties of the complex of constituents" [53]. This definition was directly related to the difficulty of defining a mode of action specific and different from that of other agricultural inputs (fertilizers and plant protection products). Another obstacle was related to the difficulty in identifying plant biostimulant active ingredients. In the last two years, attempts have been made to define the pure organic active compounds from the main classes of plant biostimulants [123]. The HS effects on plants were considered the result of "chemical priming" [51].

The mirror concept of "microalgae biostimulant" is not often used in the scientific community, despite its utility. The main effect from plant biostimulants is the increased tolerance to abiotic stress, which is directly related to the molecular priming mode of action [124]. Enhanced tolerance to stress factors, especially in commercial-scale cultivation of microalgae and/or in mixotrophic conditions, could increase the overall productivity and profitability of microalgae cultivation [125].

The microalgae biostimulants can be divided into classes similar to those of plant biostimulants, non-microbial and microbial biostimulants [126]. Only a few studies refer to non-microbial algae biostimulants as a term used for describing the applied treatments based on the observed effects, most of them related to HS. Table 3 presents these papers and the effects of non-microbial biostimulants on different microalgae strains.

The growth-promoting effect is a direct result of the biostimulant effects. In the case of terrestrial plants, the growth-promoting term was one of the initial names given to biostimulant rhizobacteria [133], which are still in use today [134]. The non-microbial plant biostimulants were also considered "plant growth promoters"—including humic acids [135]. The biostimulant effects are not limited to growth promotion. These effects are also related to the enhanced capacity of the treated photosynthetic organisms to adapt to the specific environmental conditions.

Phytohormones are among the most used microalgae growth-promoting products. Their benefits for utilization in microalgal biotechnology to increase microalgae growth and accumulation of metabolites of interest have been reviewed several times in recent years [136–138]. Recently, strigolactone analogs, phytohormones not detected in microalgae, which appear firstly in Charales [139], were demonstrated to promote microalgae growth and metabolite accumulation [140–143]. Besides phytohormones, HS and polyphenols (ferulic acid, protocatechuic acid, vanillic acid, phloroglucinol) were reported as non-microbial algae biostimulants/growth promotors. A commercial preparation of humic and fulvic acids was one of the "biochemical stimulants" used to increase biomass productivity and metabolite content in *Chlorella sorokiniana* (UTEX 2805) [130]. Phloroglucinol promotes biomass accumulation and increases fucoxanthin synthesis in the microalga *Thalassiosira pseudonana* [144]. Phloroglucinol is also a plant tissue culture growth-promotor [145,146]. Methods to enhance the oil content of oleaginous microalgae based on HS were patented—e.g., on fulvic acid [147] or melatonin [148].

Table 3. Effects of non-microbial biostimulants on different microalgae strains.

Compound	Tested Microalgae	Main Effects	Reference
Humic-like extract of anaerobic digestate (D-HL) Humic-like extract of residues from rapeseed oil production (B-HL) Humic-like extract of tomato plants (T-HL)	*Chlorella vulgaris* CCAP 211/11C *Scenedesmus quadricauda*	Increased biomass production (~25–40%) by DH-L and TH-L Increased oil accumulation (~60–90%) by DH-L and TH-L Increased unsaturated fatty acid content by B-HL Increased carbohydrate content by B-HL	Puglisi et al., 2018 [56]
Fulvic acid	*Haematococcus pluvialis* KM115647	Increased astaxanthin and lipid content	Zhao et al., 2019 [127]
Selenium and betaine	*Dunaliella salina*	Increased carotenoid and antioxidant activity	Constantinescu-Aruxandei et al., 2019 [128]
Humic acids	*Euglena pisciformis* AEW501	Increased biomass yield Higher lipid content Higher content of unsaturated fatty acids	Fan et al., 2022 [129]
Humic and fulvic acid (commercial preparation)	*Chlorella sorokiana UTEX2805*	Increased biomass yield Increased metabolite accumulation	Hunt et al., 2010 [130]
Lignosulfonate	*Euglena gracilis* NIES-48	Increased biomass yield Higher lipid content	Zhu and Wakisaka, 2021 [131]
Phenolic precursors of lignin			Zhu et al., 2021 [132]

Most published papers refer to microbial (bacterial) microalgae growth promotors/biostimulants, recently reviewed in [149]. The microbial microalgae biostimulants, mainly bacterial strains, have been used for decades to improve mixotrophic microalgae cultivation [150,151]. Such bacterial strains were described as "microalgae growth-promoting bacteria" (MGPB) two decades ago, initially to describe the synthetic mutualistic interaction of *C. vulgaris* or *C. sorokiniana* with *Azospirillum brasilense* strain Cd [152,153]. A similar artificial consortium was established between oleaginous microalgae *Ankistrodesmus* sp. strain SP2-15 and *Rhizobium* strain 1011. The microalgal strain co-cultivated with bacteria was highly efficient in lipid production (up to 112 mg L^{-1} day^{-1} compared to 87 mg L^{-1} day^{-1} for the microalgae culture alone) and accumulation of omega-3 unsaturated fatty acids [154].

Bacteria from *Rhizobium* genera were proved to be associated naturally with the microalgae. Microalgae do not usually grow well in axenic conditions—they need a phycosphere colonized by associated/symbiotic bacteria for proper development [155].

Analysis of the diversity of phycosphere bacteria from various microalgae classes revealed that the majority of these bacteria are known to be plant growth-promoting bacteria (PGPB) [156]. The illustrative examples are bacteria from *Azospirillum* and *Rhizobium* genera [126]. Plant growth-promoting (Rhizo) bacteria were reconsidered as microbial plant biostimulants [157,158]. In similar manner, microalgae growth-promoting bacteria could be reconsidered as microbial microalgae biostimulants. Nevertheless, the significant difference between the "microalgae growth-promoting bacteria" and "microbial microalgae biostimulants" is related to the effects. The biostimulant activity is not related only to growth promotion—it involves increased tolerance to (abiotic) stress and enhanced accumulation of the ingredients of interest to cultivate the microalgae [159].

There are few reports regarding HS influence on microalgae co-cultivated with biostimulant bacteria. Fulvic acids from lignite leachate promoted the growth of *Desmodesmus subspicatus* (synonym used *Scenedesmus subspicatus*) co-cultivated with bacteria [160]. Humic-like substances from landfill leachate stimulated lipid accumulation on *C. pyrenoidosa* FACHB-9 co-cultivated in consortium with bacteria [161].

Due to their amphiphilic supramolecular structures with hydrophobic pockets [19], humic acids could influence the bioavailability of the chemical exo-signals used for communication inside the microalgae microenvironment: microalgae to microalgae, microalgae to associated bacteria, bacteria to bacteria. Such exo-signals are, in general, hydrophobic compounds, e.g., acyl-homoserine-lactone (AHL), used for inter-specific quorum sensing (QS, detecting bacterial density from the same ecotype) in Gram-negative bacteria [162]. The importance of QS signals for the microalgae–bacteria interactions was recently reviewed [163].

The gap in knowledge regarding interactions between humic substances and quorum sensing molecules is significant. The data are scarce and exist only for the soil environment and wastewater anammox anaerobic treatment. Water-soluble humic substances repress the formation of QS signals in *Sinorhizobium meliloti* and promote the efficiency of symbiotic nitrogen fixation in *Medicago sativa* [164]. Fulvic acid stimulated AHL release in the anammox (anaerobic ammonium oxidation) and improved bacterial activity [165]. Further studies are needed to determine the effects of HS on QS and the underlying mechanisms, especially in microbial consortia involving microalgae. It will help to effectively utilize HS as biostimulants for microalgae cultivated in xenic conditions when large populations of associated phycosphere bacteria producing QS signals are present. HS could modulate the bioavailability of QS signals similarly to cyclodextrin—inclusion inside a hydrophobic pocket [166].

There are no data regarding the interactions between HS and the various other infochemicals present in the microalgae culture. For example, the microalgae from the *Scenedesmaceae* family respond to grazers natural cues by forming flocculating colonies [167]. Such mechanisms are of interest for microalgal biotechnology as a lower-cost harvesting process [168]. HS could support the initial formation of the extracellular matrix of polysaccharides and lipids required for colony formation and flocculation [169], reducing the metabolic costs of colony formation [170]. The surfactant agents, such as linear alkyl benzene (LAB) [171], benzalkonium bromide (BZK) [172], sodium dodecyl sulfate (SDS) [172,173], and nonionic surfactant polyoxyethylene (40) nonylphenol ether (NPE) [172] enhance microalgae colony formation in the presence of smaller amounts of grazers natural cues or even in their absence [171]. Such effect should be considered in terms of ecotoxicological risk assessment, but also for its biotechnological potential related to microalgae harvesting. HS, due to their complex structure that generates other specific chemical features and associated biological activity, could be modulator of the final inducing step of colony formation in microalgae biotechnology.

The utilization of humic substances as microalgae biostimulants benefits from the advances in knowledge related to HS structure–activity. Despite HS complexity, the various analytical techniques, combined with chemoinformatic tools, allow an accurate estimate of HS biological activity based on chemical and chemo-physical features. The ratio between HA electron-accepting capacity (EAC) and electron-donating capacity (EDC) that is related to quinone and semi-quinone and, respectively, to polyphenols and glycosylated polyphenols, predicted the plant biostimulant activity of HA in corn seedlings [174]. The low molecular weight HS, which penetrate the cell membrane more easily, modulate intracellular signals. The high molecular weight HS interact with the cell membrane receptors [175]. The ability of HS to act as eustressors (positive stressors) on rice plants is related to aromaticity, hydrophobicity, aliphaticity, and polarity [176]. A combination of different HS, of different origins, with different and well-characterized hydrophobicity and hydrophilicity, produced bioactive and environmentally friendly products [177].

The projection of the latent structure (PLS) regression, using molecular structures obtained by ^{31}P-NMR spectra of derivatized samples and ^{13}C-CPMAS-NMR spectra obtained directly on samples, predicted the biological activity of humic-like biostimulants obtained from bioeconomy side-streams [178]. Because the HS structure–activity relationship is not linear, the linear artificial neural network (ANN) based on Fourier-transform infrared (FT-IR) spectroscopy predicted better the biological activity of HS extracted from peat [176]. Fourier transform ion cyclotron resonance mass spectrometry (FT-ICR MS) was used to differentiate the molecular signature of humic acids from different sources. The degree of humification, determined according to aromaticity and degree of saturation, increases the HA from soil, river, and leonardite/oxidized lignite [179]. FT-ICR MS, in combination with ^{13}C-CPMAS NMR spectroscopy and FT-IR ATR, revealed the co-existence of anti-oxidant and pro-oxidant moieties and properties of humic acid extracted from lignite [180]. These complex chemical properties were related to the biostimulant effects on tomatoes cultivated under nutritional stress [180].

Several constraints should be considered in using HS as biostimulant for microalgae. HS have higher electron-donating capacities (EDC) in the aquatic environment than in soil [181]. The physiological window of beneficial HS doses is narrow, and inhibitory doses must be avoided. Microalgae release into their cultivation media various organic compounds with various molecular masses, including aromatic amino acids such as tryptophan and tyrosine, heterocyclic pigments such as biopterin, proteins, etc., that generate humic-like substances, algal organic matter (AOM) [182]. This dissolved organic matter that results from microalgae organisms is also called humic matter due to its aromaticity (π-π systems) related to significant fluorescence [183,184]. In several cases, accumulations of this organic matter during microalgae cultivation block water reuse. Humic acids produced from AOM are the major microalgae growth inhibitors for *E. gracilis* strain CCAP 1224/5Z [185], *S. acuminatus* strain GT-2 [186], and *N. oceanica* LARB-202-3 [187]. Filtration of the spent media through activated charcoal and oxidative degradation by ultraviolet light or ozonization effectively reduces the toxicity of the algogenic HS (organic matter—AOM) to the microalgae [185–187]. Similar treatments could be used in the situation of enhanced production of AOM under HS treatment or in the situation of biostimulant HS accumulation.

Overall, HS fulfill the criteria required for non-microbial microalgae biostimulants. HS modulate microalgae growth and mineral nutrient availability. HS increase tolerance to chemical stressors and enhance the accumulation of microalgae metabolites. These effects are proposed to result from the "chemical priming", i.e., a "preparedness" condition that promotes faster metabolic pathway activation. Similar evidences supporting the hypothesis of "chemical priming" in terrestrial plants by HS [51] were already presented in Section 2 for HS–microalgae interaction.

5. Humic Substance Interactions with Microalgae Harvesting by Flocculation

Interest in microalgae cultivation and the resulting biomass utilization has increased significantly in the last two decades. The main driving interest was initially related to the CO_2 mitigation potential combined with biofuel production [188]. The fast growth rate and the high lipid content of oleaginous microalgae represented arguments to consider microalgae as an alternative solution to fossil fuels [189]. Integration with wastewater treatment was another benefit of the microalgae-based fuels [190]. However, several problems of microalgae cultivation have prevented the development of microalgae-based biotechnologies, especially those focused on the production of bulk chemicals/ biofuels [189,191,192]. One of these issues related to the profitability of the microalgae-based product is the dewatering/harvesting step [191–193].

Efficient harvesting and dewatering represent a major bottleneck for all microalgae-based biotechnologies, including those focused on more profitable products, such as carotenoids [194]. The dilute nature of microalgae in suspension requires high energy consumption for this initial step in microalgae processing [195]. Microalgae flocculation from growing media was proposed to be a solution to reduce energy consumption for harvesting and dewatering [196]. However, the costs of flocculants reduce the profitability of this solution [195]. Affordable and natural flocculants present several advantages in terms of costs and environmental impacts [197–199].

The utilization of HA as a flocculant for microalgae exploit their capacity to bind to the microalgae cell walls and/or form a network with other microalgae flocculating agents. HA were used as flocculant together with cationic aminoclay nanoparticles. A patented composition of aminoclay (AC) and HA with 0.1–0.3 g L^{-1} HA and 3–7 g L^{-1} AC content was used to harvest oleaginous microalgae—*Ankistrodesmus* sp., *Anacystis nidulans*, *Biddulphia aurita* [200]. In cooperation with scientists from other Korean research entities, the patent authors published a short communication demonstrating that AC-HA forms a network that captures oleaginous *Chlorella* sp. microalgae [201].

Flocculation of microalgae biomass was enhanced by the utilization of the humic-like exopolymers produced by *S. acuminatus*. When the humic-like exopolymers were used, the Al^{3+} ionic coagulant concentration was almost 20 times reduced (from 77.6 to 4.5 mg L^{-1}) [202]. In other situations, the HS presence reduces the harvesting microalgae efficiency by flocculation process. HS reduces the efficiency of *T. obliquus* (synonym used *S. obliquus)* FSP-3 flocculation by ozone [203] or *C. vulgaris* 211–11b flocculation by calcium phosphate precipitation [204]. Therefore, the biotechnological process aiming to utilize HS as microalgae biostimulants should also consider the HS interference with microalgae flocculation.

6. Enhanced Biotechnological Production of Microalgae-Based High-Value Products by Humic Substances

Nowadays, the interest in microalgae cultivation is also related to the products with high added value that can be produced by microalgal biotechnology—Table 4.

Table 4. Market potential of microalgae-based high-value compounds. Reconstructed and updated, from Velea et al., 2017 [125].

High Value-Added Compounds	Market Estimation		Price Range (USD kg^{-1})
	Estimated Value (mio.US$)	Compound Annual Growth Rate—CAGR	
Plant biostimulants	3200 (2021) [a]	12.1% (2021–2026)	60–90 [a]
Carotenoids (total)	1500 (2019) [b]	4.2% (2019–2027) [b]	-
Beta-carotene	532 (2019) [a]	3.3% (2014–2019) [b]	300–1500 [b]
Lutein	314 (2019) [a]	3.6% (2014–2019) [b]	-
Astaxanthin	423 (2019) [a]	2.3% (2014–2019) [b]	200–7000 [b]
Canthaxanthin	117 (2019) [a]	3.7% (2014–2019) [b]	100–500 [d]
Omega-3 fatty acids	2100 (2020) [c]	7.4% (2020–2028) [c]	80–160 [d]

[a]—https://www.marketsandmarkets.com/Market-Reports/biostimulant-market-1081.html, accessed on 20 April 2022. [b]—https://www.marketsandmarkets.com/Market-Reports/carotenoid-market-158421566.html, accessed on 20 April 2022. [c]—https://www.grandviewresearch.com/industry-analysis/omega-3-market, accessed on 20 April 2022. [d]—Borowitzka, 2013 [205].

The potential use of humic substances as microalgae biostimulants must take into account the mixotrophic cultivation of microalgae (since parts of the HS could be used as C or N sources by microalgae and/or symbiotic associative bacteria) and to the biosynthesis of the products that are not affected by the presence of HS (since the separation of HS in the downstream process is complicated and increases production costs). The mixotrophic process for microalgae cultivation, which leads to higher biomass concentrations compared to auxotrophic grown microalgae, and the acceleration of microalgae growth (including by the use of microalgae biostimulants) are solutions that reduce the energy costs for microalgae cultivation and increase the profitability of microalgae cultivation on a large scale [124].

The products with high added value that can be obtained from microalgae have a plethora of applications within the biomedical, food, feed, and agriculture sectors. The investigation of the effects of HS on the production of these products is still in its infancy but could have a significant impact. The few studies available focused on certain compounds such as lipids, fatty acids, and carbohydrates. Still, other products such as carotenoids, polyphenols, polyamines, and proteins should be investigated more in-depth. The humic acids, applied in concentrations between 0.1 and 0.5 mg L^{-1}, were proven to stimulate the growth of microalgae *Dunaliella salina* and *Nannochloropsis salina* used in aquaculture for fish feed. The accumulation of chlorophyll *a*, carotenoids, lipids, and proteins was also stimulated by humic acids [205]. Other preliminary studies involving fulvic acid and astaxanthin [206] are promising.

Moreover, because HS were shown to be safe for many applications, there would be no need for further purification or separation. In some cases, HS were shown to have positive effects and, therefore, could even act synergistically with products from microalgae. HS were found to be antimicrobial and have prebiotic, anti-inflammatory, antioxidant, and immunomodulating activities in vertebrates [18]. HS were proven to have health-promoting effects on fish—reducing stress and fungal disease and stimulating probiotic bacteria [207]. Therefore, their use to stimulate the production of high-added-value feed ingredients (e.g., astaxanthin) is highly feasible. Trace HS will not negatively influence the value of the feeds. Humic acids and yeast-derived immunomodulator glucan were shown to have synergistic stimulation effects on the immune system [208]. These findings suggest that this synergism could apply also to similar bioactive compounds from microalgae, these vast possibilities being unexplored at the moment.

The production of plant biostimulants from microalgae biomass is one of the most recent developments in microalgal biotechnology [209]. Seaweed extracts represent a well-known and highly effective class of plant biostimulants [106]. However, the microalgae biomass is more affordable and less complicated to standardize as a raw material for plant biostimulants [210]. The sustainability of microalgae-based plant biostimulant production is higher than that of seaweed-based plant biostimulants [211].

A pioneering work of our group demonstrated the plant biostimulant effect of microalgae extracts resulting from the biomass of microalgae *Nannochloris* sp. 424-1, CCAP 251/10, cultivated mixotrophically [212].

Polyamines represent another category of endo- and exo-signals that have protectant action against abiotic stress. The *Arthrospira* (synonym used *Spirulina*) *platensis* biomass submitted to enzymatic hydrolysis for four hours generates hydrolysate with high spermine content that biostimulates field-grown lettuce [213].

The (exo)polysaccharides are other compounds from microalgae that contribute to the specific plant biostimulant action due to the activation of plant innate immunity. Their activity is related to the in situ formation of oligosaccharins [214]. Oligosaccharins are endo-signals that regulate plant growth, development, and gene expression, from the category of plant innate immunity elicitors, i.e., pathogen-associated molecular patterns (PAMPs); microbe-associated molecular patterns (MAMPs); and damage-associated molecular patterns—DAMPs [215].

The polysaccharides extracted from eukaryotic and prokaryotic microalgae (*D. salina* strain MS002, *Porphyridium* sp. strain MS081, *D. salina* strain MS067, *Phaeodactylum tricornutum* strain MS023, *Desmodesmus* sp., *A. platensis* strain MS001) and applied as leaf treatment induced the biochemical markers related to the activation of the defense pathways in tomatoes—chitinase, 1,3 beta-glucanase, phenylalanine ammonia-lyase (PAL), and peroxidase—POX [216].

The crude extract containing polysaccharides from the Moroccan strain of *C. vulgaris* and *C. sorokiniana* injected in 40-day old tomato plants determined an increase in β-1,3-glucanase activity and significantly increased the content of polyunsaturated fatty acids. The exo-polysaccharides extracted from the halophytic microalgae *D. salina* MS002, cultivated in hypersaline media, enhance tolerance of *Solanum lycopersicum* var. Jana F1 to salt stress [217].

Protein hydrolysate is another source of plant biostimulants from microalgae. The bioactive peptides resulting from enzymatically hydrolyzed microalgae proteins sustain plant hormone biosynthesis in the plant tissue, activate primary metabolism, and stimulate nutrient uptake [108]. The amino acids from microalgae protein hydrolysate, especially the glutamic and aspartic acids, modulate primary metabolism and increase nutrient use efficiency due to anaplerotic reactions of the tricarboxylic acid cycle [218]. The de-oiled microalgae biomass, resulting from the production of third-generation biodiesel, is an effective hydrolysate source that promotes plant growth and development [219]. Table 5 summarizes the active ingredients in microalgae extracts that contribute to the plant biostimulant effect.

Table 5. Active ingredients from microalgae-based microbial biostimulants.

Active Ingredients	Microalgae	Main Mechanism	Reference
Polysaccharides	*Dunaliella salina* MS002, *Porphyridium* sp. MS081, *D. salina strain* MS067, *Phaeodactylum tricornutum* MS023, *Desmodesmus* sp., *Arthrospira. platensis* MS001	Elicitation of the plant defense mechanisms and activation of secondary metabolism	Rachidi et al., 2021 [116]
Osmoprotectants—glycine-betaine and proline	*Nannochloris* sp. 424-1, CCAP 251/10	Protection of plants against hydric stress, enhanced water use efficiency	Oancea et al., 2013 [214]
Osmoprotectants—polyamines	*A. platensis*	Increased biomass yield Higher lipid content Higher content of unsaturated fatty acids	Mógor et al., 2018 [215]
Protein hydrolysate	*Chlorella vulgaris*	Activation of primary metabolism, Increased nutrient uptake and nutrient use efficiency	Maurya et al., 2016 [220]

HS components could act on terrestrial plants complementarily with the active ingredients from microalgae extract. The hormetic HS effect on microalgae could support the development of the HS-based second generation of plant biostimulants (Figure 3).

The high-throughput screening bioassay proposed in Figure 3 is multifunctional. It could be used to select a synergic combination of HS with other PB-active ingredients. Additionally, it could be used to generate PB based on HS-biostimulated microalgae, selecting the optimum HS concentrations active in microalgae and further synergizing microalgae extracts' active ingredients. A high-throughput screening assay based on microalgae application of HS and other active ingredients could select an optimal ratio between components. The hormetic effect allows fine-tuning of the optimal combination between HS and other active ingredients of plant biostimulants, including extract of microalgae. The selected combination can be further tested in a battery of plant bioassays to identify the synergistic plant biostimulant compositions. The compatibility between HS and microalgae was already proven. A combination of commercial PB based on HS (humic acids extracted from leonardite, 33% C) and lyophilized microalgae biomass *D. subspicatus* (synonym used *S. subspicatus*) in aqueous suspension has been shown to exert synergic plant biostimulant effects on onions grown in an organic farming system [174]. Therefore, microalgae stimulation by in situ application of humic substances, at a concentration from the initial range of hormetic effect, could generate a second-generation plant biostimulant. These plant biostimulants from the second generation are products based on synergic combinations of active ingredients [224].

Using humic substances as biostimulants for microalgae cultivation could be achieved in an integrated/zero-waste microalgae-based biotechnological process (Figure 4).

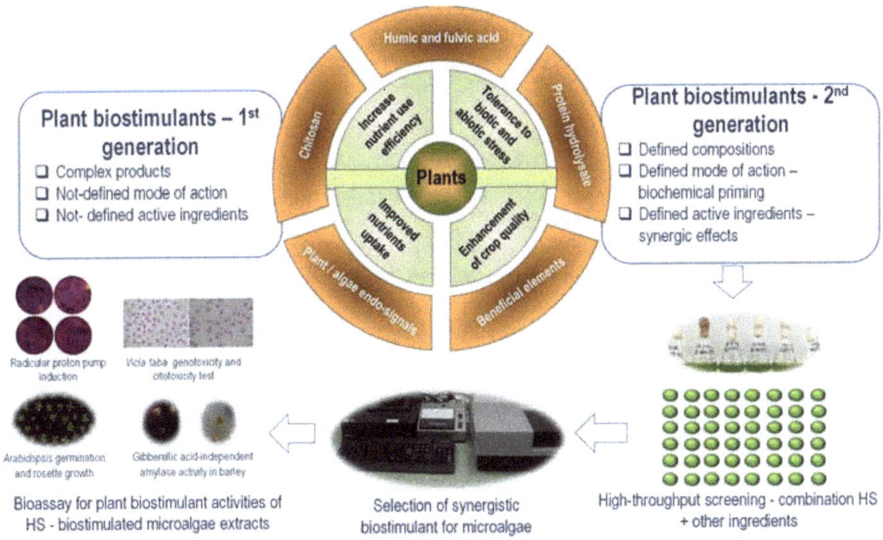

Figure 3. Development of second-generation plant biostimulants based on microalgae bioassay of interactions between humic substances and other active ingredients. The high-throughput screening on microalgae selects a combination with synergic effects that are further verified in plant biostimulant bioassay. Extracts of HS-biostimulated microalgae could be also used in association with the added HS for their activity as plant biostimulants, by using several bioassays—radicular proton pump induction [220], *Vicia faba* genotoxicity and cytotoxicity test [221], *Arabidopsis* germination and rosette growth [222], Gibberellic acid-independent amylase activity in barley [223].

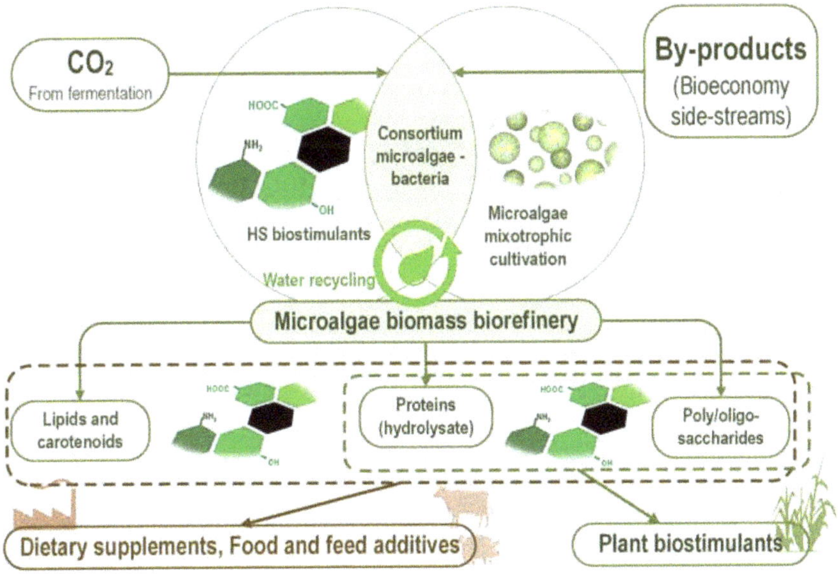

Figure 4. Using the humic substances as biostimulants for microalgae in integrated biotechnology, converting by-products from bioeconomy and carbon dioxide from fermentation process into high-value bioproducts—dietary supplements, food and feed additives, plant biostimulants. The complementary or even synergic HS interactions with components from microalgae used as active ingredients in these bioproducts underpin HS utilization as a biostimulant for microalgae.

HS promote microalgal metabolite accumulation under different stress conditions. Fulvic acids induce the accumulation of lipids in nitrogen starvation conditions in *Monoraphidium* sp. FXY-10 [225]. Melatonin and fulvic acid enhance lipid and photosynthetic pigments accumulation in *Heveochlorella* sp. Yu MK829186.1 due to modulation of nitrogen and reactive oxygen species [226]. Fulvic acids promote astaxanthin accumulation in *H. pluvialis* KM115647 under high light and nutrient starvation stress [206].

HS complement/synergize microalgae components used as active ingredients for dietary supplements or plant biostimulants in several ways. Humic acid increased water solubility and photostability of β-carotene [227]. Humic acids interact in a synergic manner with β-glucan for activation of the immune system [208], protection against liver injury induced by chemical agents [228], and binding aflatoxin B_1 in vitro [229].

7. Conclusions

Humic substances (HS) are supramolecular structures stabilized by hydrophobic interactions. In aqueous solutions/suspensions, HS generates a more dynamic supramolecular structure. Due to the dynamic structure of HS in solution/suspension, the main HS physicochemical characteristics are enhanced. The biological activities related to these characteristics are also enhanced. Amplified HS biological activities in water systems and the unicellular nature of microalgae make more evident the hormetic effects of HS on aquatic photosynthetic microorganisms.

Overall, HS fulfill the criteria required for a non-microbial microalgae biostimulant. HS increase mineral nutrient availability, microalgae growing rate, and biomass accumulation. HS increase tolerance to chemical stressors and enhance the accumulation of ingredients of interest for microalgae cultivation under abiotic stress conditions.

The HS biostimulant effect on microalgae could be exploited to improve the yield of high added-value products obtained by microalgae cultivation, such as food and feed additives, dietary supplements, and plant biostimulants.

Author Contributions: Conceptualization, D.G.P. and F.O.; methodology, D.C.-A.; data curation, C.L. and F.O.; writing—original draft preparation, D.G.P.; writing—review and editing, D.C.-A. and F.O.; visualization, F.O.; project administration, C.L.; funding acquisition, F.O. All authors have read and agreed to the published version of the manuscript.

Funding: The research leading to these results has received funding from European Regional Development Fund (ERDF), the Competitiveness Operational Programme (POC), Axis 1, project POC-A1-A1.2.3-G-2015-P_40_352, My_SMIS 105684, "Sequential processes of closing the side streams from bioeconomy and innovative (bio)products resulting from it—SECVENT", subsidiary project 1500/2020—BioLignol and subsidiary project 1882/2020—Aqua-STIM.

Institutional Review Board Statement: Not applicable.

Informed Consent Statement: Not applicable.

Data Availability Statement: Not applicable.

Conflicts of Interest: The authors declare no conflict of interest.

References

1. Lehmann, J.; Kleber, M. The contentious nature of soil organic matter. *Nature* **2015**, *528*, 60–68. [CrossRef] [PubMed]
2. Regelink, I.C.; Stoof, C.R.; Rousseva, S.; Weng, L.; Lair, G.J.; Kram, P.; Nikolaidis, N.P.; Kercheva, M.; Banwart, S.; Comans, R.N.J. Linkages between aggregate formation, porosity and soil chemical properties. *Geoderma* **2015**, *247–248*, 24–37. [CrossRef]
3. Yang, F.; Tang, C.; Antonietti, M. Natural and artificial humic substances to manage minerals, ions, water, and soil microorganisms. *Chem. Soc. Rev.* **2021**, *50*, 6221–6239. [CrossRef] [PubMed]
4. Hayes, M.H. Solvent systems for the isolation of organic components from soils. *Soil Sci. Soc. Am. J.* **2006**, *70*, 986–994. [CrossRef]
5. Olk, D.C.; Bloom, P.R.; Perdue, E.M.; McKnight, D.M.; Chen, Y.; Farenhorst, A.; Senesi, N.; Chin, Y.P.; Schmitt-Kopplin, P.; Hertkorn, N.; et al. Environmental and agricultural relevance of humic fractions extracted by alkali from soils and natural waters. *J. Environ. Qual.* **2019**, *48*, 217–232. [CrossRef]
6. Piccolo, A. The supramolecular structure of humic substances: A novel understanding of humus chemistry and implications in soil science. *Adv. Agron.* **2002**, *75*, 57–134. [CrossRef]

7. Guo, X.X.; Liu, H.T.; Wu, S.B. Humic substances developed during organic waste composting: Formation mechanisms, structural properties, and agronomic functions. *Sci. Total Environ.* **2019**, *662*, 501–510. [CrossRef]
8. Nebbioso, A.; Piccolo, A. Molecular characterization of dissolved organic matter (DOM): A critical review. *Anal. Bioanal. Chem.* **2013**, *405*, 109–124. [CrossRef]
9. Adani, F.; Spagnol, M.; Nierop, K.G. Biochemical origin and refractory properties of humic acid extracted from maize plants: The contribution of lignin. *Biogeochemistry* **2007**, *82*, 55–65. [CrossRef]
10. Shevchenko, S.M.; Bailey, G.W. Life after death: Lignin-humic relationships reexamined. *Crit. Rev. Environ. Sci. Technol.* **1996**, *26*, 95–153. [CrossRef]
11. Simpson, A.J.; Kingery, W.L.; Hayes, M.H.; Spraul, M.; Humpfer, E.; Dvortsak, P.; Kerssebaum, R.; Godejohann, M.; Hofmann, M. Molecular structures and associations of humic substances in the terrestrial environment. *Naturwissenschaften* **2002**, *89*, 84–88. [CrossRef] [PubMed]
12. Jahnel, J.B.; Frimmel, F.H. Enzymatic release of amino acids from different humic substances. *Acta Hydrochim. Hydrobiol.* **1995**, *23*, 31–35. [CrossRef]
13. Ding, G.; Mao, J.; Xing, B. Characteristics of amino acids in soil humic substances. *Commun. Soil Sci. Plant Anal.* **2006**, *32*, 1991–2005. [CrossRef]
14. Zou, J.; Huang, J.; Zhang, H.; Yue, D. Evolution of humic substances in polymerization of polyphenol and amino acid based on non-destructive characterization. *Front. Environ. Sci. Eng.* **2020**, *15*, 5. [CrossRef]
15. Young, C.C.; Chen, L.F. Polyamines in humic acid and their effect on radical growth of lettuce seedlings. *Plant Soil* **1997**, *195*, 143–149. [CrossRef]
16. Salati, S.; Papa, G.; Adani, F. Perspective on the use of humic acids from biomass as natural surfactants for industrial applications. *Biotechnol. Adv.* **2011**, *29*, 913–922. [CrossRef]
17. Li, A.; Hu, J.; Li, W.; Zhang, W.; Wang, X. Polarity based fractionation of fulvic acids. *Chemosphere* **2009**, *77*, 1419–1426. [CrossRef]
18. de Melo, B.A.; Motta, F.L.; Santana, M.H. Humic acids: Structural properties and multiple functionalities for novel technological developments. *Mater. Sci. Eng. C Mater. Biol. Appl.* **2016**, *62*, 967–974. [CrossRef]
19. Chaaban, A.A.; Lartiges, B.; El Hayek, E.; Kazpard, V.; Plisson-Chastang, C.; Vicendo, P.; Caillet, C. Complexes of humic acid with cationic surfactants support the supramolecular view of extracted humic matter. *Environ. Chem.* **2021**, *18*, 156–167. [CrossRef]
20. Gerke, J. Concepts and misconceptions of humic substances as the stable part of soil organic matter: A review. *Agronomy* **2018**, *8*, 76. [CrossRef]
21. Piccolo, A.; Cozzolino, A.; Conte, P.; Spaccini, R. Polymerization of humic substances by an enzyme-catalyzed oxidative coupling. *Naturwissenschaften* **2000**, *87*, 391–394. [CrossRef] [PubMed]
22. Spielvogel, S.; Prietzel, J.; Kgel-Knabner, I. Soil organic matter stabilization in acidic forest soils is preferential and soil type-specific. *Eur. J. Soil Sci.* **2008**, *59*, 674–692. [CrossRef]
23. Yang, F.; Antonietti, M. The sleeping giant: A polymer View on humic matter in synthesis and applications. *Prog. Polym. Sci.* **2020**, *100*, 101182. [CrossRef]
24. He, X.S.; Yang, C.; You, S.H.; Zhang, H.; Xi, B.D.; Yu, M.D.; Liu, S.J. Redox properties of compost-derived organic matter and their association with polarity and molecular weight. *Sci. Total Environ.* **2019**, *665*, 920–928. [CrossRef] [PubMed]
25. Haider, K.; Martin, J.P. Synthesis and transformation of phenolic compounds by *Epicoccum nigrum* in relation to humic acid formation. *Soil Sci. Soc. Am. J.* **1967**, *31*, 766–772. [CrossRef]
26. Boguta, P.; Skic, K.; Sokolowska, Z.; Frac, M.; Sas-Paszt, L. Chemical transformation of humic acid molecules under the influence of mineral, fungal and bacterial fertilization in the context of the agricultural use of degraded soils. *Molecules* **2021**, *26*, 4921. [CrossRef] [PubMed]
27. Wang, S.; Dou, S.; Zhang, X.; Cui, Y.J.; Wang, T. FTIR spectroscopic analysis of humic-like substances extracted from the microbial residues. *Guang Pu Xue Yu Guang Pu Fen Xi* **2015**, *35*, 3397–3401. [CrossRef]
28. Sokol, N.W.; Sanderman, J.; Bradford, M.A. Pathways of mineral-associated soil organic matter formation: Integrating the role of plant carbon source, chemistry, and point of entry. *Glob. Chang. Biol.* **2019**, *25*, 12–24. [CrossRef]
29. Pertusatti, J.; Prado, A.G. Buffer capacity of humic acid: Thermodynamic approach. *J. Colloid Interface Sci.* **2007**, *314*, 484–489. [CrossRef]
30. Rakshit, S.; Sarkar, D. Assessing redox properties of standard humic substances. *Int. J. Environ. Sci. Technol.* **2017**, *14*, 1497–1504. [CrossRef]
31. Capasso, S.; Chianese, S.; Musmarra, D.; Iovino, P. Macromolecular structure of a commercial humic acid sample. *Environments* **2020**, *7*, 32. [CrossRef]
32. Gheorghe, D.-I.; Constantinescu-Aruxandei, D.; Lupu, C.; Oancea, F. Emulsifying effect of fulvic acids from shilajit. *Chem. Proc.* **2022**, *7*, 23.
33. Kucerik, J.; Smejkalova, D.; Cechlovska, H.; Pekar, M. New insights into aggregation and conformational behaviour of humic substances: Application of high resolution ultrasonic spectroscopy. *Org. Geochem.* **2007**, *38*, 2098–2110. [CrossRef]
34. Piccolo, A. The nature of soil organic matter and innovative soil managements to fight global changes and maintain agricultural productivity. In *Carbon Sequestration in Agricultural Soils*; Springer: New York, NY, USA, 2012; pp. 1–19.
35. Verrillo, M.; Salzano, M.; Savy, D.; Di Meo, V.; Valentini, M.; Cozzolino, V.; Piccolo, A. Antibacterial and antioxidant properties of humic substances from composted agricultural biomasses. *Chem. Biol. Technol. Agric.* **2022**, *9*, 28. [CrossRef]

36. Rose, M.T.; Patti, A.F.; Little, K.R.; Brown, A.L.; Jackson, W.R.; Cavagnaro, T.R. A meta-analysis and review of plant-growth response to humic substances: Practical implications for agriculture. *Adv. Agron.* **2014**, *124*, 37–89.
37. Canellas, L.P.; Olivares, F.L.; Okorokova-Facanha, A.L.; Facanha, A.R. Humic acids isolated from earthworm compost enhance root elongation, lateral root emergence, and plasma membrane H+-ATPase activity in maize roots. *Plant Physiol.* **2002**, *130*, 1951–1957. [CrossRef]
38. Nardi, S.; Pizzeghello, D.; Muscolo, A.; Vianello, A. Physiological effects of humic substances on higher plants. *Soil Biol. Biochem.* **2002**, *34*, 1527–1536. [CrossRef]
39. Quaggiotti, S.; Ruperti, B.; Pizzeghello, D.; Francioso, O.; Tugnoli, V.; Nardi, S. Effect of low molecular size humic substances on nitrate uptake and expression of genes involved in nitrate transport in maize (*Zea mays* L.). *J. Exp. Bot.* **2004**, *55*, 803–813. [CrossRef] [PubMed]
40. Tavares, O.; Santos, L.; Ferreira, L.; Sperandio, M.; da Rocha, J.; García, A.; Dobbss, L.; Berbara, R.; de Souza, S.; Fernandes, M. Humic acid differentially improves nitrate kinetics under low-and high-affinity systems and alters the expression of plasma membrane H+-ATPases and nitrate transporters in rice. *Ann. Appl. Biol.* **2017**, *170*, 89–103. [CrossRef]
41. Jindo, K.; Soares, T.S.; Peres, L.E.P.; Azevedo, I.G.; Aguiar, N.O.; Mazzei, P.; Spaccini, R.; Piccolo, A.; Olivares, F.L.; Canellas, L.P. Phosphorus speciation and high-affinity transporters are influenced by humic substances. *J. Plant Nutr. Soil Sci.* **2016**, *179*, 206–214. [CrossRef]
42. Jannin, L.; Arkoun, M.; Ourry, A.; Laîné, P.; Goux, D.; Garnica, M.; Fuentes, M.; Francisco, S.S.; Baigorri, R.; Cruz, F. Microarray analysis of humic acid effects on *Brassica napus* growth: Involvement of N, C and S metabolisms. *Plant Soil* **2012**, *359*, 297–319. [CrossRef]
43. Zanin, L.; Tomasi, N.; Cesco, S.; Varanini, Z.; Pinton, R. Humic substances contribute to plant iron nutrition acting as chelators and biostimulants. *Front. Plant Sci.* **2019**, *10*, 675. [CrossRef] [PubMed]
44. Zanin, L.; Tomasi, N.; Zamboni, A.; Sega, D.; Varanini, Z.; Pinton, R. Water-extractable humic substances speed up transcriptional response of maize roots to nitrate. *Environ. Exp. Bot.* **2018**, *147*, 167–178. [CrossRef]
45. Trevisan, S.; Botton, A.; Vaccaro, S.; Vezzaro, A.; Quaggiotti, S.; Nardi, S. Humic substances affect Arabidopsis physiology by altering the expression of genes involved in primary metabolism, growth and development. *Environ. Exp. Bot.* **2011**, *74*, 45–55. [CrossRef]
46. Aguiar, N.O.; Olivares, F.L.; Novotny, E.H.; Canellas, L.P. Changes in metabolic profiling of sugarcane leaves induced by endophytic diazotrophic bacteria and humic acids. *Peerj* **2018**, *6*, e5445. [CrossRef] [PubMed]
47. Othibeng, K.; Nephali, L.; Ramabulana, A.T.; Steenkamp, P.; Petras, D.; Kang, K.B.; Opperman, H.; Huyser, J.; Tugizimana, F. A metabolic choreography of maize plants treated with a humic substance-based biostimulant under normal and starved conditions. *Metabolites* **2021**, *11*, 403. [CrossRef] [PubMed]
48. Zandonadi, D.B.; Santos, M.P.; Dobbss, L.B.; Olivares, F.L.; Canellas, L.P.; Binzel, M.L.; Okorokova-Facanha, A.L.; Facanha, A.R. Nitric oxide mediates humic acids-induced root development and plasma membrane H+-ATPase activation. *Planta* **2010**, *231*, 1025–1036. [CrossRef]
49. Ramos, A.C.; Dobbss, L.B.; Santos, L.A.; Fernandes, M.S.; Olivares, F.L.; Aguiar, N.O.; Canellas, L.P. Humic matter elicits proton and calcium fluxes and signaling dependent on Ca2+-dependent protein kinase (CDPK) at early stages of lateral plant root development. *Chem. Biol. Technol. Agric.* **2015**, *2*, 3. [CrossRef]
50. Canellas, L.P.; Canellas, N.O.A.; da Irineu, L.E.S.; Olivares, F.L.; Piccolo, A. Plant chemical priming by humic acids. *Chem. Biol. Technol. Agric.* **2020**, *7*, 12. [CrossRef]
51. Jindo, K.; Olivares, F.L.; Malcher, D.; Sanchez-Monedero, M.A.; Kempenaar, C.; Canellas, L.P. From lab to field: Role of humic substances under open-field and greenhouse conditions as biostimulant and biocontrol agent. *Front. Plant Sci.* **2020**, *11*, 426. [CrossRef]
52. Yakhin, O.I.; Lubyanov, A.A.; Yakhin, I.A.; Brown, P.H. Biostimulants in plant science: A global perspective. *Front. Plant Sci.* **2016**, *7*, 2049. [CrossRef] [PubMed]
53. Kim, E.; Liu, Y.; Leverage, W.T.; Yin, J.J.; White, I.M.; Bentley, W.E.; Payne, G.F. Context-dependent redox properties of natural phenolic materials. *Biomacromolecules* **2014**, *15*, 1653–1662. [CrossRef] [PubMed]
54. Stoica, R.; Oancea, F.; Minca, I.; Doncea, S.M.; Ganea, R.; Capra, L.; Senin, R.; Mateescu, M.; Ciobanu, I.; Ivan, G. Spectroscopic, textural and thermal characterization methods of biostimulants based on sodium humate. *Rev. Chim.* **2018**, *69*, 3477–3482. [CrossRef]
55. Puglisi, I.; Barone, V.; Sidella, S.; Coppa, M.; Broccanello, C.; Gennari, M.; Baglieri, A. Biostimulant activity of humic-like substances from agro-industrial waste on Chlorella vulgaris and Scenedesmus quadricauda. *Eur. J. Phycol.* **2018**, *53*, 433–442. [CrossRef]
56. Faraon, V.A.; Popa, D.G.; Tudor-Popa, I.; Mihăilă, E.G.; Oancea, F. Effect of humic acids from biomass biostimulant on microalgae growth. *Chem. Proc.* **2022**, *7*, 57.
57. Saito, S. Separation approaches towards understanding supramolecular aggregate formation of humic acid. *Anal. Sci.* **2022**, *38*, 233–234. [CrossRef]
58. Prakash, A.a.; Rashid, M. Influence of humic substances on the growth of marine phytoplankton: Dinoflagellates. *Limnol. Oceanogr.* **1968**, *13*, 598–606. [CrossRef]

59. Zheng, X.; Xu, Z.; Zhao, D.; Luo, Y.; Lai, C.; Huang, B.; Pan, X. Double-dose responses of *Scenedesmus capricornus* microalgae exposed to humic acid. *Sci. Total Environ.* **2022**, *806*, 150547. [CrossRef]
60. Toropkina, M.; Ryumin, A.; Kechaikina, I.; Chukov, S. Effect of humic acids on the metabolism of *Chlorella vulgaris* in a model experiment. *Eurasian Soil Sci.* **2017**, *50*, 1294–1300. [CrossRef]
61. Bährs, H.; Steinberg, C.E. Impact of two different humic substances on selected coccal green algae and cyanobacteria—Changes in growth and photosynthetic performance. *Environ. Sci. Pollut. Res.* **2012**, *19*, 335–346. [CrossRef]
62. Prokhotskaya, V.Y.; Steinberg, C.E. Differential sensitivity of a coccal green algal and a Cyanobacterial species to dissolved natural organic matter (NOM). *Environ. Sci. Pollut. Res.* **2007**, *14*, 11–18. [CrossRef] [PubMed]
63. Pouneva, I. Effect of humic substances on the growth of microalgal cultures. *Russ. J. Plant Physiol.* **2005**, *52*, 410–413. [CrossRef]
64. Steinberg, C.E.; Kamara, S.; Prokhotskaya, V.Y.; Manusadžianas, L.; Karasyova, T.A.; Timofeyev, M.A.; Jie, Z.; Paul, A.; Meinelt, T.; Farjalla, V.F. Dissolved humic substances–ecological driving forces from the individual to the ecosystem level? *Freshw. Biol.* **2006**, *51*, 1189–1210. [CrossRef]
65. González, P.M.; Malanga, G.; Puntarulo, S. Cellular oxidant/antioxidant network: Update on the environmental effects over marine organisms. *Open Mar. Biol. J.* **2015**, *9*, 1–12. [CrossRef]
66. Vigneault, B.; Percot, A.; Lafleur, M.; Campbell, P.G. Permeability changes in model and phytoplankton membranes in the presence of aquatic humic substances. *Environ. Sci. Technol.* **2000**, *34*, 3907–3913. [CrossRef]
67. Hasegawa, H.; Tate, Y.; Ogino, M.; Maki, T.; Begum, Z.A.; Ichijo, T.; Rahman, I.M. Laboratory culture experiments to study the effect of lignite humic acid fractions on iron solubility and iron uptake rates in phytoplankton. *J. Appl. Phycol.* **2017**, *29*, 903–915. [CrossRef]
68. Orlowska, E.; Roller, A.; Pignitter, M.; Jirsa, F.; Krachler, R.; Kandioller, W.; Keppler, B.K. Synthetic iron complexes as models for natural iron-humic compounds: Synthesis, characterization and algal growth experiments. *Sci. Total Environ.* **2017**, *577*, 94–104. [CrossRef]
69. Benderliev, K.M.; Ivanova, N.I.; Pilarski, P.S. Singlet oxygen and other reactive oxygen species are involved in regulation of release of iron-binding chelators from *Scenedesmus* cells. *Biol. Plant.* **2003**, *47*, 523–526. [CrossRef]
70. Liu, W.; Au, D.W.T.; Anderson, D.M.; Lam, P.K.S.; Wu, R.S.S. Effects of nutrients, salinity, pH and light:dark cycle on the production of reactive oxygen species in the alga *Chattonella marina*. *J. Exp. Mar. Biol. Ecol.* **2007**, *346*, 76–86. [CrossRef]
71. Lee, J.; Park, J.H.; Shin, Y.S.; Lee, B.C.; Chang, N.I.; Cho, J.; Kim, S.D. Effect of dissolved organic matter on the growth of algae, Pseudokirchneriella subcapitata, in Korean lakes: The importance of complexation reactions. *Ecotoxicol. Environ. Saf.* **2009**, *72*, 335–343. [CrossRef]
72. Perez-Perez, M.E.; Lemaire, S.D.; Crespo, J.L. Reactive oxygen species and autophagy in plants and algae. *Plant Physiol.* **2012**, *160*, 156–164. [CrossRef] [PubMed]
73. Mamaeva, A.; Taliansky, M.; Filippova, A.; Love, A.J.; Golub, N.; Fesenko, I. The role of chloroplast protein remodeling in stress responses and shaping of the plant peptidome. *New Phytol.* **2020**, *227*, 1326–1334. [CrossRef] [PubMed]
74. Fernandez, J.C.; Burch-Smith, T.M. Chloroplasts as mediators of plant biotic interactions over short and long distances. *Curr. Opin. Plant Biol.* **2019**, *50*, 148–155. [CrossRef]
75. Unal, D.; García-Caparrós, P.; Kumar, V.; Dietz, K.-J. Chloroplast-associated molecular patterns as concept for fine-tuned operational retrograde signalling. *Philos. Trans. R. Soc. B Biol. Sci.* **2020**, *375*, 20190443. [CrossRef]
76. Wang, Y.; Selinski, J.; Mao, C.; Zhu, Y.; Berkowitz, O.; Whelan, J. Linking mitochondrial and chloroplast retrograde signalling in plants. *Philos. Trans. R. Soc. B Biol. Sci.* **2020**, *375*, 20190410. [CrossRef] [PubMed]
77. Zhao, C.; Wang, Y.; Chan, K.X.; Marchant, D.B.; Franks, P.J.; Randall, D.; Tee, E.E.; Chen, G.; Ramesh, S.; Phua, S.Y.; et al. Evolution of chloroplast retrograde signaling facilitates green plant adaptation to land. *Proc. Natl. Acad. Sci. USA* **2019**, *116*, 5015–5020. [CrossRef]
78. Rea, G.; Antonacci, A.; Lambreva, M.D.; Mattoo, A.K. Features of cues and processes during chloroplast-mediated retrograde signaling in the alga Chlamydomonas. *Plant Sci.* **2018**, *272*, 193–206. [CrossRef]
79. Gilbert, M.; Bährs, H.; Steinberg, C.E.; Wilhelm, C. The artificial humic substance HS1500 does not inhibit photosynthesis of the green alga Desmodesmus armatus in vivo but interacts with the photosynthetic apparatus of isolated spinach thylakoids in vitro. *Photosynth. Res.* **2018**, *137*, 403–420. [CrossRef]
80. Allen, J. Photosynthesis of ATP-electrons, proton pumps, rotors, and poise. *Cell* **2002**, *110*, 273–276. [CrossRef]
81. Heinze, T.; Bährs, H.; Gilbert, M.; Steinberg, C.E.; Wilhelm, C. Selected coccal green algae are not affected by the humic substance Huminfeed® in term of growth or photosynthetic performance. *Hydrobiologia* **2012**, *684*, 215–224. [CrossRef]
82. Steinberg, C.E.; Timofeyev, M.A.; Menzel, R. *Dissolved Humic Substances—Interactions with Organisms*; Academic Press Oxford: Oxford, UK, 2009.
83. Peng, J.; Zhang, Y.; Li, J.; Wu, X.; Wang, M.; Gong, Z.; Gao, S. Interference of dissolved organic matter and its constituents on the accurate determination of hydrogen peroxide in water. *Sci. Rep.* **2021**, *11*, 22613. [CrossRef] [PubMed]
84. Amblard, C.; Carrias, J.-F.o.; Bourdier, G.; Maurin, N. The microbial loop in a humic lake: Seasonal and vertical variations in the structure of the different communities. *Hydrobiologia* **1995**, *300–301*, 71–84. [CrossRef]
85. Barone, M.E.; Parkes, R.; Herbert, H.; McDonnell, A.; Conlon, T.; Aranyos, A.; Fierli, D.; Fleming, G.T.A.; Touzet, N. Comparative response of marine microalgae to H_2O_2-induced oxidative stress. *Appl. Biochem. Biotechnol.* **2021**, *193*, 4052–4067. [CrossRef] [PubMed]

86. Koukal, B.; Gueguen, C.; Pardos, M.; Dominik, J. Influence of humic substances on the toxic effects of cadmium and zinc to the green alga *Pseudokirchneriella subcapitata*. *Chemosphere* **2003**, *53*, 953–961. [CrossRef]
87. Spierings, J.; Worms, I.A.; Mieville, P.; Slaveykova, V.I. Effect of humic substance photoalteration on lead bioavailability to freshwater microalgae. *Environ. Sci. Technol.* **2011**, *45*, 3452–3458. [CrossRef]
88. Shi, Z.; Xu, H.; Wang, Z.; Du, H.; Fu, X. Effects of co-exposure to copper and humic acids on microalga *Chlorella vulgaris*: Growth inhibition, oxidative stress, and extracellular secretion. *Environ. Pollut. Bioavailable* **2021**, *33*, 415–424. [CrossRef]
89. Hanachi, P.; Khoshnamvand, M.; Walker, T.R.; Hamidian, A.H. Nano-sized polystyrene plastics toxicity to microalgae *Chlorella vulgaris*: Toxicity mitigation using humic acid. *Aquat. Toxicol.* **2022**, *245*, 106123. [CrossRef]
90. Tong, M.; Li, X.; Luo, Q.; Yang, C.; Lou, W.; Liu, H.; Du, C.; Nie, L.; Zhong, Y. Effects of humic acids on biotoxicity of tetracycline to microalgae *Coelastrella* sp. *Algal Res.* **2020**, *50*, 101962. [CrossRef]
91. Zhao, J.; Li, Y.; Cao, X.; Guo, C.; Xu, L.; Wang, Z.; Feng, J.; Yi, H.; Xing, B. Humic acid mitigated toxicity of graphene-family materials to algae through reducing oxidative stress and heteroaggregation. *Environ. Sci. Nano* **2019**, *6*, 1909–1920. [CrossRef]
92. Slaveykova, V.I.; Wilkinson, K.J. Predicting the bioavailability of metals and metal complexes: Critical review of the biotic ligand model. *Environ. Chem.* **2005**, *2*, 9–24. [CrossRef]
93. Lamelas, C.; Pinheiro, J.P.; Slaveykova, V.I. Effect of humic acid on Cd(II), Cu(II), and Pb(II) uptake by freshwater algae: Kinetic and cell wall speciation considerations. *Environ. Sci. Technol.* **2009**, *43*, 730–735. [CrossRef] [PubMed]
94. Chen, Z.; Porcher, C.; Campbell, P.G.; Fortin, C. Influence of humic acid on algal uptake and toxicity of ionic silver. *Environ. Sci. Technol.* **2013**, *47*, 8835–8842. [CrossRef] [PubMed]
95. Nanayama, Y.; Sazawa, K.; Yustiawati, Y.; Syawal, M.S.; Fukushima, M.; Kuramitz, H. Effect of humic acids on the toxicity of pollutants to Chlamydomonas reinhardtii: Investigation by a microscale algal growth inhibition test. *Environ. Sci. Pollut. Res. Int.* **2021**, *28*, 211–219. [CrossRef]
96. Pandey, A.K.; Pandey, S.D.; Misra, V. Stability constants of metal-humic acid complexes and its role in environmental detoxification. *Ecotoxicol. Environ. Saf.* **2000**, *47*, 195–200. [CrossRef] [PubMed]
97. Manuel Ostera, J.; Puntarulo, S.; Malanga, G. Oxidative effects of glyphosate on the lipophobic intracellular environment in the microalgae. *Biocell* **2022**, *46*, 795–802. [CrossRef]
98. Liu, J.; Fan, J.; He, T.; Xu, X.; Ai, Y.; Tang, H.; Gu, H.; Lu, T.; Liu, Y.; Liu, G. The mechanism of aquatic photodegradation of organophosphorus sensitized by humic acid-Fe($^{3+}$) complexes. *J. Hazard Mater.* **2020**, *384*, 121466. [CrossRef]
99. Zhang, Y.; Meng, T.; Guo, X.; Yang, R.; Si, X.; Zhou, J. Humic acid alleviates the ecotoxicity of graphene-family materials on the freshwater microalgae Scenedesmus obliquus. *Chemosphere* **2018**, *197*, 749–758. [CrossRef]
100. Yin, J.; Dong, Z.; Liu, Y.; Wang, H.; Li, A.; Zhuo, Z.; Feng, W.; Fan, W. Toxicity of reduced graphene oxide modified by metals in microalgae: Effect of the surface properties of algal cells and nanomaterials. *Carbon* **2020**, *169*, 182–192. [CrossRef]
101. Ouyang, S.; Zhou, Q.; Zeng, H.; Wang, Y.; Hu, X. Natural Nanocolloids Mediate the Phytotoxicity of Graphene Oxide. *Environ. Sci. Technol.* **2020**, *54*, 4865–4875. [CrossRef]
102. Calvo, P.; Nelson, L.; Kloepper, J.W. Agricultural uses of plant biostimulants. *Plant Soil* **2014**, *383*, 3–41. [CrossRef]
103. Du Jardin, P. Plant biostimulants: Definition, concept, main categories and regulation. *Sci. Hortic.* **2015**, *196*, 3–14. [CrossRef]
104. du Jardin, P.; Xu, L.; Geelen, D. Agricultural Functions and Action Mechanisms of Plant Biostimulants (PBs): An Introduction. In *The Chemical Biology of Plant Biostimulants*; Geelen, D., Xu, L., Eds.; Wiley: Hoboken, NJ, USA, 2020; pp. 1–30.
105. Battacharyya, D.; Babgohari, M.Z.; Rathor, P.; Prithiviraj, B. Seaweed extracts as biostimulants in horticulture. *Sci. Hortic.* **2015**, *196*, 39–48. [CrossRef]
106. Khan, W.; Rayirath, U.P.; Subramanian, S.; Jithesh, M.N.; Rayorath, P.; Hodges, D.M.; Critchley, A.T.; Craigie, J.S.; Norrie, J.; Prithiviraj, B. Seaweed extracts as biostimulants of plant growth and development. *J. Plant Growth Regul.* **2009**, *28*, 386–399. [CrossRef]
107. Ertani, A.; Schiavon, M.; Muscolo, A.; Nardi, S. Alfalfa plant-derived biostimulant stimulate short-term growth of salt stressed *Zea mays* L. plants. *Plant Soil* **2013**, *364*, 145–158. [CrossRef]
108. Godlewska, K.; Pacyga, P.; Michalak, I.; Biesiada, A.; Szumny, A.; Pachura, N.; Piszcz, U. Effect of botanical extracts on the growth and nutritional quality of field-grown white head cabbage (*Brassica oleracea* var. *capitata*). *Molecules* **2021**, *26*, 1992. [CrossRef] [PubMed]
109. Canellas, L.P.; Olivares, F.L.; Aguiar, N.O.; Jones, D.L.; Nebbioso, A.; Mazzei, P.; Piccolo, A. Humic and fulvic acids as biostimulants in horticulture. *Sci. Hortic.* **2015**, *196*, 15–27. [CrossRef]
110. Bulgari, R.; Cocetta, G.; Trivellini, A.; Vernieri, P.; Ferrante, A. Biostimulants and crop responses: A review. *Biol. Agric. Hortic.* **2015**, *31*, 1–17. [CrossRef]
111. Colla, G.; Nardi, S.; Cardarelli, M.; Ertani, A.; Lucini, L.; Canaguier, R.; Rouphael, Y. Protein hydrolysates as biostimulants in horticulture. *Sci. Hortic.* **2015**, *196*, 28–38. [CrossRef]
112. Colla, G.; Rouphael, Y.; Canaguier, R.; Svecova, E.; Cardarelli, M. Biostimulant action of a plant-derived protein hydrolysate produced through enzymatic hydrolysis. *Front. Plant Sci.* **2014**, *5*, 448. [CrossRef]
113. Pichyangkura, R.; Chadchawan, S. Biostimulant activity of chitosan in horticulture. *Sci. Hortic.* **2015**, *196*, 49–65. [CrossRef]
114. Malerba, M.; Cerana, R. Recent advances of chitosan applications in plants. *Polymers* **2018**, *10*, 118. [CrossRef] [PubMed]
115. Rachidi, F.; Benhima, R.; Kasmi, Y.; Sbabou, L.; El Arroussi, H. Evaluation of microalgae polysaccharides as biostimulants of tomato plant defense using metabolomics and biochemical approaches. *Sci. Rep.* **2021**, *11*, 930.

116. Zhang, C.; Wang, W.; Zhao, X.; Wang, H.; Yin, H. Preparation of alginate oligosaccharides and their biological activities in plants: A review. *Carbohydr. Res.* **2020**, *494*, 108056. [CrossRef] [PubMed]
117. Salachna, P.; Pietrak, A. Evaluation of carrageenan, xanthan gum and depolymerized chitosan based coatings for pineapple lily plant production. *Horticulturae* **2021**, *7*, 19. [CrossRef]
118. Savvas, D.; Ntatsi, G. Biostimulant activity of silicon in horticulture. *Sci. Hortic.* **2015**, *196*, 66–81. [CrossRef]
119. Laane, H.-M. The effects of foliar sprays with different silicon compounds. *Plants* **2018**, *7*, 45. [CrossRef]
120. Bărbieru, O.-G.; Dimitriu, L.; Călin, M.; Răut, I.; Constantinescu-Aruxandei, D.; Oancea, F. Plant biostimulants based on selenium nanoparticles biosynthesized by Trichoderma strains. *Proceedings* **2019**, *29*, 95.
121. Dima, S.-O.; Neamțu, C.; Desliu-Avram, M.; Ghiurea, M.; Capra, L.; Radu, E.; Stoica, R.; Faraon, V.-A.; Zamfiropol-Cristea, V.; Constantinescu-Aruxandei, D.; et al. Plant biostimulant effects of baker's yeast vinasse and selenium on tomatoes through foliar fertilization. *Agronomy* **2020**, *10*, 133. [CrossRef]
122. García-García, A.L.; García-Machado, F.J.; Borges, A.A.; Morales-Sierra, S.; Boto, A.; Jiménez-Arias, D. Pure organic active compounds against abiotic stress: A biostimulant overview. *Front. Plant Sci.* **2020**, *11*, 5819. [CrossRef]
123. Kerchev, P.; van der Meer, T.; Sujeeth, N.; Verlee, A.; Stevens, C.V.; Van Breusegem, F.; Gechev, T. Molecular priming as an approach to induce tolerance against abiotic and oxidative stresses in crop plants. *Biotechnol. Adv.* **2020**, *40*, 107503. [CrossRef]
124. Velea, S.; Oancea, F.; Fischer, F. Heterotrophic and mixotrophic microalgae cultivation. In *Microalgae-Based Biofuels and Bioproducts, Cristina*; Gonzalez-Fernandez, R.M., Ed.; Woodhead Publishing—Elsevier: Duxford, UK, 2017; pp. 45–65.
125. Rouphael, Y.; Colla, G. Toward a sustainable agriculture through plant biostimulants: From experimental data to practical applications. *Agronomy* **2020**, *10*, 1461. [CrossRef]
126. Zhao, Y.; Xing, H.; Li, X.; Geng, S.; Ning, D.; Ma, T.; Yu, X. Physiological and metabolomics analyses reveal the roles of fulvic acid in enhancing the production of astaxanthin and lipids in *Haematococcus pluvialis* under abiotic stress conditions. *J. Agric. Food Chem.* **2019**, *67*, 12599–12609. [CrossRef] [PubMed]
127. Constantinescu-Aruxandei, D.; Vlaicu, A.; Marinas, I.C.; Vintila, A.C.N.; Dimitriu, L.; Oancea, F. Effect of betaine and selenium on the growth and photosynthetic pigment production in *Dunaliella salina* as biostimulants. *Fems Microbiol. Lett.* **2019**, *366*, fnz257. [CrossRef] [PubMed]
128. Fan, P.; Xu, P.; Zhu, Y.; Tu, X.; Song, G.; Zuo, Y.; Bi, Y. Addition of humic acid accelerates the growth of *Euglena pisciformis* AEW501 and the accumulation of lipids. *J. Appl. Phycol.* **2022**, *34*, 51–63. [CrossRef]
129. Hunt, R.W.; Chinnasamy, S.; Bhatnagar, A.; Das, K.C. Effect of biochemical stimulants on biomass productivity and metabolite content of the microalga, *Chlorella sorokiniana*. *Appl. Biochem. Biotechnol.* **2010**, *162*, 2400–2414. [CrossRef]
130. Zhu, J.; Wakisaka, M. Application of lignosulfonate as the growth promotor forfreshwater microalga *Euglena gracilis* to increase productivity of biomass and lipids. *Fuel* **2021**, *283*, 118920. [CrossRef]
131. Zhu, J.; Tan, X.; Hafid, H.S.; Wakisaka, M. Enhancement of biomass yield and lipid accumulation of freshwater microalga *Euglena gracilis* by phenolic compounds from basic structures of lignin. *Bioresour. Technol.* **2021**, *321*, 124441. [CrossRef]
132. Lugtenberg, B.; Kamilova, F. Plant-growth-promoting rhizobacteria. *Annu. Rev. Microbiol.* **2009**, *63*, 541–556. [CrossRef]
133. Kumawat, K.C.; Nagpal, S.; Sharma, P. Potential of plant growth-promoting rhizobacteria-plant interactions in mitigating salt stress for sustainable agriculture: A review. *Pedosphere* **2022**, *32*, 223–245. [CrossRef]
134. Canellas, L.P.; Olivares, F.L. Physiological responses to humic substances as plant growth promoter. *Chem. Biol. Technol. Agric.* **2014**, *1*, 3. [CrossRef]
135. Stirk, W.A.; van Staden, J. Potential of phytohormones as a strategy to improve microalgae productivity for biotechnological applications. *Biotechnol. Adv.* **2020**, *44*, 107612. [CrossRef] [PubMed]
136. Lu, Y.D.; Xu, J. Phytohormones in microalgae: A new opportunity for microalgal biotechnology? *Trends Plant Sci.* **2015**, *20*, 273–282. [CrossRef] [PubMed]
137. Han, X.; Zeng, H.; Bartocci, P.; Fantozzi, F.; Yan, Y. Phytohormones and effects on growth and metabolites of microalgae: A review. *Fermentation* **2018**, *4*, 25. [CrossRef]
138. Delaux, P.-M.; Xie, X.; Timme, R.E.; Puech-Pages, V.; Dunand, C.; Lecompte, E.; Delwiche, C.F.; Yoneyama, K.; Bécard, G.; Séjalon-Delmas, N. Origin of strigolactones in the green lineage. *New Phytol.* **2012**, *195*, 857–871. [CrossRef] [PubMed]
139. Xu, B.; Zhao, C.; Liu, J.; Zhao, Y.; Sun, S.; Wei, J. Simultaneous biogas upgrading and biogas slurry treatment by different microalgae-based technologies under various strigolactone analog (GR24) concentrations. *Bioresour. Technol.* **2022**, *351*, 127033.
140. Xu, M.; Xue, Z.; Zhao, Y.; Sun, S.; Liu, J. Enhancement of the photosynthetic and removal performance for microalgae-based technologies by co-culture strategy and strigolactone induction. *Bioresour. Technol.* **2021**, *339*, 125579. [CrossRef]
141. Song, X.; Zhao, Y.; Li, T.; Han, B.; Zhao, P.; Xu, J.-W.; Yu, X. Enhancement of lipid accumulation in *Monoraphidium* sp. QLY-1 by induction of strigolactone. *Bioresour. Technol.* **2019**, *288*, 121607. [CrossRef]
142. Shen, X.; Xue, Z.; Sun, L.; Zhao, C.; Sun, S.; Liu, J.; Zhao, Y.; Liu, J. Effect of GR24 concentrations on biogas upgrade and nutrient removal by microalgae-based technology. *Bioresour. Technol.* **2020**, *312*, 123563. [CrossRef]
143. Liu, Y.; Li, Y.X.; Zhao, H.J.; Zhu, B.L.; Xu, J.L.; Xu, F.; Liu, S.X.; Li, X.H.; Zhou, C.X. Phloroglucinol promotes fucoxanthin synthesis by activating the cis-zeatin and brassinolide pathways in *Thalassiosira pseudonana*. *Appl. Environ. Microbiol.* **2022**, *88*, e0216021. [CrossRef]
144. Da Silva, J.A.T.; Dobranszki, J.; Ross, S. Phloroglucinol in plant tissue culture. *Vitr. Cell. Dev. Biol.-Plant* **2013**, *49*, 1–16. [CrossRef]

145. Petti, C. Phloroglucinol mediated plant regeneration of *Ornithogalum dubium* as the sole "hormone-like supplement" in plant tissue culture long-term experiments. *Plants* **2020**, *9*, 929. [CrossRef] [PubMed]
146. Li, T.; Xu, J.; Yu, X.; Zhao, P.; Che, R. Method for improving oil content of oil-producing microalgae based on fulvic acid. CN105671095-A, 15 June 2016.
147. Yu, X.; Li, D.; Zhao, Y. Use of Melatonin for Increasing oil Content of Oil-Producing Microalgae and Promoting Accumulation of Oil in Microalgae. CN107475171-A, 15 December 2017.
148. Palacios, O.A.; López, B.R.; de-Bashan, L.E. Microalga Growth-Promoting Bacteria (MGPB): A formal term proposed for beneficial bacteria involved in microalgal–bacterial interactions. *Algal Res.* **2022**, *61*, 102585. [CrossRef]
149. Ramanan, R.; Kim, B.-H.; Cho, D.-H.; Oh, H.-M.; Kim, H.-S. Algae–bacteria interactions: Evolution, ecology and emerging applications. *Biotechnol. Adv.* **2016**, *34*, 14–29. [CrossRef] [PubMed]
150. Zhang, T.-Y.; Hu, H.-Y.; Wu, Y.-H.; Zhuang, L.-L.; Xu, X.-Q.; Wang, X.-X.; Dao, G.-H. Promising solutions to solve the bottlenecks in the large-scale cultivation of microalgae for biomass/bioenergy production. *Renew. Sustain. Energy Rev.* **2016**, *60*, 1602–1614. [CrossRef]
151. De-Bashan, L.E.; Moreno, M.; Hernandez, J.P.; Bashan, Y. Removal of ammonium and phosphorus ions from synthetic wastewater by the microalgae Chlorella vulgaris coimmobilized in alginate beads with the microalgae growth-promoting bacterium Azospirillum brasilense. *Water Res.* **2002**, *36*, 2941–2948. [CrossRef]
152. De-Bashan, L.E.; Hernandez, J.P.; Morey, T.; Bashan, Y. Microalgae growth-promoting bacteria as "helpers" for microalgae: A novel approach for removing ammonium and phosphorus from municipal wastewater. *Water Res.* **2004**, *38*, 466–474. [CrossRef]
153. Do Nascimento, M.; Dublan, M.d.l.A.; Ortiz-Marquez, J.C.F.; Curatti, L. High lipid productivity of an *Ankistrodesmus—Rhizobium* artificial consortium. *Bioresour. Technol.* **2013**, *146*, 400–407. [CrossRef]
154. Sapp, M.; Schwaderer, A.S.; Wiltshire, K.H.; Hoppe, H.G.; Gerdts, G.; Wichels, A. Species-specific bacterial communities in the phycosphere of microalgae? *Microb. Ecol.* **2007**, *53*, 683–699. [CrossRef]
155. Ramanan, R.; Kang, Z.; Kim, B.-H.; Cho, D.-H.; Jin, L.; Oh, H.-M.; Kim, H.-S. Phycosphere bacterial diversity in green algae reveals an apparent similarity across habitats. *Algal Res.* **2015**, *8*, 140–144. [CrossRef]
156. Ruzzi, M.; Aroca, R. Plant growth-promoting rhizobacteria act as biostimulants in horticulture. *Sci. Hortic.* **2015**, *196*, 124–134. [CrossRef]
157. Backer, R.; Rokem, J.S.; Ilangumaran, G.; Lamont, J.; Praslickova, D.; Ricci, E.; Subramanian, S.; Smith, D.L. Plant growth-promoting rhizobacteria: Context, mechanisms of action, and roadmap to commercialization of biostimulants for sustainable agriculture. *Front. Plant Sci.* **2018**, *9*, 1473. [CrossRef] [PubMed]
158. Fuentes, J.L.; Garbayo, I.; Cuaresma, M.; Montero, Z.; González-del-Valle, M.; Vílchez, C. Impact of microalgae-bacteria interactions on the production of algal biomass and associated compounds. *Mar. Drugs* **2016**, *14*, 100. [CrossRef] [PubMed]
159. Steinberg, C.E.W.; Bach, S. Growth promotion by a groundwater fulvic acid in a bacteria/algae system. *Acta Hydrochim. Hydrobiol.* **1996**, *24*, 98–100. [CrossRef]
160. Zhao, X.; Zhou, Y.; Huang, S.; Qiu, D.Y.; Schideman, L.; Chai, X.L.; Zhao, Y.C. Characterization of microalgae-bacteria consortium cultured in landfill leachate for carbon fixation and lipid production. *Bioresour. Technol.* **2014**, *156*, 322–328. [CrossRef]
161. Fuqua, C.; Greenberg, E.P. Listening in on bacteria: Acyl-homoserine lactone signalling. *Nat. Rev. Mol. Cell Biol.* **2002**, *3*, 685–695. [CrossRef]
162. Dow, L. How do quorum-sensing signals mediate algae–bacteria interactions? *Microorganisms* **2021**, *9*, 1391. [CrossRef]
163. Xu, Y.-Y.; Yang, J.-S.; Liu, C.; Wang, E.-T.; Wang, R.-N.; Qiu, X.-Q.; Li, B.-Z.; Chen, W.-F.; Yuan, H.-L. Water-soluble humic materials regulate quorum sensing in *Sinorhizobium meliloti* through a novel repressor of expR. *Front. Microbiol.* **2018**, *9*, 3194. [CrossRef]
164. Liu, L.J.; Ji, M.; Wang, F.; Tian, Z.K.; Wang, T.Y.; Wang, S.Y.; Wang, S.Y.; Yan, Z. Insight into the short-term effect of fulvic acid on nitrogen removal performance and N-acylated-L-homoserine lactones (AHLs) release in the anammox system. *Sci. Total Environ.* **2020**, *704*, 135285. [CrossRef]
165. Sandilya, A.A.; Natarajan, U.; Priya, M.H. Molecular View into the Cyclodextrin Cavity: Structure and Hydration. *ACS Omega* **2020**, *5*, 25655–25667. [CrossRef]
166. Verschoor, A.M.; Van Der Stap, I.; Helmsing, N.R.; Lürling, M.; Van Donk, E. Inducible colony formation within the Scenedesmaceae: Adaptive responses to infochemicals from two different herbivore taxa. *J. Phycol.* **2004**, *40*, 808–814. [CrossRef]
167. Roccuzzo, S.; Beckerman, A.P.; Pandhal, J. The use of natural infochemicals for sustainable and efficient harvesting of the microalgae *Scenedesmus* spp. for biotechnology: Insights from a meta-analysis. *Biotechnol. Lett.* **2016**, *38*, 1983–1990. [CrossRef] [PubMed]
168. Roccuzzo, S.; Couto, N.; Karunakaran, E.; Kapoore, R.V.; Butler, T.O.; Mukherjee, J.; Hansson, E.M.; Beckerman, A.P.; Pandhal, J. Metabolic insights into infochemicals induced colony formation and flocculation in *Scenedesmus subspicatus* unraveled by quantitative proteomics. *Front. Microbiol.* **2020**, *11*, 792. [CrossRef] [PubMed]
169. Albini, D.; Fowler, M.S.; Llewellyn, C.; Tang, K.W. Reversible colony formation and the associated costs in *Scenedesmus obliquus*. *J. Plankton Res.* **2019**, *41*, 419–429. [CrossRef]
170. Peng, Q.; Zhao, M.; Shen, G.; Gan, X.; Li, M. Linear alkylbenzene sulfonate (LAS) promotes sedimentation and lipid accumulation in *Scenedesmus obliquus*. *RSC Adv.* **2017**, *7*, 9244–9250. [CrossRef]

171. Zhu, X.X.; Wang, Z.S.; Sun, Y.F.; Gu, L.; Zhang, L.; Wang, J.; Huang, Y.; Yang, Z. Surfactants at environmentally relevant concentrations interfere the inducible defense of *Scenedesmus obliquus* and the implications for ecological risk assessment. *Environ. Pollut.* **2020**, *261*, 114131. [CrossRef]
172. Oda, Y.; Sakamoto, M.; Miyabara, Y. Colony formation in three species of the family *Scenedesmaceae* (*Desmodesmus subspicatus, Scenedesmus acutus, Tetradesmus dimorphus*) exposed to sodium dodecyl sulfate and its interference with grazing of *Daphnia galeata*. *Arch. Environ. Contam. Toxicol.* **2022**, *82*, 37–47. [CrossRef]
173. Lamar, R.T.; Monda, H.; Sleighter, R. Use of ore-derived humic acids with diverse chemistries to elucidate Structure-Activity Relationships (SAR) of humic acids in plant phenotypic expression. *Front. Plant Sci.* **2021**, *12*, 758424. [CrossRef]
174. Nardi, S.; Schiavon, M.; Francioso, O. Chemical structure and biological activity of humic substances define their role as plant growth promoters. *Molecules* **2021**, *26*, 2256. [CrossRef]
175. Van Tol de Castro, T.A.; Berbara, R.L.L.; Tavares, O.C.H.; Mello, D.F.d.G.; Pereira, E.G.; Souza, C.d.C.B.d.; Espinosa, L.M.; García, A.C. Humic acids induce a eustress state via photosynthesis and nitrogen metabolism leading to a root growth improvement in rice plants. *Plant Physiol. Biochem.* **2021**, *162*, 171–184. [CrossRef]
176. Savarese, C.; di Meo, V.; Cangemi, S.; Verrillo, M.; Savy, D.; Cozzolino, V.; Piccolo, A. Bioactivity of two different humic materials and their combination on plants growth as a function of their molecular properties. *Plant Soil* **2022**, *472*, 509–526. [CrossRef]
177. Savy, D.; Brostaux, Y.; Cozzolino, V.; Delaplace, P.; du Jardin, P.; Piccolo, A. Quantitative structure-activity relationship of humic-like biostimulants derived from agro-industrial byproducts and energy crops. *Front. Plant Sci.* **2020**, *11*, 581. [CrossRef] [PubMed]
178. Qin, S.; Xu, C.B.; Xu, Y.Z.; Bai, Y.C.; Guo, F. Molecular signatures of humic acids from different sources as revealed by ultrahigh resolution mass spectrometry. *J. Chem.* **2020**, *2020*, 7171582. [CrossRef]
179. Monda, H.; McKenna, A.M.; Fountain, R.; Lamar, R.T. Bioactivity of humic acids extracted from shale ore: Molecular characterization and structure-activity relationship with tomato plant yield under nutritional stress. *Front. Plant Sci.* **2021**, *12*, 660224. [CrossRef]
180. Aeschbacher, M.; Graf, C.; Schwarzenbach, R.P.; Sander, M. Antioxidant properties of humic substances. *Environ. Sci. Technol.* **2012**, *46*, 4916–4925. [CrossRef] [PubMed]
181. Choi, W.J.; Chae, A.N.; Song, K.G.; Park, J.; Lee, B.C. Effect of trophic conditions on microalga growth, nutrient removal, algal organic matter, and energy storage products in *Scenedesmus* (*Acutodesmus*) *obliquus* KGE-17 cultivation. *Bioprocess Biosyst. Eng.* **2019**, *42*, 1225–1234. [CrossRef]
182. Khan, S.I.; Zamyadi, A.; Rao, N.R.H.; Li, X.; Stuetz, R.M.; Henderson, R.K. Fluorescence spectroscopic characterisation of algal organic matter: Towards improved *in-situ* fluorometer development. *Environ. Sci. Water Res. Technol.* **2019**, *5*, 417–432. [CrossRef]
183. Khan, W.; Park, J.W.; Maeng, S.K. Fluorescence descriptors for algal organic matter and microalgae disintegration during ultrasonication. *J. Water Process Eng.* **2022**, *45*, 102517. [CrossRef]
184. Wu, M.C.; Du, M.; Wu, G.M.; Lu, F.M.; Li, J.; Lei, A.P.; Zhu, H.; Hu, Z.L.; Wang, J.X. Water reuse and growth inhibition mechanisms for cultivation of microalga Euglena gracilis. *Biotechnol. Biofuels* **2021**, *14*, 132. [CrossRef]
185. Sha, J.; Lu, Z.; Ye, J.; Wang, G.; Hu, Q.; Chen, Y.; Zhang, X. The inhibition effect of recycled *Scenedesmus acuminatus* culture media: Influence of growth phase, inhibitor identification and removal. *Algal Res.* **2019**, *42*, 101612. [CrossRef]
186. Zhang, X.; Lu, Z.; Wang, Y.; Wensel, P.; Sommerfeld, M.; Hu, Q. Recycling Nannochloropsis oceanica culture media and growth inhibitors characterization. *Algal Res.* **2016**, *20*, 282–290. [CrossRef]
187. Chisti, Y. Biodiesel from microalgae. *Biotechnol Adv.* **2007**, *25*, 294–306. [CrossRef]
188. Lam, M.K.; Lee, K.T. Microalgae biofuels: A critical review of issues, problems and the way forward. *Biotechnol Adv.* **2012**, *30*, 673–690. [CrossRef]
189. Christenson, L.; Sims, R. Production and harvesting of microalgae for wastewater treatment, biofuels, and bioproducts. *Biotechnol Adv.* **2011**, *29*, 686–702. [CrossRef]
190. Rawat, I.; Ranjith Kumar, R.; Mutanda, T.; Bux, F. Biodiesel from microalgae: A critical evaluation from laboratory to large scale production. *Appl. Energy* **2013**, *103*, 444–467. [CrossRef]
191. Chisti, Y. Constraints to commercialization of algal fuels. *J. Biotechnol.* **2013**, *167*, 201–214. [CrossRef]
192. Uduman, N.; Qi, Y.; Danquah, M.K.; Forde, G.M.; Hoadley, A. Dewatering of microalgal cultures: A major bottleneck to algae-based fuels. *J. Renew. Sustain. Energy* **2010**, *2*, 012701. [CrossRef]
193. Barros, A.I.; Gonçalves, A.L.; Simões, M.; Pires, J.C.M. Harvesting techniques applied to microalgae: A review. *Renew. Sustain. Energy Rev.* **2015**, *41*, 1489–1500. [CrossRef]
194. Griffiths, G.; Hossain, A.K.; Sharma, V.; Duraisamy, G. Key Targets for Improving Algal Biofuel Production. *Clean Technol.* **2021**, *3*, 711–742. [CrossRef]
195. Liber, J.A.; Bryson, A.E.; Bonito, G.; Du, Z.Y. Harvesting microalgae for food and energy products. *Small Methods* **2020**, *4*, 2000349. [CrossRef]
196. Fasaei, F.; Bitter, J.H.; Slegers, P.M.; van Boxtel, A.J.B. Techno-economic evaluation of microalgae harvesting and dewatering systems. *Algal Res.* **2018**, *31*, 347–362. [CrossRef]
197. Mathimani, T.; Mallick, N. A comprehensive review on harvesting of microalgae for biodiesel—Key challenges and future directions. *Renew. Sustain. Energy Rev.* **2018**, *91*, 1103–1120. [CrossRef]

198. Khan, S.; Naushad, M.; Iqbal, J.; Bathula, C.; Sharma, G. Production and harvesting of microalgae and an efficient operational approach to biofuel production for a sustainable environment. *Fuel* **2022**, *311*, 122543. [CrossRef]
199. Min, K.H.; Kim, D.H.; Ki, M.R.; Pack, S.P. Recent progress in flocculation, dewatering, and drying technologies for microalgae utilization: Scalable and low-cost harvesting process development. *Bioresour. Technol.* **2022**, *344*, 126404. [CrossRef] [PubMed]
200. Huh, Y.S.; Lee, Y.C. Harvesting Oleaginous Microorganisms Using Aminoclay-Induced Humic Acid. KR1567474-B1, 11 November 2015.
201. Lee, Y.C.; Oh, S.Y.; Lee, H.U.; Kim, B.; Lee, S.Y.; Choi, M.H.; Lee, G.W.; Park, J.Y.; Oh, Y.K.; Ryu, T.; et al. Aminoclay-induced humic acid flocculation for efficient harvesting of oleaginous *Chlorella* sp. *Bioresour Technol.* **2014**, *153*, 365–369. [CrossRef]
202. Yang, L.; Zhang, H.; Cheng, S.; Zhang, W.; Zhang, X. Enhanced microalgal harvesting using microalgae-derived extracellular polymeric substance as flocculation aid. *ACS Sustain. Chem. Eng.* **2020**, *8*, 4069–4075. [CrossRef]
203. Cheng, Y.L.; Juang, Y.C.; Liao, G.Y.; Tsai, P.W.; Ho, S.H.; Yeh, K.L.; Chen, C.Y.; Chang, J.S.; Liu, J.C.; Chen, W.M.; et al. Harvesting of Scenedesmus obliquus FSP-3 using dispersed ozone flotation. *Bioresour. Technol.* **2011**, *102*, 82–87. [CrossRef]
204. Beuckels, A.; Depraetere, O.; Vandamme, D.; Foubert, I.; Smolders, E.; Muylaert, K. Influence of organic matter on flocculation of Chlorella vulgaris by calcium phosphate precipitation. *Biomass Bioenergy* **2013**, *54*, 107–114. [CrossRef]
205. Borowitzka, M.A. High-value products from microalgae—Their development and commercialisation. *J. Appl. Phycol.* **2013**, *25*, 743–756. [CrossRef]
206. Mohammady, N.G.-E. Humic acid stimulation of growth and optimization of biochemical profiles on two microalgal species proposed as live feeds in aquaculture. *Int. J. Recirc. Aquac.* **2008**, *9*, 13–28. [CrossRef]
207. Zhao, Y.T.; Shang, M.M.; Xu, J.W.; Zhao, P.; Li, T.; Yu, X.Y. Enhanced astaxanthin production from a novel strain of Haematococcus pluvialis using fulvic acid. *Process. Biochem.* **2015**, *50*, 2072–2077. [CrossRef]
208. Meinelt, T.; Schreckenbach, K.; Pietrock, M.; Heidrich, S.; Steinberg, C.E. Humic substances. Part 1: Dissolved humic substances (HS) in aquaculture and ornamental fish breeding. *Environ. Sci. Pollut. Res. Int.* **2008**, *15*, 17–22. [CrossRef] [PubMed]
209. Vetvicka, V.; Baigorri, R.; Zamarreno, A.M.; Garcia-Mina, J.M.; Yvin, J.C. Glucan and humic acid: Synergistic effects on the immune system. *J. Med. Food* **2010**, *13*, 863–869. [CrossRef] [PubMed]
210. Colla, G.; Rouphael, Y. Microalgae: New source of plant biostimulants. *Agronomy* **2020**, *10*, 1240. [CrossRef]
211. González-Pérez, B.K.; Rivas-Castillo, A.M.; Valdez-Calderón, A.; Gayosso-Morales, M.A. Microalgae as biostimulants: A new approach in agriculture. *World J. Microbiol. Biotechnol.* **2022**, *38*, 1–12. [CrossRef]
212. Chiaiese, P.; Corrado, G.; Colla, G.; Kyriacou, M.C.; Rouphael, Y. Renewable sources of plant biostimulation: Microalgae as a sustainable means to improve crop performance. *Front. Plant Sci.* **2018**, *9*, 1782. [CrossRef]
213. Oancea, F.; Velea, S.; Fătu, V.; Mincea, C.; Ilie, L. Micro-algae based plant biostimulant and its effect on water stressed tomato plants. *Rom. J. Plant Prot.* **2013**, *6*, 104–117.
214. Mógor, Á.F.; Ördög, V.; Lima, G.P.P.; Molnár, Z.; Mógor, G. Biostimulant properties of cyanobacterial hydrolysate related to polyamines. *J. Appl. Phycol.* **2017**, *30*, 453–460. [CrossRef]
215. Creelman, R.A.; Mullet, J.E. Oligosaccharins, brassinolides, and jasmonates: Nontraditional regulators of plant growth, development, and gene expression. *Plant Cell* **1997**, *9*, 1211–1223. [CrossRef]
216. Abdul Malik, N.A.; Kumar, I.S.; Nadarajah, K. Elicitor and receptor molecules: Orchestrators of plant defense and immunity. *Int. J. Mol. Sci.* **2020**, *21*, 963. [CrossRef]
217. El Arroussi, H.; Benhima, R.; Elbaouchi, A.; Sijilmassi, B.; El Mernissi, N.; Aafsar, A.; Meftah-Kadmiri, I.; Bendaou, N.; Smouni, A. *Dunaliella salina* exopolysaccharides: A promising biostimulant for salt stress tolerance in tomato (*Solanum lycopersicum*). *J. Appl. Phycol.* **2018**, *30*, 2929–2941. [CrossRef]
218. Colla, G.; Hoagland, L.; Ruzzi, M.; Cardarelli, M.; Bonini, P.; Canaguier, R.; Rouphael, Y. Biostimulant action of protein hydrolysates: Unraveling their effects on plant physiology and microbiome. *Front. Plant Sci.* **2017**, *8*, 2202. [CrossRef] [PubMed]
219. Maurya, R.; Paliwal, C.; Chokshi, K.; Pancha, I.; Ghosh, T.; Satpati, G.G.; Pal, R.; Ghosh, A.; Mishra, S. Hydrolysate of lipid extracted microalgal biomass residue: An algal growth promoter and enhancer. *Bioresour. Technol.* **2016**, *207*, 197–204. [CrossRef] [PubMed]
220. Zandonadi, D.B.; Santos, M.P.; Caixeta, L.S.; Marinho, E.B.; Peres, L.E.P.; Façanha, A.R. Plant proton pumps as markers of biostimulant action. *Sci. Agric.* **2016**, *73*, 24–28. [CrossRef]
221. Klein, P.; Chauvey, L.; Kallerhoff, J.; Pinelli, E.; Morard, M.; Silvestre, J. A Tool derived from the *Vicia faba* micronucleus assay, to assess genotoxicity, cytotoxicity or biostimulation of novel compounds used in agriculture. *Agronomy* **2021**, *11*, 321. [CrossRef]
222. Ugena, L.; Hýlová, A.; Podlešáková, K.; Humplík, J.F.; Doležal, K.; Diego, N.D.; Spíchal, L. Characterization of biostimulant mode of action using novel multi-trait high-throughput screening of *Arabidopsis* germination and rosette growth. *Front. Plant Sci.* **2018**, *9*, 1327. [CrossRef]
223. Rayorath, P.; Khan, W.; Palanisamy, R.; MacKinnon, S.L.; Stefanova, R.; Hankins, S.D.; Critchley, A.T.; Prithiviraj, B. Extracts of the brown seaweed *Ascophyllum nodosum* induce gibberellic acid (GA3)-independent amylase activity in barley. *J. Plant Growth Regul.* **2008**, *27*, 370–379. [CrossRef]
224. Rouphael, Y.; Colla, G. Synergistic biostimulatory action: Designing the next generation of plant biostimulants for sustainable agriculture. *Front. Plant Sci.* **2018**, *9*, 1655. [CrossRef]
225. Che, R.Q.; Huang, L.; Xu, J.W.; Zhao, P.; Li, T.; Ma, H.X.; Yu, X.Y. Effect of fulvic acid induction on the physiology, metabolism, and lipid biosynthesis-related gene transcription of *Monoraphidium* sp. FXY-10. *Bioresour. Technol.* **2017**, *227*, 324–334. [CrossRef]

226. Cui, N.; Xiao, J.M.; Feng, Y.J.; Zhao, Y.T.; Yu, X.Y.; Xu, J.W.; Li, T.; Zhao, P. Antioxidants enhance lipid productivity in *Heveochlorella* sp. Yu. *Algal Res. Biomass Biofuels Bioprod.* **2021**, *55*, 102235. [CrossRef]
227. Martini, S.; D'Addario, C.; Bonechi, C.; Leone, G.; Tognazzi, A.; Consumi, M.; Magnani, A.; Rossi, C. Increasing photostability and water-solubility of carotenoids: Synthesis and characterization of β-carotene–humic acid complexes. *J. Photochem. Photobiol. B Biol.* **2010**, *101*, 355–361. [CrossRef]
228. Vetvicka, V.; Garcia-Mina, J.M.; Proctor, M.; Yvin, J.C. Humic Acid and glucan: Protection against liver injury induced by carbon tetrachloride. *J. Med. Food* **2015**, *18*, 572–577. [CrossRef] [PubMed]
229. Hamza, Z.K.; Hathout, A.S.; Ostroff, G.; Soto, E.; Sabry, B.A.; El-Hashash, M.A.; Hassan, N.S.; Aly, S.E. Assessment of the protective effect of yeast cell wall beta-glucan encapsulating humic acid nanoparticles as an aflatoxin B-1 adsorbent in vivo. *J. Biochem. Mol. Toxicol.* **2022**, *36*, e22941. [CrossRef] [PubMed]

marine drugs

MDPI

Review

Wide Range Applications of Spirulina: From Earth to Space Missions

Giacomo Fais [1,†], Alessia Manca [2,†], Federico Bolognesi [3,4], Massimiliano Borselli [5], Alessandro Concas [1,6], Marco Busutti [7], Giovanni Broggi [8,9], Pierdanilo Sanna [10], Yandy Marx Castillo-Aleman [10], René Antonio Rivero-Jiménez [10], Antonio Alfonso Bencomo-Hernandez [10], Yendry Ventura-Carmenate [10], Michela Altea [11], Antonella Pantaleo [2], Gilberto Gabrielli [11], Federico Biglioli [3], Giacomo Cao [1,6,12] and Giuseppe Giannaccare [5,*]

1 Interdepartmental Centre of Environmental Science and Engineering (CINSA), University of Cagliari, Via San Giorgio 12, 09124 Cagliari, Italy; faisgiacomo@gmail.com (G.F.); alessandro.concas@unica.it (A.C.); giacomo.cao@unica.it (G.C.)
2 Department of Biomedical Science, University of Sassari, Viale San Pietro 43/B, 07100 Sassari, Italy; alessia_manca@hotmail.it (A.M.); apantaleo@uniss.it (A.P.)
3 Unit of Maxillofacial Surgery, Head and Neck Department, ASST Santi Paolo e Carlo Hospital, University of Milan, Via Antonio di Rudinì 8, 20142 Milan, Italy; federico.bolognesi5@unibo.it (F.B.); federico.biglioli@unimi.it (F.B.)
4 Department of Biomedical and Neuromotor Sciences, University of Bologna, Via Zamboni 33, 40126 Bologna, Italy
5 Department of Ophthalmology, University Magna Grecia of Catanzaro, Viale Europa, 88100 Catanzaro, Italy; mborselli93@gmail.com
6 Department of Mechanical, Chemical and Materials Engineering, University of Cagliari, Via Marengo 2, 09123 Cagliari, Italy
7 Nephrology, Dialysis and Transplant Unit, IRCCS-Azienda Ospedaliero Universitaria di Bologna, University of Bologna, Via Giuseppe Massarenti 9, 40138 Bologna, Italy; marco.busutti@aosp.bo.it
8 Department of Neurosurgery, Fondazione IRCCS Istituto Neurologico Carlo Besta, University of Milan, Via Celoria 11, 20133 Milan, Italy; gbroggi@gmail.com
9 Columbus Clinic Center, Via Michelangelo Buonarroti 48, 20145 Milan, Italy
10 Abu Dhabi Stem Cells Center, Al Misaha Street, Rowdhat, Abu Dhabi, United Arab Emirates; pierdanilo.sanna@gmail.com (P.S.); yandy.castillo@adscc.ae (Y.M.C.-A.); rene.rivero@adscc.ae (R.A.R.-J.); antonio.bencomo@adscc.ae (A.A.B.-H.); yendry.ventura@adscc.ae (Y.V.-C.)
11 TOLO Green, Via San Damiano 2, 20122 Milan, Italy; m.altea@tologreen.it (M.A.); gil.gabrielli@gmail.com (G.G.)
12 Center for Advanced Studies, Research and Development in Sardinia (CRS4), Loc. Piscina Manna, Building 1, 09050 Pula, Italy
* Correspondence: giuseppe.giannaccare@unicz.it; Tel.: +39-3317186201
† These authors contributed equally to this work.

Citation: Fais, G.; Manca, A.; Bolognesi, F.; Borselli, M.; Concas, A.; Busutti, M.; Broggi, G.; Sanna, P.; Castillo-Aleman, Y.M.; Rivero-Jiménez, R.A.; et al. Wide Range Applications of Spirulina: From Earth to Space Missions. *Mar. Drugs* **2022**, *20*, 299. https://doi.org/10.3390/md20050299

Academic Editor: Carlos Almeida

Received: 22 March 2022
Accepted: 26 April 2022
Published: 28 April 2022

Abstract: *Spirulina* is the most studied cyanobacterium species for both pharmacological applications and the food industry. The aim of the present review is to summarize the potential benefits of the use of Spirulina for improving healthcare both in space and on Earth. Regarding the first field of application, Spirulina could represent a new technology for the sustainment of long-duration manned missions to planets beyond the Lower Earth Orbit (e.g., Mars); furthermore, it could help astronauts stay healthy while exposed to a variety of stress factors that can have negative consequences even after years. As far as the second field of application, Spirulina could have an active role in various aspects of medicine, such as metabolism, oncology, ophthalmology, central and peripheral nervous systems, and nephrology. The recent findings of the capacity of Spirulina to improve stem cells mobility and to increase immune response have opened new intriguing scenarios in oncological and infectious diseases, respectively.

Keywords: Spirulina; healthcare; space missions; medicine applications; microgravity effects

1. Introduction

Arthrospira platensis and *Arthrospira maxima* (also known more generically as *Spirulina* for its spiral or helical shape), which belong to the Microcoleaceae family, are the most studied cyanobacterium species in pharmacological applications and the food industry [1–4]. Cyanobacteria are Gram-negative bacteria that have played an important role in the evolution of primitive Earth and the biosphere; they have been responsible for the oxygenation of the atmosphere and oceans since the Great Oxidation Event (GOE) around 3 billion years ago [5]. These microorganisms have been classified as blue-green algae for their color due to the production of the pigment phycocyanin. Cyanobacteria have been found in various ecological niches, from freshwater to the oceans, on soils, rocks, and in environments with extreme physicochemical characteristics [6]. *Spirulina* is an obligate photoautotrophic filamentous species and has a characteristic helical shape. This cyanobacterium has a prokaryotic organization with a multilayered cell wall, ribosomes, numerous inclusions, and a lamellar photosynthetic system.

Cyanobacteria thrive naturally in alkaline waters, rich in minerals, under temperatures ranging from 35 to 40 °C. Its filaments can reach the size of 0.5 mm in length. The helical shape of the filament and the presence of gas-vacuoles inside the cells make it form floating mats. Several species of *Spirulina* are present in nature, but the most studied and used species are *Spirulina platensis* (*S. platensis*) and *Spirulina maxima* (*S. maxima*) [7]. Since the 70s, Spirulina has been considered a rich food source due to its high content of macro-and micronutrients. In fact, it is an excellent source of proteins, vitamins, fatty acids, minerals, photosynthetic pigments, and several secondary metabolites (Table 1) [8].

Table 1. *Spirulina platensis* main compounds.

Macronutrients	**% d.w.**
Proteins	50–70
Carbohydrates	15–20
Lipids	7–16
Vitamins	**mg/100 g d.w.**
Carotene	140
Biotin	0.005
Folic acid	0.01
Niacin	14
Riboflavin	4
Thiamin B1	3.5
Vitamin E	100
Vitamin B12	0.32
Vitamin K	2.2
Carbohydrates	**mg/100 g d.w.**
Glucose	54.4
Galactose	2.6
Mannose	9.3
Rhamnose	22.3
Xylose	7
Lipids	**g/100 g d.w.**
Arachidic	0.048
Gamma linolenic (GLA)	1
Myristic	0.041
Oleic	0.017
Palmitic	2

Table 1. *Cont.*

Amino acids	g/100 g d.w.
Leucine	4.94
Isoleucine	3.2
Lysine	3.02
Methionine	1.15
Phenylalaline	2.78
Threonine	2.97
Tryptophan	0.93
Valine	3.51
Phytonutrients	**g/100 g d.w.**
cis beta-carotene	0.073
Chlorophyll-a	1
Phycocyanin	12
trans beta-carotene	0.26
Minerals	**mg/100 g d.w.**
Calcium	700
Copper	1.2
Iron	100
Magnesium	400
Manganese	5
Phosphorus	800
Potassium	1400
Sodium	900
Zinc	3

Depending on how it is grown, it can contain up to 60% of dry weight (dw) proteins consisting of essential amino acids such as leucine, isoleucine, valine, tryptophan, methionine, phenylalanine, lysine, and threonine [9–12]. The protein content is of high quality, with a biological value of 75% and a high digestibility (83–90%) because *Spirulina* cells display a fragile and easily digestible murein envelope instead of cellulose walls [13–15]. Approximately 60% of the total protein content consists of phycobilisomes, composed of C-phycocyanin (C-PC) and allophycocyanin (A-PC), relevant antioxidant, anti-inflammatory, antitumor, and immunostimulant compounds, as well as ingredients for the cosmetic, food, pharmaceutical, and nutraceutical industries [16–18]. About 7–16% of the dry weight consists of lipids [19,20]. These include monounsaturated fatty acids such as oleic acid, saturated fatty acids (SFA) such as the more abundant palmitic, arachidic, and myristic acids, and the omega-6 polyunsaturated fatty acids (ω-6 PUFAs) arachidonic (AA) and γ linolenic acid (GLA), a potent immunoprotective and precursor to prostaglandins and leukotrienes [19,21–23]. Fatty acids of Spirulina also include stearidonic acid (SDA), eicosapentaenoic acid (EPA), and docosahexaenoic acid (DHA) [24]. Spirulina contains different lipophilic pigments which have different bioactive properties. Among these, the most abundant are chlorophyll-a (9–12% of the lipid fractions) and β-carotene. Xanthophylls, echinenone, myxoxanthophyll, zeaxanthin, canthaxanthin, β-cryptoxanthin, and oscillaxanthina are other terpenoids present in Spirulina [25]. Regarding vitamins, Spirulina contains provitamin A (β-carotene), vitamin B1 (thiamine), B2 (riboflavin), B3 (nicotinamide), B6 (pyridoxine), B9 (folic acid), B12 (cyanocobalamin), vitamin C, vitamin D, and vitamin E (tocopherol). Spirulina is also a source of several minerals, including iron, zinc, potassium, copper, manganese, magnesium, phosphorus, and calcium. [8,19,23,26]. It has been demonstrated that, depending on how it is grown, Spirulina carbohydrates, including glucose, fructose, sucrose, glycerol, sorbitol, and mannitol, can increase from 15 to 20% dw [26]. In addition, myo-inositol, a carbohydrate source of organic phosphorus, nitrogen, calcium spirulan (Ca-SP), and the polysaccharide immunomodulator called Immulina, can be found in Spirulina biomass [8,27]. Moreover, Spirulina species can adapt to environmental changes and grow very rapidly depending on the availability of nutrients and climatic factors [28,29]. Through photosynthesis, these strains can fix N_2 and CO_2 while producing O_2 and organic compounds such as polysaccharides, lipids, carotenoids, proteins, and vitamins, which play a key role in the food, cosmetics, and pharmaceutical industries [30].

Spirulina has been known since ancient times as a superfood. Aztecs collected it from the alkaline Lake Texcoco, Mexico, as an integral part of their diet [31,32]. The safety of Spirulina has been demonstrated by many toxicological studies both in chronic or acute administration in the liver, kidneys, reproductive system, and human physiology [33–37]. Currently, it is listed by the US Food and Drug Administration (FDA) in the category "generally recognized as safe" (GRAS) [38]. Spirulina has been declared the world's first superfood, with a complete nutritional profile [39] and the best "food of the future" for its eco-sustainability. Nowadays, Spirulina is produced industrially under controlled conditions [23]. In fact, the growth and productivity of this microorganism depend on several factors such as nutrient concentration, temperature, the light spectrum, intensity, and pH, which also influence its biochemical composition [40,41]. Furthermore, Spirulina is used in medicine to treat different health conditions thanks to bioactive compounds such as antioxidants, immuno-stimulants, anti-inflammatory, antibacterial, antiviral, antitumor, antiallergic, antidiabetic, including phenolics, phycobiliproteins, and chlorophyll [20,21,42–44]. These and other compounds from Spirulina are believed to prevent or treat a wide range of medical conditions [23,45] (Figure 1).

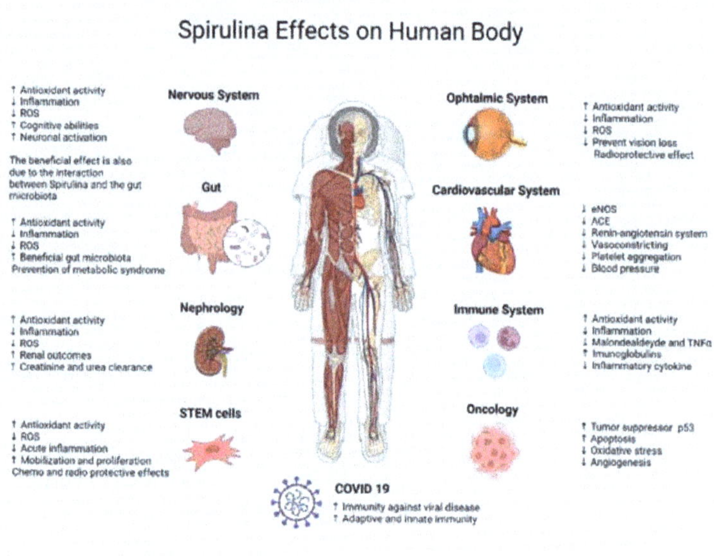

Figure 1. An overview of the effects of *Spirulina* on the various districts of the human body. Created with BioRender.com (accessed on 8 March 2022).

For these reasons, space agencies such as the National Aeronautics and Space Administration (NASA) have promoted it as a food and supplement for astronauts during their space missions since small amounts are able to provide different nutrients and protective effects [46–50]. In fact, during their stay in space, astronauts are exposed to a variety of stress factors that can have negative health consequences [51,52].

The space environment harms different districts of the astronaut's body due to several factors, such as micro-gravity and oxidative stress induced by radiation [53]. On Earth, the cells are constantly subjected to the action of Reactive Oxygen Species (ROS), normally produced as a result of cellular metabolism and exposure to environmental stresses such as X-rays, UV, pollution, xenobiotics, etc. The damage is directly proportional to the ROS produced. This production is mitigated by the activation of cellular detoxification mechanisms that physiologically protect cells. When ROS-dependent stress exceeds the body's physiological defenses, oxidative damage occurs to cells. The extraterrestrial environment determines an increase of ROS, causing damage to cellular lipid membranes, mitochondria, proteins, and DNA [54]. Microgravity and radiation affect several molecular systems and mechanisms such as repair, replication, transcription, and protein expression [55].

In this context, Spirulina, with its antioxidant activity, is able to activate cellular antioxidant enzymes and inhibit lipid peroxidation and DNA damage, thus eliminating free radicals and increasing superoxide dismutase and catalase enzyme activity. Spirulina also possesses immunomodulatory and anti-inflammatory properties that may play a preventive role against potential astronauts' pathologies [17].

In the same context, *Spirulina* has been designated to be placed in Environmental Control and Life Support Systems (ECLSS) in the Micro-Ecological Life Support System Alternative project (MELiSSA), which allows the production of water, oxygen, food, and health supplements by the total recycling of the space station crew's metabolic waste, including the exhaled CO_2 into the cabin [56–61].

To date, the effects of using Spirulina as an adjunct to classical therapies for the most common diseases on Earth are well known in the literature, while the effects of Spirulina on human health in space have not yet been studied [45,62]. In this review, we will discuss the effects of Spirulina on major diseases that can affect humans both on Earth and in space.

The aim of the present review is to summarize the potential benefits of the use of Spirulina, such as food or supplements, to improve the health of astronauts during space missions.

2. Clinical Applications on Earth

2.1. Probiotic Effect

Microbial modulation activities have been reported in vitro, suggesting that the combination of Spirulina and probiotics may represent a new strategy to enhance the growth of the beneficial gut microbiota. Spirulina can inhibit the growth of some Gram-negative (*Escherichia coli, Pseudomonas aeruginosa* and *Proteus vulgaris*) and Gram-positive (*Staphylococcus aureus, Bacillus subtilis* and *Bacillus pumilus*) bacteria [63]. In fact, *Spirulina* produces extracellular metabolites with antibacterial activity. On the contrary, low (minimum inhibitory concentrations, MIC $\geq$ 512 µg/mL) or no inhibitory effect was found against other bacteria (*Pseudomonas aeruginosa, Salmonella typhimurium*, and *Klebsiella pneumoniae*) [64]. Moreover, it has been reported that extracellular products of *Spirulina*, obtained from a culture in the late exponential stage and separated by filtration, significantly promote the in-vitro growth of lactic acid bacteria (*Lactococcus lactis, Streptococcus thermophilus, Lactobacillus casei, Lactobacillus acidophilus*, and *Lactobacillus bulgaricus*) [65]. It is largely established that drastic changes of microbiota composition occur in several gastrointestinal, immunological, and metabolic diseases [66,67]. In many microbiota-related diseases, including Inflammatory Bowel Disease (IBD), it is well known that a strong unbalanced ratio among the genera of potentially protective bacteria and normal anaerobic bacteria is present. In particular, *Bacteroides* sp., *Eubacterium* sp., and *Lactobacillus* sp. are significantly decreased [68]. Spirulina could represent an alternative strategy to symbiotic formulations [63]. The latter seems to be more effective than probiotics alone in the prevention of dysbiosis associated with immune-mediated, inflammatory, and dysmetabolic diseases. Research on the advanced medical applications of Spirulina and derived products in the treatment of infectious diseases caused by Gram-positive organisms are growing. Due to the health benefits promoted, it is gaining more and more interest, especially in of dietary supplements, where it is used as a powder or consumed as capsules or tablets [63]. This evidence suggests that Spirulina may be useful for improving animal and human health by changing the gut microbiota composition and promoting beneficial bacterial growth [68].

2.2. Antihypertensive Effect

Among the main causes of cardiovascular diseases, such as atherosclerosis, cardiac hypertrophy, heart failure, and hypertension, oxidative stress and inflammation are of primary importance. Overproduction of ROS because of oxidative stress has been observed in cardiovascular disease conditions [69]. In addition, evidence indicates that Low Density Lipoprotein (LDL) oxidation is essential for atherogenesis [70,71]. On the other hand, the microenvironment within the atherosclerotic lesion is proinflammatory. In addition to being a disorder of lipid metabolism, atherosclerosis is now recognized as a chronic inflammatory disease [72,73]. Accumulating evidence demonstrates that excessive inflammation within the arterial wall is a risk factor for cardiovascular diseases and can promote atherogenesis. Agents with antioxidant and/or anti-inflammatory activity may prove beneficial in combating cardiovascular diseases. Vascular aging is characterized by an increase in arterial stiffness and remodeling of the arterial wall with a loss of elastic properties. Silicon is an essential trace element highly present in arteries. The effects of nutritional supplementation with silicon-enriched Spirulina (SpSi) on arterial system structure and function in hypertension have been evaluated. Collagen and elastin levels were increased in association with extracellular matrix degradation decrease. The beneficial effects of SpSi supplementation evidenced here may be attributable to Spirulina enrichment and offer interesting opportunities to prevent cardiovascular risks [74]. Spirulina can be effective as a blood pressure-reducing agent by increasing endothelial nitric oxide synthase (eNOS), inhibiting angiotensin-converting enzyme (ACE), suppressing renin-angiotensin system, vasoconstricting metabolites, and platelet aggregation [74]. In the future, the use of Spirulina could have fundamental implications for hypotensive therapies in cardiology [63].

2.3. Antioxidant and Anti-Inflammatory Effects

Spirulina contains several active components, notably phycocyanin and β-carotene, with potent antioxidant and anti-inflammatory activities [64]. As anti-inflammatory activities, phycocyanin inhibits proinflammatory cytokine formation, such as TNFα, suppresses cyclooxygeanase-2 (COX-2) expression and decreases prostaglandin E(2) production [75]. Another component of Spirulina, β-carotene, has been reported to have antioxidant and anti-inflammatory activities. It was found that β-carotene is able to protect cells against singlet oxygen-mediated lipid peroxidation [76]. ROS also contribute to vascular dysfunction and remodeling through oxidative damages in endothelial cells [77]. In a recent in vitro study [78], the antioxidant and anti-inflammatory properties of four different Spirulina preparations were evaluated with a cell-free as well as a cell-based assay. It was found that Spirulina dose-dependently inactivated free superoxide radicals generated during an oxidative burst [78]. Supplements seem to affect innate immunity more effectively than acquired immunity, promoting the activity of natural killer cells. There are wide margins of use to exploit this capacity improves markers of oxidative stress and NK activity in healthy subjects and CD4+ counts in HIV+ patients. In the field of chemotherapy therapies, the effect of Spirulina against the hepatotoxicity of methotrexate has been studied. Administration of a high dose of Spirulina for 21 days prior to methotrexate treatment reduced malondialdehyde and tumor necrosis factor α [79]. In addition, the antioxidant activity of Spirulina has been associated with anti-inflammatory effects (see paragraph 3). Although the antioxidant effect of Spirulina is confirmed by some studies [17,79], the concerted modulation of antioxidant and inflammatory responses, suggested by in vitro and animal studies, requires greater confirmation in humans. Most of the studies on Spirulina are on animal samples [17,79]. From these studies, it was shown that Spirulina is an antioxidant and has a protective effect against damage to DNA. The immunostimulant effect is exerted by Spirulina in stimulating the production of immunoglobulins and the downregulation of inflammatory cytokine-producing genes [80]. Spirulina is generally considered safe for human consumption, supported by its long history of use as a food source and its favorable safety profile in animal studies. However, rare cases of side effects in humans have been reported [62].

2.4. Antiviral Effect

There are no in vivo studies providing strong evidence supporting the possible antiviral properties of Spirulina. However, Spirulina inhibits the in vitro replication of several enveloped viruses, including Herpes simplex type I, human cytomegalovirus, the measles and mumps virus, influenza A, and Human immunodeficiency virus-1 virus (HIV-1) [45,81]. L-asparaginase (L-AsnA) is an enzyme present in the composition of *Spirulina*. L-AsnA enzyme demonstrated good antiviral activity against the Coxsackie B3 (CSB3) virus in a dose-dependent manner. The antiviral mode of action is likely due to its capability of inhibiting attachment, blocking the adsorption and penetration event of the viral replication cycle with 89.24%, 72.78%, and 72.78%, respectively [82]. The ability of Spirulina-based nutraceuticals to boost immunity against viral diseases has already been reported clinically. Spirulina-based nutraceuticals boost adaptive and innate immunity, and bioactive compounds, such as ACE inhibitor peptides, phycobiliproteins, sulfated polysaccharides, and calcium spirulan, can serve as antiviral agents. The presence of these molecules indicates its potential role in resisting infection and COVID-19 disease progression (see Section 6)) [83]. A recent study has shown that the hot water extract of Spirulina inhibited the infection of herpes simplex virus type 2 (HSV-2), pseudorabies virus (PRV), human cytomegalovirus (HCMV), and HSV-1. For adenovirus, the inhibition was less than 20%, and no inhibition was found for the measles virus, subacute sclerosing panencephalitis virus (SSPE), vesicular stomatitis virus (VSV), poliovirus 1, or rotavirus SA-11. The highest antiviral activity was for HSV-2. The antiviral activity was not due to a virucidal effect. Herpesvirus infection was inhibited at the initial events (adsorption and penetration) of the viral cycle [84].

2.5. Antihistamine Effect

Spirulina exhibits anti-inflammatory properties by inhibiting the release of histamine from mast cells. In a recent study [80], individuals with allergic rhinitis were fed daily, either with placebo or Spirulina, for 12 weeks. The study showed that a high dose of Spirulina significantly reduced IL-4 levels, demonstrating the protective effects on allergic rhinitis. Spirulina consumption significantly improved the symptoms and physical findings compared with placebo, including nasal discharge, sneezing, nasal congestion, and itching. Moreover, Spirulina increases the IgA levels on the surface of mucous membranes. Two peptides, LDAVNR (P1) and MMLDF (P2), purified from enzymatic hydrolysate of Spirulina, have been reported to be effective against allergic rhinitis. It was revealed that P1 and P2 exhibited significant inhibition of mast-cell degranulation via decreasing histamine release and intracellular Ca^{2+} elevation. The inhibitory activity of P1 was found due to the blockade of calcium- and microtubule-dependent signaling pathways. Meanwhile, the inhibition of P2 was involved in the suppression of phospholipase Cγ activation and reactive oxygen species production. These findings indicate that peptides P1 and P2 from Spirulina may be promising candidates for antiallergic therapeutics, contributing to the development of bioactive food ingredients to ameliorate allergic diseases [85].

3. Applications in Ophthalmology

3.1. Therapeutic Effect of Spirulina in Corneal Inflammation

It has been established that Spirulina extract reduces alkaline burn-induced inflammation more effectively than amniotic membrane extract (AME). For this reason, it can be used for the therapy of corneal diseases involving neovascularization and inflammation. Other implications of the antioxidant effect of Spirulina can be found in the protection of the neuroepithelium. It should be highlighted that patients who assume Spirulina recorded a lower percentage of thinning of the layer of photoreceptors and death of such cells. In addition, the increase in retinal ROS levels after exposure to light has been reduced by Spirulina supplementation. Among the future implications of Spirulina should be considered the potential use in the form of a nutritional supplement to prevent vision loss related to oxidative damage [86,87].

3.2. Radioprotective Effect of Spirulina on Lacrimal Glands

The radioprotective effect of Spirulina in patients that needed radiation treatments (RAI) for other purposes has been studied. The radioprotective effect of Spirulina on lacrimal glands was evaluated with histopathological and cytopathological analysis. The evaluation was assessed before and after treatment with Spirulina, and it was found to decrease the level of oxidation after RAI [88].

3.3. The Protection of Spirulina on the Visual Function

Light-induced retinal damage is characterized by the accumulation of ROS, leading to oxidative stress and photoreceptor cell death. The light rays acting on the retina stimulate a vastness of receptors to transform the light message into electrical impulses. On the other hand, ROS are also generated. The latter may induce oxidative damage to retinal photoreceptors. In mice, it has been seen that therapy with Spirulina can reduce the production of ROS and protect visual functions from wasting [89]. Spirulina has the potential as a nutrient supplement to prevent vision loss related to oxidative damage in the future [89]. The use of natural antioxidants has emerged as a promising approach for preventing light-induced retinal damage. However, less is known about the possible efficacy of combining natural antioxidants in a multicomponent mixture. Lutein and cyanidin-3-glucoside (C3G) are particularly effective due to their antioxidant and anti-inflammatory activity. These findings suggest the rationale to formulate multicomponent blends, which may optimize the partnering compounds' bioactivity and bioavailability [90]. Age-related macular degeneration (AMD) is a significant visual impairment in older people, and there is no treatment for dry AMD. In-vitro, Spirulina decreased blue light-induced retinal pigment epithelium (RPE) cell death by inhibiting ROS production and inhibited BL-induced inflammation via regulating the NF-κB pathway, inflammatory-related gene expression, and the apoptosis pathway in RPE cells. In vivo, the administration of Spirulina inhibited blue light-induced retinal degeneration by restoring the thicknesses of the whole retina, ONL (outer nuclear layer), INL (inner nuclear layer), and PL (photoreceptor layer) by BL exposure. Therefore, Spirulina could be a potential nutraceutical approach to intercept the pathophysiological processes leading to dry AMD and advancement to wet AMD [91].

4. Applications in Oncology

4.1. Breast Cancer

High delivery efficiency, prolonged drug release, and low systemic toxicity are effective weapons for drug delivery systems to win the battle against metastatic breast cancer. It is demonstrated that Spirulina can be used as a natural carrier to build a drug-loaded system for targeted delivery and fluorescence imaging-guided chemotherapy on breast cancer lung metastases [92]. The *Spirulina's* protein C-phycocyanin (C-PC) inhibited cell proliferation and reduced the colony formation ability of MDA-MB-231 cells. Furthermore, C-PC induced cell cycle G0/G1 arrest by decreasing protein expression levels of Cyclin D1 and CDK-2 and increasing protein expression levels of p21 and p27. In addition, C-PC induced cell apoptotic by activating the cell membrane surface death receptor pathway. Besides, C-PC down-regulated the protein expression levels of cyclooxygenase-2 and further inhibited MDA-MB-231 cells migration [93]. Additionally, a study tested two nutritional supplements, namely gamma-tocotrienol (γT3) and Spirulina, for their immune-enhancing and anticancer effects in a syngeneic mouse model of breast cancer. It has been assessed that combined γT3 + Spirulina treatment did not show any synergistic anticancer effects in this study model [94]. Also, CD4, CD8, and CD56 staining were performed to investigate the effect on the immune cells' recruitment to the tumors by immunohistochemistry. This treatment combination could significantly inhibit 4T1 breast tumors' growth, decreasing the tumors' volume compared to the control and the metastatic burden [92].

4.2. Hepatocarcinoma

Most cases of hepatocellular carcinoma (HCC) are diagnosed in the advanced stages of the disease, making it the second leading cause of cancer mortality worldwide. For advanced patients, chemotherapy drugs are the best treatment option; however, their adverse effects and high cost are still the main obstacles to effective treatment. Spirulina is a rich source of nutritional and bioactive elements and potential pharmaceuticals and has an anti-proliferative effect against several tumor cell lines. It also has a prophylactic effect against the early stages of certain cancer models, including HCC. Spirulina inhibited structural and functional alterations of HCC, manifested by improving the survival rate, significantly decreasing the tumor marker AFP and the count and size of liver nodules, as well as reducing HCC [95]. This was accompanied by increased endogenous antioxidant capacity, apoptosis (Bax), and tumor suppressor protein (p53), as well as suppression of tissue levels of lipid peroxidation marker (MDA) and neoangiogenesis marker (VEGF). Spirulina has an anti-carcinogenic effect against advanced HCC exerted by activating the tumor suppressor protein p53 and apoptosis and suppressing oxidative stress and angiogenesis [95]. The L-AsnA enzyme, contained in Spirulina, in a recent study, showed an antiproliferation effect against lung cancer A549, hepatocellular carcinoma Hep-G2, and prostate cancer PC3 human cancer cell lines. Considering the antiviral and antiproliferative activity of L-AsnA against different human cell lines, it would be desirable to investigate its effects further [81].

4.3. Lung Cancer

The anticancer potential of a hot water extract of a commercial Spirulina against the human non-small-cell lung carcinoma A549 cell line has been evaluated in the literature [96]. Spirulina significantly reduced cancer cell viability and proliferation, which was accompanied by cell cycle inhibition in the G_1 phase, induction of apoptosis, and prominent morphological changes. Moreover, it has been detected that there is no cytotoxic effect of the tested Spirulina on normal skin fibroblasts. The evidence is the anticancer activity of the Spirulina against lung cancer cells and strongly supports the knowledge of the chemo-preventive properties of Spirulina [96].

5. Applications in Central and Peripheral Nervous System

It has been shown by several scientific studies that cyanobacteria, especially Spirulina, are rich in micro and macronutrients important for brain health, such as B vitamins, particularly vitamin B12, amino acids, and minerals such as iron, calcium, zinc, magnesium, manganese, and potassium [97–101]. The B vitamins can reach the brain, and in fact, like the other bioactive derivatives of cyanobacteria, they control various neuronal functions both through epigenetic mechanisms that control gene transcription neurotransmitters and through their antioxidant and anti-inflammatory abilities [102]. The beneficial effects of Spirulina on the Central Nervous System (CNS) are also derived from the interaction between the Spirulina phytocomplex and the intestinal microbiota; there is a close relationship between the nervous system and the bowel, the bowel–brain axis [103]. The microbiota can transform Spirulina into antioxidant and micronutrient molecules that are able to cross the blood–brain barrier and express their beneficial effects such as the increase in energy production and the reduction of mental fatigue with the enhancement of learning skills; the synthesis of neurotransmitters (dopamine, serotonin, glycine, glutamate/GABA); an overall improvement in cognitive functions and short- and long-term memory [104]. On the other hand, Spirulina can increase the growth of protective bacteria in the bowel, thus maintaining a good balance and intestinal integrity and reducing systemic inflammatory responses that have a negative impact on the nervous system [105,106]. Alterations of the microbiota have been demonstrated in many neurological disorders, and the use of probiotics and prebiotics such as those in the Spirulina phytocomplex can indirectly prevent the development of brain disorders [107].

In neurodegenerative diseases such as Parkinson's or Alzheimer's and other psychocognitive diseases, Spirulina plays a neuroprotective role thanks to the antioxidant capacity of its derivatives, as demonstrated by numerous in vivo studies on animals [108–112]. In Alzheimer's disease, Spirulina can prevent memory loss by reducing the deposition of β-amyloid in the brain and increasing the activity of glutathione peroxidase and catalase. The immunomodulatory and antioxidant effect of Spirulina has also been demonstrated in humans. Park et al. observed how cholesterol levels were lower in elderly patients who took Spirulina daily, while those of superoxide dismutase, IL-2, and IL-6 increased [113]. During adolescence, Spirulina can reduce stress-related disorders and the induced remodeling of limbic structures, in particular the amygdala, with the consequent lower risk of neuropsychiatric disorders in adulthood [112,113].

Further human studies have highlighted an improvement effect in cognitive abilities and neuronal activation, with an increase in memory and attention [114,115]. Finally, Spirulina is an important nutritional source in cases of severe malnutrition, especially protein malnutrition (PNM), a frequent phenomenon very common in the third world countries which causes detrimental effects on the brain development of malnourished children particularly in the hippocampus. Penton-Rol et al. [116] demonstrated that C-PC, a biliprotein derived from Spirulina with anti-inflammatory, antioxidant, and cytoprotective capacities, had neuroprotective effects on mice affected by experimental autoimmune encephalomyelitis. Such effects are due to the ability of C-PC to reduce the activation and infiltration of lymphocytes and macrophages/microglia activated in case of autoimmune encephalomyelitis. These cells are also responsible for the pathogenesis of multiple sclerosis, as they result in the production of neurotoxic molecules, proinflammatory cytokines, and present self-antigens that cause demyelination [117,118]. Penton-Rol et al. [116] demonstrated, in the animal model, that C-PC can promote significant axonal remyelination, as well as reduce the Amyloid Precursor Protein (APP), a marker protein of Alzheimer's disease. This is an encouraging result for the treatment of this neurodegenerative pathology. The data described are also encouraging for possible future clinical applications of Spirulina in the field of nerve regeneration and/or reconstructive nerve surgery [119–121].

6. Applications in COVID-19 Infection

The outbreak of the 2019 coronavirus disease (COVID-19), caused by the severe acute respiratory syndrome coronavirus 2 that has created enormous trepidation worldwide, has a mortality rate of 0.5% to 1% and is growing incessantly. Prior to the commercialization of the COVID-19 vaccine, several areas were launched in order to find better treatments for this infection. In this context, the ability of cyanobacteria-based nutraceuticals, mainly Spirulina, to increase immunity against viral diseases was studied; already widely reported clinically. Spirulina-based nutraceuticals increase adaptive and innate immunity, and bioactive compounds, such as ACE inhibitor peptides, phycobiliproteins, sulfate polysaccharides, and calcium spirulan, can serve as antiviral agents. The presence of these molecules indicates its potential role in resistance to infection and progression of the COVID-19 disease. Such cyanobacteria-based nutraceuticals could be used as immune boosters to fight the human coronavirus and other viral diseases in association with vaccines [122]. The spike protein of severe acute respiratory syndrome due to COVID-19 uses angiotensin-converting enzyme 2 as the receptor for cell entry, which is highly expressed in the gut and lungs. Nausea and diarrhea are the primary symptoms of COVID-19 even before the development of fever and respiratory symptoms. These two tissues share a relationship influencing inflammatory and immune responses via the gut–lung axis that can be responsive to probiotics through effects on commensal microbial flora. Some probiotics enhance regulatory T-cell activity and reduce pro-inflammatory cytokine production. Some probiotics have shown antiviral protective and therapeutic effects regarding upper respiratory tract disease, lessening the severity and extent of tissue damage from infection and inflammation. Evidence suggests that probiotics with anti-inflammatory or immunomodulatory properties might be predicted to have the most beneficial potential to prevent or alleviate COVID-19 symptoms [123].

7. Interaction with Stem Cells

Several studies have shown that nutraceuticals exert effects on adult stem cells. At the same time, robust evidence supports the therapeutic benefits of Spirulina in clinical settings due to its antioxidant and anti-inflammatory properties [124,125]. In the last decade, in vitro and in vivo studies have reported other effects, such as antihyperlipidemic [126], anticancer [127], anti-neurotoxic [128], and anti-type 1 diabetic [129] properties. Regarding the immune system, a review of the main immunomodulatory and anti-inflammatory properties, C-PC was one of the most abundant phycobiliproteins of Spirulina. The C-PC has been used in biomedical research as a biomarker for its fluorescence properties. It has been shown that this protein increases the release of γ-interferon in peripheral blood mononuclear cells and modulates the production of inflammatory cytokines such as tumor necrosis factor, among others. Furthermore, C-PC has immunomodulatory effects on cytokines that enhance the activation of immune cells, such as IL-6 and IL-1β, and the regulation of about 190 genes involved in immunity [80].

In addition, using Spirulina biomass for in vitro cell cultures has become feasible in biotechnological research due to its well-established nutrient-rich properties, sustainability, and ethically acceptable source. It has been considered a novel alternate supplement for fetal bovine serum (FBS). Recently, *Spirulina* animal cell culture solution showed a better growth-promoting capability than FBS, demonstrated as an effective, low-cost, and eco-friendly substitute to FBS in vitro cultures of H460 cells [130]. In preclinical studies, it was also demonstrated the chemo- and radio-protective effects of Spirulina polysaccharide on the hemopoietic system of mice and dogs [130], and the regression of tumors in animal models [124], while other papers pointed out that Spirulina was able to reduce the deleterious effects of acute inflammatory insult on neural progenitor cells functions [131]. Furthermore, other authors have demonstrated that stem cells cultivated in scaffolds associated with Spirulina biomass adhered more and had greater viability when compared with the scaffold alone [131]. Perhaps most interesting regarding the clinical use is the ability to assist with stem cells mobilization. In 2006, the effects of nutraceuticals compounds on stem cells as a viable alternative to induce the proliferation and mobilization of human bone marrow and human CD34+ and CD133+ cells were explored [132]. Another study evaluated human stem cells in vitro and in vivo of an extract from the edible cyanobacterium *Aphanizomenon flos-aqua* (AFA) [133]. This paper includes a double-blind, randomized crossover study involving 12 healthy subjects that evaluated the effects of consumption on stem cell mobilization and found that it produced a better ability for stem cells to travel to the tissues where they are most needed. The study reported that oral consumption of a supplement comprising a blend of two compounds extracted from AFA triggered an average 25% increase in the number of peripheral blood stem cells within 60 min after consumption. The magnitude of this mobilization on CD34+ cells was smaller than that triggered by Granulocyte-Colony Stimulating Factors with marketing authorization, but its safety allows for continuous use. Therefore, considering the numerous advantages of this novel approach in stem cells mobilization, the Abu Dhabi Stem Cells Center Research Team is developing basic research studies and clinical trials for further clinical application in the mobilization of hematopoietic stem cells to peripheral blood [125,134].

8. Nephrological Involvement

Renal toxicity is a restricting factor that affects the use and the dosing of many drugs and agents with disparate use in human medicine as antibiotics, chemotherapeutics, radiological contrast media, and immunosuppressive agents. Oxidative stress plays a key role in many of them. Spirulina and its antioxidant effects have been studied against several agents with known nephrotoxic potential; here, we briefly report the main results available in the current literature. Renal toxicity of oncological therapy is a major field of research, as it often brings about a reduction in drug dosage and the subsequent under-treatment of patients with renal impairment. A similar statement can be applied to antibiotics such as aminoglycosides, especially in countries where the availability of the most recent drugs is difficult and cost-biased. Spirulina seems to have the potential to significantly improve renal outcomes in these patients. In 2006, Kuhad et al. [134] evaluated renoprotective effects against Cisplatin-induced oxidative stress in the murine model. Six groups of rats were treated with a placebo, cisplatin alone, Spirulina alone, and cisplatin associated with an increasing dose of Spirulina (500, 1000, and 1500 mg/kg, from two days before and continually until three days after cisplatin administration), respectively; both biochemical and histological parameters were evaluated. Results showed a Spirulina dose-dependent attenuation of oxidative stress via a reduction of lipid peroxidation (LPO), leading to improvement in serum creatinine and urea clearance values; histology documented a reduction in the severity of morphological kidney damage. Avdagiće et al. [135] studied the efficacy of Spirulina in the reduction of gentamicin-induced nephrotoxicity in rats, demonstrating a significant reduction in plasma nitric oxide (NO) levels in rats treated with Spirulina and lighter signs of acute tubular necrosis on histology. Similar results have been reached by Hamad et al. [136], confirming beneficial effects on kidney histology and oxidative stress markers (MDA, GSH). Calcineurin inhibitors (CNI) are a class of immunosuppressant drugs currently used in several auto-immune disorders and to avoid rejection in solid organ transplantation. The most frequently used CNI's are cyclosporin A (CsA) and tacrolimus, both suffering from a known and dose-dependent nephrotoxicity; several mechanisms are involved in CNI renal damage, including oxidative stress and nitric oxide-related hemodynamic alterations caused by the activity of the inducible form of nitric oxide synthase (iNOS) in renal tissues. Khan et al. [137] found that, in rat models, the antioxidative effects of Spirulina prevented the rise in plasma creatinine and urea values as well as the severe isometric vacuolization and widening of the interstitium on histology. Furthermore, Spirulina does not interfere with the CsA metabolism. Moreover, Spirulina seems to have vasodilating properties on rat aortic rings, probably related to the cyclooxygenase-dependent product of arachidonic acid and nitric oxide (NO) [138], which could lead to improved renal perfusion; nevertheless, specific studies on effects in renal vasculature and blood flow has not been reported yet. Regarding renal drug-induced toxicity, oxidative stress plays a pivotal role in the progression of chronic kidney disease, one of the leading causes of morbidity worldwide as well as mortality and high medical costs among adults. Memije-Lazaro et al. [139] have assessed the beneficial effects of Spirulina in chronic kidney disease (CKD) rat models, obtaining significant reductions in oxidative stress and hypertension, leading to protection against cardiovascular and renal alterations. This is a brief overview of several potential benefits of Spirulina in different scenarios of kidney damage. All the studies reported were conducted in vivo in animal models (e.g., rats) and require more validation to confirm results on human tissue (both in vivo and in vitro) [81,133,139].

9. Beyond Earth: Space Sustainability

Earth resources depletion, overpopulation, climate crisis, and pandemics call for the identification of new strategies to save the future of humanity, which involve the use of renewable energies and materials, recycling of wastes, water savings, economic degrowth, etc. While these strategies can help in the short-medium term, it is well established that, in the long term, humanity should be capable of travelling and living on other planets to survive [140]. Among the planets where human life might realistically be possible in the midterm, Mars represents the strongest candidate due to its proximity to Earth. In addition, the temperatures are close to those of continental winters (~−14 °C on average on the equator), and the day duration (~25 h), average solar irradiance levels (~20 mol m^{-2} sol^{-1}), and presence of resources such as atmospheric CO_2, water, and regolith, which might be transformed and exploited in-situ to produce useful consumables [141] (Figure 2).

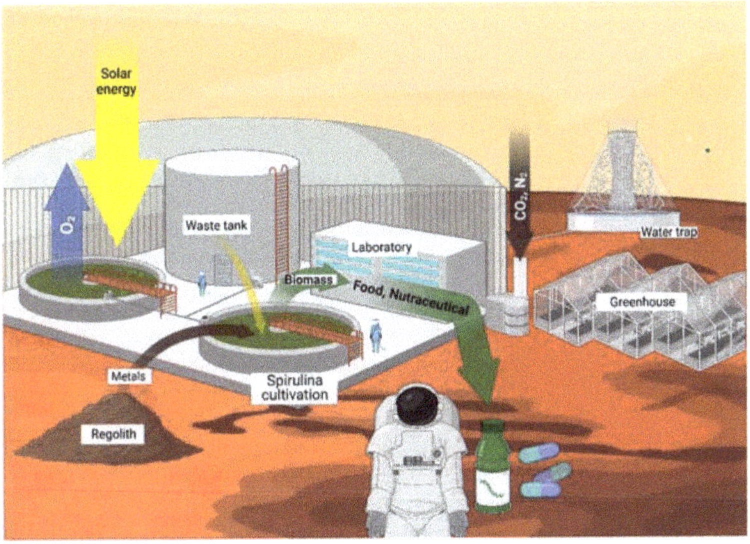

Figure 2. Rendering of ECLSS and ISRU systems for the cultivation and utilization of *Spirulina* on Mars. Created with BioRender.com (accessed on 8 March 2022).

For these reasons, the main space agencies, gathered in the International Space Exploration Coordination Group (ISECG), have listed manned missions to Mars as a common target [60].

The accomplishment of this goal necessitates the identification of new technologies for the sustainment of long-duration manned missions to planets beyond the Lower Earth Orbit (LEO) [142]. So far, several studies have been devoted to developing Environmental Control and Life Support Systems (ECLSSs), typically involving microalgae and cyanobacteria, which permit the production of water, oxygen, and food by totally recycling crew metabolic wastes, including exhausted cabin air [56–58,143]. In this regard, *Spirulina* is recognized as a crucial constituent of the MELiSSA project due to its high photosynthetic efficiency, reaching values close to 6% and protein content of 60–70% wt [142]. Moreover, *Spirulina* can use nitrogen available in the urine water through urease-catalyzed reactions that convert urea into NH_4^+ and bicarbonate, making it particularly useful for recycling astronaut's urine [144]. Moreover, it has been estimated that about ~93% of O_2 and 50–100% of the food needed by astronauts could be produced in the ECLSSs [145]. For this reason, this species is considered crucial in the framework of the new generation ECLSSs by the European Space Agency and has been successfully grown in the International Space Station (ISS) in 2017.

On the other hand, given its use on other planets, the current ECLSSs are not completely self-sustaining and thus require the integration of external inputs of supplies to produce the amount needed to meet the astronauts' needs [54,58,141]. In fact, since interplanetary trips are quite expensive [60], the integrative amounts of consumables cannot be continuously replenished from Earth but must be produced directly on Mars by exploiting the approach represented by the acronym ISRU (In Situ Resource Utilization). Therefore, the ideal strategy for sustaining long-duration manned missions beyond LEO involves the synergistic coupling of ECLSSs and ISRU technologies [56].

While the use of bio-engineering techniques involving cyanobacteria, microalgae, macroalgae, bacteria, and fungi is currently well established in the realization of bioregenerative ECLSS [56,60,142,143], very few works exist in the literature dealing with the possibility of using it to transform available in-situ resources such as regolith and atmospheric CO_2 into useful supplies on Mars. In fact, most ISRU technologies so far proposed consist of physic-chemical methods for oxygen and propellants production from Martian regolith and atmosphere but cannot contribute to the production of food [146,147]. On the other hand, in-situ food production is the main bottleneck for the manned mission since ECLSSs can contribute only to a limited extent to the needs of a crew, estimated to be around 3000 cal sol^{-1} when the latter one is composed of six members [141]. Thus, ISRU technologies for food production on Mars are needed. Albeit several works envision the cultivation of crops, cyanobacteria, algae, and fungi on Mars, only a few papers addressed the transformation of Martian resources into edible biomass under operating conditions that simulate the Martian ones.

In this regard, the few bio-ISRU technologies far proposed relies on rock weathering cyanobacteria which can photosynthetically convert N_2 and CO_2, along with S, P, Fe, and several micronutrients, available in the Mars atmosphere and the regolith, respectively, into newly formed edible biomass by relying on the water and the light available in-situ [60,148]. The use of cyanobacteria and microalgae leads to the further positive effect of producing photosynthetic oxygen, which is crucial for the crew and can integrate the amounts of oxygen produced via physic-chemical methods.

The possibility of using several cyanobacteria exposed to simulated Martian conditions (−27 °C, 0.8 kPa, pure CO_2) has been investigated by Olsson-Francis and Cockell [148]. The experimental results indicated that the strains *Anabaena cylindrica*, *Chroococcidiopsis 029*, *Gloeocapsa OU_20*, *Phormidium OU_10*, and *Leptolyngbya OU_13* were able to survive several days under Mars simulated conditions and using a regolith simulant as growth substrate [148]. Therefore, albeit neglecting the effects of microgravity, this study demonstrated that cyanobacteria could be used for ISRU purposes.

Recently Verseux et al. [60] investigated the diazotrophic growth of *Anabaena* sp. PC7983 under an artificial low-pressure (~101 hPa) atmosphere composed of N_2 (96%) and CO_2 (4%), which the authors envisioned to be produced by using the Martian atmosphere. The results showed that this strain could vigorously grow by taking C and N from such an atmosphere and other micronutrients from a Martian regolith simulant immersed in the growth medium cyanobacteria (BG-11) [60]. Growth experiments under low-pressure atmospheres (80–670 mbar) consisting of pure CO_2 to simulate Martian one were performed by Cycil et al. [144], but they found a scarce capability to survive under such low pressure due to the onset of the carbon starvation phenomena.

In the work by Billi et al. [149], the capability of the autotrophic strain *Chroococcidiopsis* sp. to tolerate perchlorate salts, typically found in Mars regolith, was investigated. The obtained experimental results showed that exposition to Mars-relevant concentrations of Mg or Ca perchlorate did not affect the growth of this strain under simulated Martian conditions, thus demonstrating that *Chroococcidiopsis* is a good candidate for bio-ISRU contexts on Mars.

To the best of our knowledge, aside from the works cited above, only a few less comprehensive papers dealing with the use of Mars environmental resources to sustain the growth of microalgae and cyanobacteria can be found in the literature. Moreover, all these papers had the common lack that the effect of microgravity (~1/3 g on Mars) was neglected during the experiments. Thus, further research activity is needed to verify the possibility of using microalgae and cyanobacteria as a potential food source in the framework of manned missions on Mars that rely on ISRU technologies.

Recently, a novel process has been proposed to grow *Spirulina* on Mars while exploiting available in-situ resources such as atmospheric CO_2 and regolith. The process, which should take place within pressurized and heated domes, also takes advantage of the urine produced by crew members to produce the growth medium wherein it is cultivated. The performances of the proposed process have been evaluated on Earth through an apparatus patented by Cao et al. [150]. It is shown that microgravity was beneficial for *Spirulina* growth because it inhibited the aggregation of cells, thus favoring the transfer of nutrients from the liquid bulk to the cells. At the same time, the high CO_2 content in the gas phase, which simulated the composition of the Martian atmosphere, was able to prevent carbon starvation phenomena, thus prolonging the growth with respect to the case where air CO_2 concentration (0.038% vol) was used. Ultimately, using the process invented by Cao et al. [150], *Spirulina* seemed to grow better under simulated microgravity conditions than on Earth. From the obtained results, it was extrapolated that a culture of 6 m^3 could be enough to meet the 40% of the protein needs of a crew consisting of six members. These encouraging results open the way to using Spirulina as a food source to be produced in situ during the future manned missions on Mars.

10. Spirulina for Astronauts' Healthcare

During their stay in space, astronauts are exposed to a variety of stress factors that can have negative health consequences even after years. With the acronym "RIDGE" (space radiation, isolation and confinement, distance from Earth, gravity fields, and hostile and closed environments), NASA indicates the main dangers of space flights. For instance, with the advent of space stations, which are habitable for long periods, it has been widely demonstrated that exposure to microgravity has several negative effects on human health. These effects may even have a long-term impact, affecting gastrointestinal diseases, including relevant effects on the composition of astronauts' microbiota [151–160], thermoregulation, heart rate, muscle tone, respiratory system, and other physiological aspects of the human body [51]. Some of the key health risks to astronauts in a microgravity environment include musculoskeletal changes such as reduced muscle strength and increased bone fragility, visual impairment, endothelial dysfunction, metabolic changes, behavioral changes due to fatigue or stress, and effects on mental well-being.

Astronauts in space are exposed to high levels of cosmic radiation, which can damage the DNA, mutations, oxidative stress, acute radiation syndromes, central nervous system injury, tissue diseases such as cardiovascular disease, and the composition of the gut microbiota. More specifically, one of the main effects of space flight at the molecular and intracellular level is the onset of oxidative stress, which causes DNA and mitochondrial damage [161,162]. Oxidative stress leads to an imbalance between the production and disposal of highly pro-oxidant radical species such as ROS and nitrogen, which have several cellular targets: cytoplasmic, mitochondrial, and nuclear, including nucleic [163,164]. Closely related to oxidative stress in mitochondrial dysregulation, identified by the reduction in the expression of genes involved in oxidative phosphorylation of the mitochondria. The latter one is, in fact, the main endogenous production site of ROS, hence its increased vulnerability to the onset of oxidative stress [162,165]. In space, there is a high production of oxidant species due to the extreme conditions to which crew members are exposed, leading to an imbalance between oxidant and antioxidant species. The most produced ROS are superoxide anion (O^{2-}), hydrogen peroxide (H_2O_2), and free radicals. An increase in ROS leads, as in radiation damage, to DNA mutations. ROS and cosmic radiation are among the major causes of ageing in space because of the damage they induce in the body. It has been demonstrated in the literature that the use of a Spirulina supplement can increase antioxidant capacity and thus counteract ROS production.

Space environments expose astronauts to a high level of oxidative stress due to extravehicular and intravehicular activities (EVA/IVA) and depend on reduced gravity, ionizing radiation, and variable atmospheric conditions, such as hyper- and hypoxia and psychophysical stress [166–169]. In addition, several studies show how astronauts experience immunodeficiency with impairment of NK cells, redox imbalance, elevated inflammation, elevated granulocytes, inhibited lymphocyte proliferation, and reduced lymphocyte functions, as well as elevated inflammation during low-Earth orbit missions [170–174].

During space missions, oxidative stress and mitochondrial dysfunction caused by exposure to cosmic rays are responsible for renal insults at multiple levels and have negative consequences on the physiology of the kidney [175]. In addition, changes in fluid balance and electrolytes occur due to microgravity conditions, which alter hemodynamics and the organ function [172,176,177]. Electrolyte imbalances increase the chance of developing kidney stones in astronauts [178,179]. Histological analysis of rats exposed to microgravity showed glomerular atrophy, interstitial edema, and degeneration of renal tubules [180].

In terms of the physiology of the human body, space flights induce effects on the cardiovascular, immune, and nervous systems, which also increases the risk of cancer. Astronauts are partially protected against cosmic radiation and from rare solar events, agents which can cause biological changes at the molecular level. [51,181,182]. At the physiological level, space flight primarily produces a displacement of biological fluids to the upper part of the body. This displacement determines an increase in the excretion and activation of compensation mechanisms by the cardiovascular system, which leads to the so-called "deconditioning of the cardiovascular system", characterized by reduced blood volume, variation in cardiac contractility, stiffening of arterial vessels, and the development of insulin resistance [183]. Microgravity reduces the heart's ability to pump blood and provide oxygen to tissues. The heart adapts to changes in blood distribution and pressure in microgravity, and the response of heart rate adaptation to changes in blood pressure is different among astronauts. In addition, arterial stiffening in space causes an increase in blood pressure that can lead to an increased risk of cardiovascular disease. The space environment can also cause the acceleration of vascular aging [184].

The extraterrestrial stressors also cause neurobiological imbalances at several levels due to molecular, biochemical, and cellular perturbations that involve the architecture of neural circuits, the expression of proteins, and synaptic plasticity, also determining neuroinflammation [185,186]. Radiation can cause deficits in cognition, learning, memory, attention, and executive functions, resulting in mood changes such as anxiety and depression [187–190]. Recent studies have linked these disorders to neuronal structural changes, variations in the microcirculation, and neuroinflammation, and it has been shown that radiation leads to an increased presence of dense fibrillar proteins and β-amyloid [191–193]. Radiation exposure reduces the glutamatergic vesicular pool in synaptosomes and reduces the expression of the N-methyle-D-aspartate (NMDA) glutamatergic receptor subunits [194], with possible repercussions on the excitatory/inhibitory state of the nervous system. Regarding the effect of microgravity on the physiology of astronauts, it has been observed that microgravity determines a variation in the architecture of the neuronal cytoskeleton, resulting in imbalances at the level of biochemical and biosynthetic pathways and altering DNA replication, RNA transcription, and protein transport [195–198].

Additionally, neurological deficits in space are closely linked to the immune dysregulation induced by space flight, which leads to a strong exacerbation of inflammatory phenomena [199]. Immune dysregulation consists of a change in the function of certain cells in the immune system (T-cells and natural killer cells) and in the expression of cytokines [159,200,201]. The resulting persistent moderate inflammation is the basis for the reactivation of latent viruses in astronauts' bodies, which increases the risk of cancer [202]. During space flights, astronauts are exposed to an increased risk of cancer due to their increased exposure to hazardous factors such as ionizing radiation, solar particle events, and the Van Allen belt. Among the various damages caused by radiation, the one caused to DNA is quite important, and although the body has repairing mechanisms, they can still create irreversible conditions leading to, among other things, the development of tumors [203–206].

Among the varieties of proteins that Spirulina contains, the study of C-PC proteins brought clinically relevant information. One of the main characteristics of C-PC is the fluorescence capacity which makes Spirulina very adaptable for diagnostic purposes, specifically in the case of tumors, e. g. breast and lung cancer. This protein also promotes the release of γ-interferon in peripheral blood mononuclear cells, thus directly stimulating the immune system. In the same way, it can stimulate the immune system and modulate the production of inflammatory cytokines such as tumor necrosis factor. The extract of Spirulina showed antioxidant activity depending on the concentration. This extract can have interesting effects on the cell cycle of lung cancer. Results showed that it could stop the growth of cancerous cells in phase G by preventing the passage in phase M, thus blocking, in fact, tumor growth [207].

Another problem encountered by astronauts during space flights is a decrease in circulating red blood cells (RBCs), leading to anemia, which is thought to be due to an inhibition of bone marrow function resulting in the decrease of erythropoietin production. It should be noted that the decrease in circulating RBCs is an acute adaptation to the hemodynamic events of cephalic fluid shift due to weightlessness [52,208]. Anemia leads to a reduced ability of red blood cells to transport oxygen and a reduction in the amount of hemoglobin, which is reflected in an impaired ability of the body to exchange gases. Studies have shown that Spirulina, due to its Vitamin B12 and iron content, can somehow counteract anemia [209].

Also, ophthalmic health in astronauts is compromised during long-term space flights, and clinical manifestations include an increased risk of cataracts, optic disc edema, globe flattening, hyperopic shifts, choroidal folds, and cotton wool spots [210]. Causes are not known but some studies suggest that vitamin B_9 (folate) and vitamin B_{12} may be involved in these dysfunctions, and proper nutrient intake may improve them [211,212]. Folate deficiency could have negative consequences during space missions. In fact, it could cause chromosomal damage due to the increased sensitivity to ionizing radiation present in space [213]. In addition, folate has been shown to eliminate a wide range of ROS efficiently [214,215]. Some research supports the idea that changes in the metabolic pathway of vitamin B_{12} could lead to ophthalmic problems such as optic neuropathy and age-related macular degeneration [212].

Space environments can influence the development of acquired immunity and immune responses. In fact, during space missions, most astronauts experience deficiencies in their immune system. Since the development of coronavirus disease 2019 (COVID-19), which is more severe for those immunocompromised, NASA has recently implemented clinical testing and monitoring to protect international space station astronauts [216,217].

Moreover, during space missions, experiments have been carried out on stem cells that demonstrate a change in their ability to differentiate when exposed to microgravity. These experiments performed on the space shuttle have shown that microgravity impairs the differentiation of stem cells [218,219].

Spirulina has already been included in the study programs of space agencies such as NASA [220,221]. Many of the problems faced by astronauts during their missions are exacerbated by the poor quality of food, which is crucial to stay healthy and protect themselves against the effects of microgravity. It is difficult and very expensive to continuously provide fresh food to crew members from Earth. Since on the ISS, it is not possible to keep fresh at room temperature, it must be rehydrated or cooked to be consumed. Lack of fresh food, such as fresh fruit and vegetables, leads to a low intake of macro- and micronutrients essential for health, and the limited amount of food available does not allow for a proper daily caloric intake. In recent years, space agencies and other companies have been investigating the possibility of producing food using the so-called ISRU, which involves microalgae to produce food and water in space to have available fresh food. Among the microalgae species currently known in the literature to be suitable for human intake is *Chlorella vulgaris* which have been declared safe for human consumption by the Food and Drug Administration (FDA) and has been placed in the GRAS category [222]. Spirulina is currently widely used not only in space but also on Earth due to its high protein content with a complete amino acid profile and high content of macro- and micro-nutrients that make it excellent for use as a food source [142,222].

11. Conclusions

Overpopulation and the depletion of the Earth resources is leading to limited consumption of natural food in the human population, thus causing a shortage of macro- and micronutrients, including vitamins and antioxidants. Along with the increase in population, the demand for quality food and the usage of food supplements has increased. People have become more aware of the link between proper nutrition and personal health. There are also many different diseases that humans can suffer, both on Earth and in space. These diseases can be caused by genetic alterations, endogenous and exogenous factors such as ionizing and cosmic radiation, changes in the composition of the microbiome, and the effects of microgravity in space environments. Specifically, proper nutrition has proven to be a valuable aid in the prevention and treatment of the most important diseases of the world. Cyanobacteria such as *Spirulina* play a key role in this contest since it is well known that its sustainable cultivation for the food and nutraceutical industry not only benefits Planet Earth but is also a potential solution to counteract malnutrition. NASA was the first space agency to conduct experiments in space using *Spirulina* as a food source for astronauts. To date, its documented effects are manifold, its high protein content, characterized by a complete amino acid profile including all the essential amino acids, and its excellent supply of vitamins such as the B vitamin complex, Vitamin C and E give it antioxidant properties. Numerous studies conducted on Spirulina have shown that its use can improve health due to its high-value composition of micro and macro nutrients. In addition, its use as a supplement to improve the health of the intestinal microbiota has anti-cancer and anti-inflammatory effects and can help strengthen the immune system. For all these positive effects on the organism, *Spirulina* has been included in the ECLSSs and ISRU technologies of the major space agencies. This brief review also shows that the intake of Spirulina by astronauts may play a crucial beneficial role in enabling a longer and safer stay of humans in space.

Author Contributions: Conceptualization, G.F., A.M., F.B. (Federico Bolognesi), G.G. (Gilberto Gabrielli), F.B. (Federico Biglioli), G.B., G.C. and G.G. (Giuseppe Giannaccare); writing—original draft preparation, G.F., A.M., A.C., M.B. (Massimiliano Borselli), M.B. (Marco Busutti), G.G. (Gilberto Gabrielli), P.S., Y.M.C.-A., R.A.R.-J., A.A.B.-H., Y.V.-C. and G.G. (Giuseppe Giannaccare); writing—review and editing, G.F., A.M., A.C., F.B. (Federico Bolognesi), M.A., M.B. (Massimiliano Borselli), M.B. (Marco Busutti) and G.G. (Giuseppe Giannaccare); visualization, G.F., A.M. and G.G. (Gilberto Gabrielli); supervision, G.G. (Gilberto Gabrielli), A.P., F.B. (Federico Biglioli), G.B. and G.G. (Giuseppe Giannaccare); funding acquisition, G.G. (Gilberto Gabrielli), G.B., G.C. and F.B. (Federico Biglioli). All authors have read and agreed to the published version of the manuscript.

Funding: This research received no external funding.

Institutional Review Board Statement: Not applicable.

Data Availability Statement: Not applicable.

Acknowledgments: The authors wish to thank TOLO Green, a renewable energy company that organized a round table in the Italy Pavilion during the EXPO 2020 Dubai where Experts in the field of Medicine reported the wide range benefits of *Spirulina*. The present manuscript is derived from the proceedings of the above mentioned round table. Two of us, G.F. and A.M, acknowledge the PhD program in Innovation Science and Technologies available at University of Cagliari and Life Science and Technologies available at University of Sassari, respectively.

Conflicts of Interest: The authors declare no conflict of interest.

References

1. Komárek, J.; Kaštovský, J.; Mareš, J.; Johansen, J.R. Taxonomic classification of cyanoprokaryotes (cyanobacterial genera) 2014, using a polyphasic approach Taxonomické hodnocení cyanoprokaryot (cyanobakteriální rody) v roce 2014 podle polyfázického přístupu. *Preslia* **2014**, *86*, 295–335.
2. Guiry, M.D.; Guiry, G.M. AlgaeBase. World-Wide Electronic Publication, National University of Ireland, Galway. 2022. Available online: https://www.algaebase.org/search/genus/detail/?genus_id=43076 (accessed on 7 March 2022).
3. Hoseini, S.M.; Khosravi-Darani, K.; Mozafari, M.R. Nutritional and Medical Applications of Spirulina Microalgae. *Mini-Rev. Med. Chem.* **2013**, *13*, 1231–1237. [CrossRef] [PubMed]
4. Banji, D.; Banji, O.J.F.; Pratusha, N.G.; Annamalai, A.R. Investigation on the role of Spirulina platensis in ameliorating behavioural changes, thyroid dysfunction and oxidative stress in offspring of pregnant rats exposed to fluoride. *Food Chem.* **2013**, *140*, 321–331. [CrossRef] [PubMed]
5. Rasmussen, B.; Fletcher, I.R.; Brocks, J.J.; Kilburn, M.R. Reassessing the first appearance of eukaryotes and cyanobacteria. *Nature* **2008**, *455*, 1101–1104. [CrossRef] [PubMed]

6. Gaysina, L.A.; Saraf, A.; Singh, P. *Cyanobacteria in Diverse Habitats*; Elsevier Inc.: Amsterdam, The Netherlands, 2018; ISBN 9780128146682.
7. Thomas, S.S. *The Role of Parry Organic Spirulina in Health Management*; Valensa International: Eustis, FL, USA, 2010.
8. Pyne, S.K.; Bhattacharjee, P.; Srivastav, P.P. Microalgae (Spirulina Platensis) and Its Bioactive Molecules: Review. *Indian J. Nutr.* **2017**, *4*, 160.
9. Soni, R.A.; Sudhakar, K.; Rana, R.S. Spirulina—From growth to nutritional product: A review. *Trends Food Sci. Technol.* **2017**, *69*, 157–171. [CrossRef]
10. Colla, L.M.; Oliveira Reinehr, C.; Reichert, C.; Costa, J.A.V. Production of biomass and nutraceutical compounds by Spirulina platensis under different temperature and nitrogen regimes. *Bioresour. Technol.* **2007**, *98*, 1489–1493. [CrossRef]
11. Vaz, B.S.; Moreira, J.B.; Morais, M.G.; Costa, J.A.V. Microalgae as a new source of bioactive compounds in food supplement. *Curr. Opin. Food Sci.* **2016**, *7*, 73–77. [CrossRef]
12. De Morais, M.G.; da Silva Vaz, B.; Greque de Morais, E.; Vieira Costa, J.A. Biologically Active Metabolites Synthesized by Microalgae. *BioMed Res. Int.* **2015**, *2015*, 835761. [CrossRef]
13. Ciferri, O. Spirulina, the edible microorganism. *Microbiol. Rev.* **1983**, *47*, 551. [CrossRef]
14. Dillon, J.C.; Phuc, A.P.; Dubacq, J.P. Nutritional Value of the Alga Spirulina. *World Rev. Nutr. Diet.* **1995**, *77*, 32–46. [CrossRef] [PubMed]
15. Dillon, J.C.; Phan, P.A. Spirulina as a source of proteins in human nutrition. *Bulletin de l'Institut Océanographique* **1993**, *12*, 103–107.
16. Braune, S.; Krüger-Genge, A.; Kammerer, S.; Jung, F.; Küpper, J.H. Phycocyanin from arthrospira platensis as potential anti-cancer drug: Review of in vitro and in vivo studies. *Life* **2021**, *11*, 91. [CrossRef] [PubMed]
17. Wu, Q.; Liu, L.; Miron, A.; Klímová, B.; Wan, D.; Kuča, K. The antioxidant, immunomodulatory, and anti-inflammatory activities of Spirulina: An overview. *Arch. Toxicol.* **2016**, *90*, 1817–1840. [CrossRef] [PubMed]
18. Liu, H.; Blankenship, R.E. On the interface of light-harvesting antenna complexes and reaction centers in oxygenic photosynthesis. *Biochim. Biophys. Acta-Bioenerg.* **2019**, *1860*, 148079. [CrossRef]
19. Mata, T.M.; Martins, A.A.; Oliveira, O.; Oliveira, S.; Mendes, A.M.; Caetano, N.S. Lipid content and productivity of arthrospira platensis and chlorella vulgaris under mixotrophic conditions and salinity stress. *Chem. Eng. Trans.* **2016**, *49*, 187–192. [CrossRef]
20. Ramadan, M.F.; Asker, M.M.S.; Ibrahim, Z.K. Functional bioactive compounds and biological activities of spirulina platensis lipids. *Czech J. Food Sci.* **2008**, *26*, 211–222. [CrossRef]
21. Kulshreshtha, A.; Anish, J.Z.; Jarouliya, U.; Bhadauriya, P.; Prasad, G.; Bisen, P. Spirulina in Health Care Management. *Curr. Pharm. Biotechnol.* **2008**, *9*, 400–405. [CrossRef]
22. Mühling, M.; Belay, A.; Whitton, B.A. Variation in fatty acid composition of Arthrospira (Spirulina) strains. *J. Appl. Phycol.* **2005**, *17*, 137–146. [CrossRef]
23. Sotiroudis, T.G.; Sotiroudis, G.T. Health aspects of Spirulina (Arthrospira) microalga food supplement. *J. Serb. Chem. Soc.* **2013**, *78*, 395–405. [CrossRef]
24. Diraman, H.; Koru, E.; Dibeklioglu, H. Fatty acid profile of Spirulina platensis used as a food supplement. *Isr. J. Aquac.-Bamidgeh* **2009**, *61*, 134–142. [CrossRef]
25. Sánchez, M.; Bernal-Castillo, J.; Rozo, C.; Rodríguez, I. SPIRULINA (ARTHROSPIRA): AN EDIBLE MICROORGANISM: A REVIEW. *Univ. Sci.* **2003**, *8*, 7–24. Available online: https://revistas.javeriana.edu.co/index.php/scientarium/article/view/4842 (accessed on 7 March 2022).
26. Matos, J.; Cardoso, C.; Bandarra, N.M.; Afonso, C. Microalgae as healthy ingredients for functional food: A review. *Food Funct.* **2017**, *8*, 2672–2685. [CrossRef] [PubMed]
27. Pugh, N.; Ross, S.A.; Elsohly, H.N.; Elsohly, M.A.; Pasco, D.S. Isolation of three high molecular weight polysaccharide preparations with potent immunostimulatory activity from Spirulina platensis, aphanizomenon flos-aquae and Chlorella pyrenoidosa. *Planta Med.* **2001**, *67*, 737–742. [CrossRef] [PubMed]
28. Sukenik, A.; Hadas, O.; Kaplan, A.; Quesada, A. Invasion of Nostocales (cyanobacteria) to subtropical and temperate freshwater lakes—physiological, regional, and global driving forces. *Front. Microbiol.* **2012**, *3*, 86. [CrossRef] [PubMed]
29. Huisman, J.; Codd, G.A.; Paerl, H.W.; Ibelings, B.W.; Verspagen, J.M.H.; Visser, P.M. Cyanobacterial blooms. *Nat. Rev. Microbiol.* **2018**, *16*, 471–483. [CrossRef] [PubMed]
30. Zahra, Z.; Choo, D.H.; Lee, H.; Parveen, A. Cyanobacteria: Review of current potentials and applications. *Environments* **2020**, *7*, 13. [CrossRef]
31. Abdulqader, G.; Barsanti, L.; Tredici, M.R. Harvest of Arthrospira platensis from Lake Kossorom (Chad) and its household usage among the Kanembu. *J. Appl. Phycol.* **2000**, *12*, 493–498. [CrossRef]
32. Ramírez, L.; Olvera, R. Uso tradicional y actual de Spirulina sp. (Arthrospira sp.). *Interciencia* **2006**, *31*, 657–663.
33. Salazar, M.; Chamorro, G.A.; Salazar, S.; Steele, C.E. Effect of Spirulina maxima consumption on reproduction and peri- and postnatal development in rats. *Food Chem. Toxicol.* **1996**, *34*, 353–359. [CrossRef]
34. Chamorro, G.; Salazar, S.; Steele, C.; Salazar, M. Reproduction and peri- and postnatal evaluation of Spirulina maxima in mice. *Toxicol. Lett.* **1996**, *88*, 67. [CrossRef]
35. Salazar, M.; Martínez, E.; Madrigal, E.; Ruiz, L.E.; Chamorro, G.A. Subchronic toxicity study in mice fed Spirulina maxima. *J. Ethnopharmacol.* **1998**, *62*, 235–241. [CrossRef]

36. Simpore, J.; Kabore, F.; Zongo, F.; Dansou, D.; Bere, A.; Pignatelli, S.; Biondi, D.M.; Ruberto, G.; Musumeci, S. Nutrition rehabilitation of undernourished children utilizing Spiruline and Misola. *Nutr. J.* **2006**, *5*, 3. [CrossRef] [PubMed]
37. Marles, R.J.; Barrett, M.L.; Barnes, J.; Chavez, M.L.; Gardiner, P.; Ko, R.; Mahady, G.B.; Dog, T.L.; Sarma, N.D.; Giancaspro, G.I.; et al. United states pharmacopeia safety evaluation of spirulina. *Crit. Rev. Food Sci. Nutr.* **2011**, *51*, 593–604. [CrossRef]
38. Tarantino LM Agency Response Letter GRASS Notice No. GRN 000127. Available online: https://www.algbiotek.com/spirulina-sertifikalar/spirulina-gras-2012.pdf (accessed on 10 March 2022).
39. Moorhead, K.; Capelli, B.; Cysewski, G.R. *SPIRULINA Nature's Superfood*; Cyanotech Corporation: Kailua-Kona, HI, USA, 2013; Volume 53, ISBN 9788578110796. Available online: https://www.terapiaclark.es/Docs/spirulina_book.pdf (accessed on 10 March 2022).
40. Milia, M.; Corrias, F.; Addis, P.; Zitelli, G.C.; Cicchi, B.; Torzillo, G.; Andreotti, V.; Angioni, A. Influence of Different Light Sources on the Biochemical Composition of Arthrospira spp. Grown in Model Systems. *Foods* **2022**, *11*, 399. [CrossRef]
41. Çelekli, A.; Yavuzatmaca, M.; Bozkurt, H. Modeling of biomass production by Spirulina platensis as function of phosphate concentrations and pH regimes. *Bioresour. Technol.* **2009**, *100*, 3625–3629. [CrossRef]
42. Mathur, M. Molécules bioactives de la spiruline: Un complément alimentaire. In *Bioactive Molecules in Food*; Mérillon, J.M., Ramawat, K., Eds.; Série de Référence en Phytochimie; Springer: Cham, Switzerland, 2018; pp. 1621–1642.
43. Hernández Lepe, M.A.; Wall-Medrano, A.; Juárez-Oropeza, M.A.; Ramos-Jiménez, A.; Hernández-Torres, R.P. Spirulina Y Su Efecto Hipolipemiante Y Antioxidante En Humanos: Una Revisión Sistemática [Spirulina and Its Hypolipidemic And Antioxidant Effects In Humans: A Systematic Review] (in Spanish). *Nutr. Hosp.* **2015**, *32*, 494–500.
44. Mysliwa-Kurdziel, B.; Solymosi, K. Phycobilins and Phycobiliproteins Used in Food Industry and Medicine. *Mini-Rev. Med. Chem.* **2017**, *17*, 1173–1193. [CrossRef]
45. Karkos, P.D.; Leong, S.C.; Karkos, C.D.; Sivaji, N.; Assimakopoulos, D.A. Spirulina in clinical practice: Evidence-based human applications. Evidence-based Complement. *Altern. Med.* **2011**, *2011*, 4–7. [CrossRef]
46. Tranquille, N.; Emeis, J.J.; de Chambure, D.; Binot, R.; Tamponnet, C. Spirulina acceptability trials in rats. A study for the "Melissa" life-support system. *Adv. Sp. Res.* **1994**, *14*, 167–170. [CrossRef]
47. Soni, R.A.; Sudhakar, K.; Rana, R.S.; Baredar, P. Food Supplements Formulated with Spirulina. *Algae* **2021**, 201–226. [CrossRef]
48. Jung, F.; Krüger-Genge, A.; Waldeck, P.; Küpper, J.H. Spirulina platensis, a super food? *J. Cell. Biotechnol.* **2019**, *5*, 43–54. [CrossRef]
49. Ciani, M.; Lippolis, A.; Fava, F.; Rodolfi, L.; Niccolai, A.; Tredici, M.R. Microbes: Food for the Future. *Foods* **2021**, *10*, 971. [CrossRef] [PubMed]
50. Ramírez-Rodrigues, M.M.; Estrada-Beristain, C.; Metri-Ojeda, J.; Pérez-Alva, A.; Baigts-Allende, D.K. Spirulina platensis protein as sustainable ingredient for nutritional food products development. *Sustainability* **2021**, *13*, 6849. [CrossRef]
51. Afshinnekoo, E.; Scott, R.T.; MacKay, M.J.; Pariset, E.; Cekanaviciute, E.; Barker, R.; Gilroy, S.; Hassane, D.; Smith, S.M.; Zwart, S.R.; et al. Fundamental Biological Features of Spaceflight: Advancing the Field to Enable Deep-Space Exploration. *Cell* **2020**, *183*, 1162–1184. [CrossRef]
52. Garrett-Bakelman, F.E.; Darshi, M.; Green, S.J.; Gur, R.C.; Lin, L.; Macias, B.R.; McKenna, M.J.; Meydan, C.; Mishra, T.; Nasrini, J.; et al. The NASA twins study: A multidimensional analysis of a year-long human spaceflight. *Science* **2019**, *364*, eaau8650. [CrossRef]
53. Stein, T.P.; Leskiw, M.J. Oxidant damage during and after spaceflight. *Am. J. Physiol.-Endocrinol. Metab.* **2000**, *278*, 375–382. [CrossRef]
54. Tominaga, H.; Kodama, S.; Matsuda, N.; Suzuki, K.; Watanabe, M. Involvement of reactive oxygen species (ROS) in the induction of genetic instability by radiation. *J. Radiat. Res.* **2004**, *45*, 181–188. [CrossRef]
55. Yatagai, F.; Honma, M.; Dohmae, N.; Ishioka, N. Biological effects of space environmental factors: A possible interaction between space radiation and microgravity. *Life Sci. Sp. Res.* **2019**, *20*, 113–123. [CrossRef]
56. Concas, A.; Corrias, G.; Orrù, R.; Licheri, R.; Pisu, M.; Cao, G. Remarks on ISRU and ISFR technologies for manned missions on moon and mars. *Eurasian Chem. J.* **2012**, *14*, 243–248. [CrossRef]
57. Poughon, L.; Laroche, C.; Creuly, C.; Dussap, C.G.; Paille, C.; Lasseur, C.; Monsieurs, P.; Heylen, W.; Coninx, I.; Mastroleo, F.; et al. Limnospira indica PCC8005 growth in photobioreactor: Model and simulation of the ISS and ground experiments. *Life Sci. Sp. Res.* **2020**, *25*, 53–65. [CrossRef] [PubMed]
58. Alemany, L.; Peiro, E.; Arnau, C.; Garcia, D.; Poughon, L.; Cornet, J.F.; Dussap, C.G.; Gerbi, O.; Lamaze, B.; Lasseur, C.; et al. Continuous controlled long-term operation and modeling of a closed loop connecting an air-lift photobioreactor and an animal compartment for the development of a life support system. *Biochem. Eng. J.* **2019**, *151*, 107323. [CrossRef]
59. Verseux, C.; Baqué, M.; Lehto, K.; De Vera, J.P.P.; Rothschild, L.J.; Billi, D. Sustainable life support on Mars—The potential roles of cyanobacteria. *Int. J. Astrobiol.* **2016**, *15*, 65–92. [CrossRef]
60. Verseux, C.; Heinicke, C.; Ramalho, T.P.; Determann, J.; Duckhorn, M.; Smagin, M.; Avila, M. A Low-Pressure, N_2/CO_2 Atmosphere Is Suitable for Cyanobacterium-Based Life-Support Systems on Mars. *Front. Microbiol.* **2021**, *12*, 611798. [CrossRef]
61. Hendrickx, L.; De Wever, H.; Hermans, V.; Mastroleo, F.; Morin, N.; Wilmotte, A.; Janssen, P.; Mergeay, M. Microbial ecology of the closed artificial ecosystem MELiSSA (Micro-Ecological Life Support System Alternative): Reinventing and compartmentalizing the Earth's food and oxygen regeneration system for long-haul space exploration missions. *Res. Microbiol.* **2006**, *157*, 77–86. [CrossRef]

62. Deng, R.; Chow, T.-J. Hypolipidemic, Antioxidant, and Antiinflammatory Activities of Microalgae Spirulina. *Cardiovasc. Ther.* **2010**, *28*, e33–e45. [CrossRef]
63. Yousefi, R.; Saidpour, A.; Mottaghi, A. The effects of Spirulina supplementation on metabolic syndrome components, its liver manifestation and related inflammatory markers: A systematic review. *Complement. Ther. Med.* **2019**, *42*, 137–144. [CrossRef]
64. Kaushik, P.; Chauhan, A. In vitro antibacterial activity of laboratory grown culture of Spirulina platensis. *Indian J. Microbiol.* **2008**, *48*, 348–352. [CrossRef]
65. Parada, J.L.; Zulpa de Caire, G.; Zaccaro de Mulé, M.C.; Storni de Cano, M.M. Lactic acid bacteria growth promoters from Spirulina platensis. *Int. J. Food Microbiol.* **1998**, *45*, 225–228. [CrossRef] [PubMed]
66. Davis, C.D. The Gut Microbiome and Its Role in Obesity. *Nutr. Today* **2016**, *51*, 167–174. [CrossRef]
67. Schroeder Bjoern, O.; Bäckhed, F. Signals from the gut microbiota to distant organs in physiology and disease. *Nat. Med.* **2016**, *22*, 1079–1089. [CrossRef] [PubMed]
68. Wang, Z.K.; Yang, Y.S.; Chen, Y.; Yuan, J.; Sun, G.; Peng, L.H. Intestinal microbiota pathogenesis and fecal microbiota transplantation for inflammatory bowel disease. *World J. Gastroenterol.* **2014**, *20*, 14805–14820. [CrossRef] [PubMed]
69. Dhalla, N.S.; Temsah, R.M.; Netticadan, T. Role of oxidative stress in cardiovascular diseases. *J. Hypertens.* **2000**, *18*, 655–673. [CrossRef] [PubMed]
70. Steinberg, D. Low density lipoprotein oxidation and its pathobiological significance. *J. Biol. Chem.* **1997**, *272*, 20963–20966. [CrossRef]
71. Chisolm, G.M.; Steinberg, D. The oxidative modification hypothesis of atherogenesis: An overview. *Free. Radic. Biol. Med.* **2000**, *28*, 1815–1826. [CrossRef]
72. Lusis, A.J. Atherosclerosis. *Nature* **2000**, *407*, 233–241. [CrossRef]
73. Glass, C.K.; Witztum, J.L. Atherosclerosis. the road ahead. *Cell* **2001**, *104*, 503–516. [CrossRef]
74. Arthur-Ataam, J.; Bideaux, P.; Charrabi, A.; Sicard, P.; Fromy, B.; Liu, K.; Eddahibi, S.; Pasqualin, C.; Jouy, N.; Richard, S.; et al. Dietary supplementation with silicon-enriched spirulina improves arterial remodeling and function in hypertensive rats. *Nutrients* **2019**, *11*, 2574. [CrossRef]
75. Romay, C.; Delgado, R.; Remirez, D.; González, R.; Rojas, A. Effects of phycocyanin extract on tumor necrosis factor-alpha and nitrite levels in serum of mice treated with endotoxin. *Arzneimittel-Forschung* **2001**, *51*, 733–736. [CrossRef]
76. Schafer, F.Q.; Wang, H.P.; Kelley, E.E.; Cueno, K.L.; Martin, S.M.; Buettner, G.R. Comparing beta-carotene, vitamin E and nitric oxide as membrane antioxidants. *Biol. Chem.* **2002**, *383*, 671–681. [CrossRef]
77. Yung, L.M.; Leung, F.P.; Yao, X.; Chen, Z.Y.; Huang, Y. Reactive oxygen species in vascular wall. *Cardiovasc. Hematol. Disord. Drug Targets* **2006**, *6*, 1–19. [CrossRef] [PubMed]
78. Dartsch, P.C. Antioxidant potential of selected Spirulina platensis preparations. *Phytother. Res. PTR* **2008**, *22*, 627–633. [CrossRef]
79. Grosshagauer, S.; Kraemer, K.; Somoza, V. The True Value of Spirulina. *J. Agric. Food Chem.* **2020**, *68*, 4109–4115. [CrossRef] [PubMed]
80. Mao, T.K.; van de Water, J.; Gershwin, M.E. Effects of a Spirulina-based dietary supplement on cytokine production from allergic rhinitis patients. *J. Med. Food* **2005**, *8*, 27–30. [CrossRef] [PubMed]
81. Jeong, Y.; Choi, W.Y.; Park, A.; Lee, Y.J.; Lee, Y.; Park, G.H.; Lee, S.J.; Lee, W.K.; Ryu, Y.K.; Kang, D.H. Marine cyanobacterium Spirulina maxima as an alternate to the animal cell culture medium supplement. *Sci. Rep.* **2021**, *11*, 4906. [CrossRef]
82. Abd El-Baky, H.H.; El-Baroty, G.S. *Spirulina maxima* L-asparaginase: Immobilization, Antiviral and Antiproliferation Activities. *Recent Pat Biotechnol.* **2020**, *14*, 154–163. [CrossRef] [PubMed]
83. Ratha, S.K.; Renuka, N.; Rawat, I.; Bux, F. Prospective options of algae-derived nutraceuticals as supplements to combat COVID-19 and human coronavirus diseases. *Nutrition* **2021**, *83*, 111089. [CrossRef]
84. Hernández-Corona, A.; Nieves, I.; Meckes, M.; Chamorro, G.; Barron, B.L. Antiviral activity of Spirulina maxima against herpes simplex virus type 2. *Antiviral Res.* **2002**, *56*, 279–285. [CrossRef] [PubMed]
85. Vo, T.S.; Ngo, D.H.; Kang, K.H.; Park, S.J.; Kim, S.K. The role of peptides derived from Spirulina maxima in downregulation of FcεRI-mediated allergic responses. *Mol. Nutr. Food Res.* **2014**, *58*, 2226–2234. [CrossRef]
86. Yang, L.; Wang, Y.; Zhou, Q.; Chen, P.; Wang, Y.; Wang, Y.; Liu, T.; Xie, L. Inhibitory effects of polysaccharide extract from Spirulina platensis on corneal neovascularization. *Mol. Vis.* **2009**, *15*, 1951–1961.
87. Yang, L.L.; Zhou, Q.J.; Wang, Y.; Gao, Y.; Wang, Y.Q. Comparison of the therapeutic effects of extracts from Spirulina platensis and amnion membrane on inflammation-associated corneal neovascularization. *Int. J. Ophthalmol.* **2012**, *5*, 32–37. [CrossRef] [PubMed]
88. Atilgan, H.I.; Akbulut, A.; Yazihan, N.; Yumusak, N.; Singar, E.; Koca, G.; Korkmaz, M. The Cytokines-Directed Roles of Spirulina for Radioprotection of Lacrimal Gland. *Ocul. Immunol. Inflamm.* **2022**, 1–6. [CrossRef] [PubMed]
89. Okamoto, T.; Kawashima, H.; Osada, H.; Toda, E.; Homma, K.; Nagai, N.; Imai, Y.; Tsubota, K.; Ozawa, Y. Dietary spirulina supplementation protects visual function from photostress by suppressing retinal neurodegeneration in mice. *Transl. Vis. Sci. Technol.* **2019**, *8*, 20. [CrossRef] [PubMed]
90. Amato, R.; Canovai, A.; Melecchi, A.; Pezzino, S.; Corsaro, R.; Dal Monte, M.; Rusciano, D.; Bagnoli, P.; Cammalleri, M. Dietary Supplementation of Antioxidant Compounds Prevents Light-Induced Retinal Damage in a Rat Model. *Biomedicines* **2021**, *9*, 1177. [CrossRef] [PubMed]

91. Cho, H.M.; Jo, Y.D.; Choung, S.Y. Protective Effects of *Spirulina maxima* against Blue Light-Induced Retinal Damages in A2E-Laden ARPE-19 Cells and Balb/c Mice. *Nutrients* **2022**, *14*, 401. [CrossRef]
92. Kefayat, A.; Ghahremani, F.; Safavi, A.; Hajiaghababa, A.; Moshtaghian, J. Spirulina extract enriched for Braun-type lipoprotein (Immulina®) for inhibition of 4T1 breast tumors' growth and metastasis. *Phyther. Res.* **2020**, *34*, 368–378. [CrossRef]
93. Jiang, L.; Wang, Y.; Liu, G.; Liu, H.; Zhu, F.; Ji, H.; Li, B. C-Phycocyanin exerts anti-cancer effects via the MAPK signaling pathway in MDA-MB-231 cells. *Cancer Cell Int.* **2018**, *18*, 12. [CrossRef]
94. Subramaiam, H.; Chu, W.L.; Radhakrishnan, A.K.; Chakravarthi, S.; Selvaduray, K.R.; Kok, Y.Y. Evaluating anticancer and immunomodulatory effects of spirulina (Arthrospira) platensis and gamma-tocotrienol supplementation in a syngeneic mouse model of breast cancer. *Nutrients* **2021**, *13*, 2320. [CrossRef]
95. Mahmoud, Y.I.; Shehata, A.M.M.; Fares, N.H.; Mahmoud, A.A. Spirulina inhibits hepatocellular carcinoma through activating p53 and apoptosis and suppressing oxidative stress and angiogenesis. *Life Sci.* **2021**, *265*, 118827. [CrossRef]
96. Czerwonka, A.; Kaławaj, K.; Sławińska-Brych, A.; Lemieszek, M.K.; Bartnik, M.; Wojtanowski, K.K.; Zdzisińska, B.; Rzeski, W. Anticancer effect of the water extract of a commercial Spirulina (Arthrospira platensis) product on the human lung cancer A549 cell line. *Biomed. Pharmacother.* **2018**, *106*, 292–302. [CrossRef]
97. Kennedy, D.O. B vitamins and the brain: Mechanisms, dose and efficacy—A review. *Nutrients* **2016**, *8*, 68. [CrossRef] [PubMed]
98. Rosa, G.P.; Tavares, W.R.; Sousa, P.M.C.; Pagès, A.K.; Seca, A.M.L.; Pinto, D.C.G.A. Seaweed secondary metabolites with beneficial health effects: An overview of successes in in vivo studies and clinical trials. *Mar. Drugs* **2020**, *18*, 8. [CrossRef] [PubMed]
99. Pereira, L.; Valado, A. The Seaweed Diet in Prevention and Treatment of the. *Mar. Drugs* **2021**, *19*, 128. [CrossRef] [PubMed]
100. Miranda, M.S.; Cintra, R.G.; Barros, S.B.M.; Mancini-Filho, J. Antioxidant activity of the microalga Spirulina maxima. *Braz. J. Med. Biol. Res.* **1998**, *31*, 1075–1079. [CrossRef] [PubMed]
101. Cornish, M.L.; Critchley, A.T.; Mouritsen, O.G. Consumption of seaweeds and the human brain. *J. Appl. Phycol.* **2017**, *29*, 2377–2398. [CrossRef]
102. Makkar, R.; Behl, T.; Bungau, S.; Zengin, G.; Mehta, V.; Kumar, A.; Uddin, M.S.; Ashraf, G.M.; Abdel-Daim, M.M.; Arora, S.; et al. Nutraceuticals in neurological disorders. *Int. J. Mol. Sci.* **2020**, *21*, 4424. [CrossRef] [PubMed]
103. Morais, L.H.; Schreiber, H.L.; Mazmanian, S.K. The gut microbiota–brain axis in behaviour and brain disorders. *Nat. Rev. Microbiol.* **2021**, *19*, 241–255. [CrossRef]
104. Sorrenti, V.; Castagna, D.A.; Fortinguerra, S.; Buriani, A.; Scapagnini, G.; Willcox, D.C. Spirulina microalgae and brain health: A scoping review of experimental and clinical evidence. *Mar. Drugs* **2021**, *19*, 293. [CrossRef]
105. Hu, J.; Li, Y.; Pakpour, S.; Wang, S.; Pan, Z.; Liu, J.; Wei, Q.; She, J.; Cang, H.; Zhang, R.X. Dose Effects of Orally Administered Spirulina Suspension on Colonic Microbiota in Healthy Mice. *Front. Cell. Infect. Microbiol.* **2019**, *9*, 243. [CrossRef]
106. Neyrinck, A.M.; Taminiau, B.; Walgrave, H.; Daube, G.; Cani, P.D.; Bindels, L.B.; Delzenne, N.M. Spirulina protects against hepatic inflammation in aging: An effect related to the modulation of the gut microbiota? *Nutrients* **2017**, *9*, 633. [CrossRef]
107. Westfall, S.; Lomis, N.; Kahouli, I.; Dia, S.Y.; Singh, S.P.; Prakash, S. Microbiome, probiotics and neurodegenerative diseases: Deciphering the gut brain axis. *Cell. Mol. Life Sci.* **2017**, *74*, 3769–3787. [CrossRef] [PubMed]
108. Hwang, J.H.; Lee, I.-T.; Jeng, K.C.; Wang, M.F.; Hou, R.C.W.; Wu, S.M.; Chan, Y.C. Spirulina prevents memory dysfunction, reduces oxidative stress damage and augments antioxidant activity in senescence-accelerated mice. *J. Nutr. Sci. Vitaminol.* **2011**, *57*, 186–191. [CrossRef] [PubMed]
109. Pabon, M.M.; Jernberg, J.N.; Morganti, J.; Contreras, J.; Hudson, C.E.; Klein, R.L.; Bickford, P.C. A Spirulina-Enhanced Diet Provides Neuroprotection in an α-Synuclein Model of Parkinson's Disease. *PLoS ONE* **2012**, *7*, e90972. [CrossRef]
110. Lima, F.A.V.; Joventino, I.P.; Joventino, F.P.; de Almeida, A.C.; Neves, K.R.T.; do Carmo, M.R.; Leal, L.K.A.M.; de Andrade, G.M.; de Barros Viana, G.S. Neuroprotective Activities of Spirulina platensis in the 6-OHDA Model of Parkinson's Disease Are Related to Its Anti-Inflammatory Effects. *Neurochem. Res.* **2017**, *42*, 3390–3400. [CrossRef]
111. Koh, E.J.; Kim, K.J.; Song, J.H.; Choi, J.; Lee, H.Y.; Kang, D.H.; Heo, H.J.; Lee, B.Y. Spirulina maxima extract ameliorates learning and memory impairments via inhibiting GSK-3β phosphorylation induced by intracerebroventricular injection of amyloid-β 1-42 in mice. *Int. J. Mol. Sci.* **2017**, *18*, 2401. [CrossRef]
112. Moradi-Kor, N.; Ghanbari, A.; Rashidipour, H.; Bandegi, A.R.; Yousefi, B.; Barati, M.; Kokhaei, P.; Rashidy-Pour, A. Therapeutic effects of spirulina platensis against adolescent stress-induced oxidative stress, brain-derived neurotrophic factor alterations and morphological remodeling in the amygdala of adult female rats. *J. Exp. Pharmacol.* **2020**, *12*, 75–85. [CrossRef]
113. Park, H.J.; Lee, Y.J.; Ryu, H.K.; Kim, M.H.; Chung, H.W.; Kim, W.Y. A randomized double-blind, placebo-controlled study to establish the effects of spirulina in elderly Koreans. *Ann. Nutr. Metab.* **2008**, *52*, 322–328. [CrossRef]
114. Reid, S.N.S.; Ryu, J.K.; Kim, Y.; Jeon, B.H. The Effects of Fermented Laminaria japonica on Short-Term Working Memory and Physical Fitness in the Elderly. Evidence-based Complement. *Altern. Med.* **2018**, *2018*, 8109621. [CrossRef]
115. Haskell-Ramsay, C.F.; Jackson, P.A.; Dodd, F.L.; Forster, J.S.; Bérubé, J.; Levinton, C.; Kennedy, D.O. Acute post-prandial cognitive effects of Brown seaweed extract in humans. *Nutrients* **2018**, *10*, 85. [CrossRef]
116. Pentón-Rol, G.; Lagumersindez-Denis, N.; Muzio, L.; Bergami, A.; Furlan, R.; Fernández-Massó, J.R.; Nazabal-Galvez, M.; Llópiz-Arzuaga, A.; Herrera-Rolo, T.; Veliz-Rodriguez, T.; et al. Comparative Neuroregenerative Effects of C-Phycocyanin and IFN-Beta in a Model of Multiple Sclerosis in Mice. *J. Neuroimmune Pharmacol.* **2016**, *11*, 153–167. [CrossRef]

117. Heppner, F.L.; Greter, M.; Marino, D.; Falsig, J.; Raivich, G.; Hövelmeyer, N.; Waisman, A.; Rülicke, T.; Prinz, M.; Priller, J.; et al. Experimental autoimmune encephalomyelitis repressed by microglial paralysis. *Nat. Med.* **2005**, *11*, 146–152. [CrossRef] [PubMed]
118. Takeuchi, O.; Akira, S. Pattern Recognition Receptors and Inflammation. *Cell* **2010**, *140*, 805–820. [CrossRef] [PubMed]
119. Monk, C.; Georgieff, M.K.; Osterholm, E.A. Research Review: Maternal prenatal distress and poor nutrition—Mutually influencing risk factors affecting infant neurocognitive development. *J. Child Psychol. Psychiatry Allied Discip.* **2013**, *54*, 115–130. [CrossRef]
120. Sinha, S.; Patro, N.; Tiwari, P.K.; Patro, I.K. Maternal Spirulina supplementation during pregnancy and lactation partially prevents oxidative stress, glial activation and neuronal damage in protein malnourished F1 progeny. *Neurochem. Int.* **2020**, *141*, 104877. [CrossRef] [PubMed]
121. Patro, N.; Sharma, A.; Kariaya, K.; Patro, I. Spirulina platensis protects neurons via suppression of glial activation and peripheral sensitization leading to restoration of motor function in collagen-induced arthritic rats. *Indian J. Exp. Biol.* **2011**, *49*, 739–748. [PubMed]
122. Tzachor, A.; Rozen, O.; Khatib, S.; Jensen, S.; Avni, D. Photosynthetically Controlled Spirulina, but Not Solar Spirulina, Inhibits TNF-α Secretion: Potential Implications for COVID-19-Related Cytokine Storm Therapy. *Mar. Biotechnol.* **2021**, *23*, 149–155. [CrossRef]
123. Baindara, P.; Chakraborty, R.; Holliday, Z.M.; Mandal, S.M.; Schrum, A.G. Oral probiotics in coronavirus disease 2019: Connecting the gut-lung axis to viral pathogenesis, inflammation, secondary infection and clinical trials. *New Microbes New Infect.* **2021**, *40*, 100837. [CrossRef]
124. Arab, S.; Ghasemi, S.; Ghanbari, A.; Bahraminasab, M.; Satari, A.; Mousavi, M.; Dehcheshme, H.G.; Asgharzade, S. Chemo-preventive effect of spirulina microalgae on an animal model of glioblastoma via down-regulation of PI3K/AKT/mTOR and up-regulation of miR-34a/miR-125B expression. *Phyther. Res.* **2021**, *35*, 6452–6461. [CrossRef]
125. Díaz Domínguez, G.; Marsán Suárez, V.; del Valle Pérez, L.O. Principales propiedades inmunomoduladoras y antinflamatorias de la ficobiliproteína C-ficocianina. Rev. Cuba. *Hematol. Inmunol. Hemoter.* **2016**, *32*, 447–454.
126. Ponce-Canchihuamán, J.C.; Pérez-Méndez, O.; Hernández-Müoz, R.; Torres-Durán, P.V.; Juárez-Oropeza, M.A. Protective effects of Spirulina maxima on hyperlipidemia and oxidative-stress induced by lead acetate in the liver and kidney. *Lipids Health Dis.* **2010**, *9*, 35. [CrossRef]
127. Oh, S.H.; Ahn, J.; Kang, D.H.; Lee, H.Y. The Effect of Ultrasonificated Extracts of Spirulina maxima on the Anticancer Activity. *Mar. Biotechnol.* **2011**, *13*, 205–214. [CrossRef] [PubMed]
128. Koh, E.J.; Seo, Y.J.; Choi, J.; Lee, H.Y.; Kang, D.H.; Kim, K.J.; Lee, B.Y. Spirulina maxima extract prevents neurotoxicity via promoting activation of BDNF/CREB signaling pathways in neuronal cells and mice. *Molecules* **2017**, *22*, 1363. [CrossRef] [PubMed]
129. Lee, J.; Park, A.; Kim, M.J.; Lim, H.J.; Rha, Y.A.; Kang, H.G. Spirulina extract enhanced a protective effect in type 1 diabetes by anti-apoptosis and anti-ROS production. *Nutrients* **2017**, *9*, 1363. [CrossRef] [PubMed]
130. Bachstetter, A.D.; Jernberg, J.; Schlunk, A.; Vila, J.L.; Hudson, C.; Cole, M.J.; Shytle, R.D.; Tan, J.; Sanberg, P.R.; Sanberg, C.D.; et al. Spirulina promotes stem cell genesis and protects against LPS induced declines in neural stem cell proliferation. *PLoS ONE* **2010**, *5*, e10496. [CrossRef] [PubMed]
131. Steffens, D.; Lersch, M.; Rosa, A.; Scher, C.; Crestani, T.; Morais, M.G.; Costa, J.A.V.; Pranke, P. A new biomaterial of nanofibers with the microalga spirulina as scaffolds to cultivate with stem cells for use in tissue engineering. *J. Biomed. Nanotechnol.* **2013**, *9*, 710–718. [CrossRef]
132. Bickford, P.C.; Tan, J.; Douglas Shytle, R.; Sanberg, C.D.; El-Badri, N.; Sanberg, P.R. Nutraceuticals synergistically promote proliferation of human stem cells. *Stem Cells Dev.* **2006**, *15*, 118–123. [CrossRef]
133. Jensen, G.S.; Hart, A.N.; Zaske, L.A.M.; Drapeau, C.; Gupta, N.; Schaeffer, D.J.; Cruickshank, J.A. Mobilization of human CD34+CD133+ and CD34+CD133- stem cells in vivo by consumption of an extract from Aphanizomenon flos-aquae-related to modulation of CXCR4 expression by an L-selectin ligand? Cardiovasc. *Revascularization Med.* **2007**, *8*, 189–202. [CrossRef]
134. Kuhad, A.; Tirkey, N.; Pilkhwal, S.; Chopra, K. Renoprotective effect of Spirulina fusiformis on cisplatin-induced oxidative stress and renal dysfunction in rats. *Ren. Fail.* **2006**, *28*, 247–254. [CrossRef]
135. Avdagić, N.; Ćosović, E.; Nakaš-Ićindič, E.; Mornjaković, Z.; Začiragić, A.; Hadžović-Džuvo, A. Spirulina platensis protects against renal injury in rats with gentamicin-induced acute tubular necrosis. *Bosn. J. Basic Med. Sci.* **2008**, *8*, 331–336. [CrossRef]
136. Hamad, E.M.; Mousa, H.M.; Ashoush, I.S.; Abdel-Salam, A.M. Nephroprotective effect of camel milk and spirulina platensis in gentamicin-induced nephrotoxicity in rats. *Int. J. Pharmacol.* **2018**, *14*, 559–565. [CrossRef]
137. Khan, M.; Shobha, J.C.; Mohan, I.K.; Rao Naidu, M.U.; Prayag, A.; Kutala, V.K. Spirulina attenuates cyclosporine-induced nephrotoxicity in rats. *J. Appl. Toxicol.* **2006**, *27*, 511–518. [CrossRef] [PubMed]
138. Paredes-Carbajal, M.C.; Torres-Duràn, P.V.; Dìaz-Zagoya, J.C.; Mascher, D.; Juárez-Oropeza, M.A. Effects of dietary spirulina maxima on endothelium dependent vasomotor responses of rat aortic rings. *Life Sci.* **1997**, *61*, 211–219. [CrossRef]
139. Memije-Lazaro, I.N.; Blas-Valdivia, V.; Franco-Colín, M.; Cano-Europa, E. Arthrospira maxima (Spirulina) and C-phycocyanin prevent the progression of chronic kidney disease and its cardiovascular complications. *J. Funct. Foods* **2018**, *43*, 37–43. [CrossRef]
140. Kovic, M. Risks of space colonization. *Futures* **2021**, *126*, 102638. [CrossRef]
141. Nangle, S.N.; Wolfson, M.Y.; Hartsough, L.; Ma, N.J.; Mason, C.E.; Merighi, M.; Nathan, V.; Silver, P.A.; Simon, M.; Swett, J.; et al. The case for biotech on Mars. *Nat. Biotechnol.* **2020**, *38*, 401–407. [CrossRef]

142. Revellame, E.D.; Aguda, R.; Chistoserdov, A.; Fortela, D.L.; Hernandez, R.A.; Zappi, M.E. Microalgae cultivation for space exploration: Assessing the potential for a new generation of waste to human life-support system for long duration space travel and planetary human habitation. *Algal Res.* **2021**, *55*, 102258. [CrossRef]
143. Belz, S.; Buchert, M.; Bretschneider, J.; Nathanson, E.; Fasoulas, S. Physicochemical and biological technologies for future exploration missions. *Acta Astronaut.* **2014**, *101*, 170–179. [CrossRef]
144. Cycil, L.M.; Hausrath, E.M.; Ming, D.W.; Adcock, C.T.; Raymond, J.; Remias, D.; Ruemmele, W.P. Investigating the Growth of Algae Under Low Atmospheric Pressures for Potential Food and Oxygen Production on Mars. *Front. Microbiol.* **2021**, *12*, 733244. [CrossRef]
145. Yang, C.; Liu, H.; Li, M.; Yu, C.; Yu, G. Treating urine by Spirulina platensis. *Acta Astronaut.* **2008**, *63*, 1049–1054. [CrossRef]
146. Rapp, D. *Use of Extraterrestrial Resources for Human Space Missions to Moon or Mars*; Elsevier: Amsterdam, The Netherlands, 2018. [CrossRef]
147. De Man, P. In-Situ Resource Utilization: Legal Aspects. In *Promoting Productive Cooperation Between Space Lawyers and Engineers*; Pecujlic, A.N., Tugnoli, M., Eds.; IGI Global: Hershey, PA, USA, 2019; pp. 211–224. [CrossRef]
148. Olsson-Francis, K.; Cockell, C.S. Use of cyanobacteria for in-situ resource use in space applications. *Planet. Space Sci.* **2010**, *58*, 1279–1285. [CrossRef]
149. Billi, D.; Gallego Fernandez, B.; Fagliarone, C.; Chiavarini, S.; Rothschild, L.J. Exploiting a perchlorate-tolerant desert cyanobacterium to support bacterial growth for in situ resource utilization on Mars. *Int. J. Astrobiol.* **2021**, *20*, 29–35. [CrossRef]
150. Cao, G.; Concas, A.; Fais, G.; Gabrielli, G.; Manca, A.; Pantaleo, A. Process and Kit to Investigate Microgravity Effect on Animal/Vegetable Cells under Extraterrestrial Cultivation Conditions and Cultivation Process thereof to Sustain Manned Space Missions 2021. PCT/IB2012/053754, 2021. Available online: https://patentscope2.wipo.int/search/en/detail.jsf?docId=WO2013014606 (accessed on 10 March 2022).
151. Nickerson, C.A.; Ott, C.M.; Wilson, J.W.; Ramamurthy, R.; Pierson, D.L. Microbial Responses to Microgravity and Other Low-Shear Environments. *Microbiol. Mol. Biol. Rev.* **2004**, *68*, 345–361. [CrossRef] [PubMed]
152. Wilson, J.W.; Ott, C.M.; Ramamurthy, R.; Quick, L.; Porwollik, S.; Cheng, P.; Mcclelland, M.; Ho, K.; Tsaprailis, G.; Radabaugh, T.; et al. Space flight alters bacterial gene expression and virulence and reveals a role for global regulator Hfq. *Proc. Natl. Acad. Sci. USA* **2007**, *104*, 16299–16304. [CrossRef] [PubMed]
153. Wilson, J.W.; Ott, C.M.; Quick, L.; Davis, R.; zu Bentrup, K.H.; Crabbé, A.; Richter, E.; Sarker, S.; Barrila, J.; Porwollik, S.; et al. Media ion composition controls regulatory and virulence response of Salmonella in spaceflight. *PLoS ONE* **2008**, *3*, e3923. [CrossRef] [PubMed]
154. Barrila, J.; Radtke, A.L.; Crabbé, A.; Sarker, S.F.; Herbst-Kralovetz, M.M.; Ott, C.M.; Nickerson, C.A. Organotypic 3D cell culture models: Using the rotating wall vessel to study host-pathogen interactions. *Nat. Rev. Microbiol.* **2010**, *8*, 791–801. [CrossRef] [PubMed]
155. Bailey, M.T.; Dowd, S.E.; Galley, J.D.; Hufnagle, A.R.; Allen, R.G.; Lyte, M. Exposure to a social stressor alters the structure of the intestinal microbiota: Implications for stressor-induced immunomodulation. *Brain. Behav. Immun.* **2011**, *25*, 397–407. [CrossRef] [PubMed]
156. Castro, S.L.; Nelman-Gonzalez, M.; Nickerson, C.A.; Ott, C.M. Induction of attachment-independent biofilm formation and repression of hfq expression by low-fluid-shear culture of Staphylococcus aureus. *Appl. Environ. Microbiol.* **2011**, *77*, 6368–6378. [CrossRef]
157. Crabbé, A.; Schurr, M.J.; Monsieurs, P.; Morici, L.; Schurr, J.; Wilson, J.W.; Ott, C.M.; Tsaprailis, G.; Pierson, D.L.; Stefanyshyn-Piper, H.; et al. Transcriptional and proteomic responses of Pseudomonas aeruginosa PAO1 to spaceflight conditions involve Hfq regulation and reveal a role for oxygen. *Appl. Environ. Microbiol.* **2011**, *77*, 1221–1230. [CrossRef]
158. Turroni, S.; Magnani, M.; KC, P.; Lesnik, P.; Vidal, H.; Heer, M. Gut Microbiome and Space Travelers' Health: State of the Art and Possible Pro/Prebiotic Strategies for Long-Term Space Missions. *Front. Physiol.* **2020**, *11*, 553929. [CrossRef]
159. Crucian, B.; Sams, C. Immune system dysregulation during spaceflight: Clinical risk for exploration-class missions. *J. Leukoc. Biol.* **2009**, *86*, 1017–1018. [CrossRef]
160. Siddiqui, R.; Akbar, N.; Khan, N.A. Gut microbiome and human health under the space environment. *J. Appl. Microbiol.* **2021**, *130*, 14–24. [CrossRef] [PubMed]
161. Bradbury, P.; Wu, H.; Choi, J.U.; Rowan, A.E.; Zhang, H.; Poole, K.; Lauko, J.; Chou, J. Modeling the Impact of Microgravity at the Cellular Level: Implications for Human Disease. *Front. Cell Dev. Biol.* **2020**, *8*, 96. [CrossRef] [PubMed]
162. Da Silveira, W.A.; Fazelinia, H.; Rosenthal, S.B.; Laiakis, E.C.; Kim, M.S.; Meydan, C.; Kidane, Y.; Rathi, K.S.; Smith, S.M.; Stear, B.; et al. Comprehensive Multi-omics Analysis Reveals Mitochondrial Stress as a Central Biological Hub for Spaceflight Impact. *Cell* **2020**, *183*, 1185–1201.e20. [CrossRef]
163. Powers, S.K.; Ji, L.L.; Kavazis, A.N.; Jackson, M.J. Reactive oxygen species: Impact on skeletal muscle. *Compr. Physiol.* **2011**, *1*, 941–969. [CrossRef] [PubMed]
164. Moreno-Villanueva, M.; Wong, M.; Lu, T.; Zhang, Y.; Wu, H. Interplay of space radiation and microgravity in DNA damage and DNA damage response. *NPJ Microgravity* **2017**, *3*, 14. [CrossRef] [PubMed]
165. Richter, C. Reactive oxygen and DNA damage in mitochondria. *Mutat. Res. DNAging* **1992**, *275*, 249–255. [CrossRef]
166. Goodwin, T.J.; Christofidou-Solomidou, M. Oxidative Stress and Space Biology: An Organ-Based Approach. *Int. J. Mol. Sci.* **2018**, *19*, 959. [CrossRef]

167. Schimmerling, W. Space and radiation protection: Scientific requirements for space research. *Radiat. Environ. Biophys.* **1995**, *34*, 133–137. [CrossRef]
168. Belobrajdic, B.; Melone, K.; Diaz-Artiles, A. Planetary extravehicular activity (EVA) risk mitigation strategies for long-duration space missions. *NPJ Microgravity* **2021**, *7*, 16. [CrossRef]
169. Buonanno, M.; De Toledo, S.M.; Howell, R.W.; Azzam, E.I. Low-dose energetic protons induce adaptive and bystander effects that protect human cells against DNA damage caused by a subsequent exposure to energetic iron ions. *J. Radiat. Res.* **2015**, *56*, 502–508. [CrossRef]
170. Bigley, A.B.; Agha, N.H.; Baker, F.L.; Spielmann, G.; Kunz, H.E.; Mylabathula, P.L.; Rooney, B.V.; Laughlin, M.S.; Mehta, S.K.; Pierson, D.L.; et al. NK cell function is impaired during long-duration spaceflight. *J. Appl. Physiol.* **2019**, *126*, 842–853. [CrossRef] [PubMed]
171. Crucian, B.; Stowe, R.P.; Mehta, S.; Quiriarte, H.; Pierson, D.; Sams, C. Alterations in adaptive immunity persist during long-duration spaceflight. *NPJ Microgravity* **2015**, *1*, 15013. [CrossRef] [PubMed]
172. Martinez, E.M.; Yoshida, M.C.; Candelario, T.L.T.; Hughes-Fulford, M. Spaceflight and simulated microgravity cause a significant reduction of key gene expression in early T-cell activation. *Am. J. Physiol. Regul. Integr. Comp. Physiol.* **2015**, *308*, R480–R488. [CrossRef] [PubMed]
173. Pecaut, M.J.; Mao, X.W.; Bellinger, D.L.; Jonscher, K.R.; Stodieck, L.S.; Ferguson, V.L.; Bateman, T.A.; Mohney, R.P.; Gridley, D.S. Is spaceflight-induced immune dysfunction linked to systemic changes in metabolism? *PLoS ONE* **2017**, *12*, e0174174. [CrossRef]
174. Paul, A.M.; Cheng-Campbell, M.; Blaber, E.A.; Anand, S.; Bhattacharya, S.; Zwart, S.R.; Crucian, B.E.; Smith, S.M.; Meller, R.; Grabham, P.; et al. Beyond Low-Earth Orbit: Characterizing Immune and microRNA Differentials following Simulated Deep Spaceflight Conditions in Mice. *iScience* **2020**, *23*, 101747. [CrossRef]
175. Pavlakou, P.; Dounousi, E.; Roumeliotis, S.; Eleftheriadis, T.; Liakopoulos, V. Oxidative Stress and the Kidney in the Space Environment. *Int. J. Mol. Sci.* **2018**, *19*, 3176. [CrossRef]
176. Roumelioti, M.-E.; Glew, R.H.; Khitan, Z.J.; Rondon-Berrios, H.; Argyropoulos, C.P.; Malhotra, D.; Raj, D.S.; Agaba, E.I.; Rohrscheib, M.; Murata, G.H.; et al. Fluid balance concepts in medicine: Principles and practice. *World J. Nephrol.* **2018**, *7*, 1–28. [CrossRef]
177. Drummer, C.; Gerzer, R.; Baisch, F.; Heer, M. Body fluid regulation in micro-gravity differs from that on Earth: An overview. *Pflugers Arch.* **2000**, *441*, R66–R72. [CrossRef]
178. Whitson, P.A.; Pietrzyk, R.A.; Jones, J.A.; Nelman-Gonzalez, M.; Hudson, E.K.; Sams, C.F. Effect of potassium citrate therapy on the risk of renal stone formation during spaceflight. *J. Urol.* **2009**, *182*, 2490–2496. [CrossRef]
179. Pietrzyk, R.A.; Jones, J.A.; Sams, C.F.; Whitson, P.A. Renal stone formation among astronauts. *Aviat. Space. Environ. Med.* **2007**, *78* (Suppl. S4), A9–A13.
180. Liakopoulos, V.; Leivaditis, K.; Eleftheriadis, T.; Dombros, N. The kidney in space. *Int. Urol. Nephrol.* **2012**, *44*, 1893–1901. [CrossRef] [PubMed]
181. Barcellos-Hoff, M.H.; Blakely, E.A.; Burma, S.; Fornace, A.J.; Gerson, S.; Hlatky, L.; Kirsch, D.G.; Luderer, U.; Shay, J.; Wang, Y.; et al. Concepts and challenges in cancer risk prediction for the space radiation environment. *Life Sci. Sp. Res.* **2015**, *6*, 92–103. [CrossRef] [PubMed]
182. Hu, S.; Kim, M.H.Y.; McClellan, G.E.; Cucinotta, F.A. Modeling the acute health effects of astronauts from exposure to large solar particle events. *Health Phys.* **2009**, *96*, 465–476. [CrossRef] [PubMed]
183. Patel, S. The effects of microgravity and space radiation on cardiovascular health: From low-Earth orbit and beyond. *IJC Heart Vasc.* **2020**, *30*, 100595. [CrossRef] [PubMed]
184. Melissa Gaskill Vascular Studies in Space: Good Heart, Everyone's; 2021. p. 166. Available online: https://www.nasa.gov/sites/default/files/atoms/files/benefits-for-humanity_third.pdf (accessed on 7 March 2022).
185. Meyers, C.A. Neurocognitive dysfunction in cancer patients. *Oncology (Williston Park)* **2000**, *14*, 75–85. [PubMed]
186. Jandial, R.; Hoshide, R.; Waters, J.D.; Limoli, C.L. Space-brain: The negative effects of space exposure on the central nervous system. *Surg. Neurol. Int.* **2018**, *9*, 9. [CrossRef]
187. Andersen, B.L.; Tewfik, H.H. Psychological reactions to radiation therapy: Reconsideration of the adaptive aspects of anxiety. *J. Pers. Soc. Psychol.* **1985**, *48*, 1024–1032. [CrossRef]
188. Makale, M.T.; McDonald, C.R.; Hattangadi-Gluth, J.A.; Kesari, S. Mechanisms of radiotherapy-associated cognitive disability in patients with brain tumours. *Nat. Rev. Neurol.* **2017**, *13*, 52–64. [CrossRef]
189. Tang, Y.; Luo, D.; Rong, X.; Shi, X.; Peng, Y. Psychological disorders, cognitive dysfunction and quality of life in nasopharyngeal carcinoma patients with radiation-induced brain injury. *PLoS ONE* **2012**, *7*, e36529. [CrossRef]
190. Wellisch, D.K.; Kaleita, T.A.; Freeman, D.; Cloughesy, T.; Goldman, J. Predicting major depression in brain tumor patients. *Psychooncology* **2002**, *11*, 230–238. [CrossRef]
191. Parihar, V.K.; Allen, B.D.; Caressi, C.; Kwok, S.; Chu, E.; Tran, K.K.; Chmielewski, N.N.; Giedzinski, E.; Acharya, M.M.; Britten, R.A.; et al. Cosmic radiation exposure and persistent cognitive dysfunction. *Sci. Rep.* **2016**, *6*, 34774. [CrossRef] [PubMed]
192. Cherry, J.D.; Liu, B.; Frost, J.L.; Lemere, C.A.; Williams, J.P.; Olschowka, J.A.; O'Banion, M.K. Galactic cosmic radiation leads to cognitive impairment and increased aβ plaque accumulation in a mouse model of Alzheimer's disease. *PLoS ONE* **2012**, *7*, e53275. [CrossRef] [PubMed]

193. Parihar, V.K.; Pasha, J.; Tran, K.K.; Craver, B.M.; Acharya, M.M.; Limoli, C.L. Persistent changes in neuronal structure and synaptic plasticity caused by proton irradiation. *Brain Struct. Funct.* **2015**, *220*, 1161–1171. [CrossRef] [PubMed]
194. Machida, M.; Lonart, G.; Britten, R.A. Low (60 cGy) doses of (56)Fe HZE-particle radiation lead to a persistent reduction in the glutamatergic readily releasable pool in rat hippocampal synaptosomes. *Radiat. Res.* **2010**, *174*, 618–623. [CrossRef] [PubMed]
195. Gaboyard, S.; Blanchard, M.-P.; Travo, C.; Viso, M.; Sans, A.; Lehouelleur, J. Weightlessness affects cytoskeleton of rat utricular hair cells during maturation in vitro. *Neuroreport* **2002**, *13*, 2139–2142. [CrossRef] [PubMed]
196. He, J.; Zhang, X.; Gao, Y.; Li, S.; Sun, Y. Effects of altered gravity on the cell cycle, actin cytoskeleton and proteome in Physarum polycephalum. *Acta Astronaut.* **2008**, *63*, 915–922. [CrossRef]
197. Huang, Y.; Dai, Z.-Q.; Ling, S.-K.; Zhang, H.-Y.; Wan, Y.-M.; Li, Y.-H. Gravity, a regulation factor in the differentiation of rat bone marrow mesenchymal stem cells. *J. Biomed. Sci.* **2009**, *16*, 87. [CrossRef]
198. Sarkar, P.; Sarkar, S.; Ramesh, V.; Hayes, B.E.; Thomas, R.L.; Wilson, B.L.; Kim, H.; Barnes, S.; Kulkarni, A.; Pellis, N.; et al. Proteomic analysis of mice hippocampus in simulated microgravity environment. *J. Proteome Res.* **2006**, *5*, 548–553. [CrossRef]
199. Clément, G.R.; Boyle, R.D.; George, K.A.; Nelson, G.A.; Reschke, M.F.; Williams, T.J.; Paloski, W.H. Challenges to the central nervous system during human spaceflight missions to Mars. *J. Neurophysiol.* **2020**, *123*, 2037–2063. [CrossRef]
200. Tauber, S.; Hauschild, S.; Paulsen, K.; Gutewort, A.; Raig, C.; Hürlimann, E.; Biskup, J.; Philpot, C.; Lier, H.; Engelmann, F.; et al. Signal transduction in primary human T lymphocytes in altered gravity during parabolic flight and clinostat experiments. *Cell. Physiol. Biochem.* **2015**, *35*, 1034–1051. [CrossRef]
201. Tauber, S.; Hauschild, S.; Crescio, C.; Secchi, C.; Paulsen, K.; Pantaleo, A.; Saba, A.; Buttron, I.; Thiel, C.S.; Cogoli, A.; et al. Signal transduction in primary human T lymphocytes in altered gravity—Results of the MASER-12 suborbital space flight mission. *Cell Commun. Signal.* **2013**, *11*, 32. [CrossRef]
202. Mehta, S.K.; Laudenslager, M.L.; Stowe, R.P.; Crucian, B.E.; Feiveson, A.H.; Sams, C.F.; Pierson, D.L. Latent virus reactivation in astronauts on the international space station. *NPJ Microgravity* **2017**, *3*, 11. [CrossRef] [PubMed]
203. Cortés-Sánchez, J.L.; Callant, J.; Krüger, M.; Sahana, J.; Kraus, A.; Baselet, B.; Infanger, M.; Baatout, S.; Grimm, D. Cancer studies under space conditions: Finding answers abroad. *Biomedicines* **2022**, *10*, 25. [CrossRef] [PubMed]
204. Furukawa, S.; Nagamatsu, A.; Nenoi, M.; Fujimori, A.; Kakinuma, S.; Katsube, T.; Wang, B.; Tsuruoka, C.; Shirai, T.; Nakamura, A.J.; et al. Space Radiation Biology for "Living in Space". *Biomed Res. Int.* **2020**, *2020*, 4703286. [CrossRef] [PubMed]
205. Jeggo, P.; Löbrich, M. Radiation-induced DNA damage responses. *Radiat. Prot. Dosim.* **2006**, *122*, 124–127. [CrossRef]
206. Santivasi, W.L.; Xia, F. Ionizing radiation-induced DNA damage, response, and repair. *Antioxid. Redox Signal.* **2014**, *21*, 251–259. [CrossRef]
207. Tajvidi, E.; Nahavandizadeh, N.; Pournaderi, M.; Pourrashid, A.Z.; Bossaghzadeh, F.; Khoshnood, Z. Study the antioxidant effects of blue-green algae Spirulina extract on ROS and MDA production in human lung cancer cells. *Biochem. Biophys. Rep.* **2021**, *28*, 101139. [CrossRef]
208. Leach, C.S.; Johnson, P.C. Influence of spaceflight on erythrokinetics in man. *Science* **1984**, *225*, 216–218. [CrossRef]
209. Selmi, C.; Leung, P.S.C.; Fischer, L.; German, B.; Yang, C.Y.; Kenny, T.P.; Cysewski, G.R.; Gershwin, M.E. The effects of Spirulina on anemia and immune function in senior citizens. *Cell. Mol. Immunol.* **2011**, *8*, 248–254. [CrossRef]
210. Mader, T.H.; Gibson, C.R.; Pass, A.F.; Kramer, L.A.; Lee, A.G.; Fogarty, J.; Tarver, W.J.; Dervay, J.P.; Hamilton, D.R.; Sargsyan, A.; et al. Optic disc edema, globe flattening, choroidal folds, and hyperopic shifts observed in astronauts after long-duration space flight. *Ophthalmology* **2011**, *118*, 2058–2069. [CrossRef]
211. Zwart, S.R.; Gibson, C.R.; Mader, T.H.; Ericson, K.; Ploutz-Snyder, R.; Heer, M.; Smith, S.M. Vision Changes after Spaceflight Are Related to Alterations in Folate– and Vitamin B-12–Dependent One-Carbon Metabolism. *J. Nutr.* **2012**, *142*, 427–431. [CrossRef] [PubMed]
212. Ata, F.; Bint, I.; Bilal, A.; Javed, S.; Shabir Chaudhry, H.; Sharma, R.; Fatima Malik, R.; Choudry, H.; Bhaskaran Kartha, A. Optic neuropathy as a presenting feature of vitamin B-12 deficiency: A systematic review of literature and a case report. *Ann. Med. Surg.* **2020**, *60*, 316–322. [CrossRef] [PubMed]
213. Beetstra, S.; Thomas, P.; Salisbury, C.; Turner, J.; Fenech, M. Folic acid deficiency increases chromosomal instability, chromosome 21 aneuploidy and sensitivity to radiation-induced micronuclei. *Mutat. Res.* **2005**, *578*, 317–326. [CrossRef] [PubMed]
214. Moens, A.L.; Vrints, C.J.; Claeys, M.J.; Timmermans, J.P.; Champion, H.C.; Kass, D.A. Mechanisms and potential therapeutic targets for folic acid in cardiovascular disease. *Am. J. Physiol.—Heart Circ. Physiol.* **2008**, *294*, 1971–1977. [CrossRef] [PubMed]
215. Zwart, S.R.; Gibson, C.R.; Gregory, J.F.; Mader, T.H.; Stover, P.J.; Zeisel, S.H.; Smith, S.M. Astronaut ophthalmic syndrome. *FASEB J.* **2017**, *31*, 3746–3756. [CrossRef]
216. Akiyama, T.; Horie, K.; Hinoi, E.; Hiraiwa, M.; Kato, A.; Maekawa, Y.; Takahashi, A.; Furukawa, S. How does spaceflight affect the acquired immune system? *NPJ Microgravity* **2020**, *6*, 14. [CrossRef]
217. Makedonas, G.; Mehta, S.K.; Scheuring, R.A.; Haddon, R.; Crucian, B.E. SARS-CoV-2 Pandemic Impacts on NASA Ground Operations to Protect ISS Astronauts. *J. Allergy Clin. Immunol. Pract.* **2020**, *8*, 3247–3250. [CrossRef]
218. Blaber, E.A.; Finkelstein, H.; Dvorochkin, N.; Sato, K.Y.; Yousuf, R.; Burns, B.P.; Globus, R.K.; Almeida, E.A.C. Microgravity Reduces the Differentiation and Regenerative Potential of Embryonic Stem Cells. *Stem Cells Dev.* **2015**, *24*, 2605–2621. [CrossRef]
219. Blaber, E.; Sato, K.; Almeida, E.A.C. Stem cell health and tissue regeneration in microgravity. *Stem Cells Dev.* **2014**, *23*, 73–78. [CrossRef]

220. Tadros, M.G. *Characterization of Spirulina Biomass for CELSS Diet Potential*; NASA Contract: NCC 2-501; US Gov.: Washington, DC, USA, 1988; 53p. Available online: https://ntrs.nasa.gov/citations/19890016190 (accessed on 24 April 2022).
221. Oguchi, M.; Nitta, K.; Hatayaina, S.; Technology, S. Food production and gas exchange system using blue-green alga (Spirulina) for CELSS. *Adv. Space Res.* **1987**, *7*, 7–10. [CrossRef]
222. García, J.L.; de Vicente, M.; Galán, B. Microalgae, old sustainable food and fashion nutraceuticals. *Microb. Biotechnol.* **2017**, *10*, 1017–1024. [CrossRef] [PubMed]

marine drugs

Review

Carotenoid Production from Microalgae: Biosynthesis, Salinity Responses and Novel Biotechnologies

Yuanyuan Ren [1,2,3], Han Sun [2,3], Jinquan Deng [2,3], Junchao Huang [2,3,*] and Feng Chen [2,3,*]

1 Institute for Food and Bioresource Engineering, College of Engineering, Peking University, Beijing 100871, China; 1701111648@pku.edu.cn
2 Shenzhen Key Laboratory of Marine Microbiome Engineering, Institute for Advanced Study, Shenzhen University, Shenzhen 518060, China; sunhanias@163.com (H.S.); 1900392002@email.szu.edu.cn (J.D.)
3 Institute for Innovative Development of Food Industry, Shenzhen University, Shenzhen 518060, China
* Correspondence: huangjc65@szu.edu.cn (J.H.); sfchen@szu.edu.cn (F.C.)

Abstract: Microalgae are excellent biological factories for high-value products and contain biofunctional carotenoids. Carotenoids are a group of natural pigments with high value in social production and human health. They have been widely used in food additives, pharmaceutics and cosmetics. Astaxanthin, β-carotene and lutein are currently the three carotenoids with the largest market share. Meanwhile, other less studied pigments, such as fucoxanthin and zeaxanthin, also exist in microalgae and have great biofunctional potentials. Since carotenoid accumulation is related to environments and cultivation of microalgae in seawater is a difficult biotechnological problem, the contributions of salt stress on carotenoid accumulation in microalgae need to be revealed for large-scale production. This review comprehensively summarizes the carotenoid biosynthesis and salinity responses of microalgae. Applications of salt stress to induce carotenoid accumulation, potentials of the Internet of Things in microalgae cultivation and future aspects for seawater cultivation are also discussed. As the global market share of carotenoids is still ascending, large-scale, economical and intelligent biotechnologies for carotenoid production play vital roles in the future microalgal economy.

Keywords: carotenoids; microalgae; salt stress; seawater cultivation; Internet of Things

Citation: Ren, Y.; Sun, H.; Deng, J.; Huang, J.; Chen, F. Carotenoid Production from Microalgae: Biosynthesis, Salinity Responses and Novel Biotechnologies. *Mar. Drugs* **2021**, *19*, 713. https://doi.org/10.3390/md19120713

Academic Editor: Carlos Almeida

Received: 16 November 2021
Accepted: 10 December 2021
Published: 20 December 2021

1. Introduction

Carotenoids are a class of terpenoid pigments with C40 backbones and are health-promoting for human daily diets. Up until now, over 1100 carotenoids have been discovered, and they exist in various species [1], especially in aquatic creatures, microorganisms and terrestrial plants. Carotenoids are very common in our daily lives; tomato fruit is rich in lycopene (pink red), maize corn is abundant with zeaxanthin (yellow) and carrots and *Dunaliella salina* are well known for producing β-carotene (orange) [1]. Carotenoids ingested by humans can work as precursors of Vitamin A [2], reduce free radicals [2] and repair damaged retina [3]. Moreover, they can also reduce the risks of some diseases as they have anti-cancer, anti-inflammatory and anti-obesity properties [4]. Hence, carotenoids have been widely used in the food, feed, cosmetic and pharmaceutical industries.

With the development of modern biotechnologies and market concern for food safety, the demand for carotenoids from natural sources is increasing remarkably. Compared with plant-derived carotenoids, those from microorganisms are more efficient, have lower cost and are not limited by regions and seasons. However, the carotenoids market is still mainly occupied by chemically synthetic products (80–90%), with a much lower portion of natural sources (10–20%) [5]. Thus, choosing low-cost, profitable and safe microorganisms for carotenoid production is drawing extensive interest.

Microalgae are diverse and abundant. They can adapt to different cultivation conditions, grow rapidly and accumulate a high amount of desirable bioproducts, such as fatty acids, proteins and carotenoids. The global algae market is expected to reach 970 million U.S. dollars by the end of 2025 [6] and that of carotenoids is expected to reach 2 billion U.S. dollars by 2026, mainly including food and beverages (26.1%), pharmaceuticals (9.2%), cosmetics (6.5%) and dietary supplements (23.5%) [7]. Some species, such as *Chlorella* and *Spirulina*, have GRAS status and are well accepted as health foods [8].

Microalgae have versatile metabolic modes and tend to accumulate different metabolites. Different trophic modes (autotrophy, mixotrophy, heterotrophy), culture conditions (light, pH, dissolved oxygen) and nutrient availability (repletion or depletion) will lead carbon skeletons to have significantly different cell compositions. For primary metabolites, light and adequate nutrients are essential for active growth. On the contrary, adverse stresses, such as nutrient deficiency, high light and salt stress, are key factors that arouse the defense mechanisms of microalgae and accumulate secondary metabolites to survive.

Salt stress is a common stress factor in natural environments, especially for freshwater microalgae. Salt stress can lead to oxidative damage, chlorophyll degradation and inhibit photosynthesis and growth [9]. To adapt to salt stress, microalgae have evolved a survival strategy to guarantee the balance of growth and stress responses. During the adaptation to stress, some microalgae may become cysts and accumulate secondary metabolites, such as astaxanthin and triacylglycerol [10]. However, serious salt stress can be fatal to microalgae when the cells are in low concentration or poor viability. Thus, two-stage cultivation, balancing cell density and metabolites accumulation, is a popular strategy in the practical production of high-value compounds from microalgae [11]. Due to the shortage of freshwater resources, overcoming the salt tolerance problem of freshwater microalgae is also of invaluable significance for microalgae cultivation in seawater.

This review comprehensively introduces the production of high-value carotenoids and their beneficial effects on human health. The latest research results are summarized, and biotechnology topics are focused on, such as yields of microalgae-derived carotenoids and microalgal cultivation strategies. Meanwhile, the contribution of salt stress on the growth and carotenoid accumulation of microalgae is also discussed. Moreover, considering modern advanced technologies, the application prospects of the Internet of Things in innovative biotechnology are discussed. In summary, this article aims to provide readers with comprehensive knowledge of microalgae-derived carotenoids and their related biotechnologies, coupled with salt-stress treatment or endurance.

2. Health-Promoting Carotenoids from Microalgae and Their Biofunctions

In the global market for carotenoids, β-carotene will reach a value of 620 million U.S. dollars by the end of 2026, lutein will reach 357.7 million U.S. dollars by the end of 2024 and astaxanthin will surpass 800 million in 2026 [1]. In addition, other health-promoting carotenoids, such as fucoxanthin [12] and zeaxanthin [13], are also drawing customers' attention.

Microalgal biomass can be harvested regardless of seasons and districts. Other biofunctional metabolites in microalgae can also be processed as value-added products to increase revenue. Carotenoid production from microalgae requires less labor and an easy harvesting process. Compared to terrestrial plants, microalgae have higher photosynthetic efficiency and a superior growth rate. Hyperaccumulating strain selection and the culture condition optimization required to achieve high biomass and maximum carotenoid productivities are the bottlenecks to be resolved at present. In this section, we comprehensively describe carotenoid biosynthesis in microalgae and list several microalgae-derived carotenoids and their biofunctions.

2.1. Carotenoid Biosynthesis in Microalgae

Carotenoids in microalgae can be categorized as the primary ones related to photosynthesis, and the secondary ones accumulated under adversity [14]. In this review, we mainly focus on functional microalgal carotenoids, including lutein, β-carotene, astaxanthin, zeaxanthin and fucoxanthin.

The biosynthetic pathways of carotenoids involve several intermediates and key enzymes (shown in Figure 1) [15,16]. First, through the methylerythritol phosphate pathway (MEP) in the chloroplast, isopentenyl pyrophosphate (IPP, C5) and its isomer dimethylallyl diphosphate (DMAPP, C5) are catalyzed at a ratio of 3:1 by geranyl diphosphate synthase (GPPS) and geranylgeranyl pyrophosphate synthase (GGPPS) to geranylgeranyl pyrophosphate (GGPP, C20). In *H. pluvialis*, this step from C5 to C20 can be catalyzed by GGPPS only [10]. Phytoene synthase (PSY) condenses two molecules of GGPPS into (15Z)-phytoene (C40), which is desaturated and isomerized into lycopene by phytoene desaturases (PDS), ζ-carotene isomerase (Z-ISO), ζ-carotene desaturase (ZDS) and carotene isomerase (CrtISO).

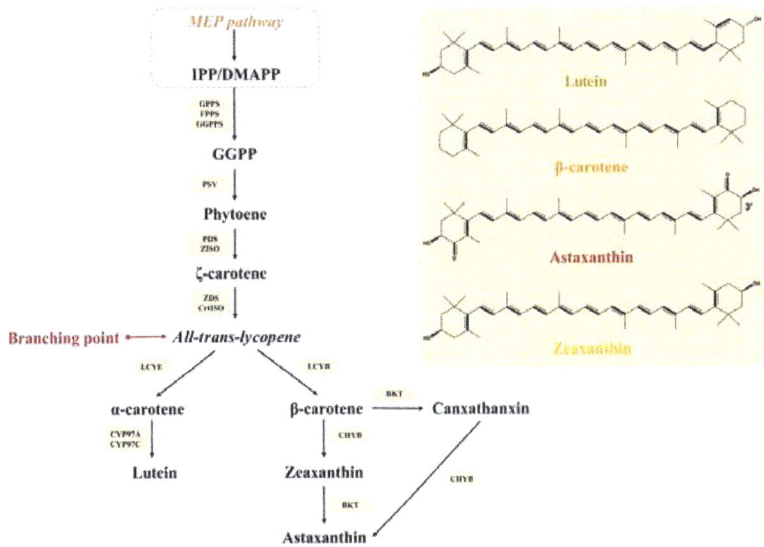

Figure 1. Biosynthesis pathways of carotenoids in microalgae and their chemical structures. MEP: methylerythritol phosphate; IPP: isopentenyl pyrophosphate; DMAPP: dimethylallyl diphosphate; GPPS: geranyl diphosphate synthase; GGPPS: geranylgeranyl pyrophosphate synthase; GGPP: geranylgeranyl pyrophosphate; PSY: phytoene synthase; Z-ISO: ζ-carotene isomerase; ZDS: ζ-carotene desaturase; CrtISO: carotene isomerase; LCYE: lycopene epsilon cyclase; CYP97A: cytochrome P450 beta hydroxylase; CYP97C: cytochrome P450 epsilon hydroxylase; LCYB: lycopene β-cyclase; BKT: β-carotene ketolase; CHYB: β-carotene hydroxylase.

Lycopene may flow to two branches, α-branch and β-branch. Lycopene ε-cyclase (LCYE) and lycopene β-cyclase (LCYB) catalyze lycopene to α-carotene, and then cytochrome P450 beta hydroxylase (CYP97A) and cytochrome P450 epsilon hydroxylase (CYP97C) convert α-carotene to lutein. As for the β-branch, lycopene is converted by LCYB to β-carotene. For the microalgae capable of astaxanthin accumulation, such as *Haematococcus pluvialis* and *Chromochloris zofingiensis*, β-carotene may also flow to different branches. β-carotene can first be catalyzed by β-carotene ketolase (BKT) to canthaxanthin and then to astaxanthin by β-carotene hydroxylase (CHYB). β-carotene can also flow to zeaxanthin first by CHYB and then to astaxanthin by BKT [15]. This also reflects the diversity of microalgae metabolism. Up until now, the exact biosynthesis pathway of fucoxanthin is still unclear.

2.2. Health-Promoting Carotenoids and Their Production from Microalgae

2.2.1. Lutein

Lutein, (3*R*, 3′*R*, 6′*R*)-β, ε-carotene-3, 3′-diol, is a natural antioxidant and has drawn interest for its health-promoting functions. Human metabolism cannot synthesize lutein, and lutein uptake in a suggested dose (6 mg day^{-1}) has been proved to be beneficial for human health. Lutein has potentials in free radical scavenging for skin health and can also prevent age-related macular degeneration (AMD) and Alzheimer's Disease (AD) [3,17]. Absorbed lutein can accumulate in human retina, filter blue light and, thus, protect eyesight. Biofunctional carotenoids and their natural sources, biofunctions and recommended doses are listed in Table 1.

Table 1. Natural carotenoids and their natural sources, biofunctions and recommended doses.

Carotenoids	Natural Sources	Biofunctions	Recommended Dose	Ref.
Lutein	Marigold flower *; Yolk; Broccoli; Microalgae Orange-yellow fruits; Leafy green vegetables;	Antioxidant; Filter blue light; Prevent AMD; Prevent AD	6 mg day^{-1}	[3,17]
Astaxanthin	Shrimp; Salmon; Crabs; Microalgae (*Haematococcus pluvialis* *) *Phaffia rhodozyma*	Antioxidant; Anti-aging; Anti-inflammatory; Anti-hypertensive; Anti-cancer;	4–12 mg day^{-1}	[10,18]
β-carotene	Pumpkin; Mango; Carrots; Microalgae (*Dunaliella salina* *)	Vitamin A precursor; Antioxidant; Anti-cancer; Anti-cardiovascular; Immune enhancement	600 μg RE [1]/day	[19]
Zeaxanthin	Marigold flower *; Maize; Orange peppers; Microalgae; Scallions	Filter blue light; Improve visual acuity; Anti-cancer; Anti-inflammatory; Anti-allergy Against UV, skin redness	2 mg day^{-1}	[13,20]
Fucoxanthin	Macroalgae *; Microalgae	Anti-cancer; Anti-hypertensive; Anti-inflammatory; Anti-obesity	–	[12,21]

Footnotes: DW, dry weight; AMD, age-related macular degeneration; AD, Alzheimer's Disease. * This symbol represents the main source of a certain carotenoid. [1] RE, retinol equivalent.

Orange-yellow fruits like mango, broccoli and other green leafy vegetables are dietary sources of lutein [22].The marigold flower is the main source of natural lutein and lutein esters. Lutein contents in different species of marigold petals range from 17 to 570 mg/100 g [23]. However, there are some drawbacks of this source, such as mandatory harvesting in specific seasons and time-consuming petal separation. Other sources containing lutein also have such disadvantages as low concentration (corn residues, leafy green vegetables) and low bioavailability (egg yolk, crustaceans) [24].

The production of lutein from microalgae may avoid these troubles. Microalgae can accumulate considerable biomass concentrations and accumulate lutein under suitable culture conditions. As a primary xanthophyll carotenoid, lutein is an antenna pigment in the light-harvesting complex (LHC) of microalgal photosynthetic apparatus. When light intensity is too high, lutein can reduce oxidative damage by non-photochemical quenching (NPQ) [25]. Thus, illumination is a key environmental factor for lutein accumulation.

Compared with marigold-originated lutein, lutein in microalgae exists in free form. The feasibility and economic competitiveness of using microalgae to replace the marigold flower to produce lutein has been widely discussed [23,24], including harvesting cost, extraction and processing methods and bioavailability analysis. Up until now, many microalgae subgenera have been investigated for lutein production, such as *Scenedesmus*, *Chlorella*, *Coccomyxa*, *Parachlorella* and *Tetraselmis* [25].

To achieve both high content and productivity of carotenoid from microalgae, various cultivation factors are globally considered, including trophic mode, carbon source, fed-batch, semi-continuous operation, light intensities and light/dark cycles, etc. [25]. Mixotrophic is popular in lutein production from various microalgae capable of glucose uptake, such as *Chlorella minutissima* [26], *Chlorella sorokiniana* MB-1-M12 [27] and *Chlorella* sp. GY-H4 [28]. The lutein production of *Chlorella* sp. GY-H4 under 20 g/L glucose (10.5 mg/L/day, 63 mg/L) was higher than that under 10 g/L glucose (7.3 mg/L/day, 44 mg/L) [28]. By adding NaAc, *Chlorella sorokiniana* MB-1-M12 can accumulate 7.39 mg/g DW of lutein [27]. Two-stage cultivation, such as the mixotrophic-photoautotrophic process of *Chlorella sorokiniana* FZU60, also achieved an efficient production at 8.25 mg/L/day [29]. Moreover, food waste [28], poultry litter waste [30] and wastewater [31] have been applied to microalgae cultivation for lower cost. A recently isolated microalga, *Parachlorella sp.* JD-076, is high in lutein productivity at 25 mg/L/day via regulation of illumination and CO_2 supply [32]. Results of satisfactory lutein production from microalgae are summarized in Table 2 with details of contents and productivities.

Table 2. Microalgae-derived carotenoids, content and their biofunctions.

Carotenoid	Microalgae	Content	Productivity/Yield	Ref.
Lutein	*Chromochloris zofingiensis bkt1* (mutant)	13.81 mg/g DW	33.97 mg/L	[33]
	Parachlorella sp. JD-076	11.87 mg/g DW	25.0 mg/L/day	[32]
	Chlorella sorokiniana FZU60	11.22 mg/g DW	8.25 mg/L/day	[29]
	Chlorella vulgaris UTEX 265	9.82 mg/g DW	11.98 mg/g/day	[34]
	Chlorella vulgaris CS-41	9.0 mg/g DW	1.56 mg/L/day	[35]
	Scenedesmus sp.	7.47 mg/g DW	19.70 mg/L/day	[36]
	Chlorella sp. GY-H4	8.9 mg/g DW	10.50 mg/L/day	[28]
	Chlorella sorokiniana MB-1-M12	7.39 mg/g DW	3.43 mg/L/day	[27]
	Chlorella minutissima MCC-27	7.05 mg/g DW	6.34 mg/L/day	[26]
Astaxanthin	*Haematococcus pluvialis*	5% DW	65.8 mg/m^2/day	[37]
	Chromochloris zofingiensis	6.5 mg/g DW	0.8 mg/L/day	[38]
	Coelastrum sp. G1-C1 (mutant)	–	28.32 mg/L	[39]
β-carotene	*Dunaliella salina* *	13% DW	–	[40]
	Chromochloris zofingiensis bkt1 (mutant)	7.18 mg/g DW	34.64 ± 1.39 mg/L	[33]
Zeaxanthin	*Nannochloropsis oceanica* CCNM 1081 *	30.2 mg/g DW	–	[41]
	Coelastrella sp. M60	13.15 mg/g DW	0.72 mg/L/day	[42]
	Chromochloris zofingiensis bkt1 (mutant)	7.00 mg/g DW	36.79 ± 2.23 mg/L	[33]
	Dunaliella salina zea1 (mutant) *	5.9 mg/g DW	–	[43]
	Chlorella ellipsoidea *	4.26 mg/g DW	–	[44]
Fucoxanthin	*Mallomonas sp.*	26.6 mg/g DW	–	[45]
	Isochrysis zhanjiangensis *	23.29 mg/g DW	2.94 mg/L/day	[46]
	Odontella aurita *	18.47 mg/g DW	7.96 mg/L/day	[47]
	Isochrysis aff. *Galbana* *	18.23 mg/g DW	–	[48]
	Tisochrysis lutea *	16.39 mg/g DW	9.81 mg/L/day	[49]
	Phaeodactylum tricornutum *	16.33 mg/g DW	–	[50]

* indicates seawater microalgae species.

2.2.2. Astaxanthin

Astaxanthin (3, 3′-dihydroxy-β, β′-carotene-4, 4′-dione) is a natural pigment existing in various species of aquacultures [18]. It is acknowledged as the strongest natural antioxidant, which is 65 times that of vitamin C and 10 times higher than other carotenoids [51]. Astaxanthin has been widely applied in functional foods, pharmaceuticals and cosmetics, with its outstanding potential in free radical scavenging and its anti-aging, anti-inflammatory, anti-hypertensive and anti-cancer properties [10].

Currently, the astaxanthin market is mostly occupied by chemically synthesized products. It is a mixture of three stereoisomers, namely (*3R,3′R*), (*3R, 3′S*) and (*3S, 3′S*) astaxanthin at the ratio of 1:2:1 [52], which is structurally heterogeneous and inefficient for biological uptake. For now, microalgae-derived astaxanthin only occupies 1% of the current astaxanthin market. Additionally, the production cost of natural astaxanthin (USD 1800/kg) is much higher than that of the chemically synthesized one (USD 1000/kg) [53]. Thus, improving the productivity of astaxanthin by microalgae is essential for commercialization.

Microalgae-derived astaxanthin has been successfully approved by the Food and Drug Administration (FDA) for direct human consumption [8]. In microalgae, astaxanthin is a secondary carotenoid accumulated under adversity. It can mediate cellular redox imbalance for microalgae to survive oxidative stress [14]. As for the current main source of natural astaxanthin, *Haematococcus pluvialis* can accumulate up to 5% DW of astaxanthin with attached cultivation [37] and has been used for biotechnology industry for astaxanthin production with an annual yield of 300 tons of biomass [54]. Under adverse growth conditions, *H. pluvialis* transforms from green and motile cells to red cysts cells with high astaxanthin content [52]. Many pieces of research have exploited strategies for boosting astaxanthin accumulation, such as nutrient deficiency, high light, adding extra chemicals and some combined strategies [10].

Chromochloris zofingiensis is another promising candidate for astaxanthin production. In comparison to *H. pluvialis*, *C. zofingiensis* can take up glucose as a carbon source and achieve high biomass concentration in mixotrophy [55]. It has drawn a lot attention for its broad capability for producing carotenoids, lipids and exopolysaccharides under different conditions [56]. Astaxanthin content in *C. zofingiensis* can reach 6.5 mg/g DW [38], and robust biotechnological traits of *C. zofingiensis* may be competitive as a better organism for large-scale astaxanthin production. Results of satisfactory astaxanthin production from microalgae are summarized in Table 2 with details of contents and productivities.

2.2.3. β-Carotene

β-carotene ($C_{40}H_{56}$) is a pure hydrocarbon carotenoid and commonly exists in plants (pumpkins, mango, carrots, etc.), fungi (*Phaffia rhodozyma*) and algae [57]. It is lipid-soluble and serves as the precursor of Vitamin A in human body, benefiting the treatment of night blindness [2]. Currently, β-carotene is widely used in foods and medicines for its excellent antioxidant, anti-cardiovascular and immune enhancement properties [57]. Although chemically synthetic β-carotene is low in cost, natural β-carotene is more acceptable to consumers.

β-carotene is a primary carotenoid in microalgae, responsible for transferring light energy to chlorophylls to expand the light absorbing spectrum [40]. Moreover, it is also the first ever commercialized high-value product from microalgae. β-carotene from microalgae has been approved by the FDA for direct human consumption [8].

The marine microalga *D. salina* can accumulate β-carotene up to 13% DW [58] when induced by different stress factors, such as high light, high temperature, nutrient deprivation and high salinity [59], which is the main source of natural β-carotene. As *D. salina* is capable of growing in high salinity, cultivation of *D. salina* in seawater with high yield of β-carotene is profitable, which may also avoid microorganism contamination. In addition, many species of microalgae can synthesize β-carotene. *Chlorella zofingiensis*, *Spirulina platensis* and *Caulerpa taxifolia* accumulate β-carotene at an average yield of 0.1–2% DW [40]. Studies have shown that the β-carotene content of *Spirulina* is 10 times that in carrots [57]. A mutant *C. zofingiensis bkt1* can accumulate high amounts of three carotenoids, as lutein (13.81 mg/g DW), β-carotene (7.18 mg/g DW) and zeaxanthin (7.00 mg/g DW) under high light and salt stress [33]. Results of satisfactory β-carotene production from microalgae are summarized in Table 2 with details of contents and productivities.

Moreover, the profiles of chemically synthesized and microalgae-derived β-carotene are different. The synthetic β-carotene only has (*all-E*)-isomer. In contrast, β-carotene from *D. salina* mainly contains three isomers, (*all-E*)-β-carotene (~42%), (9Z)-β-carotene (~41%), (15Z)-β-carotene (~10%) and others (~6%) [60]. As the (9Z)-β-carotene plays a key role in antioxidation, microalgae have potentials in producing low-cost and safer natural β-carotene for stronger biofunctions.

2.2.4. Zeaxanthin

Zeaxanthin is a xanthophyll pigment existing in human eyes and skin. Together with lutein, zeaxanthin accumulates in the cornea as a macular pigment, protecting the retinal membrane from blue light and improving visual acuity [24]. Zeaxanthin also has potentials as an antioxidant, anti-inflammatory and for preventing neurological disease [20]. Maize, orange peppers and cooked scallions are good choices for dietary intake, while their contents are relatively as low at 16.3, 16.7 and 24.9 μg/g, respectively [13,22].

Currently, the marigold flower is applied for commercial production of zeaxanthin. Zeaxanthin content in different species of marigold petals ranges from 10 to 300 μg/g [61]. As the global number of people suffering from AMD [13] increases, the demand for natural sources with higher productivity of zeaxanthin is surging. The marine green alga *Chlorella ellipsoidea* produces zeaxanthin (4.26 mg/g DW) nine-fold over that produced by red pepper [44]. Additionally, microalgae-derived zeaxanthin partly exists in free form, rather than as mono-/di- esters in flowers and fruits of plants. Thus, microalgae are more valuable sources for processing zeaxanthin with higher productivity and bioavailability.

The production of zeaxanthin from microalgae is attractive and varies depending on microalgae species and cultivation factors. The favorable light condition for zeaxanthin varies in microalgae species. The zeaxanthin content of *C. zofingiensis bkt1* (5.69 ± 0.45 mg/g DW) was strongly enhanced by high light as compared with that of low light control (0.34 ± 0.04 mg/g DW) [33]. Under high light, combined with N-starvation, C. *zofingiensis bkt1* can accumulate 7.00 ± 0.82 mg/g DW. However, high light is not always preferred for microalgae to accumulate zeaxanthin. A mutant *D. salinazea1* can accumulate 5.9 mg/g DW of zeaxanthin under continuous low light (100 μmol photons m^{-2} s^{-1}), higher than that transferred to high light (4.18 mg/g DW) [43]. As *D. salina* grows well in hypersaline medium, cultivation of this strain with seawater is potentially profitable. Results of satisfactory zeaxanthin production from microalgae are summarized in Table 2 with details of contents and productivities.

Additionally, downstream processing of zeaxanthin needs more attention, such as extraction and separation. Solvent systems have different efficiencies in zeaxanthin extraction from microalgae, and the maximum extraction of zeaxanthin (30.2 mg/g DW) was in chloroform: methanol (1:2) [41]. Pressurized liquid extraction was also applied to extract zeaxanthin from *C. ellipsoidea* [41]. More importantly, zeaxanthin shares similar chemical structures to lutein. Huang et al. developed an efficient method for the separation of zeaxanthin and lutein by ultra-high performance liquid chromatography (UHPLC) equipped with a Waters YMC Carotenoid C30 column [33].

2.2.5. Fucoxanthin

Fucoxanthin is a natural carotenoid that occupies nearly 10% of total natural carotenoid production [62]. The global production of fucoxanthin was about 500 tons in 2016 and was estimated to increase further with an annual rate of 5.3% [63]. As shown in Figure 2, it has a unique molecular structure with an allenic bond, a conjugated carbonyl, a 5,6-monoepoxide and acetyl groups. Fucoxanthin and its derivatives have shown potential anti-cancer [64], anti-inflammatory [65] and anti-obesity effects [46].

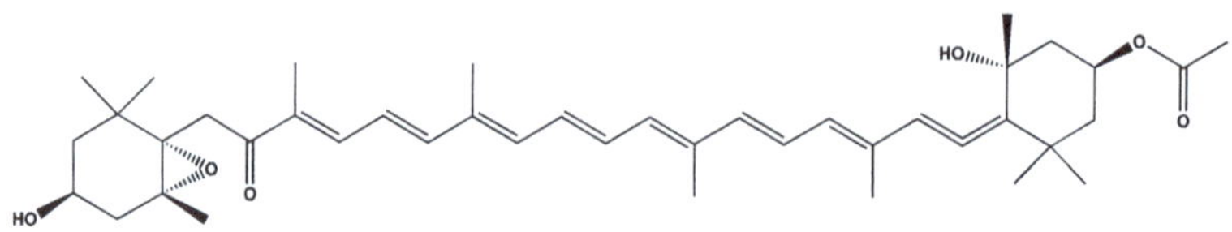

Figure 2. The chemical structure of (*all-E*)-fucoxanthin.

Currently, the main natural sources of fucoxanthin rely on some macroalgae species, such as *Laminaria japonica* and *Undaria pinnatifid*, which are common in Asian diets for iodine supplement [66]. However, the fucoxanthin contents in these macroalgae are relatively low (0.02–0.58 mg/g fresh weight) and not feasible for commercialization. Microalgae grow faster than macroalgae with higher productivity of carotenoids. Fucoxanthin is a primary light-harvesting carotenoid that transfers energy to LHC with high efficiency (>80%) [67] and may also protect microalgae from high light with its internal antioxidant property [47]. Up until now, research about producing fucoxanthin from microalgae is still relatively lacking.

Microalgae are potential candidates for fucoxanthin production. Fucoxanthin mainly exists in heterokont and haptophyte groups of algae (>20,000 species, especially in Synurophyceae (up to 26.6 mg/g DW), diatoms (up to 21.67 mg/g DW) and Prymnesiophyceae (up to 18.23 mg/g DW) [45]. Under low light intensity, carbon skeletons are diverted into carotenoids, especially fucoxanthin in *Isochrysis zhanjiangensis*, which can produce 23.29 mg/g DW of fucoxanthin [46]. The marine diatom *Odontella aurita* can accumulate 18.47 mg/g DW of fucoxanthin in a nitrogen-replete medium under low light [47]. Moreover, the purified fucoxanthin from *O. aurita* is (*all-E*)-fucoxanthin, which owns strong antioxidant ability and bioavailability. Maria et al. isolated a novel strain *Mallomonas sp.* (Synurophyceae) with the highest known content of fucoxanthin at 26.6 mg/g DW [45]. The highest ever reported fucoxanthin productivity was found in *Tisochrysis lutea* with 9.81 mg/L/day under batch culture with continuous chemostat dilution [49].

Diatom *Phaeodactylum tricornutum* also has a high fucoxanthin content at 16.33 mg/g DW [50]. As *P. tricornutum* is a model system for investigations and its genome has been sequenced, further exploitations of key genes related to fucoxanthin, fucoxanthin hyper-accumulating mutants and genetic modifications of diatoms for fucoxanthin production are expected in the near future. Results of satisfactory fucoxanthin production from microalgae are summarized in Table 2 with details of contents and productivities.

3. Salt-Stress Treatment for Carotenoid Production from Microalgae

Microalgae have flexible metabolic systems, which means that external environmental changes will affect biochemical components of microalgal cells. Stress factors, such as high light intensity, salinity, nitrogen starvation and high/low temperatures, will provoke metabolic changes in microalgae and lead to different cell compositions.

Though carotenoid production could be boosted by several environmental abiotic stress factors, salt-stress treatment is much more attractive because it is cheap, easy to operate and has great commercialization value for seawater culture of microalgae. Here, we elucidate the contributions of salinity in microalgal culture and production of carotenoids and summarize the current strategies of salt-stress treatment in carotenoid production.

3.1. *Microalgal Responses to Salt Stress*

Salt stress is one of the major abiotic stresses decreasing productivity of agriculture all over the earth [68,69]. Salinity inhibits plant growth, development, seed germination and yields. In contrast, plants have developed strategies to adapt to high salt concentrations, such as regulating stress hormones and growth hormones to balance growth and stress responses [69]. Currently, research on salt stress is more in-depth and comprehensive in higher plants, but there are a few studies on microalgae.

Due to the fast growth rate of microalgae and their richness of high-value products, large-scale microalgae cultivation is imperative at present [6]. Microalgae are widely distributed in the ocean and freshwater, while salt stress is still a threat to the large-scale production of freshwater microalgae. As freshwater resources are becoming scarce today, the cultivation of freshwater microalgae with seawater is of great significance. On the other hand, appropriate salt stress shows promoting effects on secondary metabolite accumulation (such as secondary carotenoids and lipids) [70]. Therefore, it is of great significance to study the response of microalgae to salt stress. The possible responses in microalgae cells caused by salt stress are summarized in Figure 3.

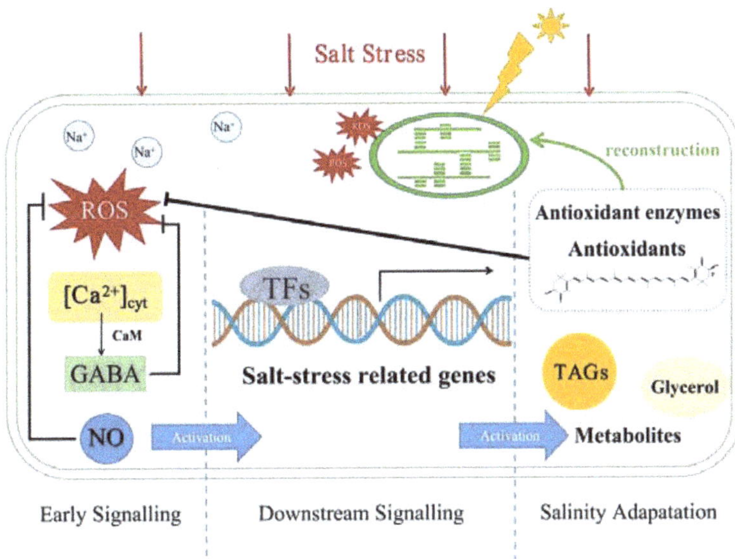

Figure 3. Possible mechanisms of responses of microalgae induced by salt stress.

Salt stress can affect microalgae on multiple levels, as physiological characteristics, gene expression and metabolic pathways. Salt stress can reduce cell growth, decrease chlorophyll content, inhibit photosynthesis and cause morphological changes in microalgae [9]. Under salt stress, the cell wall of *H. pluvialis* thickens, the cell volume becomes larger and gradually becomes immobile cysts [10]. These morphological changes are related to a series of in-depth signaling and downstream changes in both genetic and metabolic aspects.

3.1.1. Early Signaling

Ca^{2+} is considered as a universal second messenger for the primary stress signals [71], acting in real-time in response to the imbalance of cell ion homeostasis under NaCl stress in plants [72]. Similar to plants, salt stress can elicit a temporary increase in cytosolic Ca^{2+} concentration ($[Ca^{2+}]_{cyt}$), which can regulate the activity of downstream effector proteins, such as calmodulin (CaM), Ca^{2+}-dependent protein kinases and CaM-dependent protein kinases [73]. Moreover, the decarboxylation of glutamate to γ-aminobutyric acid (GABA) was regulated by the Ca^{2+}/CaM protein.

GABA is a recently identified endogenous signaling molecule in plants, participating in cell growth and enhancing abiotic stress tolerance [74]. It is a non-protein amino acid derived from the decarboxylation of glutamate, and exogenous GABA can enhance plant resistance to abiotic stress by activating the GABA bypass pathways and TCA cycle. Under salt stress, GABA helps to maintain C/N balance and even acts as a scavenger of toxic

ROS [75]. It is also related to nitric oxide (NO) accumulation under stress conditions as it amplifies NO stress signaling [74].

NO is an important molecule involved in plant growth, development and tolerance to abiotic stress. It plays important roles in resistance to drought, temperature (high or low), UV-B and heavy metal stress [76]. NO can also act as a signal in activating antioxidant enzyme defense against oxidative stress induced by salt stress [76]. By applying NO donors, such as sodium nitroprusside, plants under stress conditions showed restoration of chlorophyll to recover their damaged photosynthetic system [77].

Reactive oxygen species (ROS) are also second messengers induced by salinity stress, which is also associated with Ca^{2+} signaling [73]. ROS is a stress indicator in response to abiotic stress of microalgae, which can regulate cell growth and metabolites synthesis. Toxic ROS will lead to lipid peroxidation, membrane deterioration, DNA and protein damage [68]. To eliminate excessive ROS, antioxidant enzymes (such as superoxide dismutase and catalase) and antioxidants (such as carotenoids) in microalgae are two essential mechanisms evolved by organisms [78].

3.1.2. Downstream Signaling

Salinity-induced signals may then have effects on the gene expression of microalgae cells, which is related to salt concentrations [79]. Low salt concentration can promote the growth of some microalgae, such as cultures of *Scenedesmus sp.* [80], *Botryococcus braunii* [81] and *H. pluvialis* [79]. Treatment of *H. pluvialis* with low-dose NaCl (12.5 mg/L) showed a promoting effect on biomass concentration (28% higher) and productivity (from $0.15\ d^{-1}$–$0.22\ d^{-1}$). These effects were related to upregulation of growth-related genes, such as *rbcL*, *rbcS* and nitrate reductase gene (*NR*).

Under high-dose salt stress, both ionic and osmotic homeostasis need to be maintained. As for the halotolerant microalga *D. salina*, osmoregulation under osmotic stresses can be divided into two mechanisms. One is to maintain intracellular Na^+ and K^+ concentrations by the plasma membrane electron transport (redox) system, as Na^+-ATPase and K^+ carriers, and the other is to regulate glycerol concentrations inside to maintain the water potential inside and outside of cells [82]. Accumulation of glycerol as a soluble substance is a strategy for microalgae to keep osmotic homeostasis under salt stress, which has been demonstrated in *C. reinhardtii* [83] and *C. zofingiensis* [15]. Salt stress can also upregulate the genes related to starch catabolism (like pyruvate kinase, PK) and downregulate the genes for gluconeogenesis (like phosphoenolpyruvate carboxykinase, PEPCK), providing more building blocks for storage lipids (fatty acids and triacylglycerol, TAGs) and carotenoids in *C. zofingiensis* [15]. Salt-induced acetyl-coenzyme A carboxylase (ACCase) expression for fatty acid synthesis was also demonstrated in *Chlamydomonas* sp. [84], *Chlorella* sp. [85] and *Nitzschia* sp. [86].

Omics approaches like genomic, transcriptomic and metabolomic are also applied to reveal the changes occurring under salt stress. By analyzing differential expressed genes (DEGs) under stress, transcription factors (TFs), such as myeloblastosis (MYB), WRKY and basic helix–loop–helix (bHLH), are demonstrated to play important roles in regulating gene expressions under salinity [68]. Responses of microalgae to both low-dose and high-dose salinity are shown in Table 3.

Table 3. Overview of the effects of salinity on microalgae at different doses.

Responses	Low-Dose NaCl	High-Dose NaCl
Physiology	Growth ↑; Photosynthesis ↑	Growth ↓; Chlorophyll content↓; Photosynthesis ↓
Morphology	No significant changes	Cell size ↑; Cell wall ↑ Color change
Main carbon sinks	Carbohydrate ↑	Carbohydrate (providing building blocks) ↓ TAGs ↑
Gene expression	• *nr* gene (for N assimilation) ↑ • Photosynthetic enzyme genes (*rbcL*, *rbcS*) ↑	• TFs (WRKY, MYB, bHLH...) • Antioxidant enzymes genes (SOD, CAT) ↑ • Secondary metabolites genes ↑ • Fatty acid synthesis genes ↑ • Starch catabolism genes (PK) ↑ • Gluconeogenesis gene (PEPCK) ↓
Metabolites	Lutein ↑	Secondary carotenoids ↑ TFA ↑

3.2. Salt Stress Strategies for Carotenoid Accumulation

Carotenoid production could be affected by environmental abiotic stress. Under salt stress, carotenoids will accumulate and protect cells as antioxidants to increase the surviving possibility of microalgae. In addition, optimal salt condition varies among different microalgal species. After NaCl treatment (1%, 0.17 M) for 10 days, the astaxanthin content of *H. pluvialis* climbed from 3.53 mg/g to 17.7 mg/g [87]. *C. zofingiensis* CCAP 211/14 can tolerate moderate NaCl concentration of 100 mM, and a significant enhancement of astaxanthin content was observed under 200 mM NaCl treatment [88]. The optimal condition to obtain the highest amount of fucoxanthin (79.40 ± 0.95 mg/g DW) of *Tisochrysis lutea* was determined as 36.27 g/L salt addition [64].

As the optimal conditions for cell growth are commonly different from those for secondary metabolite accumulation, two-stage cultivation has been widely applied to microalgae cultivation, a potent strategy to balance cell growth and metabolite accumulation. Acidophilic eukaryotic microalga *Coccomyxa onubensis* can endure moderate salt stress and serves as a potential resource for lutein production. Its growth rate and biomass productivity were significantly boosted under salt treatment (100 mM NaCl). By adding 500 mM NaCl, the lutein content was significantly enhanced by 47% to 7.80 mg/g DW, though the cell growth was inhibited [89]. Thus, cultures containing 100 mM salt can be applied at the first stage for higher biomass, then extra salt was added to induce lutein accumulation at the second stage.

Moreover, salt stress can also be coupled with other treatments to increase the carotenoid accumulation in microalgae. Light induction and nutrient starvation are widely used methods to boost carotenoid production. Combined with high light, salinity treatment increased the astaxanthin yield of *C. zofingiensis* 7.53-fold compared with the control [70]. *D. salina* was able to adapt to NaCl ranging from 0.05 to 5.5 M and the β-carotene content of *D. salina* achieved 13% of DW under salt stress, combined with high light at high temperature under nutrient deficiency [40].

The addition of chemicals can also boost carotenoid accumulation under salt stress. Photocatalyst TiO_2 can enhance zeaxanthin accumulation under salt stress in *Coelastrella sp.* by increasing oxidative stress [42]. The maximum zeaxanthin (13.2 ± 4.4 mg/g DW) was achieved under high salinity (3% NaCl) and N-starvation at 40 °C after TiO_2 treatment. An 0.25 mM amount of γ-aminobutyric acid (GABA) could facilitate astaxanthin productivity by 3.24-fold in *H. pluvialis* under high light with salinity treatment (2 g/L) [90]. Melatonin (MT) addition enhanced the expression of carotenogenic genes of *H. pluvialis* and induced astaxanthin accumulation by 1.20-fold under N-starvation and salt stress (1 g/L) [91]. NaCl treatment could also amplify the effect of linoleic acid (LA) on boosting astaxanthin accumulation in *Chlorella sorokiniana*, and LA could increase astaxanthin content by 1.25-fold in the presence of 20% NaCl (*w*/*v*) [92]. Strategies with salinity treatment to boost carotenoid accumulation in microalgae are summarized in Table 4.

Table 4. Salt treatment and salinity-combined conditions for carotenoids production.

Stress Conditions	Microalgae	Carotenoids	Fold Change	Ref.
100–500 mM NaCl (Two-stage)	*Coccomyxa onubensis*	Lutein	0.47-fold	[89]
200 mM NaCl	*C. zofingiensis* CCAP 211/14	Astaxanthin	1.23-fold	[88]
36.27 g/L NaCl	*Tisochrysis lutea*	Fucoxanthin	–	[64]
2% NaCl (*w*/*v*)	*Chromochloris zofingiensis bkt1*	Zeaxanthin Lutein β-carotene	1.38-fold 0.22-fold 0.36-fold	[33]
High light + NaCl	*Chromochloris zofingiensis*	Astaxanthin	7.53-fold	[70]
LA + NaCl (20%)	*Chlorella sorokiniana*	Astaxanthin	1.25-fold	[92]
GABA + high light + NaCl	*Haematococcus pluvialis*	Astaxanthin	3.24-fold	[90]
MT+ N-starvation + NaCl	*Haematococcus pluvialis*	Astaxanthin	1.20-fold	[91]
TiO_2 + N-starvation + NaCl	*Coelastrella sp.*	Zeaxanthin Astaxanthin	0.51-fold 1.16-fold	[42]

In summary, moderate salt stress can increase biomass productivity and induce accumulation of carotenoids, depending on microalgae species, while strict stress will be toxic and even fatal to microalgae. As both high cell density and carotenoid productivity are desirable for biotechnological goals, an optimal salt concentration should be chosen cautiously according to different cultivation goals and microalgae species. Two-stage or multi-stage cultivation and strategies combining salinity with other stress factors also have potential in microalgae cultivation for carotenoids.

4. Potential Applications of Internet of Things (IoT) in Carotenoids Production

The Internet of Things (IoT) is a huge ecosystem that connects things, machines and humans, anytime and anywhere. As the third wave of the world information industry, the IoT has attracted thriving research interest from various fields, such as the food supply chain [93], healthcare applications [94] and precision agriculture [95]. The number of IoT devices worldwide is predicted to reach 75 billion by 2025 [93]. Considering its successful experiences in agriculture and its advantages as a real-time, efficient and intelligent comprehensive data system of works, the IoT also shows competitiveness in microalgae biorefinery [96].

Microalgae biorefinery can be divided into two consecutive categories, the upstream processing and downstream processing [97]. The upstream processing mainly focuses on the high-density cultivation of selected microalgae and the regulation and optimization of the cultivation process, such as choosing the suitable trophic mode and light conditions. The downstream process mainly includes the harvest of microalgae, the separation and extraction of key products and subsequent processing to make microalgae-derived products. The IoT, as an ubiquitous, huge network, can be perfectly combined with traditional biotechnologies in microalgae biorefinery [96]. As shown in Figure 4, both upstream and downstream events are connected and collected, and all information will then be processed by the IoT platform. Then, how can the IoT be applied in the production of microalgae-derived carotenoids?

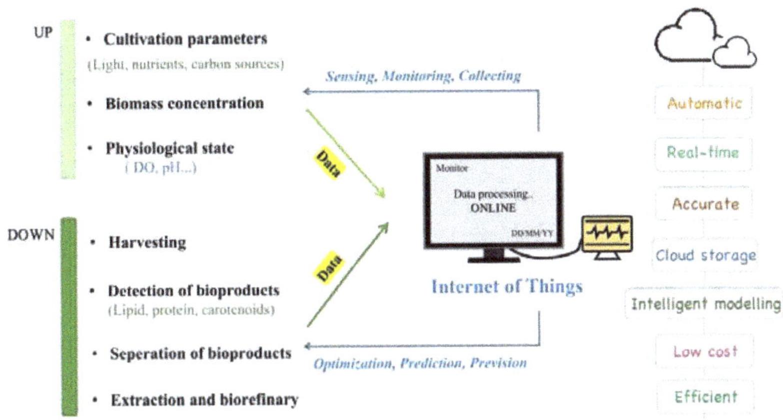

Figure 4. Characteristics of IoT and their potential utilizations in microalgae biorefinery. UP, upstream processing; DOWN, downstream processing.

For the upstream processing of carotenoid production, sensors are indispensable for data monitoring. Traditional detection techniques (such as measurement of biomass, cell size, etc.) are time-consuming, prone to causing culture contamination and cannot get data in time. In comparison, online sensors can monitor microalgae growth to gain real-time data, calculate unmeasured variables and even predict growth models.

For example, *H. pluvialis* is relatively sensitive to culture environment; its growth and cell composition can go far away from expectations without careful tuning. Possible contamination during sampling can disturb its growth. In addition, nutrient deficiency could drive it into a cyst structure and cause it to begin accumulating astaxanthin and other secondary metabolites [10]. In this case, innovative online sensors can monitor its growth in several aspects. Optical density (OD) sensors [98,99] can determine real-time biomass and dissolved oxygen (DO), CO_2 sensors [100] can screen metabolic activity, pH sensors and nitrate and phosphorous sensors [101] can guarantee a suitable environment and nutrient sufficiency; these can all be applied to ensure high-density cultivation of *H. pluvialis*. Moreover, the IoT can also act as the base of a decision support system. By combining the data from sensors with simulation models, the IoT platform can predict growth trends and provide cultivation suggestions to maximize biomass production [102]. Afterwards, automation units of the IoT are essential to achieve condition debugging at the right time after data analysis, such as a suitable feeding strategy [103]. As for cultivation with salt-stress treatments, the IoT will facilitate picking the right timepoint to conduct salinity pressure to adequate microalgal cell density.

For the downstream processing of carotenoid production, it is particularly critical to distinguish microalgal cells from other contaminants in the culture medium. Since traditional turbidimeters detect both microalgal cells and pollutants, the measurement results are often not accurate enough [104]. As fluorescence only comes from chlorophyll pigment of alive cells, convenient chlorophyll fluorescence sensors have been designed and widely applied in in situ detection of microalgal cells with high efficiency and accuracy [105,106].

Take the astaxanthin production from *C. zofingiensis* as an example. *C. zofingiensis* can synthesize astaxanthin in adversity and store it in lipid droplets [55]. As it can take up glucose for rapid growth, it is more likely to be contaminated by bacteria during the actual cultivation process. Chlorophyll fluorescence sensors can visibly separate microalgae from bacteria and cell debris, which can be applied in quantification of microalgae. Additionally, bioproducts content can be non-invasively detected *in situ*, such as with a triacylglycerides (TAG) observer [107]. As for cultivation with salt-stress treatments, the IoT will facilitate in distinguishing cell debris under salt stress to accurately harvest and quantify microalgal cells, and a TAG observer can also display astaxanthin content to achieve preliminary evaluation of productivities. The IoT, as an intelligent system, can also choose the best extraction method after the target product has been separated, which saves time while ensuring purity and extraction efficiency.

5. Future Prospects

5.1. Genetic Modifications of Microalgae for Salt Tolerance and Carotenoid Accmulation

Microalgae are widespread and large in number, and only a small part of them have been studied. Therefore, there are still numerous novel microalgae species that may be able to tolerate salinity and accumulate carotenoids. Here, we mainly focus on three methods to get expected microalgae strains: (1) adaptive evolution, (2) random mutagenesis and (3) targeted genetic engineering.

Different to strain isolation, adaptive evolution can obtain an expected strain with a target phenotype. Adaptive evolution has been widely utilized for strain improvement to stress tolerance in various microalgae species, such as *C. reinhardtii* for 200 mM salt tolerance [108] and *Chlorella sp.* for high phenol concentration [109]. It was performed for improving the tolerance of a freshwater strain *Chlorella sp.* AE10 to 30 g/L salt for 138 days (46 cycles). The genes of the resulting strain *Chlorella sp.* S30 related to Calvin–Benson cycle, C4-dicarboxylic acid cycle and crassulacean acid metabolism (CAM) pathways were upregulated, which was beneficial for CO_2 fixation under salt stress [85]. However, salinity acclimation of microalgae involves different functional genes and numerous pathways, which makes the exploitation of specific genes in salt tolerance more difficult.

Chemical mutagens, such as N-methyl-N′-nitro-N-nitrosoguanidine (MNNG) and ethyl methane sulfonate (EMS), are effective in mutant generation and may lead to mutants with enhanced carotenoid accumulation [110]. When genes in the carotenoid synthesis pathway (shown in Figure 1) are mutated, carotenoid profiles in microalgae may change. By chemical mutagenesis with EMS, carotenoid content in *Coelastrum sp.* C1-G1 was increased about 2-fold over its mother strain [39]. The Astaxanthin-overproduction strain *H. pluvialis* MT 2877 is a mutant by MNNG, which produced 4-fold astaxanthin over the WT strain [111]. *C. zofingiensis bkt1*, a chemical mutagen by MNNG, can accumulate high amounts of three essential carotenoids, e.g., zeaxanthin (7.00 $\pm$ 0.82 mg/g), lutein (13.81 $\pm$ 1.23 mg/g) and β-carotene (7.18 $\pm$ 0.72 mg/g) under different cultivation conditions [33], which can serve as a competent option for large-scale cultivation. *Chlorella sp.* was irradiated by ^{137}Se-γ ray and domesticated with a seawater culture medium (salinity 3% wt.) under 15% CO_2 stress. The biomass yield of the mutant was increased by 25% with 54.9% DW of lipids [112]. Although random mutagenesis could lead to novel traits, unpredictable traits may also show up, and the mutated genes still need further verification, such as gene sequencing, knockout and retro-complementation.

Targeted genetic engineering could generate specific insertions, deletions or substitutions in the host genome while avoiding random results [113]. Up until now, a lot of genes related to salt responses have been recognized, such as *nhaP* encoding a Na^+/H^+ antiporter, *codA* encoding choline oxidase to synthesize compatible solutes and *Dps* encoding DNA-binding proteins (Dps). Their introduction or overexpression strains have shown enhanced salt-tolerant abilities and potentials for seawater cultivation, which have been comprehensively reviewed [114].

Genetic engineering of microalgae for carotenoid hyperaccumulation is now a mature technology [115]. Genetic engineering of the green algae *C. zofingiensis* with a modified norflurazon-resistant endogenous *pds* gene resulted in up to 54.3% higher astaxanthin [116]. Transformation and expression of *dxs* and *psy* genes in *P. tricornutum* can increase fucoxanthin content by 2.4-fold and 1.8-fold, respectively [117]. *H. pluvialis* can also act as a gene donor. Transgenic *D. salina* with *bkt* and *chyb* genes from *H. pluvialis* was capable of astaxanthin biosynthesis with a better tolerance to high light [118].

At present, several microalgal genomes sequences are available in the public databases, such as *C. reinhardtii*, *Chlorella variabilis*, *Nanochloropsis graditana*, *C. zofingiensis*, etc. [119].

This offers worthy opportunities for precise engineering technology. Novel molecular tools have been applied to modify microalgal traits, including advanced genome editing systems, such as clustered regularly interspaced palindromic repeats and associated proteins (CRISPR-Cas), transcription activator-like effector nuclease (TALEN), zinc-finger nuclease (ZFN) and RNA interference (RNAi) [10,116]. For example, knocking down zeaxanthin epoxidase gene (*zep*) in *C. reinhardtii* by using DNA-free CRISPR-Cas9 RNP-mediated mutagenesis can increase zeaxanthin content by 56-fold [120].

5.2. Co-Production of Carotenoids with Value-Added Products from Microalgae

Lipids and several carotenoids (such as astaxanthin) are secondary metabolites of microalgae under adversity, which means certain stimulating conditions may promote the accumulation of two products simultaneously. In *H. pluvialis*, astaxanthin esterified by FAs is stored in lipid bodies, where TAGs are also positioned. Additionally, the biosynthesis of FAs and TAG has been demonstrated to be correlated with astaxanthin biosynthesis in *H. pluvialis* [121]. By adding 6 μM Cu^{2+}, the astaxanthin and lipid content was increased by 66.87 and 34.99%, respectively, with carotenogenic and lipogenic genes being upregulated at transcriptional level [122].

Microalgae also contain biofunctional unsaturated fatty acids. Omega-3 polyunsaturated fatty acids (Omega-3 PUFA) from microalgae, such as eicosapentaenoic acid (EPA. C20:5) and docosahexaenoic acid (DHA, C22:6), have been proved to have various health effects with dietary intake, lowering the risks of neurodegenerative diseases and cardiovascular disorders [123]. *Nannochloropsis oceanica* oil is rich in EPA, while its zeaxanthin content (30.2 mg/g DW) is relatively high among microalgae species [41], making it a potential species for co-production of zeaxanthin and EPA. *O. aurita* also contains high concentrations of EPA (28% of total fatty acids), which has been confirmed for large-scale culture in open ponds [47], making it competent for producing both fucoxanthin (18.47 mg/g DW) and EPA.

In addition, co-production of fucoxanthin and DHA is feasible in microalgae. The marine algae genus *Isochrysis* is rich in DHA, and it allows easy extraction of DHA and fucoxanthin as it has no cell wall. Sun et al. screened 16 strains of *Isochrysis* for DHA and fucoxanthin production, and *Isochrysis* CCMP1324 in semi-continuous cultivation excels with productivities as 9.05 and 7.96 mg/L/day of DHA and fucoxanthin, respectively [124]. *I. zhanjiangensis* accumulated both high content of fucoxanthin (23.29 mg/g DW) and stearidonic acid (SDA, 17.16% of total fatty acid), which is the common precursor of EPA and DHA [46].

Additionally, both cultivation strategies and efficient extraction methods are worth noting. As zeaxanthin and lipid are soluble in organic solvents, choosing an efficient extraction system, such as hexane: methanol (3:2), can realize the best extraction efficacy for both EPA (36.4 mg/g DW) and zeaxanthin (28.2 mg/g DW), which also avoid the usage of toxic chloroform [41]. As most fucoxanthin was mostly in the hydroalcoholic phase (over 99%), lipid and fucoxanthin can be separated by this biphasic system after the same extraction process with pure ethanol solvent [50].

The residual biomass after carotenoid extraction of microalgae can be further processed to harvest other by-products. Under nutrient starvation, *D. salina* accumulated β-carotene and triglycerides up to 30–60% DW [125]. After carotenoid extraction, the residuals can be employed for co-production of biodiesel after transesterification. In addition, the carbohydrate in *D. salina* can also be employed for bioethanol production [126].

6. Conclusions

In this review, we comprehensively introduced health-promoting carotenoids in microalgae, summarized the carotenoid biosynthetic pathways and discussed microalgal responses to salt stress. Advances in applying the Internet of Things (IoT) in modern biotechnology and co-production of carotenoids with other high-value products were also considered in detail. Through in-depth research on the effects of salt stress on microalgae, it is believed that the cultivation of freshwater microalgae in seawater will no longer be an obstacle in the future. Moreover, it is also important to exploit more potential salt-tolerant microalgae species for high-yield production of biofunctional carotenoids.

Author Contributions: Conceptualization, resources and supervision, F.C. and J.H.; writing—original draft preparation, Y.R. and J.D.; writing—review and editing, H.S., J.H. and F.C. All authors have read and agreed to the published version of the manuscript.

Funding: This research was funded by the National Natural Science Foundation of China (Project 31471717), Key Realm R&D Program of Guangdong Province (No. 2018B020206001), Guangdong Province Zhujiang Talent Program (No. 2019ZT08H476) and the Science and Technology Innovation Commission of Shenzhen (No. KQTD20180412181334790).

Institutional Review Board Statement: Not applicable.

Informed Consent Statement: Not applicable.

Data Availability Statement: Not applicable.

Conflicts of Interest: The authors declare no conflict of interest.

References

1. Liu, C.; Hu, B.; Cheng, Y.; Guo, Y.; Yao, W.; Qian, H. Carotenoids from fungi and microalgae: A review on their recent production, extraction, and developments. *Bioresour. Technol.* **2021**, *337*, 125398. [CrossRef]
2. Qiang, S.; Su, A.P.; Li, Y.; Chen, Z.; Hu, C.Y.; Meng, Y.H. Elevated β-Carotene Synthesis by the Engineered Rhodobacter sphaeroides with Enhanced CrtY Expression. *J. Agric. Food Chem.* **2019**, *67*, 9560–9568. [CrossRef]
3. Roberts, R.L.; Green, J.; Lewis, B. Lutein and zeaxanthin in eye and skin health. *Clin. Dermatol.* **2009**, *27*, 195–201. [CrossRef]
4. Le Goff, M.; Le Ferrec, E.; Mayer, C.; Mimouni, V.; Lagadic-Gossmann, D.; Schoefs, B.; Ulmann, L. Microalgal carotenoids and phytosterols regulate biochemical mechanisms involved in human health and disease prevention. *Biochimie* **2019**, *167*, 106–118. [CrossRef]
5. Saini, R.K.; Keum, Y.-S. Carotenoid extraction methods: A review of recent developments. *Food Chem.* **2018**, *240*, 90–103. [CrossRef]
6. Kumar, L.; Bharadvaja, N. A review on microalgae biofuel and biorefinery: Challenges and way forward. *Energy Sources Part A Recover. Util. Environ. Eff.* **2020**, *42*, 1–24. [CrossRef]
7. Liu, Z.; Berg, C.V.D.; Weusthuis, R.A.; Dragone, G.; Mussatto, S.I. Strategies for an improved extraction and separation of lipids and carotenoids from oleaginous yeast. *Sep. Purif. Technol.* **2020**, *257*, 117946. [CrossRef]
8. Gong, M.; Bassi, A. Carotenoids from microalgae: A review of recent developments. *Biotechnol. Adv.* **2016**, *34*, 1396–1412. [CrossRef] [PubMed]
9. Sinetova, M.A.; Sidorov, R.A.; Medvedeva, A.A.; Starikov, A.Y.; Markelova, A.G.; Allakhverdiev, S.I.; Los, D.A. Effect of salt stress on physiological parameters of microalgae Vischeria punctata strain IPPAS H-242, a superproducer of eicosapentaenoic acid. *J. Biotechnol.* **2021**, *331*, 63–73. [CrossRef]
10. Ren, Y.; Deng, J.; Huang, J.; Wu, Z.; Yi, L.; Bi, Y.; Chen, F. Using green alga *Haematococcus pluvialis* for astaxanthin and lipid co-production: Advances and outlook. *Bioresour. Technol.* **2021**, *340*, 125736. [CrossRef] [PubMed]
11. Liyanaarachchi, V.C.; Premaratne, M.; Ariyadasa, T.U.; Nimarshana, P.; Malik, A. Two-stage cultivation of microalgae for production of high-value compounds and biofuels: A review. *Algal Res.* **2021**, *57*, 102353. [CrossRef]
12. Martin, L.J. Fucoxanthin and Its Metabolite Fucoxanthinol in Cancer Prevention and Treatment. *Mar. Drugs* **2015**, *13*, 4784–4798. [CrossRef]
13. Bourdon, L.; Jensen, A.A.; Kavanagh, J.M.; McClure, D.D. Microalgal production of zeaxanthin. *Algal Res.* **2021**, *55*, 102266. [CrossRef]
14. Nisar, N.; Li, L.; Lu, S.; Khin, N.C.; Pogson, B.J. Carotenoid Metabolism in Plants. *Mol. Plant* **2015**, *8*, 68–82. [CrossRef]
15. Mao, X.; Zhang, Y.; Wang, X.; Liu, J. Novel insights into salinity-induced lipogenesis and carotenogenesis in the oleaginous astaxanthin-producing alga Chromochloris zofingiensis: A multi-omics study. *Biotechnol. Biofuels* **2020**, *13*, 1–24. [CrossRef] [PubMed]
16. Sun, T.; Li, L. Toward the 'golden' era: The status in uncovering the regulatory control of carotenoid accumulation in plants. *Plant Sci.* **2019**, *290*, 110331. [CrossRef]
17. Bhat, I.; Mamatha, B.S. Genetic factors involved in modulating lutein bioavailability. *Nutr. Res.* **2021**, *91*, 36–43. [CrossRef]
18. Fang, N.; Wang, C.; Liu, X.; Zhao, X.; Liu, Y.; Liu, X.; Du, Y.; Zhang, Z.; Zhang, H. De novo synthesis of astaxanthin: From organisms to genes. *Trends Food Sci. Technol.* **2019**, *92*, 162–171. [CrossRef]
19. Ye, X.; Al-Babili, S.; Klöti, A.; Zhang, J.; Lucca, P.; Beyer, P.; Potrykus, I. Engineering the Provitamin A (β-Carotene) Biosynthetic Pathway into (Carotenoid-Free) Rice Endosperm. *Science* **2000**, *287*, 303–305. [CrossRef]

20. Bouyahya, A.; El Omari, N.; Hakkur, M.; El Hachlafi, N.; Charfi, S.; Balahbib, A.; Guaouguaou, F.-E.; Rebezov, M.; Maksimiuk, N.; Shariati, M.A.; et al. Sources, health benefits, and biological properties of zeaxanthin. *Trends Food Sci. Technol.* **2021**, *118*, 519–538. [CrossRef]
21. Muradian, K.; Vaiserman, A.; Min, K.-J.; Fraifeld, V. Fucoxanthin and lipid metabolism: A minireview. *Nutr. Metab. Cardiovasc. Dis.* **2015**, *25*, 891–897. [CrossRef] [PubMed]
22. Butnariu, M.; Rodino, S.; Petrache, P.; Butu, M. Determination and quantification of maize zeaxanthin stability. *Dig. J. Nanomater. Bios.* **2014**, *9*, 745–755.
23. Lin, J.-H.; Lee, D.-J.; Chang, J.-S. Lutein production from biomass: Marigold flowers versus microalgae. *Bioresour. Technol.* **2014**, *184*, 421–428. [CrossRef]
24. Fernández-Sevilla, J.M.; Fernandez, F.G.A.; Grima, E.M. Biotechnological production of lutein and its applications. *Appl. Microbiol. Biotechnol.* **2010**, *86*, 27–40. [CrossRef] [PubMed]
25. Zheng, H.; Wang, Y.; Li, S.; Nagarajan, D.; Varjani, S.; Lee, D.-J.; Chang, J.-S. Recent advances in lutein production from microalgae. *Renew. Sustain. Energy Rev.* **2021**, *153*, 111795. [CrossRef]
26. Dineshkumar, R.; Subramanian, G.; Dash, S.K.; Sen, R. Development of an optimal light-feeding strategy coupled with semi-continuous reactor operation for simultaneous improvement of microalgal photosynthetic efficiency, lutein production and CO_2 sequestration. *Biochem. Eng. J.* **2016**, *113*, 47–56. [CrossRef]
27. Chen, J.-H.; Chen, C.-Y.; Hasunuma, T.; Kondo, A.; Chang, C.-H.; Ng, I.-S.; Chang, J.-S. Enhancing lutein production with mixotrophic cultivation of *Chlorella sorokiniana* MB-1-M12 using different bioprocess operation strategies. *Bioresour. Technol.* **2019**, *278*, 17–25. [CrossRef] [PubMed]
28. Wang, X.; Zhang, M.-M.; Sun, Z.; Liu, S.-F.; Qin, Z.-H.; Mou, J.; Zhou, Z.-G.; Lin, C.S.K. Sustainable lipid and lutein production from *Chlorella mixotrophic* fermentation by food waste hydrolysate. *J. Hazard. Mater.* **2020**, *400*, 123258. [CrossRef] [PubMed]
29. Ma, R.; Zhang, Z.; Ho, S.-H.; Ruan, C.; Li, J.; Xie, Y.; Shi, X.; Liu, L.; Chen, J. Two-stage bioprocess for hyper-production of lutein from microalga *Chlorella sorokiniana* FZU60: Effects of temperature, light intensity, and operation strategies. *Algal Res.* **2020**, *52*, 102119. [CrossRef]
30. De Bhowmick, G.; Sen, R.; Sarmah, A.K. Consolidated bioprocessing of wastewater cocktail in an algal biorefinery for enhanced biomass, lipid and lutein production coupled with efficient CO_2 capture: An advanced optimization approach. *J. Environ. Manag.* **2019**, *252*, 109696. [CrossRef]
31. Chen, J.-H.; Kato, Y.; Matsuda, M.; Chen, C.-Y.; Nagarajan, D.; Hasunuma, T.; Kondo, A.; Dong, C.-D.; Lee, D.-J.; Chang, J.-S. A novel process for the mixotrophic production of lutein with *Chlorella sorokiniana* MB-1-M12 using aquaculture wastewater. *Bioresour. Technol.* **2019**, *290*, 121786. [CrossRef]
32. Heo, J.; Shin, D.-S.; Cho, K.; Cho, D.-H.; Lee, Y.J.; Kim, H.-S. Indigenous microalga *Parachlorella* sp. JD-076 as a potential source for lutein production: Optimization of lutein productivity via regulation of light intensity and carbon source. *Algal Res.* **2018**, *33*, 1–7. [CrossRef]
33. Huang, W.; Lin, Y.; He, M.; Gong, Y.; Huang, J. Induced High-Yield Production of Zeaxanthin, Lutein, and β-Carotene by a Mutant of *Chlorella zofingiensis*. *J. Agric. Food Chem.* **2018**, *66*, 891–897. [CrossRef] [PubMed]
34. Gong, M.; Bassi, A. Investigation of *Chlorella vulgaris* UTEX 265 Cultivation under Light and Low Temperature Stressed Conditions for Lutein Production in Flasks and the Coiled Tree Photo-Bioreactor (CTPBR). *Appl. Biochem. Biotechnol.* **2017**, *183*, 652–671. [CrossRef]
35. McClure, D.D.; Nightingale, J.K.; Luiz, A.; Black, S.; Zhu, J.; Kavanagh, J.M. Pilot-scale production of lutein using *Chlorella vulgaris*. *Algal Res.* **2019**, *44*, 101707. [CrossRef]
36. Přibyl, P.; Pilný, J.; Cepák, V.; Kaštánek, P. The role of light and nitrogen in growth and carotenoid accumulation in *Scenedesmus* sp. *Algal Res.* **2016**, *16*, 69–75. [CrossRef]
37. Wan, M.; Hou, D.; Li, Y.; Fan, J.; Huang, J.; Liang, S.; Wang, W.; Pan, R.; Wang, J.; Li, S. The effective photoinduction of *Haematococcus pluvialis* for accumulating astaxanthin with attached cultivation. *Bioresour. Technol.* **2014**, *163*, 26–32. [CrossRef] [PubMed]
38. Liu, J.; Sun, Z.; Gerken, H.; Liu, Z.; Jiang, Y.; Chen, F. *Chlorella zofingiensis* as an Alternative Microalgal Producer of Astaxanthin: Biology and Industrial Potential. *Mar. Drugs* **2014**, *12*, 3487–3515. [CrossRef]
39. Tharek, A.; Yahya, A.; Salleh, M.M.; Jamaluddin, H.; Yoshizaki, S.; Hara, H.; Iwamoto, K.; Suzuki, I.; Mohamad, S.E. Improvement and screening of astaxanthin producing mutants of newly isolated Coelastrum sp. using ethyl methane sulfonate induced mutagenesis technique. *Biotechnol. Rep.* **2021**, *32*, e00673. [CrossRef]
40. Rammuni, M.; Ariyadasa, T.U.; Nimarshana, P.; Attalage, R. Comparative assessment on the extraction of carotenoids from microalgal sources: Astaxanthin from H. pluvialis and β-carotene from D. salina. *Food Chem.* **2018**, *277*, 128–134. [CrossRef]
41. Mitra, M.; Mishra, S. A comparative analysis of different extraction solvent systems on the extractability of eicosapentaenoic acid from the marine eustigmatophyte Nannochloropsis oceanica. *Algal Res.* **2019**, *38*, 101387. [CrossRef]
42. Pushpalatha, S.; Sangeetha, R.; Ariraman, S.; Ashokkumar, B.; Varalakshmi, P. Photocatalyst (TiO_2) as an enhancer: An at-tempt to enhance the production of carotenoids and lipids with the combined oxidative stresses in *Coelastrella* sp. *MClean Technol. Environ. Policy* **2020**, *23*, 41–53. [CrossRef]
43. Jin, E.; Feth, B.; Melis, A. A mutant of the green alga *Dunaliella salina* constitutively accumulates zeaxanthin under all growth conditions. *Biotechnol. Bioeng.* **2002**, *81*, 115–124. [CrossRef] [PubMed]

44. Koo, S.Y.; Cha, K.H.; Song, D.-G.; Chung, D.; Pan, C.-H. Optimization of pressurized liquid extraction of zeaxanthin from *Chlorella ellipsoidea*. *Environ. Boil. Fishes* **2011**, *24*, 725–730. [CrossRef]
45. Petrushkina, M.; Gusev, E.; Sorokin, B.; Zotko, N.; Mamaeva, A.; Filimonova, A.; Kulikovskiy, M.; Maltsev, Y.; Yampolsky, I.; Guglya, E.; et al. Fucoxanthin production by heterokont microalgae. *Algal Res.* **2017**, *24*, 387–393. [CrossRef]
46. Li, Y.; Sun, H.; Wu, T.; Fu, Y.; He, Y.; Mao, X.; Chen, F. Storage carbon metabolism of Isochrysis zhangjiangensis under different light intensities and its application for co-production of fucoxanthin and stearidonic acid. *Bioresour. Technol.* **2019**, *282*, 94–102. [CrossRef]
47. Xia, S.; Wang, K.; Wan, L.; Li, A.; Hu, Q.; Zhang, C. Production, Characterization, and Antioxidant Activity of Fucoxanthin from the Marine Diatom Odontella aurita. *Mar. Drugs* **2013**, *11*, 2667–2681. [CrossRef]
48. Kim, S.M.; Kang, S.-W.; Kwon, O.N.; Chung, D.; Pan, C.-H. Fucoxanthin as a major carotenoid in Isochrysis aff. galbana: Characterization of extraction for commercial application. *J. Korean Soc. Appl. Biol. Chem.* **2012**, *55*, 477–483. [CrossRef]
49. Gao, F.; Cabanelas, I.I.T.; Wijffels, R.H.; Barbosa, M.J. Process optimization of fucoxanthin production with Tisochrysis lutea. *Bioresour. Technol.* **2020**, *315*, 123894. [CrossRef]
50. Kim, S.M.; Jung, Y.-J.; Kwon, O.-N.; Cha, K.H.; Um, B.-H.; Chung, D.; Pan, C.-H. A Potential Commercial Source of Fucoxanthin Extracted from the Microalga Phaeodactylum tricornutum. *Appl. Biochem. Biotechnol.* **2012**, *166*, 1843–1855. [CrossRef]
51. Zhao, T.; Yan, X.; Sun, L.; Yang, T.; Hu, X.; He, Z.; Liu, F.; Liu, X. Research progress on extraction, biological activities and delivery systems of natural astaxanthin. *Trends Food Sci. Technol.* **2019**, *91*, 354–361. [CrossRef]
52. Shah, M.; Mahfuzur, R.; Liang, Y.; Cheng, J.J.; Daroch, M. Astaxanthin-Producing Green Microalga *Haematococcus pluvialis*: From Single Cell to High Value Commercial Products. *Front. Plant Sci.* **2016**, *7*, 531. [CrossRef]
53. Jannel, S.; Caro, Y.; Bermudes, M.; Petit, T. Novel Insights into the Biotechnological Production of *Haematococcus pluvialis*-Derived Astaxanthin: Advances and Key Challenges to Allow Its Industrial Use as Novel Food Ingredient. *J. Mar. Sci. Eng.* **2020**, *8*, 789. [CrossRef]
54. Torres-Tiji, Y.; Fields, F.J.; Mayfield, S.P. Microalgae as a future food source. *Biotechnol. Adv.* **2020**, *41*, 107536. [CrossRef] [PubMed]
55. Sun, H.; Ren, Y.; Fan, Y.; Lu, X.; Zhao, W.; Chen, F. Systematic metabolic tools reveal underlying mechanism of product biosynthesis in Chromochloris zofingiensis. *Bioresour. Technol.* **2021**, *337*, 125406. [CrossRef]
56. Gorgich, M.; Martins, A.A.; Mata, T.M.; Caetano, N.S. Composition, cultivation and potential applications of *Chlorella zofingiensis*—A comprehensive review. *Algal Res.* **2021**, *60*, 102508. [CrossRef]
57. Wang, L.; Liu, Z.; Jiang, H.; Mao, X. Biotechnology advances in β-carotene production by microorganisms. *Trends Food Sci. Technol.* **2021**, *111*, 322–332. [CrossRef]
58. Raposo, M.F.D.J.; De Morais, A.M.M.B.; Morais, R. Carotenoids from Marine Microalgae: A Valuable Natural Source for the Prevention of Chronic Diseases. *Mar. Drugs* **2015**, *13*, 5128–5155. [CrossRef] [PubMed]
59. Lamers, P.P.; Janssen, M.; De Vos, R.C.; Bino, R.J.; Wijffels, R.H. Exploring and exploiting carotenoid accumulation in *Dunaliella salina* for cell-factory applications. *Trends Biotechnol.* **2008**, *26*, 631–638. [CrossRef] [PubMed]
60. Roso, G.R. The Bioeconomy of Microalgal Carotenoid-Rich Oleoresins Produced in Agroindustrial Biorefineries. *J. Biosens. Bioelectron.* **2015**, *6*. [CrossRef]
61. Rodrigues, D.B.; Mercadante, A.; Mariutti, L.R.B. Marigold carotenoids: Much more than lutein esters. *Food Res. Int.* **2018**, *119*, 653–664. [CrossRef]
62. Matsuno, T. Aquatic animal carotenoids. *Fish. Sci.* **2001**, *67*, 771–783. [CrossRef]
63. Leong, Y.K.; Chen, C.-Y.; Varjani, S.; Chang, J.-S. Producing fucoxanthin from algae—Recent advances in cultivation strategies and downstream processing. *Bioresour. Technol.* **2021**, *344*, 126170. [CrossRef] [PubMed]
64. Mohamadnia, S.; Tavakoli, O.; Faramarzi, M.A. Enhancing production of fucoxanthin by the optimization of culture media of the microalga Tisochrysis lutea. *Aquaculture* **2020**, *533*, 736074. [CrossRef]
65. Shiratori, K.; Ohgami, K.; Ilieva, I.; Jin, X.-H.; Koyama, Y.; Miyashita, K.; Yoshida, K.; Kase, S.; Ohno, S. Effects of fucoxanthin on lipopolysaccharide-induced inflammation in vitro and in vivo. *Exp. Eye Res.* **2005**, *81*, 422–428. [CrossRef] [PubMed]
66. Kanazawa, K.; Ozaki, Y.; Hashimoto, T.; Das, S.K.; Matsushita, S.; Hirano, M.; Okada, T.; Komoto, A.; Mori, N.; Nakatsuka, M. Commercial-scale Preparation of Biofunctional Fucoxanthin from Waste Parts of Brown Sea Algae Laminalia japonica. *Food Sci. Technol. Res.* **2008**, *14*, 573–582. [CrossRef]
67. Kajikawa, T.; Okumura, S.; Iwashita, T.; Kosumi, D.; Hashimoto, H.; Katsumura, S. Stereocontrolled Total Synthesis of Fucoxanthin and Its Polyene Chain-Modified Derivative. *Org. Lett.* **2012**, *14*, 808–811. [CrossRef]
68. Arif, Y.; Singh, P.; Siddiqui, H.; Bajguz, A.; Hayat, S. Salinity induced physiological and biochemical changes in plants: An omic approach towards salt stress tolerance. *Plant Physiol. Biochem.* **2020**, *156*, 64–77. [CrossRef]
69. Yu, Z.; Duan, X.; Luo, L.; Dai, S.; Ding, Z.; Xia, G. How Plant Hormones Mediate Salt Stress Responses. *Trends Plant Sci.* **2020**, *25*, 1117–1130. [CrossRef]
70. Kou, Y.; Liu, M.; Sun, P.; Dong, Z.; Liu, J. High light boosts salinity stress-induced biosynthesis of astaxanthin and lipids in the green alga Chromochloris zofingiensis. *Algal Res.* **2020**, *50*, 101976. [CrossRef]
71. Gong, Z.; Xiong, L.; Shi, H.; Yang, S.; Herrera-Estrella, L.R.; Xu, G.; Chao, D.-Y.; Li, J.; Wang, P.-Y.; Qin, F.; et al. Plant abiotic stress response and nutrient use efficiency. *Sci. China Life Sci.* **2020**, *63*, 635–674. [CrossRef] [PubMed]
72. Choi, W.-G.; Toyota, M.; Kim, S.-H.; Hilleary, R.; Gilroy, S. Salt stress-induced Ca2+ waves are associated with rapid, long-distance root-to-shoot signaling in plants. *Proc. Natl. Acad. Sci. USA* **2014**, *111*, 6497–6502. [CrossRef] [PubMed]

73. Wheeler, G.L.; Brownlee, C. Ca2+ signalling in plants and green algae—Changing channels. *Trends Plant Sci.* **2008**, *13*, 506–514. [CrossRef]
74. Khanna, R.R.; Jahan, B.; Iqbal, N.; Khan, N.A.; AlAjmi, M.F.; Rehman, T.; Khan, M.I.R. GABA reverses salt-inhibited photosynthetic and growth responses through its influence on NO-mediated nitrogen-sulfur assimilation and antioxidant system in wheat. *J. Biotechnol.* **2020**, *325*, 73–82. [CrossRef] [PubMed]
75. Barbosa, J.M.; Singh, N.K.; Cherry, J.H.; Locy, R.D. Nitrate uptake and utilization is modulated by exogenous γ-aminobutyric acid in Arabidopsis thaliana seedlings. *Plant Physiol. Biochem.* **2010**, *48*, 443–450. [CrossRef]
76. Siddiqui, M.H.; Al-Whaibi, M.H.; Basalah, M.O. Role of nitric oxide in tolerance of plants to abiotic stress. *Protoplasma* **2010**, *248*, 447–455. [CrossRef] [PubMed]
77. Procházková, D.; Haisel, D.; Wilhelmová, N.; Pavlíková, D.; Száková, J. Effects of exogenous nitric oxide on photosynthesis. *Photosynthetica* **2013**, *51*, 483–489. [CrossRef]
78. Zhao, Y.; Wang, H.-P.; Han, B.; Yu, X. Coupling of abiotic stresses and phytohormones for the production of lipids and high-value by-products by microalgae: A review. *Bioresour. Technol.* **2018**, *274*, 549–556. [CrossRef]
79. Li, Q.; You, J.; Qiao, T.; Zhong, D.-B.; Yu, X. Sodium chloride stimulates the biomass and astaxanthin production by *Haematococcus pluvialis* via a two-stage cultivation strategy. *Bioresour. Technol.* **2021**, *344*, 126214. [CrossRef] [PubMed]
80. Elloumi, W.; Jebali, A.; Maalej, A.; Chamkha, M.; Sayadi, S. Effect of Mild Salinity Stress on the Growth, Fatty Acid and Carotenoid Compositions, and Biological Activities of the Thermal Freshwater Microalgae *Scenedesmus* sp. *Biomolecules* **2020**, *10*, 1515. [CrossRef]
81. Rao, A.R.; Dayananda, C.; Sarada, R.; Shamala, T.; Ravishankar, G. Effect of salinity on growth of green alga Botryococcus braunii and its constituents. *Bioresour. Technol.* **2007**, *98*, 560–564. [CrossRef] [PubMed]
82. Lv, H.; Kim, M.; Park, S.; Baek, K.; Oh, H.; Polle, J.E.; Jin, E. Comparative transcriptome analysis of short-term responses to salt and glycerol hyperosmotic stress in the green alga *Dunaliella salina*. *Algal Res.* **2020**, *53*, 102147. [CrossRef]
83. Morales-Sánchez, D.; Kim, Y.; Terng, E.L.; Peterson, L.; Cerutti, H. A multidomain enzyme, with glycerol-3-phosphate dehydrogenase and phosphatase activities, is involved in a chloroplastic pathway for glycerol synthesis in Chlamydomonas reinhardtii. *Plant J.* **2017**, *90*, 1079–1092. [CrossRef] [PubMed]
84. Kato, Y.; Ho, S.-H.; Vavricka, C.J.; Chang, J.-S.; Hasunuma, T.; Kondo, A. Evolutionary engineering of salt-resistant Chlamydomonas sp. strains reveals salinity stress-activated starch-to-lipid biosynthesis switching. *Bioresour. Technol.* **2017**, *245*, 1484–1490. [CrossRef] [PubMed]
85. Li, X.; Yuan, Y.; Cheng, D.; Gao, J.; Kong, L.; Zhao, Q.; Wei, W.; Sun, Y. Exploring stress tolerance mechanism of evolved freshwater strain *Chlorella* sp. S30 under 30 g/L salt. *Bioresour. Technol.* **2018**, *250*, 495–504. [CrossRef]
86. Cheng, R.; Feng, J.; Zhang, B.-X.; Huang, Y.; Cheng, J.; Zhang, C.-X. Transcriptome and Gene Expression Analysis of an Oleaginous Diatom Under Different Salinity Conditions. *BioEnergy Res.* **2013**, *7*, 192–205. [CrossRef]
87. Gao, Z.; Meng, C.; Chen, Y.C.; Ahmed, F.; Mangott, A.; Schenk, P.M.; Li, Y. Comparison of astaxanthin accumulation and biosynthesis gene expression of three *Haematococcus pluvialis* strains upon salinity stress. *Environ. Boil. Fishes* **2014**, *27*, 1853–1860. [CrossRef]
88. Del Campo, J.A.; Rodríguez, H.; Moreno, J.; Vargas, M.Á.; Rivas, J.; Guerrero, M.G. Accumulation of astaxanthin and lutein in *Chlorella zofingiensis* (Chlorophyta). *Appl. Microbiol. Biotechnol.* **2004**, *64*, 848–854. [CrossRef]
89. Bermejo, E.; Ruiz-Domínguez, M.C.; Cuaresma, M.; Vaquero, I.; Ramos-Merchante, A.; Vega, J.M.; Vílchez, C.; Garbayo, I. Production of lutein, and polyunsaturated fatty acids by the acidophilic eukaryotic microalga *Coccomyxa onubensis* under abiotic stress by salt or ultraviolet light. *J. Biosci. Bioeng.* **2018**, *125*, 669–675. [CrossRef]
90. Li, Q.; Zhao, Y.; Ding, W.; Han, B.; Geng, S.; Ning, D.; Ma, T.; Yu, X. Gamma-aminobutyric acid facilitates the simultaneous production of biomass, astaxanthin and lipids in *Haematococcus pluvialis* under salinity and high-light stress conditions. *Bioresour. Technol.* **2020**, *320*, 124418. [CrossRef] [PubMed]
91. Xing, H.; Zhao, Y.; Li, T.; Han, B.; Zhao, P.; Yu, X. Enhancing astaxanthin and lipid coproduction in *Haematococcus pluvialis* by the combined induction of plant growth regulators and multiple stresses. *Bioresour. Technol.* **2021**, *344*, 126225. [CrossRef] [PubMed]
92. Khalili, Z.; Jalili, H.; Noroozi, M.; Amrane, A. Effect of linoleic acid and methyl jasmonate on astaxanthin content of Scenedesmus acutus and *Chlorella sorokiniana* under heterotrophic cultivation and salt shock conditions. *Environ. Boil. Fishes* **2019**, *31*, 2811–2822. [CrossRef]
93. Maroli, A.; Narwane, V.S.; Gardas, B.B. Applications of IoT for achieving sustainability in agricultural sector: A comprehensive review. *J. Environ. Manag.* **2021**, *298*, 113488. [CrossRef] [PubMed]
94. Kashani, M.H.; Madanipour, M.; Nikravan, M.; Asghari, P.; Mahdipour, E. A systematic review of IoT in healthcare: Applications, techniques, and trends. *J. Netw. Comput. Appl.* **2021**, *192*, 103164. [CrossRef]
95. Akhter, R.; Sofi, S.A. Precision agriculture using IoT data analytics and machine learning. *J. King Saud Univ.—Comput. Inf. Sci.* **2021**. [CrossRef]
96. Wang, K.; Khoo, K.S.; Leong, H.Y.; Nagarajan, D.; Chew, K.W.; Ting, H.Y.; Selvarajoo, A.; Chang, J.-S.; Show, P.L. How does the Internet of Things (IoT) help in microalgae biorefinery? *Biotechnol. Adv.* **2021**, 107819. [CrossRef]
97. Tan, X.B.; Lam, M.K.; Uemura, Y.; Lim, J.W.; Wong, C.Y.; Lee, K.T. Cultivation of microalgae for biodiesel production: A review on upstream and downstream processing. *Chin. J. Chem. Eng.* **2018**, *26*, 17–30. [CrossRef]

98. Krujatz, F.; Fehse, K.; Jahnel, M.; Gommel, C.; Schurig, C.; Lindner, F.; Bley, T.; Weber, J.; Steingroewer, J. MicrOLED-photobioreactor: Design and characterization of a milliliter-scale Flat-Panel-Airlift-photobioreactor with optical process monitoring. *Algal Res.* **2016**, *18*, 225–234. [CrossRef]
99. Barbosa, R.C.; Soares, J.; Martins, M.A. Low-cost and versatile sensor based on multi-wavelengths for real-time estimation of microalgal biomass concentration in open and closed cultivation systems. *Comput. Electron. Agric.* **2020**, *176*, 105641. [CrossRef]
100. Haigh-Flórez, D.; Cano-Raya, C.; Bedoya, M.; Orellana, G. Rugged fibre-optic luminescent sensor for CO_2 determination in microalgae photoreactors for biofuel production. *Sens. Actuators B Chem.* **2015**, *221*, 978–984. [CrossRef]
101. Martin, J.; Dannenberg, A.; Detrell, G.; Ewald, S.F. Noninvasive process control of a microalgae-based system for automated treatment of polluted agricultural ground water transferred from the development of a biological Life Support Systems. In Proceedings of the International Conference on Environmental Systems; 2020. Available online: https://www.ices.space/conference-proceedings-2/ (accessed on 10 December 2021).
102. Giannino, F.; Esposito, S.; Diano, M.; Cuomo, S.; Toraldo, G. A predictive Decision Support System (DSS) for a microalgae production plant based on Internet of Things paradigm. *Concurr. Comput. Pract. Exp.* **2018**, *30*, e4476. [CrossRef]
103. Havlik, I.; Scheper, T.; Reardon, K.F. Monitoring of Microalgal Processes. In *Microalgae Biotechnology*; Posten, C., Feng Chen, S., Eds.; Springer International Publishing: Cham, Switzerland, 2016; pp. 89–142.
104. Benson, B.C.; Gutierrez-Wing, M.T.; Rusch, K.A. Optimization of the lighting system for a Hydraulically Integrated Serial Turbidostat Algal Reactor (HISTAR): Economic implications. *Aquac. Eng.* **2009**, *40*, 45–53. [CrossRef]
105. Dufková, K.; Barták, M.; Morkusová, J.; Elster, J.; Hájek, J. Screening of growth phases of Antarctic algae and cyanobacteria cultivated on agar plates by chlorophyll fluorescence imaging. *Czech Polar Rep.* **2019**, *9*, 170–181. [CrossRef]
106. Takahashi, T. Routine Management of Microalgae Using Autofluorescence from Chlorophyll. *Molecules* **2019**, *24*, 4441. [CrossRef]
107. Mairet, F.; Moisan, M.; Bernard, O. Estimation of neutral lipid and carbohydrate quotas in microalgae using adaptive interval observers. *Bioprocess Biosyst. Eng.* **2013**, *37*, 51–61. [CrossRef] [PubMed]
108. Perrineau, M.M.; Zelzion, E.; Gross, J.; Price, D.C.; Boyd, J.; Bhattacharya, D. Evolution of salt tolerance in a laboratory reared population of Chlamydomonas reinhardtii. *Environ. Microbiol.* **2014**, *16*, 1755–1766. [CrossRef]
109. Wang, L.; Xue, C.; Wang, L.; Zhao, Q.; Wei, W.; Sun, Y. Strain improvement of *Chlorella* sp. for phenol biodegradation by adaptive laboratory evolution. *Bioresour. Technol.* **2016**, *205*, 264–268. [CrossRef]
110. Kamath, B.S.; Vidhyavathi, R.; Sarada, R.; Ravishankar, G. Enhancement of carotenoids by mutation and stress induced carotenogenic genes in *Haematococcus pluvialis* mutants. *Bioresour. Technol.* **2008**, *99*, 8667–8673. [CrossRef] [PubMed]
111. Hu, Z.; Li, Y.; Sommerfeld, M.; Chen, F.; Hu, Q. Enhanced protection against oxidative stress in an astaxanthin-overproduction *Haematococcus mutant* (Chlorophyceae). *Eur. J. Phycol.* **2008**, *43*, 365–376. [CrossRef]
112. Cheng, J.; Lu, H.; Huang, Y.; Li, K.; Huang, R.; Zhou, J.; Cen, K. Enhancing growth rate and lipid yield of *Chlorella* with nuclear irradiation under high salt and CO_2 stress. *Bioresour. Technol.* **2016**, *203*, 220–227. [CrossRef]
113. Muñoz, C.F.; Südfeld, C.; Naduthodi, M.I.; Weusthuis, R.A.; Barbosa, M.J.; Wijffels, R.H.; D'Adamo, S. Genetic engineering of microalgae for enhanced lipid production. *Biotechnol. Adv.* **2021**, *52*, 107836. [CrossRef]
114. Cui, J.; Sun, T.; Chen, L.; Zhang, W. Engineering salt tolerance of photosynthetic cyanobacteria for seawater utilization. *Biotechnol. Adv.* **2020**, *43*, 107578. [CrossRef] [PubMed]
115. Sproles, A.E.; Fields, F.J.; Smalley, T.N.; Le, C.H.; Badary, A.; Mayfield, S.P. Recent advancements in the genetic engineering of microalgae. *Algal Res.* **2020**, *53*, 102158. [CrossRef]
116. Liu, J.; Sun, Z.; Gerken, H.; Huang, J.; Jiang, Y.; Chen, F. Genetic engineering of the green alga *Chlorella zofingiensis*: A modified norflurazon-resistant phytoene desaturase gene as a dominant selectable marker. *Appl. Microbiol. Biotechnol.* **2014**, *98*, 5069–5079. [CrossRef] [PubMed]
117. Eilers, U.; Bikoulis, A.; Breitenbach, J.; Büchel, C.; Sandmann, G. Limitations in the biosynthesis of fucoxanthin as targets for genetic engineering in Phaeodactylum tricornutum. *Environ. Boil. Fishes* **2015**, *28*, 123–129. [CrossRef]
118. Chen, Y.; Bi, C.; Zhang, J.; Hou, H.; Gong, Z. Astaxanthin biosynthesis in transgenic *Dunaliella salina* (Chlorophyceae) enhanced tolerance to high irradiation stress. *S. Afr. J. Bot.* **2020**, *133*, 132–138. [CrossRef]
119. Brar, A.; Kumar, M.; Soni, T.; Vivekanand, V.; Pareek, N. Insights into the genetic and metabolic engineering approaches to enhance the competence of microalgae as biofuel resource: A review. *Bioresour. Technol.* **2021**, *339*, 125597. [CrossRef] [PubMed]
120. Baek, K.; Yu, J.; Jeong, J.; Sim, S.J.; Bae, S.; Jin, E. Photoautotrophic production of macular pigment in a Chlamydomonas reinhardtii strain generated by using DNA-free CRISPR-Cas9 RNP-mediated mutagenesis. *Biotechnol. Bioeng.* **2017**, *115*, 719–728. [CrossRef]
121. Chen, G.; Wang, B.; Han, D.; Sommerfeld, M.; Lu, Y.; Chen, F.; Hu, Q. Molecular mechanisms of the coordination between astaxanthin and fatty acid biosynthesis in *Haematococcus pluvialis* (Chlorophyceae). *Plant J.* **2014**, *81*, 95–107. [CrossRef]
122. Guo, H.; Li, T.; Zhao, Y.; Yu, X. Role of copper in the enhancement of astaxanthin and lipid coaccumulation in *Haematococcus pluvialis* exposed to abiotic stress conditions. *Bioresour. Technol.* **2021**, *335*, 125265. [CrossRef] [PubMed]
123. Meyer, B. Are we consuming enough long chain omega-3 polyunsaturated fatty acids for optimal health? *Prostaglandins Leukot. Essent. Fat. Acids* **2011**, *85*, 275–280. [CrossRef] [PubMed]
124. Sun, Z.; Wang, X.; Liu, J. Screening of Isochrysis strains for simultaneous production of docosahexaenoic acid and fucoxanthin. *Algal Res.* **2019**, *41*, 101545. [CrossRef]

125. Ye, Z.-W.; Jiang, J.-G.; Wu, G.-H. Biosynthesis and regulation of carotenoids in *Dunaliella*: Progresses and prospects. *Biotechnol. Adv.* **2008**, *26*, 352–360. [CrossRef] [PubMed]
126. Doan, Q.C.; Moheimani, N.R.; Mastrangelo, A.J.; Lewis, D.M. Microalgal biomass for bioethanol fermentation: Implications for hypersaline systems with an industrial focus. *Biomass Bioenergy* **2012**, *46*, 79–88. [CrossRef]

marine drugs

Review

Microalgae as Sustainable Biofactories to Produce High-Value Lipids: Biodiversity, Exploitation, and Biotechnological Applications

Tomásia Fernandes [1,2] and Nereida Cordeiro [1,2,*]

1 Laboratory of Bioanalysis, Biomaterials, and Biotechnology (LB3), Faculty of Exact Sciences and Engineering, University of Madeira, Campus Universitário da Penteada, 9020-105 Funchal, Portugal; tomasia.fernandes@uma.pt

2 Interdisciplinary Centre of Marine and Environmental Research (CIIMAR), University of Porto, 4450-208 Matosinhos, Portugal

* Correspondence: ncordeiro@staff.uma.pt

Abstract: Microalgae are often called "sustainable biofactories" due to their dual potential to mitigate atmospheric carbon dioxide and produce a great diversity of high-value compounds. Nevertheless, the successful exploitation of microalgae as biofactories for industrial scale is dependent on choosing the right microalga and optimum growth conditions. Due to the rich biodiversity of microalgae, a screening pipeline should be developed to perform microalgal strain selection exploring their growth, robustness, and metabolite production. Current prospects in microalgal biotechnology are turning their focus to high-value lipids for pharmaceutic, nutraceutic, and cosmetic products. Within microalgal lipid fraction, polyunsaturated fatty acids and carotenoids are broadly recognized for their vital functions in human organisms. Microalgal-derived phytosterols are still an underexploited lipid resource despite presenting promising biological activities, including neuroprotective, anti-inflammatory, anti-cancer, neuromodulatory, immunomodulatory, and apoptosis inductive effects. To modulate microalgal biochemical composition, according to the intended field of application, it is important to know the contribution of each cultivation factor, or their combined effects, for the wanted product accumulation. Microalgae have a vital role to play in future low-carbon economy. Since microalgal biodiesel is still costly, it is desirable to explore the potential of oleaginous species for its high-value lipids which present great global market prospects.

Keywords: sustainability; industrial valorization; carbon dioxide fixation; biological activities; polyunsaturated fatty acids; phytosterol; carotenoids

Citation: Fernandes, T.; Cordeiro, N. Microalgae as Sustainable Biofactories to Produce High-Value Lipids: Biodiversity, Exploitation, and Biotechnological Applications. *Mar. Drugs* **2021**, *19*, 573. https://doi.org/10.3390/md19100573

Academic Editor: Carlos Almeida

Received: 23 September 2021
Accepted: 11 October 2021
Published: 14 October 2021

1. Introduction

Microalgae use light energy and inorganic nutrients to produce oxygen and biomass rich in a diversity of value-added compounds (Figure 1) [1]. They can thrive in almost all environments and can be found in oceans, brackish water, freshwater, rocks, and soils [2]. Microalgal biotechnological historical data go back to the Aztec population, who harvested *Arthrospira* from lake Texcoco for food purposes [2,3]. *Arthrospira* has also been collected by local people surrounding Lake Chad and consumed as a nutritional supplement called "dihe" [2]. *Nostoc* species have been used by the Chinese as a food delicacy and for their properties for hundreds of years [2].

Arthrospira platensis, *Aphanizomenon flosaquae* var. flosaquae, *Chlorella luteoviridis*, *Chlorella pyrenoidosa*, *Chlorella vulgaris*, and *Auxenochlorella protothecoides* have been on the market as a food or food ingredient and consumed to a significant degree before 15 May 1997 in the European Union market; thus, its access to the market is not subject to the Novel Food Regulation (EU) 2015/2283 [4]. Dried *Tetraselmis chuii*, *Odontella aurita*, and astaxanthin-rich oleoresin from *Haematococcus pluvialis* are microalgal products approved as a novel food and, as the microalgae listed previously, are within the list of microalgae that can be commercialized in the EU [5]. Through Figure 2a, it is possible to visualize that in most European countries, algae farmers mainly produce microalgal species belonging to Cyanobacteria and Chlorophyta phyla, except for Belgium (BE), Norway (NO), and Sweden (SE). This is consistent with the microalgae approved for human consumption which belong to Chlorophyta and Cyanophyta phyla, except for *Odontella aurita*, which belongs to Bacillariophyta phylum.

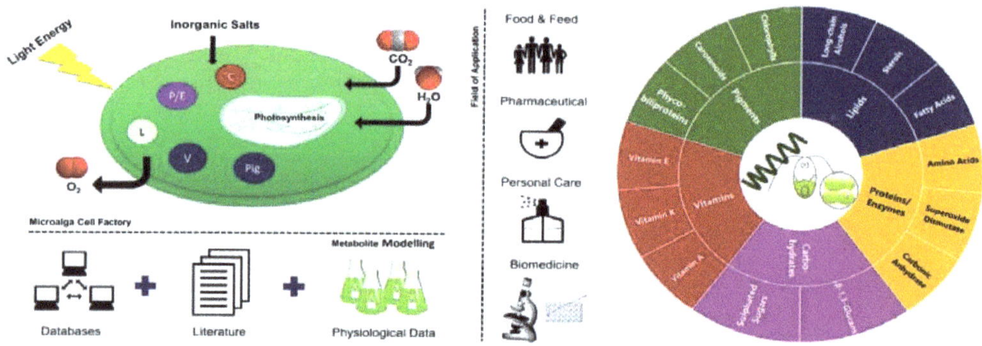

Figure 1. Uncovering the potential of a microalga as a biofactory. C—carbohydrates; P/E—protein/enzymes; L—lipids; V—vitamins; Pig.—pigments.

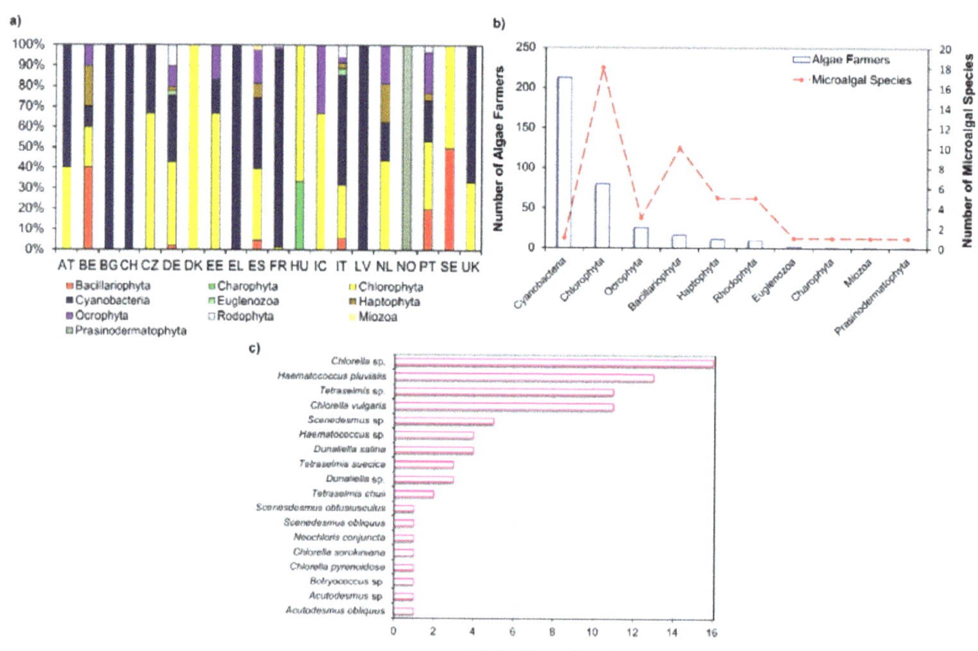

Figure 2. Microalgae production in Europe: (**a**) relative abundance of microalgae at phylum level produced by algae farmers (AT—Austria; BE—Belgium; BG—Bulgaria; CH—Switzerland; CZ—Czech Republic; DE—Germany; DK—Denmark; EE—Estonia; EL—Greece; ES—Spain; FO—Faroe Islands; FR—France; GR—Greenland; HU—Hungary; IC—Iceland; IE—Ireland; IT—Italy; LV—Latvia; NL—the Netherlands; NO—Norway; PT—Portugal; SE—Sweden; UK—the United Kingdom). (**b**) Number of algae farmers against main microalgae phyla and diversity of species exploited. (**c**) Diversity of Chlorophyta species produced by algae farmers. Based on EMODnet database [6].

Health, energy, and human nutrition are the three main applications for microalgal products [7]. However, the energy production from microalgae is experiencing a slow growth compared to other segments [7]. Through Figure 3a, it is possible to observe that both patents and research activities presented an exponential increase between 2004 and 2014. This trend continued for research activities, in contrast to patent publications. These waves of microalgae-related research and development activities are mainly associated with energy-driven trends, and having as the driving force the high crude oil price [8]. After 2015, this driving force was lost due to the development and popularization of electric cars [8]. Figure 3b shows that research activities were mainly focused on the biological sciences before 1990s. From this decade forward, biotechnology-applied microbiology appeared in the main research categories meeting the slight increase in patent and research publications.

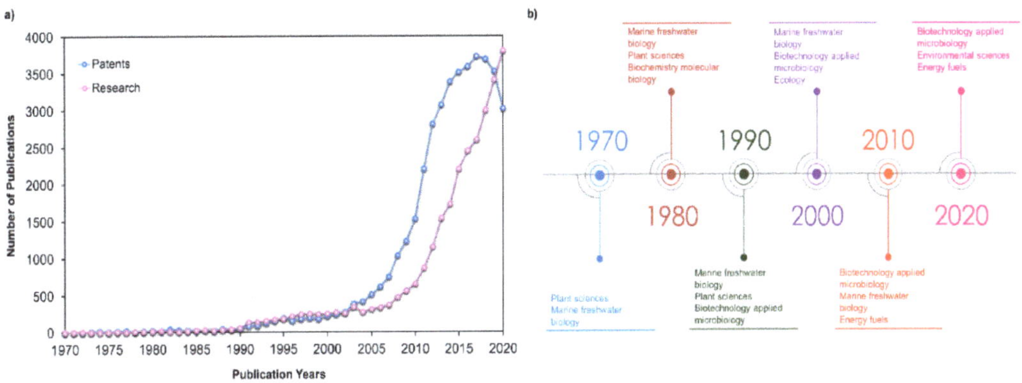

Figure 3. Patenting and research activity in the microalgal field: (**a**) numbers of microalgae-related patents and research publications against publication years; (**b**) timeline with the main research activities categories according to Web of Science. The information used to construct these plots using "microalgae" as topic can be found in Espacenet [9] and Web of Science databases [10].

Through Figure 2b, it is possible to visualize that the divisions comprising the greatest diversity of microalgal species industrially exploited are Chlorophyta (18) and Bacillariophyta (10) phyla. According to Griffiths et al. [11] most microalgal species considered for biofuel production are either Chlorophyta or Bacillariophyta, which may explain the previous observation. In the north-west European algae strategic initiatives, the bioenergy (e.g., biodiesel) market is the most mentioned. The energy demand is growing worldwide, especially in the rapidly developing countries such as China and India [12]. Biofuels currently account only 1.9% of global transport fuel consumption and are projected to achieve a threefold increase over the next 20 years [13]. Brazil's ethanol, derived from agricultural crops, is the most price-competitive biofuel in the world, which reflects the large investment of governmental agencies in research and technology that allowed the improvement of production processes, which in turn lowered biofuel manufacturing costs [14]. However, the food or fuel controversy derived from using agricultural crops for biofuel prompt the search for alternative sources as microalgae. Although these microorganisms have several advantages such as that they have a rapid growth rate, an ability to readily adapt to a wide range of climatic conditions, and that they do not compete for arable lands, the high initial capital investment and high biofuel production costs makes its exploitation for bioenergy purposes not feasible [15]. According to Moshood et al. [13], an economically efficient policy assistance may be required to overcome the main challenges found for microalgal biofuel production.

Ongoing efforts have been performed to reduce microalgal production costs; these include the evaluation of reactor design approaches, the development of biorefinery approaches, and the use of low-cost inputs for microalgal production. The use of wastewater for microalgal cultivation can offset the cost of freshwater and nutrients delivering environmental benefits such as the recycling of resources and reducing the nutrient discharges responsible for eutrophication in water bodies [16,17]. Numerous studies have evaluated the feasibility of microalgae cultivation for lipid production using wastewater from different sources: municipal [18–20], industrial [21], mining activity [22], landfill leachate [23], agricultural [24], and fish farm effluents [25]. Nevertheless, further research is needed to solve the drawbacks of the simultaneous microalgal biomass/product production and wastewater treatment, namely the everchanging chemical content of wastewater; the fact that not all metal ions and contaminants can be reduced using microalgae; and the need of wastewater pre-treatment (e.g., anaerobic digestate) to reduce organic compounds concentration, water turbidity, and low transparency, which affect microalgal growth [21,26].

According to the International Energy Agency [27], global energy-related carbon dioxide (CO_2) emissions remained at 31.5 Gt, despite the decline in 2020. CO_2 contributes up to 68% of the total greenhouse gases, the accumulation of which in the atmosphere has been considered as the main driver of climate changes [28]. Presenting higher carbon dioxide fixation efficiencies than terrestrial plants, microalgae have a pivotal role to play in future low-carbon economy [28,29]. In the literature, it is estimated that producing 280 t of microalgal dry biomass per ha per year using 9% of the incoming solar energy fixes roughly 513 t of CO_2 [30]. In this sense, microalgae are often called "sustainable biofactories" due to their dual potential to mitigate/bioremediate atmospheric CO_2 and produce a wide array of high-value compounds, which can be further enhanced through induced changes in its growth conditions.

Current prospects of algal biotechnology are turning their focus to high-value lipids production, namely polyunsaturated fatty acids (PUFA), which can be used in dietary supplements, functional food, pharmaceutical, and infant formula segments [31]. Some companies already produce ω3-PUFA with microalgal origin as dietary supplements or food ingredients (Oceans Alive (USA), Blue Biotech (Germany), Flora Health (USA)—*Nannochloropsis*; InnovalG (France)—*Odontella*) [32]. The global market size for ω3-PUFA in 2020 was estimated at US $16.2 billion, and its 2027 value projections are expected to reach US $36.9 billion at a compound annual growth rate (CAGR) of 12.5% over the forecast period of 2020–2027 [33].

Within Chlorophyta phylum, it is possible to observe that most algae farmers have been focusing on the production of oleaginous microalgae species, namely from *Chlorella*, *Tetraselmis*, *Botryococcus*, *Scenedesmus* genera, which are more specialized for biofuel production [7]. *Chlorella* sp., *Haematococcus pluvialis*, *Tetraselmis* sp., and *Chlorella vulgaris* are on the first line of microalgal species produced by algae farmers (Figure 2c). These microalgae have in common its versatility which allows to apply them in food, feed, energy production, and as a source of high-value molecules, which may reflect the efforts of algal farmers to target more than one market [34–36]. Furthermore, species belonging to *Haematococcus* and *Dunaliella* genera are mainly used for pigments exploitation, namely for astaxanthin and β-carotene production, respectively [7]. Protein is on the first line of development in human nutrition and health sectors, followed by pigments and lipids [7]. Accounting for the large number of oleaginous species already produced by algae farmers and the potentialities of microalgal lipids for high-value market, this review outlines the potential of microalgal high-value lipids for dietary supplements, cosmetics, and pharmaceutics, along with its health-promoting activities and optimization strategies.

2. Species Selection and Exploitation

Microalgae present a rich biodiversity comprising 40,000–50,000 described species with an estimation of nearly 800,000 existing species [37]. Chlorophyta comprises 6952 algal species [38], and Bacillariophyta is the most diversified group within microalgae, with more than 10,000 diatom species being described [39]. Despite this great diversity, only a few microalgal species have been exploited for biotechnological applications, with only 18 species of Chlorophyta and 10 species of Baccilariophyta phyla being produced by European algae farmers (Figure 2b). A smaller number of species is recorded for the other phyla: Cyanobacteria, Ocrophyta, Haptophyta, Rhodophyta, Euglenozoa, Charophyta, Miozoa, and Prasinodermatophyta.

The industrial production of microalgae heavily depends on biomass productivity, which is the most significant factor that can reduce the production cost levels [40]. To increase microalgae productivity, several strategies have been developed, namely the exploitation of the cultivation conditions to direct the metabolism towards desired product accumulation, selection and breeding of strains with increased biomass productivity, and genetic modification [40].

The selection of the right microalga considering specific culture conditions and desired product content could be performed through an exhaustive screen of scientific data through databases and literature, followed by physiological data collection (Figure 1). With respect to the product synthesized by microalgae, they can be applied in several fields from bioenergy (low-value market) to cosmetics and pharmaceutics (high-value market).

When exploiting the great biodiversity of microalgae as new natural sources of high-value phytochemicals, a screening pipeline should be developed including the following key aspects: *(i) growth*—for the successful improvement and progression of microalgae-based industries, the discovery and improvement of new fast-growing strains is essential [41]. Thus, the study of the maximum growth rate, maximum cell density, and dry cell biomass at different growth conditions, along with their amenability for heterotrophic/mixotrophic growth, are critical features to predict the feasibility of microalgae for large-scale production [42]; *(ii) robustness*—the selected microalgae should withstand variable local climatic conditions and be resistant to possible infections (e.g., other algae strains, grazers, bacteria, or viruses) in order to prevent large-scale crashes [43]. Nevertheless, strategies have been developed to combat and prevent contamination in microalgae cultivation such as the use of extreme conditions to create an unfavorable environment for the competitive organisms of the microalgae [44,45], and the use of allelopathic approaches to control microalgae cultivation [46]; *(iii) metabolite production*—in the production of microalgae for food and health purposes, the potential toxicity (phycotoxins) of some species as well as desired product (high-value products) accumulation should be considered.

2.1. *Phycotoxins*

Phycotoxins are causative agents of seafood-borne poisoning syndromes (e.g., ciguatera fish poisoning) in humans [47]. From the wide microalgal diversity, only around 200 species are health-threatening, with the main toxic microalgae belonging to dinoflagellates and diatoms groups [48]. Moreover, contamination of microalgae-based products resultant from unsuitable location of cultivation ponds (e.g., inflows of effluents containing pollutants) is a concern for human health [49]. Therefore, standard guidelines provided by international regulatory organizations, such as the Food and Drug Administration (United States of America) and European Food Safety Authority (Europe), guarantee that microalgae-based industries operate in conformity with safety requirements [48]. Through Table S1, is possible to have a brief insight on the regulations and directives applied in the European Union.

2.2. *High-Value Products*

Microalgae can be produced targeting different fields of application such as food, feed, pharmaceutical, personal care, and biomedicine (Figure 1). This versatility is derived from its ability to synthesize a multiplicity of metabolites distributed among pigments, vitamins, carbohydrates, lipids, and proteins/enzymes. Moreover, several biological properties attributed to microalgae, namely anti-inflammatory [50,51], anti-pyretic [51], and anti-cancer [52] activities, have shown its potential to high-value compound production.

Using the Cosmetic Ingredient Database (CosIng) [53], it is possible to see that several substances with microalgal origin have already been authorized, namely the oils from

Odontella aurita, *Nannochloropsis oceanica*, *Chlorella minutisssima*, *Chlorella protothecoides*, and *Haematococcus pluvialis*. These can be incorporated into cosmeceuticals functioning as skin-conditioning emollient, skin protection, and anti-oxidants in cosmetic products [53]. Moreover, when analyzing research activities performed on microalgal production for food and health purposes, two main groups of metabolites stand out: fatty acids and carotenoids (Figure 4). These are mainly related to the lipid-soluble fraction of microalgae.

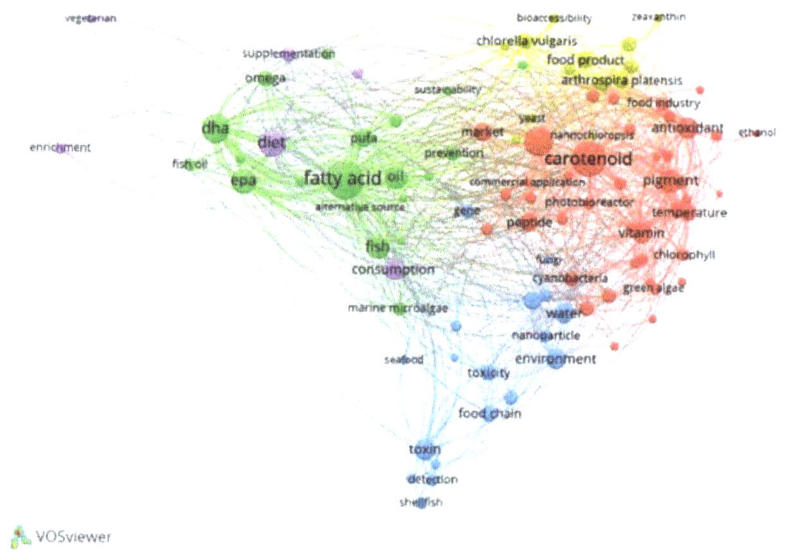

Figure 4. Concept's network obtained with VOSviewer software [54] for the research on "microalgae AND food AND health" in Web of Science database [10].

3. Polyunsaturated Fatty Acids (PUFA) Exploitation from Autotrophic Microalgae

PUFA are broadly known for their vital functions in human organisms [55]. For instance, docosahexaenoic acid (DHA, C22ω3) is enriched in human milk, plasma, and sperm, and along with arachidonic acid (AA, C20:4ω6) is concentrated in the membrane lipids of gray matter and in the visual elements of retina [56]. There is a lot of evidence that an adequate supply of these fatty acids improves visual acuity and infant cognitive development [56]. Moreover, several therapeutic properties have been attributed to the consumption of these fatty acids such as reduced risk of arthritis and cardiovascular diseases [57]. In contrast to microalgae, mammals do not have the ability to convert oleic acid (C18:1ω9) to the precursors of long-chain polyunsaturated fatty acids (LC-PUFA) biosynthesis pathway, and they poorly synthesize C20-C22 PUFA from dietary linoleic acid (LA, C18:2ω6) and α-linolenic acid (ALA, C18:3ω3).

3.1. PUFA—Synthesis by Microalgae

LC-PUFA biosynthesis pathways by microalgae are initiated by Δ12 desaturation of C18:1ω9, producing LA, which might be further desaturated by a ω3-desaturase generating ALA (see Figure 5) [58]. The ω3-pathway is initiated with the Δ6 desaturation of LA and leads to the synthesis of ω3-LC-PUFA eicosapentaenoic acid (EPA, C20:5ω3) and DHA, whereas the ω6-pathway initiates with the Δ6 desaturation of ALA and produces the ω6-LC-PUFA, AA [58]. However, some EPA-producing eustigmatophytes, such as *Nannochloropsis* sp. and *Monodus subterraneus*, are thought to preferentially synthesize EPA via the ω6-pathway by the action of a ω3-desaturase, which catalyzes the conversion of AA to EPA [58,59].

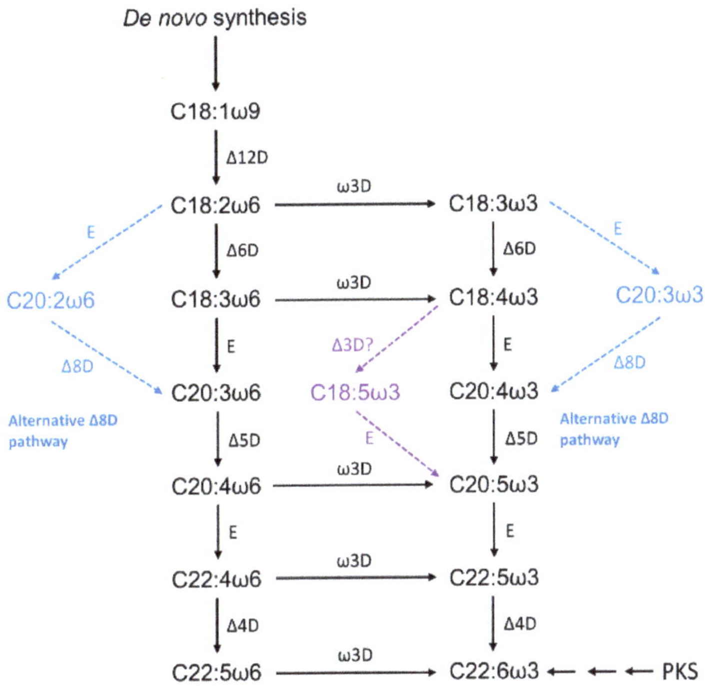

Figure 5. Biosynthesis of long-chain polyunsaturated fatty acids in microalgae [55,58,60]. ΔxD—"front-end" desaturase, adds a double bond at position x from carboxyl end; ωyD—"methyl-end" desaturase, adds a double bond at position y from methyl end; E—elongase, which catalyzes carbon chain extension; PKS—polyketide synthase.

As with other organisms, the microalgal fatty acid composition is known to vary among the different phylogenetic groups [58]. In Figure 5, is possible to visualize an alternative route for LC-PUFA biosynthesis: Δ8 desaturase pathway. According to Khozin-Goldberg [55], this pathway is known to exist in some microalgae, namely in haptophytes *Isochrysis galbana*, *Pavlova salina*, and *Emiliana huxleyi*, and the Euglenophyte *Euglena gracilis*. For the DHA-producing haptophyta *Isochrysis galbana*, a gene encoding a C22-Δ4 polyunsaturated fatty acid specific desaturase has been isolated and characterized [61]. The Trebouxiophyceae *Lobosphaera incisa* is a rare case in which AA is the major product of LC-PUFA biosynthesis in microalgae [55].

3.2. *PUFA Role in Human Health*

From microalgal lipids, PUFA are the most studied for their pharmacological potential. In human health, C20-C22 PUFA play important roles in many physiological and pathological processes [57]. Moreover, most of PUFA health-benefits are due to their key roles as lipid mediators in inflammatory processes and as important compounds for growth and development. EPA and DHA are parent compounds of specialized pro-resolving lipid mediators (protectins, resolvins, and maresins) that act as inflammatory brakes and promote the return of the affected site to homeostasis (Figure 6) [62,63]. As with ω3-PUFA, AA-derived lipoxins and their carbon-15 position epimers have beneficial effects on inflammation and resolution [62].

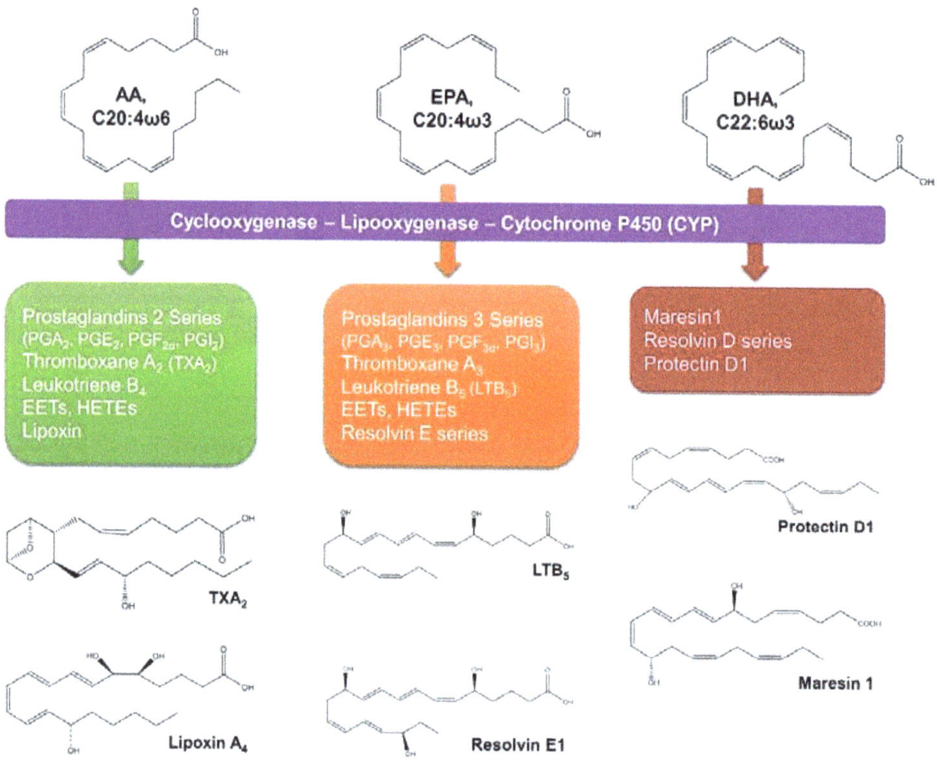

Figure 6. Lipid mediators of inflammatory process derived from arachidonic acid (AA), eicosapentaenoic acid (EPA), and docosahexaenoic acid (DHA) [63,64]. EETs—epoxyeicosatrienoic acids; HETEs—hydroxyeicosatetraenoic acids; PG—prostaglandin.

EPA competitively inhibits the utilization of AA by cyclooxygenasse/lipoxygenases to less pro-inflammatory mediators [65]. In contrast with eicosanoid products from AA (prostaglandin E2, thromboxane A2, and leukotriene B4), EPA-derived eicosanoids (thromboxane A3, and leukotriene B5) are weak inducers of inflammation and have attenuated platelet-aggregating and vasoconstriction abilities [65]. In mammalian cells, ω6- and ω3-fatty acids are not interconvertible because they lack ω3-desaturase; therefore, their balance in the diet is important [65]. In the Western diet, the greatest amounts of ω6-fatty acids (ω6:ω3 of 20:1) drive higher levels of AA eicosanoids products (e.g., prostaglandins, thromboxanes, leukotrienes, and hydroxy fatty acids), causing an imbalance between pro- and anti-inflammatory molecules and shifting the physiological state to one that is proinflammatory, prothrombotic, and proaggregatory [65]. Thus, a balanced intake of ω6- and ω3-fatty acids is crucial.

3.3. *Microalgae—PUFA Enhancement Strategies*

Although the largest share of the EPA/DHA oil market comes from wild fish, the declining fish stocks and susceptibility to contamination by pollutants (such as mercury) have turned the attention of PUFA exploitation to microalgae [66]. In this sense, the oxidative stability, sustainability, suitability for vegetarians, and the absence of fishy taste/smell are some of the advantages that make microalgae a feasible source for PUFA commercialization [66]. However, the immature production process is one of the weaknesses that must be surpassed for making PUFA exploitation from microalgae a feasible process [66].

New competitors of PUFA exploitation from microalgae are DHA producers belonging to thraustochytrids (*Traustochytrium* sp. and *Aurantiochytrium*) [66]. These unicellular heterothrophic organisms have already been authorized for food/feed/nutraceuticals and are marketed by DSM/Evonik and Source-Omega companies [66]. However, some disadvantages from the exploitation of these organisms for ω3-production (mostly produce DHA; the production chain increases pressure on arable land—since heterotrophic organisms need a sugar input to grow, and produces CO_2) are opportunities for the production of ω3-fatty acids from microalgae [66].

Since the microalgae composition is known to vary with growth conditions, the study of strategies for microalgae PUFA enhancement is crucial to overcome the challenge of the undeveloped production process of phototrophic PUFA exploitation from microalgae. Although several strategies have been proposed to enhance microalgae lipids production, most of these have been projected for biofuel production. However, nowadays, more attention has been given towards PUFA production. Conventional approaches for enhancing microalgae lipid accumulation include nutrient stress (e.g., alterations in nitrogen, phosphorus, and carbon supply) or changes in cultivation conditions (e.g., light, and temperature) [67]. The main advantage of nutrient stress is its easy applicability at both lab and large-scale cultivation, while cultivation conditions such as light have high operational costs and are not easy to control within open cultivation systems [67]. Within nutrient regime alterations, nitrate limitation is a commonly employed strategy to enhance microalgal lipids quantity [67].

As essential constituent of proteins, nucleotides, vitamins, and coenzymes, any changes in the nitrogen source and concentrations can trigger growth changes and biochemical remodeling in microalgae species [68–70]. Although the study of Huang et al. [71] had the purpose of improving lipids properties of the microalgal strains *Tetraselmis subcordiformis* SHOU-S05, *Nannochloropsis oculata* SHOU-S14, and *Pavlova viridis* SHOU-S16, the insights on the fatty acid composition at different nitrogen concentrations enables to see some trends with respect to PUFA accumulation. Therefore, towards high nitrogen supplementations, the lipid content decreased, whereas the PUFA proportion increased for *N. oculata* and *P. viridis*. For *T. subcordiformis*, the highest PUFA percentage was registered at lower nitrogen supplementations as with the highest lipid content. As with Sukenik [72], this seems to suggest that the growth conditions for the maximization of PUFA production are similar to the conditions required to maximize biomass production for *N. oculata* and *P. viridis*.

Other cultivation conditions affecting microalgal growth and chemical diversity are the salinity, light intensity, and photoperiod. Mitra et al. [73] studied the effects of these factors in *Nannochloropsis gaditana* CCNM1032 strain, and concluded that the most positive factor for fatty acid enhancement was the photoperiod. In this study, maximal EPA productivities were achieved at 60 μmol photons m^{-2} s^{-1} and at a photoperiod regime of 18 h:6 h (light:dark); this observation was made for 1L cultures.

Two-stage cultivation and combined nutrient and abiotic stresses are novel approaches used to enhance the microalgal biochemical composition. In the two-stage cultivation, microalgae are first grown to gain higher biomass and are then exposed to different cultivation conditions to trigger the accumulation of desired product content. For *Nannochloropsis gaditana* IMTE1, a two-stage cultivation was studied by Xiao et al. [74]. This microalgal strain was firstly grown in batch culture for 6 days, washed, and then transferred for a chemostat culture with a fixed dilution rate and adjusted nitrate concentrations. In these cultivation conditions, the highest biomass (897.10 mg dw L^{-1}; dw—dry weight) and EPA content (2.62% dw) were obtained at high nitrogen supplementations.

When combining nutrient and abiotic stresses, it is important to know the importance of each factor for the desired product accumulation, as well as its synergistic effects [67]. With the purpose of optimizing ω3-fatty acid production by *Pavlova lutheri*, Carvalho and Malcata [75] studied the combined effects of the dilution rate, light intensity, and CO_2 concentration under continuous mode. The optimum conditions for EPA and DHA production were found in cultures supplied with 0.5% CO_2, at a dilution rate of 0.297 d^{-1} and a light intensity of 120 µE m^{-2} s^{-1} [75]. Other combined nutrient and abiotic stresses are summarized in Table 1 for the following strains: *Chlamydomonas reinharditii* CC124, *Nannochloropsis gaditana* CCNM1032, *Phaeodactylum tricornutum* CS-29C, and *Chaetoceros muelleri* CS-176. The major disadvantage of the enounced novel approaches is that large-scale trials are required.

Table 1. Some strategies used for microalgae lipid enhancement and its impact on polyunsaturated fatty acids accumulation.

Microalga	Strain	Factors Used	Biomass	Lipid	PUFA Content	Notes	Ref.
Chlamydomonas reinhardtii	CC124	Phosphorus supplementation under nitrogen deficiency			PUFA: 17.15–45.23 µg mg^{-1}; DHA: 0.09–0.17 µg mg^{-1}		[76]
Chlamydomonas reinhardtii	CC124	Acetate input (1, 2, and 4 g L^{-1} sodium acetate)	1.08–2.49 g L^{-1}		PUFA: 28.84–51.58 µg mg^{-1}; DHA: 0.03–0.09 µg mg^{-1}		[76]
Heterochlorella luteoviridis	BE002	Temperature (22, 27 and 32 °C) and $NaNO_3$ content (12, 24, 36, 48 or 60 mg L^{-1} of N-NO_3)	0.48 g L^{-1} d^{-1}	82.5–99.1 mg g^{-1}	PUFA: 34.4–40.7% TFA	Biomass productivity obtained at higher nitrogen conditions	[77]
Nannochloropsis oceanica	IMET1	Nitrogen-deficiency stress (60, 120, and 2200 µmol L^{-1} NO_3^{2-})	319.10–897.10 mg L^{-1}	34.04–56.17% dw	EPA: 1.77–2.62% dw	The highest EPA amount was observed at 2200 µmol L^{-1} NO_3^{2-}, in contrast to the lipid content	[74]
Tetraselmis subcordiformis	SHOU-S05	Nitrogen supplementation (0, 0.22, 0.44, 0.88 and 1.76 mmol N·L^{-1})		13.40–29.77%	PUFA: 57.97–62.59% TFA EPA: 2.92–3.85% TFA	The highest values of PUFA and EPA were obtained at 0.22 mmol N L^{-1}	[71]
Nannochloropsis oculata	SHOU-S14			22.5–35.85%	PUFA: 46.10–53.69% TFA EPA: 29.34–35.51% TFA	The highest values of PUFA and EPA were obtained at 1.76 mmol N L^{-1}	[71]
Pavlova viridis	SHOU-S16			26.45–32.10%	PUFA: 26.94–41.28% TFA EPA: 9.52–15.71% TFA DHA: 2.39–7.17% TFA	The highest values of PUFA and EPA were obtained at 1.76 mmol N L^{-1}	[71]
Nannochloropsis gaditana	CCNM1032	Salinity (20, 30, 35, and 40 g L^{-1}), light intensity (60 and 150 µmol photons m^{-2} s^{-1}), and photoperiod (24/0, 18/6, 12/12, 6/18 and 0/24 light/dark hour)	45.01 mg L^{-1} d^{-1}	14.63 mg L^{-1} d^{-1}	EPA: 19.13–37.83% TFA	Biomass and lipid productivities were obtained at a salinity gradient of 20 g L^{-1}	[73]
Phaeodactylum tricornutum	CS-29C	Nitrogen source (nitrate, ammonium, and urea) and ultraviolet (UV) radiation (UV-A: 315–400 nm; UV-B: 280–315 nm)			PUFA: 34.89–48.85% TFA EPA: 18.86–23.42% TFA DHA: 1.49–2.52% TFA		[78]
Chaetoceros muelleri	CS-176				PUFA: 29.26–36.76% TFA EPA: 9.61–14.23% TFA DHA: 0.75–1.42% TFA		[78]
Pavlova lutheri	SMBA60	CO_2 concentrations (0–2% *v/v*), light intensity (75 and 120 µmol photons m^{-2} s^{-1}) and cultivation mode (batch and continuous)	0.900 g L^{-1}	132.5 mg L^{-1}	EPA: 3.61 mg L^{-1} d^{-1} DHA: 1.29 mg L^{-1} d^{-1}	Values obtained at 0.5% (*v/v*) CO_2, a dilution rate of 0.297 d^{-1}, and a light intensity of 120 µmol photons m^{-2} s^{-1}	[75]

PUFA—polyunsaturated fatty acids; DHA—docosahexaenoic acid; EPA—eicosapentaenoic acid; TFA—total fatty acids; dw—dry weight.

4. Sterols as an Underexploited Lipid Resource from Microalgae

When linking microalgae with food and human health, the predominant terms are fatty acids and carotenoids (Figure 4). Nevertheless, some microalgae are high-level producers of phytosterols, which have been playing a key role in the functional food market [79,80]. In addition, mixtures of phytosterols can function as skin conditioning in cosmetic products (creams and lipstick), and in pharmaceutics, they are gaining interest for the production of therapeutic steroids [53,81].

Plant-derived phytosterols have been added to food products for their ability to reduce serum cholesterol levels and prevent coronary heart diseases [79,82]. Cyanophya *Nostoc commune* var. *sphaeroides* lipid extracts have been found to inhibit the expression of key regulatory genes involved in cholesterol and fatty acid biosynthetic pathways; this property could contribute to lower serum cholesterol as well as triglyceride concentrations [83]. Phytosterols occur in four common forms: as free sterols (FS), as fatty acid esters (sterol is esterified to fatty acid; SE), as steryl glycosides (bound to sugar with a glycosidic bond; SG), and as acylated steryl glycosides (sugar moiety is acylated with a fatty acid; ASG) [84]. Due to the poor solubility of free phytosterols, major phytosterol/phytostanol products being marketed are in their conjugated forms (SE, SG, and ASG) as it is easier to add them into food products [84]. Table 2 summarizes information with respect to some brands that commercialize phytosterol fortified food products (yoghurts, spreads, soft cheese, and drinks), and other phytosterol-based products (supplements and paste). Although the raw materials used for phytosterols isolation for the food industry are tall oil (fat-soluble by-product obtained from trees) and vegetable oil, the scarcity of land resources has pulled the attention of scientific and industrial communities towards the search for new sustainable natural sources of phytosterols [85,86]. The concentration of phytosterols in different vegetable oils was estimated by Yang et al. [87] and ranged between 142.64 (camellia oil) and 1891.82 (rice bran oil) mg 100 g^{-1}.

Table 2. Phytosterols marketed as low cholesterol agents.

Manufacturer	Brand	Products	Source	Ref.
Raisio group	Benecol	Soft cheese Yoghurt drinks Yoghurts Spreads	Plant phytostanol esters	[88]
Upfield	Flora ProActiv	Spreads Drinks	Plant sterols	[89]
Goodman Fielder	Logicol	Spread	Plant sterols	[90]
Archer Daniels Midland (ADM)	CardioAid	Powder Paste soluble in oils and fats	Plant sterols	[91]
Cargill	CoroWise	Dietary foods * Beverages Supplements	Plant sterols (phytosterols and steryl esters)	[92]
Lipofoods	Lipophytol	Water-dispersible powder	Plant sterols (from soy or pine tree origin)	[93]

* Foods with plant sterols and commercialized include emulsified sterols (ES200); fine particle sterols (FP100); granular phytosterols (FP300); steryl esters (SE-C100); water dispersible steryl esters (WDSE-33).

The potential of Chlorophytes *Dunaliella tertiolecta* and *Dunaliella salina* as sources of phytosterols was studied by Francavilla et al. [94]; in its study, the highest yields of total sterols were 1.3% and 0.89% dw for *D. tertiolecta* and *D. salina*, respectively. Another promising microalgal strain for phytosterol production, investigated by Ahmed et al. [79], was *Pavlova lutheri*, with the phytosterol content reaching up to 5.1% dw. Comparing these values with the ones previously mentioned for vegetable oils, it is possible to recognize that microalgae have the potential to become a useful alternative source of phytosterols to use as functional ingredients. Microalgal-derived phytosterols are mainly distributed among four groups: 4-desmethyl-Δ5-sterols; 4-desmethyl-Δ7-sterols; 4-methylsterols; and di-hydroxylated sterols [80], with 4-desmethyl-Δ5-sterols being the predominant phytosterol in microalgae [80,95]. According to Moreau et al. [84], 4-desmethyl sterols and stanols have been shown to inhibit the uptake of cholesterol from the intestine, resulting in a decrease of serum cholesterol levels. This once again strengthens the potential of using microalgae-phytosterols as a novel industrial application.

4.1. Sterol Synthesis by Microalgae

For microalgae, there are two distinct and compartmentalized pathways for isoprenoid synthesis; these are (i) the mevalonic acid (MVA) pathway, in the cytosol; and (ii) the 1-deoxy-D-xylulose-5-phosphate/2-C-methyl-D-erythritol-4-phosphate (DOXP/MEP) pathway, in the plastid [80,96]. In general, microalgae possess both DOXP/MEP and MVA pathways [96]. Exceptions include Cyanophytes, Chlorophytes, as well as the Bacillariophyta *Haslea ostrearia*, and the Rhodophyta *Cyanidioschyzon merolae*, which produce sterols only from the MEP/DOXP pathway [97,98]. Figure 7 shows a generalized overview of sterol biosynthesis pathway for microalgae. Therefore, sterol biosynthetic pathway can be split into three main stages: (i) biosynthesis of isoprenoid precursors—isopentenyl pyrophosphate (IPP) and its isomer dimethylallyl pyrophosphate (DMAPP); (ii) biosynthesis of polyprenyl pyrophosphates such as farnesyl pyrophosphate (FPP); and (iii) squalene (precursor of all phytosterols) formation—dimerization of FPP [97]. According to Sasso et al. [97], there is evidence for the transport of IPP and polyprenyl pyrophosphates, such as FPP and geranyl pyrophosphate (GPP), across plastid membranes. Therefore, in algae such as Chlorophytes, in which sterols are only produced through MEP/DOXP pathway, IPP synthesized in the plastids could be exported to cytosol for the formation of sterols [99].

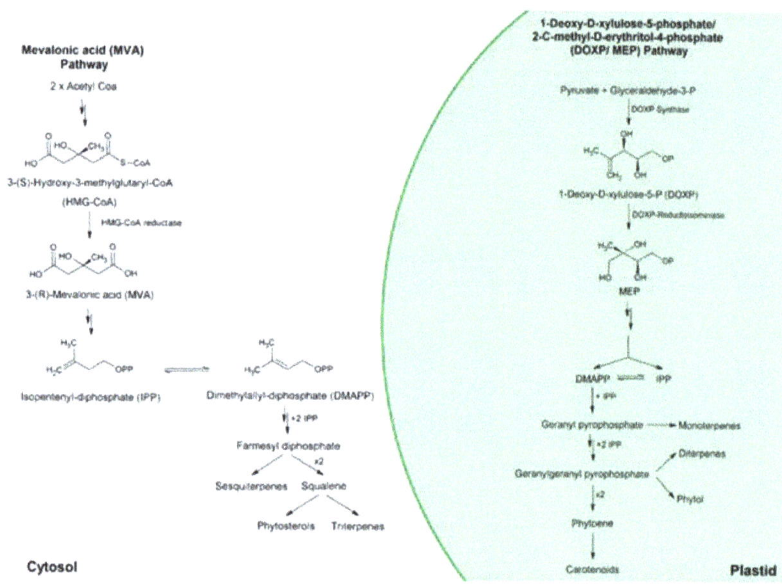

Figure 7. Generalized sterol biosynthesis pathway for microalgae [96,97].

4.2. Microalgae-Derived Phytosterols Biological Activities

Phytosterols synthesized by microalgae have shown interesting biological activities including neuroprotective, anti-inflammatory, anti-cancer, neuromodulatory, immunomodulatory, and apoptosis inductive effects. From Table S2, it is possible to visualize that the predominant biological activity studied for microalgal-derived phytosterols was anti-inflammatory. The anti-inflammatory potential of phytosterols can be assessed in vitro through determination of nitric oxide (NO), prostaglandins (PG) and cytokines production, and/or expression of nitric oxide synthase (iNOS), and cyclooxygenase (COX-2), after cell treatment with inflammation stimulation agents such as concanavalin A (Con A) and lipopolysaccharide (LPS) [100–104]. Promising results were obtained for a sterol rich fraction of *N. oculata* which was found to inhibit NO production, and down-regulate LPS-stimulated protein levels of inducible iNOS and COX-2 in a dose-dependent manner [102]. Moreover, a study performed for *D. tertiolecta*, testing several mixtures of phytosterols, showed that ergosterol and a mix of ergosterol and 7-dehydroporiferasterol suppressed a highly pro-inflammatory cytokine (tumor necrosis factor alpha; TNF-α), a pleiotropic cytokine (interleukin (IL)-6), and increased the levels of an anti-inflammatory cytokine (IL-10), showing the anti-inflammatory potential of both sterols and suggesting that phytosterol anti-inflammatory properties might depend on the existence of a synergistic effect of these molecules [101].

Besides anti-inflammatory activity, the ability of phytosterols to cross the blood–brain barrier and act as acetylcholinesterase enzymes inhibitors has sparked the attention of neurodegenerative diseases research [105]. Fagundes et al. [105] studied the neuroprotective potential of *Phormidium autumnale* phytosterol-rich extracts and determined the anti-cholinergic, antioxidant, and anti-inflammatory capacities. In this study, the phytosterol-rich extract demonstrated higher in vitro neuroprotective activity than non-enriched extract, exhibiting a moderate–high anticholinergic potential, and showing to be an effective lipoxygenase inhibitor. This was further supported through molecular docking simulation, which showed the specificity of stigmasterol interaction with acetylcholinesterase active sites. Moreover, a previous study has found in vivo neuromodulatory activity of *D. tertiolecta*—derived phytosterols in selective brain areas of rats [85].

Through Table S2, it is possible to visualize that the biological studies with respect to phytosterols were mainly performed in vitro presenting some limitations, namely, these types of studies only give partial information on bio-functionality, and a lack of systemic factors [106]. However, they provide fast and inexpensive screening of bioactivities, have high sensitivity, and are easy to perform, manage, and interpret [106]. Moreover, it is crucial to highlight that most of the studies summarized in Table S2 are mainly performed in sterol-rich fractions and that future research targeting the potential of the different types of microalgae-derived phytosterols, including their functional activity and synergistic effects, is crucial for gaining in-depth knowledge of microalgae sterols potential.

4.3. Strategies for Sterol Enhancement

To boost phytosterol accumulation in microalgae, it is crucial to understand its trigger mechanisms. Table 3 summarizes some strategies already employed for inducing microalgae phytosterol accumulation. For Haptophyta *P. lutheri*, the effects of nutrient-induced changes, salinity, ultraviolet-C (UV-C) radiation, and sampling days have been assessed aiming phytosterol production [79,82]. From these variables, the most effective were UV-C radiation and sampling days [79,82]. Although the UV-C radiation equipment is simple and easy for operation and maintenance, this physical stressor has a great disadvantage for large-scale production, which is the significant cell damage connected with UV-C radiation mutagenic factor (as it attacks an organism's deoxyribonucleic acid (DNA)) [107–109]. The mutagenic factor also poses a concern since little is known about the stability of modified algal strains or whether they can potentially take any environmental risks [110].

Table 3. Some strategies evaluated for microalgae sterol enhancement.

Microalga	Variables Studied	Total Sterols	Major Sterols	Observations	Ref.
Diacronema lutheri (syn. Pavlova lutheri)	UV-C radiation (50–250 mJ m^{-2}) Hydrogen Peroxide (H_2O_2: 1–500 μM)	9.9–20.3 mg g^{-1} dw 19.5–30.9 mg g^{-1} dw	Poriferasterol Clionasterol 4-α-methylporiferast-22-enol Methylpavlovol Epicampesterol	↑TS was found at 100 mJ cm^{-2} No significant increase of TS due to H_2O_2	[82]
	Combined effects of sampling days (2, 4, 6, 12, 14, and 16), and salinity (15, 25, 35, and 45‰)	20.29–51.86 mg g^{-1} dw		Significant differences were observed between sampling days but not for the different salinities	[79]
Dunaliella salina *Dunaliella tertiolecta*	Salinity (0.6, 1.4 and 2.1 M NaCl)	0.89% dw 1.3% dw	7-Dehydroporiferasterol Ergosterol	Good yields of TS were found at lower salt concentrations (0.6 M)	[94]

Table 3. *Cont.*

Microalga	Variables Studied	Total Sterols	Major Sterols	Observations	Ref.
Scenedesmus quadricauda	Combined effects of light intensity (30, 60, 140, 230, and 490 µmol photons m^{-2} s^{-1}), and phosphorus (1–50 µM)	8–13 µg mg C^{-1}	Fungisterol Chondrillasterol 22-Dihydrochondrillasterol	In the high-P TS increased with light intensity	[111]
Cryptomonas ovata		7–8 µg mg C^{-1}	Epibrassicasterol Stigmasterol	No significant changes in TS	
Cycotella meneghiniana SAG 1020-1a		5–8 µg mg C^{-1}	24-Methylene-cholesterol 22-Dihydrobrassicasterol		
Chlamydomonas globosa		3–4 µg mg C^{-1}	Ergosterol Fungisterol		
Prorocentrum donghaiense	Temperature (15, 20, and 25 °C) N:P supply (10:1, 24:1, and 63:1 molar ratios) Growth phase (exponential and stationary growth phases)	Brassicasterol: 0.03–0.12 pg cell^{-1} Dinosterol: 0.15–1.54 pg cell^{-1}	Brassicasterol Dinosterol	Growth phase changes showed the most pronounced effects, while temperature and nutrient deficiency had moderate effects on sterol contents	[112]
Prorocentrum minimum		Brassicasterol: 0.04–0.20 pg cell^{-1} Dinosterol: 0.28–1.83 pg cell^{-1}			
Karenia mikimotoi		Brassicasterol: 0.07–1.56 pg cell^{-1} Dinosterol: 0.20–1.30 pg cell^{-1}			
Thalassiosira pseudonana CCMP1335	Rapid cooling (18 to 4 °C) Salinity (10, 17, 25, 30, 39, 47, 53 and 61‰)		24-Methylenecholesta-5,24(24′)-dien-3β-ol Fucosterol Isofucosterol Cholesterol		[113]
Phaeodactylum tricornutum CCMP632			24-Methylenechol esta-5,24(24′)-dienol Fucosterol Isofucosterol Cholesterol	Shifts its sterol content at a reduced temperature	
Chaetoceros muelleri CCMP1316			Brassicasterol Campesterol Cholesterol	Rapid cooling did not significantly change sterols relative abundance	

According to Ahmed et al. [79] and Ahmed and Schenck [82], *P. lutheri* did not increased its sterol content when subjected to peroxide hydrogen input, variations in nitrogen and phosphorus concentrations, and changes in salinity. Still, previous studies have described salinity-induced changes as an effective tool to induce phytosterol accumulation in microalgae. Francavilla et al. [94] studied the effect of different salinity concentrations for two Chlorophytas (*D. tertiolecta* and *D. salina*) using a different approach from the study performed to Haptophyta *P. lutheri*, and good yields of total sterols were observed at a lower salt concentration [94]. Thus, the different observations made in both studies might be derived from (i) species-specific differences and/or (ii) the different approaches used to impose salinity-induced changes.

Besides the growth conditions, the specific growth phase at which microalgal biomass is harvested can influence lipid yields and composition for specific purposes [70,114]. The effect of growth phase on the sterol content of dinoflagellate species (*Prorocentrum donghaiense, Prorocentrum minimum, Karenia mikimotoi*) in batch cultures was assessed by Chen et al. [112]. In this study, the sterol content of two microalgae was susceptible to changes in the growth phase (from exponential to stationary phase), with the greatest increase being verified for dinosterol (168%) in *P. minimum*, and for brassicasterol (423%) in *K. mikimotoi*. When assessing product accumulation over the growth phase, it is also important to take into consideration that there are some growth phases in which the production of stress-associated molecules (e.g., carotenoids, lipids) is still balanced with good amounts of growth-associated ones (e.g., proteins) [115]. This point is especially important for co-exploitation of high-value microalgae products.

Interactive effects of nutrient and abiotic factors have been reported to enhance the production of phytosterol in some freshwater microalgal species [111,116]. For instance, varying light intensities in both high- and low-phosphorus environments was shown to affect sterol accumulation for Chlorophyta *Scenedesmus quadricauda* and *Chlamydomonas globosa*, and Bacillariophyta *Cyclotella meneghiniana*, with sterol contents increasing with light intensity under high phosphorus [111]. Another strategy applied by Piepho et al. [116] was varying the temperature with phosphorus, as well as the silicate supply. For this study,

C. meneghiniana increased its sterol content from low to high temperature, and this was even higher in the high-phosphorus treatment. Simultaneous effects of nutrient supply and abiotic factors in microalgae highlight the importance of investigating more than just one environmental factor when inducing the accumulation of a desired product for a field of application.

5. Carotenoids

Some studies have explored the co-production of carotenoids with lipids [117] and fatty acids [118,119] in microalgae. Carotenoids are lipid-soluble compounds synthesized by microalgae and can be divided into carotenes (hydrocarbons) and xanthophylls (oxygenated hydrocarbons) [34,120]. As structural components of light-harvesting complexes, these compounds play key roles within microalgal cells, namely in the protection against excess irradiance, chlorophyll triplets, and reactive oxygen species [121]. Through Figure 8, it is possible to visualize that the global market for carotenoids was USD 1.7 billion in 2020. In microalgae, astaxanthin, β-carotene, and lutein are among the key carotenoids with high market potential [34,120,122]. Astaxanthin presents the highest value for the global market size, USD 663.89 million in 2020, and value projections for 2027, USD 977.74 million (Figure 7). *Haematococcus pluvialis* and *Dunaliella salina* are the most popular microalgal species exploited for the commercial production of astaxanthin and β-carotene, respectively [34]. In *Haematococcus*, astaxanthin is majorly esterified to palmitic acid (C16:0) and unsaturated fatty acids of the C18 family (C18:1, C18:2, and C18:3) [120,123].

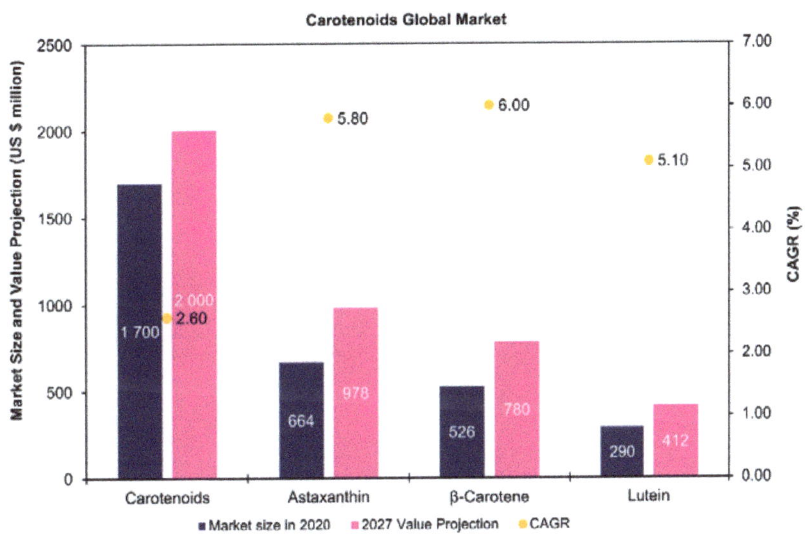

Figure 8. Carotenoids global market prospects—market size in 2020, 2027 value projection, and compound annual growth rate (CAGR). The values presented for carotenoids and lutein CAGR correspond to the forecast period of 2020–2027, while for astaxanthin and β-carotene, CAGR values correspond to the forecast period of 2021–2027. Data for carotenoids and lutein were collected from StrategyR [33], whereas data for astaxanthin and β-carotene were collected from Global Market Insights [124].

Carotenoids are extensively used in food, feed, nutraceuticals, and cosmetics [125]. The consumption of a diet rich in carotenoids is often associated with positive effects on skin health, cancer, cardiovascular, neuronal, and gastrointestinal protection, and vision and immune system enhancement [125]. As with essential fatty acids and phytosterols, carotenoids cannot be synthesized by humans, which, in turn, must obtain these through their diet [125]. The beneficial effects of carotenoids to human health are thought to be derived from its potent anti-oxidant activity and the provitamin A activity of some carotenoids, which can be converted to retinal (e.g., α- and β- carotenes) [125].

Currently, astaxanthin is the best biological antioxidant, presenting a free-radical scavenging capacity 65 times more powerful than ascorbic acid (vitamin C) and 54 times stronger than β-carotene [126]. This carotenoid is best known for its use in aquaculture, namely for giving the pinkish-red color of salmonids, shrimps, lobsters, and crayfishes. In aquaculture this carotenoid is also used for its positive impacts on organisms' immune-system and fertility [127]. Astaxanthin has a wide range of applications beyond aquaculture, namely in the food, cosmetic, and pharmaceutical industries [126]. Astaxanthin has several health-promoting properties, and it is used for anti-tumor therapies and prevention, treatment of neural damage interrelated with age-related macular degeneration, Alzheimer and Parkinson diseases [127].

Although natural astaxanthin has a notably higher antioxidant capacity and safety for human consumption, synthetically derived astaxanthin has a low production cost [128]. This makes astaxanthin from natural sources only account for less than 5% of the commercialized astaxanthin [128]. Given the metabolic plasticity of microalgae and its high growth rates, efforts have been displaced for reducing the costs of naturally derived astaxanthin. From astaxanthin producers, *H. pluvialis* is one of the richest sources of natural astaxanthin, accumulating up to 4% of astaxanthin on a dry weight basis. Table 4 summarizes different strategies already applied to *H. pluvialis* strains for prompting astaxanthin accumulation [128].

Table 4. Strategies applied for enhancing astaxanthin production for *Haematococcus pluvialis* strains.

Strain	Medium	Induction Stage		Factors Studied	Observations	Ref.
		Green-Phase Cells	Red-Phase Cells			
H. pluvialis Flotow 1844em. Wille K-0084	mBG-11	N-replete 5 days 75 μmol photons m^{-2} s^{-1}	N-free 2×10^5 cells ml^{-1} 350 μmol photons m^{-2} s^{-1} 48 or 96 h	High light intensity time exposure: 48 h—0, 6, 12, 24, 36, 48 h 96 h—0, 24, 48, 72, 96 h	Astaxanthin dominated carotenoid composition: 92.6% Car and 97.7% Car, after 24 and 48 h respectively	[129]
H. pluvialis IPPAS H-2018 (former BM1)	BG-11	N-replete 40 μmol photons m^{-2} s^{-1} 60 μmol photons m^{-2} s^{-1}	N-free 480 μmol photons m^{-2} s^{-1}	Green-phase cells: CO_2 concentrations (5, 10, 20%); Growth phase (exponential and stationary). Red-phase cells: CO_2 concentrations (5, 10, 20%)	5% CO_2 resulted in a higher astaxanthin productivity.	[130]
H. pluvialis NIES 14	BG-11	N-replete	N-free	Organic carbon: 0.5% (*v/v*) of methanol, ethanol, glycol, acetaldehyde, isopropanol, and glycerol Ethanol treatment: 1, 2, 3 and 5% (*v/v*) Light Intensity: low (25 μmol photons m^{-2} s^{-1}) and high (150 μmol photons m^{-2} s^{-1})	Astaxanthin productivity reached 11.26 mg L^{-1} d^{-1} at 3% (*v/v*) ethanol	[131]
H. pluvialis LUGU (KM115647.1)	BG-11	N-replete 30 μmol photons m^{-2} s^{-1}	N-free 250 μmol photons m^{-2} s^{-1}	Cultivation mode (batch and batch-fed) Concentration of SA (0.25, 0.5, 1.0, 2.0 and 4.0 mM) on fed-batch operation (7, 9, and 11 days)	Maximum values of astaxanthin (35.88 mg g^{-1}) and lipid (54.79%) contents were obtained after supplementation of SA on day 7	[132]
H. pluvialis NIES-144	NIES-C	50 μmol photons m^{-2} s^{-1}	250 μmol photons m^{-2} s^{-1}	Cultivation mode (batch and semi-continuous Large-scale cultivation	Induction stage lasted 8 and 20 days; For semi-continuous cultivation light intensity was constant (250 μmol photons m^{-2} s^{-1}); Semi-continuous process produced 700.4 mg L^{-1} of astaxanthin over 60 days	[128]
H. pluvialis LUGU	BG-11	N-replete 30 μmol photons m^{-2} s^{-1}	N-free 250 μmol photons m^{-2} s^{-1}	Walnut shell extracts (WSE) concentrations (10, 15, and 20%)	The highest astaxanthin (29.53 mg g^{-1}) and lipid (51.75%) occurred with 15% of WSE	[133]

Car—carotenoids; N—nitrogen; SA—succinic acid.

As can be seen in Table 4, most studies concerning astaxanthin accumulation by *H. pluvialis* have an induction stage from green- to red-phase. This distinction is associated with *H. pluvialius* cellular morphologies. Therefore, macrozooids (zoospores), microzooids, and palmella cellular morphologies are usually called "green-phase cells", while hematocysts (aplanospores) are referred as "red-phase cells" [126]. The first predominate in favorable growth conditions, whereas the second occur under unfavorable environmental or culture conditions [126]. Since *H. pluvialis* accumulate large amounts of carotenoids, especially astaxanthin, under the red-phase, stress inputs such as nitrogen depletion and high light intensity are often applied to *H. pluvialis* to prompt astaxanthin production [126].

Several factors have already being studied for *H. pluvialis* in both green- and red-phases with the aim to prompt astaxanthin production, namely the effect of high light intensity exposure time, growth phase, cultivation mode, and the use of alternative substrates such as succinic acid, ethanol, and walnut shell extracts, as detailed in Table 4. Nonetheless, commercial production of astaxanthin from *H. pluvialis* still faces challenges such as the high operating costs of the mass cultivation of microalgae [128].

Besides *H. pluvialis* (5.8–22.7 mg g^{-1} dw), there are other astaxanthin-producing microalgae, such as *Scenedesmus vacuolatus* (1.5–2.7 mg g^{-1} dw), *Scotiellopsis oocystiformis* (6.4–10.9 mg g^{-1} dw), *Chlorella zofingiensis* (3.5–6.8 mg g^{-1} dw), *Neochloris wimmeri* (5.1–19.3 mg g^{-1} dw), and *Protosiphon botryoides* (12.9–14.3 mg g^{-1} dw).

The carotenoids database [134] at the access date of 8 October 2021 presented information on 1204 natural compounds distributed among 722 organisms from all domains of life. Xantophylls can be classified based on their chemical modifications as hydroxyl (lutein, zeaxanthin), epoxide (violaxanthin, neoxanthin), keto (astaxanthin), or carbonyl groups (canthaxanthin, capsanthin) [135]. Besides their health benefits and industrial applications, carotenoids can be useful biomarkers for microalgal distinction. Table 5 summarizes the main carotenoids and xantophylls that can be found for each algal phylum.

Table 5. Main carotenoids and xanthophylls of the algal phyla [136].

Phylum	Carotenoids	Xantophylls
Cyanobacteria	β-carotene	Myxoxanthin, zeaxanthin
Prochlorophyta	β-carotene	Zeaxanthin
Glaucophyta	β-carotene	Zeaxanthin
Rhodophyta	α- and β-carotene	Lutein
Cryptophyta	α-, β-, and ε-carotene	Alloxanthin
Ocrophyta	α-, β-, and ε-carotene	Fucoxanthin, violaxanthin
Haptophyta	α- and β-carotene	Fucoxanthin
Dinophyta	β-carotene	Peridinin, fucoxanthin, diadinoxanthin, dinoxanthin, gyroxanthin
Euglenophyta	β- and **Y**-carotene	Diadinoxanthin
Chlorarachniophyta	absent	Lutein, neoxanthin, violaxanthin
Chlorophyta	absent	Lutein, prasinoxanthin

As can be seen in Figure 7, in microalgae, carotenoids are biosynthesized using the isoprenoid precursor's IPP and DMAPP, which are further condensed to yield geranyl pyrophosphate. This molecule is elongated yielding geranylgeranyl pyrophosphate, which, in turn, can be dimerized to generate phytoene, the precursor of carotenoids [97]. Carotenoid accumulation often occurs when microalgal is exposed to some stress factors (e.g., nitrogen deficiency [137], ultraviolet-A radiation [118], salinity [138,139]) and growth is arrested [34]. Thus, for prompting carotenoids overproduction, for industrial scale, a two-stage cultivation strategy is often applied to first obtain higher biomass and then to trigger carotenoids accumulation [34].

6. Lipid Characterization

Carbon fixed by microalgae can be allocated into diverse metabolites. This distribution is dependent on microalgae metabolic and cellular organization, which in turn varies across distinct phylogenetic classes [140]. Thus, to improve microalgal biomass/product accumulation, several questions must be solved, namely how flux to desired product accumulation is controlled, and which interactions occur between and within metabolic pathways [141].

Metabolic profiling is a promising approach for identifying and quantifying the intracellular metabolic fluxes of microalgae, under different cultivation conditions [129].

Studies monitoring chemical diversity often screen a specific biological activity, a targeted compound class or individual molecule [142]. This constrains the development of microalgae-based industries once it overshadows the diversity of compounds produced by microalgae [143]. Several methods have been developed to allow a quantitative and simultaneous analysis of many groups of metabolites on complex mixtures by capillary electrophoresis mass spectroscopy (CE-MS), gas chromatography–mass spectroscopy (GC-MS), liquid chromatography–mass spectroscopy (LC-MS), nuclear magnetic resonance spectroscopy (NMR), and Fourier transform ion cyclotron resonance–mass spectroscopy (FTICR-MS) [144,145]. These strategies allow to save time, with respect to laborious isolation and quantification procedures, and give prompt information for complex mixtures, constituting a major advantage for microalgal-based industries [145].

Lipidomic comprises the detailed identification and quantification of lipid classes [146]. GC-MS has been widely used for the determination of the fatty acid compositions of microalgal lipids, which, in turn, are the basis of microalgae lipidomic studies [146]. However, the challenge with microalgal lipidomic studies lies in addressing its vast complexity and chemical heterogeneity [147]. GC-MS methods are recognized by their high detection sensitivity, accuracy, and excellent reproducibility; nevertheless, sample pretreatment (e.g., hydrolysis, derivatization) needs to be performed for samples [148]. A trimethylsilyl 2,2,2-trifluoro-N-trimethylsilylethanimidate (BSTFA) derivatization with GC-MS has been used as a simple, fast, low-cost, and powerful tool to gain in-depth knowledge on unknown but relevant lipids [149]. Through this non-target approach, it is possible to simultaneously analyze fatty acids, sterols, monoglycerides, aliphatic alcohols, glycosyl sterols, and other lipid-soluble molecules, such as α-tocopherols, without prior knowledge of the sample composition [150].

7. Conclusions

Until this stage of algal development, the production of bioenergy from microalgae is still not feasible. Thus, to take advantage of already exploited oleaginous microalgal species, algae farmers are turning their focus towards high-value lipids production for health and food sectors. Nevertheless, the great biodiversity of microalgae remains to be explored, holding back the opportunities that come from the wide diversity of compounds amongst microalgae taxa. Within lipids, the already recognized ω3-PUFA health-promoting properties are turning the focus of microalgal-based industries towards autotrophic ω3-PUFA production. However, more knowledge on the production strategies to prompt ω3-PUFA accumulation is lacking. Phytosterols are an unexplored lipid resource, which could pose an opportunity to explore them as food additives, as functional food, or dietary supplements. Lipid-soluble compounds carotenoids are the most extensively used in food, feed, nutraceuticals, and cosmetics. For high-value lipid exploitation, increasing the knowledge on new, simple, and cost-effective strategies to increase the production of these molecules is necessary. Additionally, the use of high-throughput methods that allow the identification and quantification of a wide array of lipid components are needed. The promising biological activities of microalgal-derived phytosterols show that the future of microalgal high-value lipids should not be restricted to fatty acids.

Supplementary Materials: The following are available online at https://www.mdpi.com/article/10.3390/md19100573/s1: Table S1. Insight on the legislation available in the European Union to regulate the food market, Table S2. Biological activities studied for microalgae-derived phytosterols in cell culture experiments and animal models.

Author Contributions: Conceptualization, T.F. and N.C.; writing—original draft preparation, T.F.; writing—review and editing, T.F. and N.C. All authors have read and agreed to the published version of the manuscript.

Funding: This research was partially supported by the Foundation for Science and Technology-FCT, through UIDB/04423/2020 and UIDP/04423/2020, and by the European Territorial Cooperation Programme PCT-MAC 2014-2020, through the REBECA-CCT (MAC/1.1.B/269) project. Tomásia Fernandes was financially supported by a doctoral grant from the Regional Agency for the Development of Research, Technology, and Innovation of Madeira-ARDITI, Project M1420-09-5369-FSE-000002.

Data Availability Statement: The data presented in this study are available on request from the corresponding author.

Conflicts of Interest: The authors declare no conflict of interest.

References

1. Dolganyuk, V.; Belova, D.; Babich, O.; Prosekov, A.; Ivanova, S.; Katserov, D.; Patyukov, N.; Sukhikh, S. Microalgae: A promising source of valuable bioproducts. *Biomolecules* **2020**, *10*, 1153. [CrossRef]
2. Borowitzka, M.A. Biology of microalgae. In *Microalgae in Health and Disease Prevention*; Levine, I., Fleurence, J., Eds.; Elsevier: Amsterdam, The Netherlands, 2018; pp. 23–72, ISBN 978-0-12-811405-6.
3. Pulz, O.; Gross, W. Valuable products from biotechnology of microalgae. *Appl. Microbiol. Biotechnol.* **2004**, *65*, 635–648. [CrossRef]
4. European Comission. EU Novel Food Catalogue. Available online: https://ec.europa.eu/food/safety/novel_food/catalogue/search/public/index.cfm (accessed on 29 June 2021).
5. European Comission. Commission implementing Regulation (EU) 2017/2470 of 20 December 2017 establishing the Union list of novel foods in accordance with Regulation (EU) 2015/2283 of the European Parliament and of the Council on novel foods. *Off. J. Eur. Union* **2017**, *351*, 1–188.
6. The European Marine Observation and Data Network. EMODnet Human Activities, Algae Production. Available online: https://www.emodnet-humanactivities.eu/search-results.php?dataname=Microalgae (accessed on 29 June 2021).
7. *World Intellectual Property Organization (WIPO) Patent Landscape Report: Microalgae-Related Technologies*; WIPO: Geneva, Switzerland, 2016.
8. Li, D.; Du, W.; Fu, W.; Cao, X. A Quick Look Back at the Microalgal Biofuel Patents: Rise and Fall. *Front. Bioeng. Biotechnol.* **2020**, *8*, 1035. [CrossRef] [PubMed]
9. European Patent Office. Espacenet Patent Search. Available online: https://worldwide.espacenet.com/ (accessed on 28 July 2021).
10. Clarivate. Web of Science. Available online: https://www.webofscience.com/wos/woscc/basic-search (accessed on 28 July 2021).
11. Griffiths, M.; Dicks, R.; Richardson, C.; Harrison, S. Advantages and Challenges of Microalgae as a Source of Oil for Biodiesel. In *Biodiesel Feedstocks and Processing Technologies*; Stoytcheva, M., Montero, G., Eds.; IntechOpen: London, UK, 2011; pp. 177–200, ISBN 978-953-307-713-0.
12. Mondal, M.; Goswami, S.; Ghosh, A.; Oinam, G.; Tiwari, O.N.; Das, P.; Gayen, K.; Mandal, M.K.; Halder, G.N. Production of biodiesel from microalgae through biological carbon capture: A review. *3 Biotech* **2017**, *7*, 99. [CrossRef]
13. Moshood, T.D.; Nawanir, G.; Mahmud, F. Microalgae biofuels production: A systematic review on socioeconomic prospects of microalgae biofuels and policy implications. *Environ. Challenges* **2021**, *5*, 100207. [CrossRef]
14. Sorda, G.; Banse, M.; Kemfert, C. An overview of biofuel policies across the world. *Energy Policy* **2010**, *38*, 6977–6988. [CrossRef]
15. Rafa, N.; Ahmed, S.F.; Badruddin, I.A.; Mofijur, M.; Kamangar, S. Strategies to Produce Cost-Effective Third-Generation Biofuel From Microalgae. *Front. Energy Res.* **2021**, *9*, 1–11. [CrossRef]
16. Al-Jabri, H.; Das, P.; Khan, S.; Thaher, M.; AbdulQuadir, M. Treatment of wastewaters by microalgae and the potential applications of the produced biomass—A review. *Water* **2021**, *13*, 27. [CrossRef]
17. Gao, F.; Yang, H.L.; Li, C.; Peng, Y.Y.; Lu, M.M.; Jin, W.H.; Bao, J.J.; Guo, Y.M. Effect of organic carbon to nitrogen ratio in wastewater on growth, nutrient uptake and lipid accumulation of a mixotrophic microalgae Chlorella sp. *Bioresour. Technol.* **2019**, *282*, 118–124. [CrossRef] [PubMed]
18. Jiang, L.; Luo, S.; Fan, X.; Yang, Z.; Guo, R. Biomass and lipid production of marine microalgae using municipal wastewater and high concentration of CO_2. *Appl. Energy* **2011**, *88*, 3336–3341. [CrossRef]
19. Zhang, C.; Zhang, Y.; Zhuang, B.; Zhou, X. Strategic enhancement of algal biomass, nutrient uptake and lipid through statistical optimization of nutrient supplementation in coupling Scenedesmus obliquus-like microalgae cultivation and municipal wastewater treatment. *Bioresour. Technol.* **2014**, *171*, 71–79. [CrossRef] [PubMed]
20. Álvarez-Díaz, P.D.; Ruiz, J.; Arbib, Z.; Barragán, J.; Garrido-Pérez, M.C.; Perales, J.A. Freshwater microalgae selection for simultaneous wastewater nutrient removal and lipid production. *Algal Res.* **2017**, *24*, 477–485. [CrossRef]
21. Debowski, M.; Zielinski, M.; Kisielewska, M.; Kazimierowicz, J.; Dudek, M.; Swica, I.; Rudnicka, A. The cultivation of lipid-rich microalgae biomass as anaerobic digestate valorization technology-A pilot-scale study. *Processes* **2020**, *8*, 517. [CrossRef]
22. Li, T.; Lin, G.; Podola, B.; Melkonian, M. Continuous removal of zinc from wastewater and mine dump leachate by a microalgal biofilm PSBR. *J. Hazard. Mater.* **2015**, *297*, 112–118. [CrossRef]
23. Zhao, X.; Zhou, Y.; Huang, S.; Qiu, D.; Schideman, L.; Chai, X.; Zhao, Y. Characterization of microalgae-bacteria consortium cultured in landfill leachate for carbon fixation and lipid production. *Bioresour. Technol.* **2014**, *156*, 322–328. [CrossRef]
24. Yang, Y.; Xu, J.; Vail, D.; Weathers, P. Ettlia oleoabundans growth and oil production on agricultural anaerobic waste effluents. *Bioresour. Technol.* **2011**, *102*, 5076–5082. [CrossRef]

25. Tossavainen, M.; Lahti, K.; Edelmann, M.; Eskola, R.; Lampi, A.M.; Piironen, V.; Korvonen, P.; Ojala, A.; Romantschuk, M. Integrated utilization of microalgae cultured in aquaculture wastewater: Wastewater treatment and production of valuable fatty acids and tocopherols. *J. Appl. Phycol.* **2019**, *31*, 1753–1763. [CrossRef]
26. Lavrinovičs, A.; Juhna, T. Review on Challenges and Limitations for Algae-Based Wastewater Treatment. *Constr. Sci.* **2018**, *20*, 17–25. [CrossRef]
27. *International Energy Agency Global Energy Review 2021: Assessing the Effects of Economic Recoveries on Global Energy Demand and CO2 Emissions in 2021*; International Energy Agency: Paris, France, 2021.
28. Bhola, V.; Swalaha, F.; Ranjith Kumar, R.; Singh, M.; Bux, F. Overview of the potential of microalgae for CO_2 sequestration. *Int. J. Environ. Sci. Technol.* **2014**, *11*, 2103–2118. [CrossRef]
29. Saxon, R.J.; Rad-Menéndez, C.; Campbell, C.N. Patent depositing of algal strains. *Appl. Phycol.* **2020**, *263*, 1–8. [CrossRef]
30. Bilanovic, D.; Andargatchew, A.; Kroeger, T.; Shelef, G. Freshwater and marine microalgae sequestering of CO_2 at different C and N concentrations Response surface methodology analysis. *Energy Convers. Manag.* **2009**, *50*, 262–267. [CrossRef]
31. Ratledge, C.; Cohen, Z. Microbial and algal oils: Do they have a future for biodiesel or as commodity oils? *Lipid Technol.* **2008**, *20*, 155–160. [CrossRef]
32. Rao, A.R.; Ravishankar, G.A. Global microalgal-based products for industrial applications. In *Handbook of Algal Technologies and Phytochemicals: Volume II: Phycoremediation, Biofuels and Global Biomass Production*; Ravishankar, G.A., Ambati, R.R., Eds.; CRC Press: Boca Raton, FL, USA, 2019; pp. 267–278.
33. Global Industry Analysts Inc. StrategyR: Influencer Driven. Available online: https://www.strategyr.com/ (accessed on 14 September 2021).
34. Minhas, A.K.; Hodgson, P.; Barrow, C.J.; Adholeya, A. A review on the assessment of stress conditions for simultaneous production of microalgal lipids and carotenoids. *Front. Microbiol.* **2016**, *7*, 546. [CrossRef]
35. Figueroa-Torres, G.; Bermejo-Padilla, E.; Pittman, J.K.; Theodoropoulos, C. *Microalgae Strain Catalogue a Strain Selection Guide for Microalgae Users: Cultivation and Chemical Characteristics for High Added-Value Products*; University of Manchester: Manchester, UK, 2021; pp. 1–86.
36. Sprujit, J. *Output WP2A7.01: Inventory of North-West European Algae Initiatives*; Public Output Report of the EnAlgae Project; Swansea University: Swansea, UK, 2015.
37. Morançais, M.; Mouget, J.-L.; Dumay, J. Chapter 7 Proteins and pigments. In *Microalgae in Health and Disease Prevention*; Levine, I., Fleurence, J., Eds.; Elsevier: London, UK, 2018; pp. 145–175, ISBN 9780128114056.
38. Guiry, M. Phylum: Chlorophyta. Available online: https://www.algaebase.org/browse/taxonomy/?id=97241 (accessed on 5 June 2021).
39. Barra, L.; Chandrasekaran, R.; Corato, F.; Brunet, C. The challenge of ecophysiological biodiversity for biotechnological applications of marine microalgae. *Mar. Drugs* **2014**, *12*, 1641–1675. [CrossRef] [PubMed]
40. Chauton, M.S.; Reitan, K.I.; Norsker, N.H.; Tveterås, R.; Kleivdal, H.T. A techno-economic analysis of industrial production of marine microalgae as a source of EPA and DHA-rich raw material for aquafeed: Research challenges and possibilities. *Aquaculture* **2015**, *436*, 95–103. [CrossRef]
41. Steinrücken, P.; Erga, S.R.; Mjøs, S.A.; Kleivdal, H.; Prestegard, S.K. Bioprospecting North Atlantic microalgae with fast growth and high polyunsaturated fatty acid (PUFA) content for microalgae-based technologies. *Algal Res.* **2017**, *26*, 392–401. [CrossRef] [PubMed]
42. Geada, P.; Vasconcelos, V.; Vicente, A.; Fernandes, B. Microalgal biomass cultivation. In *Algal Green Chemistry: Recent Progress in Biotechnology*; Rastogi, R.P., Madamwar, D., Pandey, A., Eds.; Elsevier: Amsterdam, The Netherlands, 2017; pp. 257–284, ISBN 9780444640413.
43. Duong, V.T.; Li, Y.; Nowak, E.; Schenk, P.M. Microalgae isolation and selection for prospective biodiesel production. *Energies* **2012**, *5*, 1835–1849. [CrossRef]
44. von Alvensleben, N.; Stookey, K.; Magnusson, M.; Heimann, K. Salinity Tolerance of Picochlorum atomus and the Use of Salinity for Contamination Control by the Freshwater Cyanobacterium Pseudanabaena limnetica. *PLoS ONE* **2013**, *8*, e63569. [CrossRef] [PubMed]
45. Vadlamani, A.; Viamajala, S.; Pendyala, B.; Varanasi, S. Cultivation of Microalgae at Extreme Alkaline pH Conditions: A Novel Approach for Biofuel Production. *ACS Sustain. Chem. Eng.* **2017**, *5*, 7284–7294. [CrossRef]
46. Bacellar Mendes, L.B.; Vermelho, A.B. Allelopathy as a potential strategy to improve microalgae cultivation. *Biotechnol. Biofuels* **2013**, *6*, 1. [CrossRef] [PubMed]
47. Rico, M.; González, A.G.; Santana-Casiano, M.; González-Dávila, M.; Pérez-Almeida, N.; Miguel Tangil, M.S. Production of primary and secondary metabolites using algae. In *Prospects and Challenges in Algal Biotechnology*; Tripathi, B.N., Kumar, D., Eds.; Springer: Singapore, 2017; pp. 311–326, ISBN 9789811019500.
48. Morocho-Jácome, A.L.; Ruscinc, N.; Martinez, R.M.; de Carvalho, J.C.M.; Santos de Almeida, T.; Rosado, C.; Costa, J.G.; Velasco, M.V.R.; Baby, A.R. (Bio)Technological aspects of microalgae pigments for cosmetics. *Appl. Microbiol. Biotechnol.* **2020**, *104*, 9513–9522. [CrossRef]
49. Rzymski, P.; Niedzielski, P.; Kaczmarek, N.; Jurczak, T.; Klimaszyk, P. The multidisciplinary approach to safety and toxicity assessment of microalgae-based food supplements following clinical cases of poisoning. *Harmful Algae* **2015**, *46*, 34–42. [CrossRef]

50. Tabarzad, M.; Atabaki, V.; Hosseinabadi, T. Anti-inflammatory Activity of Bioactive Compounds from Microalgae and Cyanobacteria by Focusing on the Mechanisms of Action. *Mol. Biol. Rep.* **2020**, *47*, 6193–6205. [CrossRef]
51. Somchit, M.N.; Mohamed, N.A.; Ahmad, Z.; Amiruddin, Z.; Shamsuddin, L.; Sofian, M.; Fauzee, O.; Kadir, A.A. Anti-inflammatory and anti-pyretic properties of Spirulina platensis and Spirulina lonar: A comparative study. *Pak. J. Pharm. Sci.* **2014**, *27*, 1277–1280.
52. Ahmad, R.R.; Adzahar, N.S.; Basri, D.F.; Latif, E.S.; Sallehudin, N.J. In Vitro and in Vivo Cytotoxic Effects of Chlorella Against Various types of Cancer. *IIUM Med. J. Malaysia* **2021**, *20*, 149–158. [CrossRef]
53. European Comission. CosIng: European Commission Database for Information on Cosmetic Substances and Ingredients. Available online: https://ec.europa.eu/growth/tools-databases/cosing/index.cfm?fuseaction=search.simple (accessed on 28 May 2021).
54. Centre for Science and Technology Studies Leiden University. VOSviewer: Visualizing Scientific Landscapes. Available online: https://www.vosviewer.com/ (accessed on 5 June 2021).
55. Khozin-Goldberg, I.; Iskandarov, U.; Cohen, Z. LC-PUFA from photosynthetic microalgae: Occurrence, biosynthesis, and prospects in biotechnology. *Appl. Microbiol. Biotechnol.* **2011**, *91*, 905–915. [CrossRef]
56. Ryan, A.S.; Astwood, J.D.; Gautier, S.; Kuratko, C.N.; Nelson, E.B.; Salem, N. Effects of long-chain polyunsaturated fatty acid supplementation on neurodevelopment in childhood: A review of human studies. *Prostaglandins Leukot. Essent. Fat. Acids* **2010**, *82*, 305–314. [CrossRef]
57. Harwood, J.L. Algae: Critical sources of very long-chain polyunsaturated fatty acids. *Biomolecules* **2019**, *9*, 708. [CrossRef] [PubMed]
58. Khozin-Goldberg, I. Lipid metabolism in microalgae. In *Developments in Applied Phycology: The Physiology of Microalgae*; Borowitzka, M.A., Raven, J.A., Eds.; Springer: Cham, Switzerland, 2016; pp. 413–485, ISBN 9783319249438.
59. Mühlroth, A.; Li, K.; Røkke, G.; Winge, P.; Olsen, Y.; Hohmann-Marriott, M.F.; Vadstein, O.; Bones, A.M. Pathways of lipid metabolism in marine algae, co-expression network, bottlenecks and candidate genes for enhanced production of EPA and DHA in species of chromista. *Mar. Drugs* **2013**, *11*, 4662–4697. [CrossRef] [PubMed]
60. Remize, M.; Brunel, Y.; Silva, J.L.; Berthon, J.Y.; Filaire, E. Microalgae n-3 PUFAs Production and Use in Food and Feed Industries. *Mar. Drugs* **2021**, *19*, 113. [CrossRef]
61. Shi, T.; Yu, A.; Li, M.; Ou, X.; Xing, L.; Li, M. Identification of a novel C22-Δ4-producing docosahexaenoic acid (DHA) specific polyunsaturated fatty acid desaturase gene from Isochrysis galbana and its expression in Saccharomyces cerevisiae. *Biotechnol. Lett.* **2012**, *34*, 2265–2274. [CrossRef] [PubMed]
62. Buckley, C.; Gilroy, D.; Serhan, C. Pro-Resolving lipid mediators and Mechanisms in the resolution of acute inflammation. *Immunity* **2014**, *40*, 315–327. [CrossRef] [PubMed]
63. Miyata, J.; Arita, M. Role of omega-3 fatty acids and their metabolites in asthma and allergic diseases. *Allergol. Int.* **2015**, *64*, 27–34. [CrossRef] [PubMed]
64. Das, U.N. Essential fatty acids: Biochemistry, physiology and pathology. *Biotechnol. J.* **2006**, *1*, 420–439. [CrossRef]
65. Simopoulos, A.P. An increase in the Omega-6/Omega-3 fatty acid ratio increases the risk for obesity. *Nutrients* **2016**, *8*, 128. [CrossRef] [PubMed]
66. van der Voort, M.P.J.; Spruijt, J.; Potters, J.; de Wolf, P.L.; Elissen, H.J.H. *Socio-Economic Assessment of Algae-Based PUFA Production*; Public Output Report of the PUFAChain Project; PUFAChain: Göttingen, Germany, 2017.
67. Singh, P.; Kumari, S.; Guldhe, A.; Misra, R.; Rawat, I.; Bux, F. Trends and novel strategies for enhancing lipid accumulation and quality in microalgae. *Renew. Sustain. Energy Rev.* **2016**, *55*, 1–16. [CrossRef]
68. Fernandes, T.; Fernandes, I.; Andrade, C.A.P.; Ferreira, A.; Cordeiro, N. Marine microalgae monosaccharide fluctuations as a stress response to nutrients inputs. *Algal Res.* **2017**, *24*, 340–346. [CrossRef]
69. Fernandes, T.; Fernandes, I.; Andrade, C.A.P.; Cordeiro, N. *Changes in Fatty acid Biosynthesis in Marine Microalgae as a Response to Medium Nutrient Availability*; Elsevier: Amsterdam, The Netherlands, 2016; Volume 18, pp. 314–320.
70. Fidalgo, J.P.; Cid, A.; Torres, E.; Sukenik, A.; Herrero, C. Effects of nitrogen source and growth phase on proximate biochemical composition, lipid classes and fatty acid profile of the marine microalga Isochrysis galbana. *Aquaculture* **1998**, *166*, 105–116. [CrossRef]
71. Huang, X.; Huang, Z.; Wen, W.; Yan, J. Effects of nitrogen supplementation of the culture medium on the growth, total lipid content and fatty acid profiles of three microalgae (Tetraselmis subcordiformis, Nannochloropsis oculata and Pavlova viridis). *J. Appl. Phycol.* **2013**, *25*, 129–137. [CrossRef]
72. Sukenik, A. Ecophysiological considerations in the optimization of eicosapentaenoic acid production by Nannochloropsis sp. (Eustigmatophyceae). *Bioresour. Technol.* **1991**, *35*, 263–269. [CrossRef]
73. Mitra, M.; Patidar, S.K.; George, B.; Shah, F.; Mishra, S. A euryhaline nannochloropsis gaditana with potential for nutraceutical (EPA) and biodiesel production. *Algal Res.* **2015**, *8*, 161–167. [CrossRef]
74. Xiao, Y.; Zhang, J.; Cui, J.; Feng, Y.; Cui, Q. Metabolic profiles of Nannochloropsis oceanica IMET1 under nitrogen-deficiency stress. *Bioresour. Technol.* **2013**, *130*, 731–738. [CrossRef]
75. Carvalho, A.P.; Malcata, F.X. Optimization of ω-3 fatty acid production by microalgae: Crossover effects of CO_2 and light intensity under batch and continuous cultivation modes. *Mar. Biotechnol.* **2005**, *7*, 381–388. [CrossRef]

76. Yang, L.; Chen, J.; Qin, S.; Zeng, M.; Jiang, Y.; Hu, L.; Xiao, P.; Hao, W.; Hu, Z.; Lei, A.; et al. Growth and lipid accumulation by different nutrients in the microalga Chlamydomonas reinhardtii. *Biotechnol. Biofuels* **2018**, *11*, 1–12. [CrossRef]
77. Menegol, T.; Diprat, A.B.; Rodrigues, E.; Rech, R. Effect of temperature and nitrogen concentration on biomass composition of heterochlorella luteoviridis. *Food Sci. Technol.* **2017**, *37*, 28–37. [CrossRef]
78. Liang, Y.; Beardall, J.; Heraud, P. Effects of nitrogen source and UV radiation on the growth, chlorophyll fluorescence and fatty acid composition of Phaeodactylum tricornutum and Chaetoceros muelleri (Bacillariophyceae). *J. Photochem. Photobiol. B Biol.* **2006**, *82*, 161–172. [CrossRef] [PubMed]
79. Ahmed, F.; Zhou, W.; Schenk, P.M. Pavlova lutheri is a high-level producer of phytosterols. *Algal Res.* **2015**, *10*, 210–217. [CrossRef]
80. Luo, X.; Su, P.; Zhang, W. Advances in microalgae-derived phytosterols for functional food and pharmaceutical applications. *Mar. Drugs* **2015**, *13*, 4231–4254. [CrossRef] [PubMed]
81. Fernandes, P.; Cabral, J.M.S. Phytosterols: Applications and recovery methods. *Bioresour. Technol.* **2007**, *98*, 2335–2350. [CrossRef] [PubMed]
82. Ahmed, F.; Schenk, P.M. UV–C radiation increases sterol production in the microalga Pavlova lutheri. *Phytochemistry* **2017**, *139*, 25–32. [CrossRef] [PubMed]
83. Rasmussen, H.E.; Blobaum, K.R.; Park, Y.K.; Ehlers, S.J.; Lu, F.; Lee, J.Y. Lipid extract of Nostoc commune var. sphaeroides Kützing, a blue-green alga, inhibits the activation of sterol regulatory element binding proteins in HepG2 cells. *J. Nutr.* **2008**, *138*, 476–481. [CrossRef]
84. Moreau, R.A.; Whitaker, B.D.; Hicks, K.B. Phytosterols, phytostanols, and their conjugates in foods: Structural diversity, quantitative analysis, and health-promoting uses. *Prog. Lipid Res.* **2002**, *41*, 457–500. [CrossRef]
85. Francavilla, M.; Colaianna, M.; Zotti, M.; Morgese, M.G.; Trotta, P.; Tucci, P.; Schiavone, S.; Cuomo, V.; Trabace, L. Extraction, Characterization and In Vivo Neuromodulatory Activity of Phytosterols from Microalga Dunaliella Tertiolecta. *Curr. Med. Chem.* **2012**, *19*, 3058–3067. [CrossRef]
86. Jones, P.J.H.; Ntanios, F.Y.; Raeini-Sarjaz, M.; Vanstone, C.A. Cholesterol-lowering efficacy of a sitostanol-containing phytosterol mixture with a prudent diet in hyperlipidemic men. *Am. J. Clin. Nutr.* **1999**, *69*, 1144–1150. [CrossRef]
87. Yang, R.; Xue, L.; Zhang, L.; Wang, X.; Qi, X.; Jiang, J.; Yu, L.; Wang, X.; Zhang, W.; Zhang, Q.; et al. Phytosterol contents of edible oils and their contributions to estimated phytosterol intake in the Chinese diet. *Foods* **2019**, *8*, 334. [CrossRef]
88. Benecol Limited. Benecol Products. Available online: https://benecol.co.uk/products/#/feed (accessed on 18 May 2021).
89. Upfield. ProActiv Products. Available online: https://www.pro-activ.com/en-ie/products (accessed on 18 May 2021).
90. Goodman Fielder. Logicol: Creating a Better Life Everyday. Available online: https://goodmanfielder.com/portfolio/logicol/ (accessed on 18 May 2021).
91. ADM. CardioAid Plant Sterols: Protecting the Heart Using Nature. Available online: https://www.adm.com/products-services/health-wellness/product-solutions/nature-based-health-solutions/cardioaid (accessed on 18 May 2021).
92. Cargill. CoroWiseTM Plant Sterols. Available online: https://www.cargill.com/food-bev/ap/corowise (accessed on 18 May 2021).
93. The Lubrizol Corporation. LIPOPHYTOLTM Microcapsules. Available online: https://www.lipofoods.com/en/products/lipophytol.html (accessed on 18 May 2021).
94. Francavilla, M.; Trotta, P.; Luque, R. Phytosterols from Dunaliella tertiolecta and Dunaliella salina: A potentially novel industrial application. *Bioresour. Technol.* **2010**, *101*, 4144–4150. [CrossRef]
95. Volkman, J.K. Sterols in microalgae. In *Developments in Applied Phycology: The Physiology of Microalgae*; Borowitzka, M.A., Beardall, J., Raven, J.A., Eds.; Springer: Cham, Switzerland, 2016; pp. 485–505.
96. Lopes, G.; Sousa, C.; Valentão, P.; Andrade, P.B. *Sterols in Algae and Health*; Hernández-Ledesma, B., Herrero, M., Eds.; John Wiley & Sons: Chicago, IL, USA, 2013; ISBN 9781118412893.
97. Sasso, S.; Pohnert, G.; Lohr, M.; Mittag, M.; Hertweck, C. Microalgae in the postgenomic era: A blooming reservoir for new natural products. *FEMS Microbiol. Rev.* **2012**, *36*, 761–785. [CrossRef]
98. Fagundes, M.B.; Wagner, R. Sterols biosynthesis in algae. In *Biosynthesis*; Zepka, L.Q., Do Nascimento, T.C., Jacob-Lopes, E., Eds.; IntechOpen: London, UK, 2021; pp. 137–144, ISBN 9781626239777.
99. Scodelaro Bilbao, P.G.; Garelli, A.; Díaz, M.; Salvador, G.A.; Leonardi, P.I. Crosstalk between sterol and neutral lipid metabolism in the alga Haematococcus pluvialis exposed to light stress. *Biochim. Biophys. Acta Mol. Cell Biol. Lipids* **2020**, *1865*, 158767. [CrossRef] [PubMed]
100. Ciliberti, M.G.; Francavilla, M.; Intini, S.; Albenzio, M.; Marino, R.; Santillo, A.; Caroprese, M. Phytosterols from Dunaliella tertiolecta reduce cell proliferation in sheep fed flaxseed during post partum. *Mar. Drugs* **2017**, *15*, 216. [CrossRef] [PubMed]
101. Caroprese, M.; Albenzio, M.; Ciliberti, M.G.; Francavilla, M.; Sevi, A. A mixture of phytosterols from Dunaliella tertiolecta affects proliferation of peripheral blood mononuclear cells and cytokine production in sheep. *Vet. Immunol. Immunopathol.* **2012**, *150*, 27–35. [CrossRef]
102. Sanjeewa, K.K.A.; Fernando, I.P.S.; Samarakoon, K.W.; Lakmal, H.H.C.; Kim, E.A.; Kwon, O.N.; Dilshara, M.G.; Lee, J.B.; Jeon, Y.J. Anti-inflammatory and anti-cancer activities of sterol rich fraction of cultured marine microalga nannochloropsis oculata. *Algae* **2016**, *31*, 277–287. [CrossRef]
103. Wang, L.; Jeon, Y.J.; Kim, J. Il In vitro and in vivo anti-inflammatory activities of a sterol-enriched fraction from freshwater green alga, spirogyra sp. *Fish. Aquat. Sci.* **2020**, *23*, 1–9. [CrossRef]

104. Ciliberti, M.G.; Albenzio, M.; Francavilla, M.; Neglia, G.; Esposito, L.; Caroprese, M. Extracts from microalga Chlorella sorokiniana exert an anti-proliferative effect and modulate cytokines in sheep peripheral blood mononuclear cells. *Animals* **2019**, *9*, 45. [CrossRef] [PubMed]
105. Fagundes, M.B.; Alvarez-Rivera, G.; Mendiola, J.A.; Bueno, M.; Sánchez-Martínez, J.D.; Wagner, R.; Jacob-Lopes, E.; Zepka, L.Q.; Ibañez, E.; Cifuentes, A. Phytosterol-rich compressed fluids extracts from Phormidium autumnale cyanobacteria with neuroprotective potential. *Algal Res.* **2021**, *55*, 102264. [CrossRef]
106. Fini, M.; Giardino, R. In vitro and in vivo tests for the biological evaluation of candidate orthopedic materials: Benefits and limits. *J. Appl. Biomater. Biomech.* **2003**, *1*, 155–163. [CrossRef]
107. Sydney, T.; Marshall-Thompson, J.A.; Kapoore, R.V.; Vaidyanathan, S.; Pandhal, J.; Fairclough, J.P.A. The effect of high-intensity ultraviolet light to elicit microalgal cell lysis and enhance lipid extraction. *Metabolites* **2018**, *8*, 65. [CrossRef]
108. Barrado-Moreno, M.M.; Beltrán-Heredia, J.; Martín-Gallardo, J. Degradation of microalgae from freshwater by UV radiation. *J. Ind. Eng. Chem.* **2017**, *48*, 1–4. [CrossRef]
109. Tao, Y.; Zhang, X.; Au, D.W.T.; Mao, X.; Yuan, K. The effects of sub-lethal UV-C irradiation on growth and cell integrity of cyanobacteria and green algae. *Chemosphere* **2010**, *78*, 541–547. [CrossRef]
110. Borowitzka, M.A.; Vonshak, A. Scaling up microalgal cultures to commercial scale. *Eur. J. Phycol.* **2017**, *52*, 407–418. [CrossRef]
111. Piepho, M.; Martin-Creuzburg, D.; Wacker, A. Simultaneous effects of light intensity and phosphorus supply on the sterol content of phytoplankton. *PLoS ONE* **2010**, *5*, e15828. [CrossRef]
112. Chen, M.; Bi, R.; Chen, X.; Ding, Y.; Zhang, H.; Li, L.; Zhao, M. Stoichiometric and sterol responses of dinoflagellates to changes in temperature, nutrient supply and growth phase. *Algal Res.* **2019**, *42*, 101609. [CrossRef]
113. Jaramillo-Madrid, A.C.; Ashworth, J.; Ralph, P.J. Levels of diatom minor sterols respond to changes in temperature and salinity. *J. Mar. Sci. Eng.* **2020**, *8*, 85. [CrossRef]
114. da Costa, F.; Le Grand, F.; Quéré, C.; Bougaran, G.; Cadoret, J.P.; Robert, R.; Soudant, P. Effects of growth phase and nitrogen limitation on biochemical composition of two strains of Tisochrysis lutea. *Algal Res.* **2017**, *27*, 177–189. [CrossRef]
115. Gifuni, I.; Pollio, A.; Safi, C.; Marzocchella, A.; Olivieri, G. Current bottlenecks and challenges of the microalgal biorefinery. *Trends Biotechnol.* **2019**, *37*, 242–252. [CrossRef] [PubMed]
116. Piepho, M.; Martin-Creuzburg, D.; Wacker, A. Phytoplankton sterol contents vary with temperature, phosphorus and silicate supply: A study on three freshwater species. *Eur. J. Phycol.* **2012**, *47*, 138–145. [CrossRef]
117. Chen, L.; Zhang, L.; Liu, T. Concurrent production of carotenoids and lipid by a filamentous microalga Trentepohlia arborum. *Bioresour. Technol.* **2016**, *214*, 567–573. [CrossRef]
118. Huang, J.J.H.; Cheung, P.C.K. +UVA treatment increases the degree of unsaturation in microalgal fatty acids and total carotenoid content in Nitzschia closterium (Bacillariophyceae) and Isochrysis zhangjiangensis (Chrysophyceae). *Food Chem.* **2011**, *129*, 783–791. [CrossRef] [PubMed]
119. Lamers, P.P.; Janssen, M.; De Vos, R.C.H.; Bino, R.J.; Wijffels, R.H. Carotenoid and fatty acid metabolism in nitrogen-starved Dunaliella salina, a unicellular green microalga. *J. Biotechnol.* **2012**, *162*, 21–27. [CrossRef]
120. Nethravathy, M.U.; Mehar, J.G.; Mudliar, S.N.; Shekh, A.Y. Recent Advances in Microalgal Bioactives for Food, Feed, and Healthcare Products: Commercial Potential, Market Space, and Sustainability. *Compr. Rev. Food Sci. Food Saf.* **2019**, *18*, 1882–1897. [CrossRef]
121. Masojídek, J.; Torzillo, G.; Koblížek, M. Photosynthesis in microalgae. In *Handbook of Microalgal Culture: Applied Phycology and Biotechnology*; Richmond, A., Hu, Q., Eds.; Wiley Blackwell: Hoboken, NJ, USA, 2013; pp. 21–36, ISBN 9780470673898.
122. Le Goff, M.; Le Ferrec, E.; Mayer, C.; Mimouni, V.; Lagadic-Gossmann, D.; Schoefs, B.; Ulmann, L. Microalgal carotenoids and phytosterols regulate biochemical mechanisms involved in human health and disease prevention. *Biochimie* **2019**, *167*, 106–118. [CrossRef]
123. Holtin, K.; Kuehnle, M.; Rehbein, J.; Schuler, P.; Nicholson, G.; Albert, K. Determination of astaxanthin and astaxanthin esters in the microalgae Haematococcus pluvialis by LC-(APCI)MS and characterization of predominant carotenoid isomers by NMR spectroscopy. *Anal. Bioanal. Chem.* **2009**, *395*, 1613–1622. [CrossRef]
124. Global Market Insights Inc. Global Market Insights: Insights to Innovation. Available online: https://www.gminsights.com/ (accessed on 14 September 2021).
125. Saini, R.K.; Keum, Y.S. Microbial platforms to produce commercially vital carotenoids at industrial scale: An updated review of critical issues. *J. Ind. Microbiol. Biotechnol.* **2019**, *46*, 657–674. [CrossRef] [PubMed]
126. Shah, M.M.R.; Liang, Y.; Cheng, J.J.; Daroch, M. Astaxanthin-producing green microalga Haematococcus pluvialis: From single cell to high value commercial products. *Front. Plant Sci.* **2016**, *7*, 531. [CrossRef] [PubMed]
127. Panis, G.; Carreon, J.R. Commercial astaxanthin production derived by green alga Haematococcus pluvialis: A microalgae process model and a techno-economic assessment all through production line. *Algal Res.* **2016**, *18*, 175–190. [CrossRef]
128. Cho, S.J.; Sung, Y.J.; Lee, J.S.; Yu, B.S.; Sim, S.J. Robust cyst germination induction in Haematococcus pluvialis to enhance astaxanthin productivity in a semi-continuous outdoor culture system using power plant flue gas. *Bioresour. Technol.* **2021**, *338*, 125533. [CrossRef]
129. Recht, L.; Töpfer, N.; Batushansky, A.; Sikron, N.; Gibon, Y.; Fait, A.; Nikoloski, Z.; Boussiba, S.; Zarka, A. Metabolite profiling and integrative modeling reveal metabolic constraints for carbon partitioning under nitrogen starvation in the green algae Haematococcus pluvialis. *J. Biol. Chem.* **2014**, *289*, 30387–30403. [CrossRef] [PubMed]

130. Chekanov, K.; Schastnaya, E.; Solovchenko, A.; Lobakova, E. Effects of CO_2 enrichment on primary photochemistry, growth and astaxanthin accumulation in the chlorophyte Haematococcus pluvialis. *J. Photochem. Photobiol. B Biol.* **2017**, *171*, 58–66. [CrossRef]
131. Wen, Z.; Liu, Z.; Hou, Y.; Liu, C.; Gao, F.; Zheng, Y.; Chen, F. Ethanol induced astaxanthin accumulation and transcriptional expression of carotenogenic genes in Haematococcus pluvialis. *Enzyme Microb. Technol.* **2015**, *78*, 10–17. [CrossRef]
132. Yu, C.; Wang, H.P.; Qiao, T.; Zhao, Y.; Yu, X. A fed-batch feeding with succinic acid strategy for astaxanthin and lipid hyper-production in Haematococcus pluviualis. *Bioresour. Technol.* **2021**, *340*, 125648. [CrossRef]
133. Yu, C.; Li, X.; Han, B.; Zhao, Y.; Geng, S.; Ning, D.; Ma, T.; Yu, X. Simultaneous improvement of astaxanthin and lipid production of Haematococcus pluvialis by using walnut shell extracts. *Algal Res.* **2021**, *54*, 102171. [CrossRef]
134. Yabuzaki, J. Carotenoids Database. Available online: http://carotenoiddb.jp/ (accessed on 8 October 2021).
135. Rodriguez-Concepcion, M.; Avalos, J.; Bonet, M.L.; Boronat, A.; Gomez-Gomez, L.; Hornero-Mendez, D.; Limon, M.C.; Meléndez-Martínez, A.J.; Olmedilla-Alonso, B.; Palou, A.; et al. A global perspective on carotenoids: Metabolism, biotechnology, and benefits for nutrition and health. *Prog. Lipid Res.* **2018**, *70*, 62–93. [CrossRef] [PubMed]
136. Barsanti, L.; Gualtieri, P. Chapter 1 General Overview. In *Algae: Anatomy, Biochemistry, and Biotechnology*; Barsanti, L., Gualtieri, P., Eds.; CRC Press Taylor & Francis: Boca Raton, FL, USA, 2006; pp. 1–6, ISBN 9780849314674.
137. Jo, S.W.; Hong, J.W.; Do, J.M.; Na, H.; Kim, J.J.; Park, S.I.; Kim, Y.S.; Kim, I.S.; Yoon, H.S. Nitrogen deficiency-dependent abiotic stress enhances carotenoid production in indigenous green microalga Scenedesmus rubescens KNUA042, for use as a potential resource of high value products. *Sustain.* **2020**, *12*, 5445. [CrossRef]
138. Fu, W.; Paglia, G.; Magnúsdóttir, M.; Steinarsdóttir, E.A.; Gudmundsson, S.; Palsson, B.T.; Andrésson, Ó.S.; Brynjólfsson, S. Effects of abiotic stressors on lutein production in the green microalga Dunaliella salina. *Microb. Cell Fact.* **2014**, *13*, 1–9. [CrossRef] [PubMed]
139. Paliwal, C.; Pancha, I.; Ghosh, T.; Maurya, R.; Chokshi, K.; Vamsi Bharadwaj, S.V.; Ram, S.; Mishra, S. Selective carotenoid accumulation by varying nutrient media and salinity in Synechocystis sp. CCNM 2501. *Bioresour. Technol.* **2015**, *197*, 363–368. [CrossRef]
140. Fernandes, T.; Fernandes, I.; Andrade, C.A.P.; Cordeiro, N. Marine microalgae growth and carbon partitioning as a function of nutrient availability. *Bioresour. Technol.* **2016**, *214*, 541–547. [CrossRef]
141. Rawsthorne, S. Carbon flux and fatty acid synthesis in plants. *Prog. Lipid Res.* **2002**, *41*, 182–196. [CrossRef]
142. Martin, A.C.; Pawlus, A.D.; Jewett, E.M.; Wyse, D.L.; Angerhofer, C.K.; Hegeman, A.D. Evaluating solvent extraction systems using metabolomics approaches. *RSC Adv.* **2014**, *4*, 26325–26334. [CrossRef]
143. Fernandes, T.; Martel, A.; Cordeiro, N. Exploring Pavlova pinguis chemical diversity: A potentially novel source of high value compounds. *Sci. Rep.* **2020**, *10*, 339. [CrossRef]
144. Arora, N.; Pienkos, P.T.; Pruthi, V.; Poluri, K.M.; Guarnieri, M.T. Leveraging algal omics to reveal potential targets for augmenting TAG accumulation. *Biotechnol. Adv.* **2018**, *36*, 1274–1292. [CrossRef]
145. Aguilera-Sáez, L.M.; Abreu, A.C.; Camacho-Rodríguez, J.; González-López, C.V.; Del Carmen Cerón-García, M.; Fernández, I. NMR Metabolomics as an Effective Tool to Unravel the Effect of Light Intensity and Temperature on the Composition of the Marine Microalgae Isochrysis galbana. *J. Agric. Food Chem.* **2019**, *67*, 3879–3889. [CrossRef] [PubMed]
146. Da Costa, E.; Silva, J.; Mendonça, S.H.; Abreu, M.H.; Domingues, M.R. Lipidomic approaches towards deciphering glycolipids from microalgae as a reservoir of bioactive lipids. *Mar. Drugs* **2016**, *14*, 101. [CrossRef]
147. Gupta, V.; Thakur, R.S.; Reddy, C.R.K.; Jha, B. Central metabolic processes of marine macrophytic algae revealed from NMR based metabolome analysis. *RSC Adv.* **2013**, *3*, 7037–7047. [CrossRef]
148. Nguyen, H.T.; Ramli, A.; Kee, L.M. A Review on Methods Used in Analysis of Microalgae Lipid Composition. *J. Japan Inst. Energy* **2017**, *96*, 532–537. [CrossRef]
149. Sun, A.; Tao, X.; Sun, H.; Chen, J.; Li, L.; Wang, Y. Analysis of polyphenols in apple pomace using gas chromatography-mass spectrometry with derivatization. *Int. J. Food Prop.* **2014**, *17*, 1818–1827. [CrossRef]
150. Santos, S.A.O.; Vilela, C.; Freire, C.S.R.; Abreu, M.H.; Rocha, S.M.; Silvestre, A.J.D. Chlorophyta and Rhodophyta macroalgae: A source of health promoting phytochemicals. *Food Chem.* **2015**, *183*, 122–128. [CrossRef] [PubMed]

marine drugs

MDPI

Review

Antioxidant Compounds from Microalgae: A Review

Noémie Coulombier [1,*], Thierry Jauffrais [2] and Nicolas Lebouvier [3]

1 ADECAL Technopole, 1 Bis Rue Berthelot, 98846 Nouméa, New Caledonia, France
2 Ifremer, UMR 9220 ENTROPIE, RBE/LEAD, 101 Promenade Roger Laroque, 98897 Nouméa, New Caledonia, France; Thierry.Jauffrais@ifremer.fr
3 ISEA, EA7484, Campus de Nouville, Université de Nouvelle Calédonie, 98851 Nouméa, New Caledonia, France; nicolas.lebouvier@unc.nc
* Correspondence: noemie.coulombier@adecal.nc

Abstract: The demand for natural products isolated from microalgae has increased over the last decade and has drawn the attention from the food, cosmetic and nutraceutical industries. Among these natural products, the demand for natural antioxidants as an alternative to synthetic antioxidants has increased. In addition, microalgae combine several advantages for the development of biotechnological applications: high biodiversity, photosynthetic yield, growth, productivity and a metabolic plasticity that can be orientated using culture conditions. Regarding the wide diversity of antioxidant compounds and mode of action combined with the diversity of reactive oxygen species (ROS), this review covers a brief presentation of antioxidant molecules with their role and mode of action, to summarize and evaluate common and recent assays used to assess antioxidant activity of microalgae. The aim is to improve our ability to choose the right assay to assess microalgae antioxidant activity regarding the antioxidant molecules studied.

Keywords: reactive oxygen species; ascorbic acid; glutathione; tocopherols; phenolic compounds; carotenoids

Citation: Coulombier, N.; Jauffrais, T.; Lebouvier, N. Antioxidant Compounds from Microalgae: A Review. *Mar. Drugs* **2021**, *19*, 549. https://doi.org/10.3390/md19100549

Academic Editor: Carlos Almeida

Received: 14 September 2021
Accepted: 24 September 2021
Published: 28 September 2021

1. Introduction

The demand for natural products isolated from microalgae has increased over the last decade and has drawn attention from the food, cosmetic and nutraceutical industries. Microalgae are eukaryotic unicellular cells that combine several advantages for the development of biotechnological applications: high biodiversity, photosynthetic yield, growth, productivity and a metabolic plasticity that can be orientated using culture conditions [1,2]. Some of these metabolites are molecules of interest such as pigments (e.g., carotenoids), polyunsaturated fatty acids (PUFAs, e.g., the omega-3 or -6 fatty acids), polysaccharides, vitamins and sterols which can be introduced as dietary supplements in human nutrition and animal feed e.g., [3,4]. In addition, most of them are bioactive molecules with anti-inflammatory, antibacterial, anti-UV, antifungal, anticancer, and/or antioxidant activities which may bring added value to cosmetics, nutraceuticals or food products e.g., [5–9].

The demand for natural antioxidants as an alternative to synthetic antioxidants has increased [6,10]. Indeed, many synthetic antioxidants (e.g., butylated hydroxyanisole (BHA), butylated hydroxytoluene (BHT)) are considered to have a carcinogenic and/or toxic effect on animal models [11–14]. Although, most natural antioxidants currently available on the market are derived from terrestrial plants, microalgae are being more and more considered as a potential source of natural antioxidant compounds by the food industry [15–17] and by the cosmetic and nutraceutical industries [4,18].

Regarding the wide diversity of antioxidant compounds and mode of action combined with the diversity of ROS, this review first covers a global presentation of antioxidant molecules with their role and mode of action, to finally summarize and evaluate common and recent assays used to assess antioxidant activity of microalgae. The aim of this review is to improve our ability to choose the right assay to assess microalgae antioxidant activity regarding the antioxidant molecules studied. It also emphasizes and discusses the potential use of microalgae by the food industry for their antioxidant activity.

2. Antioxidant and Reactive Oxygen Species (ROS)

An antioxidant is defined as "a substance that, when present at low concentrations compared with those of an oxidizable substrate, significantly delays or prevents oxidation of that substrate" [19]. Antioxidant molecules produced by microalgae are used to protect the cell against reactive oxygen species (ROS) produced in response to biotic or abiotic stressors. Indeed, irradiance, UV, temperature, pH, metals, and nutrient can directly influence the production of antioxidant molecules in response to their availability, either through an excess or a limitation [7,20–26].

Antioxidants used for ROS detoxification have enzymatic and nonenzymatic origins with intracellular or extracellular mode of action (e.g., singlet O_2 quencher, radical scavenger, electron donor, hydrogen donor, peroxide decomposer, enzyme inhibitor, gene expression regulation, synergist, and metal-chelating agents) [27].

In microalgae, ROS are produced by electron transport chains in chloroplasts and mitochondria, by the activity of some enzymes such as peroxidases and oxidases and also by the activity of some photosensitizers such as the chlorophyll [28]. The reactive oxygen species are therefore essentially generated in the chloroplasts and mitochondria but also in the peroxisomes [29]. More generally, ROS refer to O_2 derivatives that are more reactive than O_2 itself. This includes free radicals that contain at least one unpaired electron, as well as nonradical molecules [30]. Briefly, the activation of O_2, in its stable state triplet oxygen ($^3O_2$), takes place (i) either by a transfer of energy large enough to reverse the spin of one of the electrons, which leads to the formation of singlet oxygen ($^1O_2$), or (ii) by an electron transfer that leads to the sequential reduction of $^3O_2$ to superoxide radical ($O_2^{-\bullet}$), hydrogen peroxide (H_2O_2) and hydroxyl radical ($OH^\bullet$).

In plants and algae, singlet oxygen $^1O_2$ is produced under high light by chloroplasts in the reaction center of the photosystem II (PSII) and to a lower extent in the antenna complex [31]. In the antenna complex, triplet-excited chlorophyll (^{3}Chl *) is formed from singlet-excited chlorophyll (^{1}Chl *) by intersystem conversion [32]. The chlorophyll in the triplet state has a longer lifespan than in the singlet state and can react with $^3O_2$ to form the highly reactive $^1O_2$ [33]. The singlet oxygen is responsible for extensive cell damage (e.g., protein, lipid and nucleic acid oxidation, chloroplasts and thylakoids membranes disruption and photoinhibition) around the production area [34,35]. The reaction center of PSII is thus particularly threatened. The superoxide radical ($O_2^{-\bullet}$) generation takes place in the chloroplast during photosynthesis, in the mitochondria during oxidative phosphorylation and in cell membranes through the activity of the NADPH oxidase [30]. The superoxide radical is poorly reactive because it lacks the ability to modify macromolecules and is quickly transformed into hydrogen peroxide (H_2O_2) [34]. However, its protonated form is the precursor of much more reactive radicals [30]. The hydrogen peroxide is formed by disproportionation of the $O_2^{-\bullet}$ a redox reaction that can be spontaneous or catalyzed by the superoxide dismutase (SOD). The hydrogen peroxide is also poorly reactive; however, it remains particularly toxic, as it can cross membranes, diffuse throughout the cell and oxidize sulfhydryl groups, causing the deactivation of essential enzymes [36]. It can also react with DNA and more specifically with some transition metals (e.g., iron and copper) inducing the formation of highly reactive hydroxyl radicals by the Haber–Weiss reaction [36,37]. The hydroxyl radical ($OH^\bullet$) is formed in the same cell compartments as the H_2O_2, i.e., in the stroma of the chloroplasts using the H_2O_2 generated by the photosystems, but needs the presence of reduced metal of transition [30]. The hydroxyl radicals can induce lipid peroxidation, protein and nucleic acid denaturation. In addition, there are no enzymes that can detoxify these radicals; in excess, it might lead to cell death [38], and lipid peroxidation may also generate other very reactive free radicals (e.g., the perhydroxyl $HO_2^\bullet$, alkyl radical, reactive aldehydes malondialdehyde

(MDA) and 4-hydroxy-2-nonenal (HNE)) [33,35]. Thus, the lipid-rich membranes and their functions are particularly affected by lipid peroxidation mainly through a decrease in membrane fluidity, an increase in their permeability and by enzyme, protein, ion channel and membrane receptor inactivation, which could lead to cell damage [33].

3. The Antioxidants Molecules of Microalgae

3.1. Ascorbic Acid

Ascorbic acid or vitamin C (**1**) is one of the most abundant water-soluble antioxidants synthesized by plants (Figure 1). It is mainly present in the cytosol and chloroplasts where it can directly neutralize superoxide and hydroxyl radicals as well as singlet oxygen by electron transfer, in addition to its role in the detoxification of hydrogen peroxide during the ascorbate-glutathione cycle [39]. Ascorbic acid is also involved in the protection of the photosynthetic apparatus through its participation in the regeneration of carotenoids of the xanthophyll cycle (cofactor of violaxanthin de-epoxidase) and α-tocopherol linked to membranes [39]. It has been shown that ascorbate can also have a pro-oxidant action by the reduction of transition metals (Fe^{3+} to Fe^{2+} and Cu^{2+} to Cu^{+}) which can reduce hydrogen peroxide to hydroxyl radical by the Fenton reaction [40].

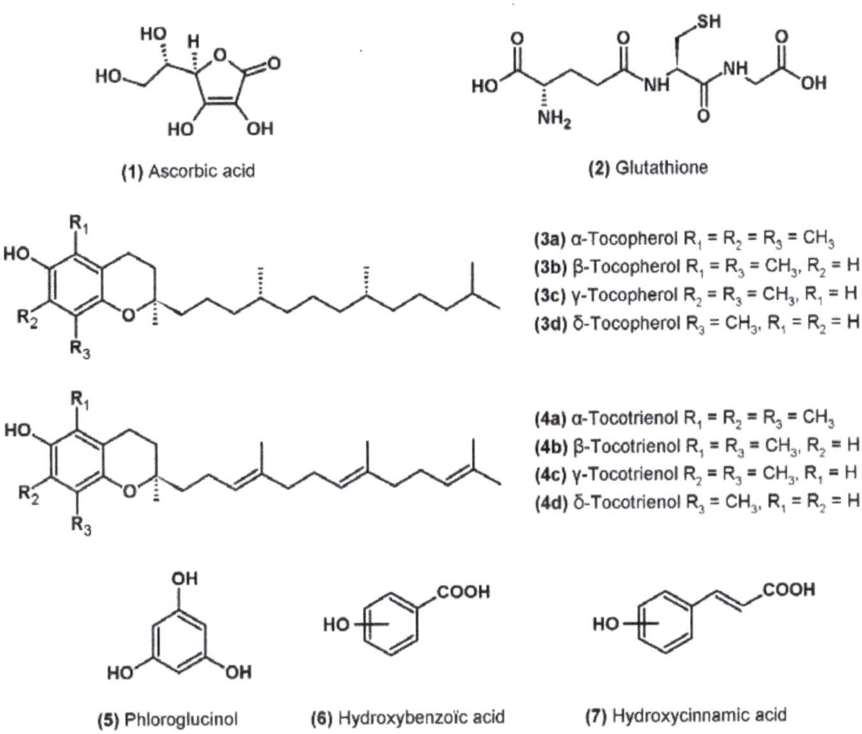

Figure 1. Molecular structure of ascorbic acid, glutathione, tocopherols and phenolic compounds.

3.2. Glutathione

Glutathione (**2**) is a water-soluble tripeptide (L-γ-glutamyl-L-cysteinylglycine) present in all cellular compartments that play a crucial role in the antioxidant response (Figure 1). In addition to its role as a cofactor in the neutralization of hydrogen peroxide by glutathione peroxidase and in the regeneration of ascorbate in reduced form via the ascorbate-glutathione cycle, glutathione can directly deactivate superoxide and hydroxyl radicals as well as singlet oxygen. In addition, like ascorbate, glutathione participates in the regeneration of α-tocopherol in its reduced form [37].

3.3. Tocopherols

Tocopherols or vitamin E are fat-soluble molecules only synthesized by photosynthetic organisms and located in the lipid bilayers of membranes, mainly in those of chloroplasts [41]. The name "vitamin E" groups together four natural forms of tocopherols (α-, β-, γ- and δ-) (**3a–d**) to which are added the four forms of tocotrienols (α-, β-, γ- and δ-) (**4a–d**) (Figure 1). Tocopherols and tocotrienols consist of a chromanol ring and a hydrophobic phytyl side chain, tocotrienols differing from tocopherols by the presence of three double bonds on the side chain [41].

Tocopherols and tocotrienols have the capacity to neutralize lipid peroxyl radicals by giving a hydrogen atom from the hydroxyle group of the chromanol ring, thus making it possible to stop the chain reaction of lipid peroxidation [41]. The reaction results in the formation of a hydroperoxide, which can be neutralized by the action of glutation peroxidase, and of a tocopheroxyl radical (for tocopherols) or tocotrienoxyl (for tocotrienols), which are less reactive. Tocopherols and tocotrienols can then be regenerated by the action of ascorbate and glutathione at the interface of the membrane and cytosol or by coenzyme Q (UQH_2) in the membrane [41]. Tocopherols can also deactivate singlet oxygen by two mechanisms: a physical quenching by charge transfer and a chemical reaction resulting in the formation of tocopherol quinone by irreversible opening of the chromanol ring [42].

3.4. Phenolic Compounds

Phenolic compounds are a large family of molecules: more than 8000 phenolic structures have been described to date in the plant kingdom [43]. These molecules contain at least one aromatic ring carrying one or more hydroxyl groups (Figure 1). The main families of compounds are phenolic acids, tocopherols described above, flavonoids and tannins as well as stilbenes and lignans [43]. Phenolics are an important class of antioxidants in higher plants and macroalgae but have only recently been studied in microalgae. However, the total content of phenolic compounds has been shown to contribute to the antioxidant activity of microalgae extracts [10,44–47]. The main molecules identified to date in microalgae are phloroglucinol (**5**) and phenolic acids derived from hydroxybenzoic acid (**6**) and hydroxycinnamic acid (**7**). Several studies have also shown the presence of weak concentrations of flavonoids e.g., [8,47–54]. All of these molecules are found in higher plants where their concentration is generally higher than in microalgae [55].

Phenolic acids can neutralize ROS primarily by hydrogen atom transfer. The antioxidant activity of the different molecules is directly linked to their chemical structure such as the number of hydroxyl groups or their position on the aromatic cycle [55]. The reaction results in the formation of a phenoxyl radical which is stabilized by the delocalization of the single electron around the aromatic ring (resonance stabilization). Phenolic acids also have the ability to inactivate radicals by monoelectronic transfer, and some can chelate the transition metals involved in the Fenton reaction thus preventing the formation of the highly reactive hydroxyl radical [55,56].

Among the pigments, we can also mention marennine, a blue-green pigment produced by *Haslea ostrearia*, which shows particularly interesting anti-free radical and antioxidant properties [57].

3.5. Carotenoids

Carotenoids are the most common pigments in nature, and more than 750 molecules have been described in algae, higher plants, bacteria and fungi [58] (Figure 2). They are fat-soluble molecules belonging to the terpenoids family containing a central chain with a system of conjugated double bonds, which can carry cyclic end groups. Carotenoids are separated into two groups: carotenes which contain only carbon and hydrogen atoms, and xanthophylls which contain at least one oxygen atom (hydroxyl, epoxy, ketone functions, for example) [59].

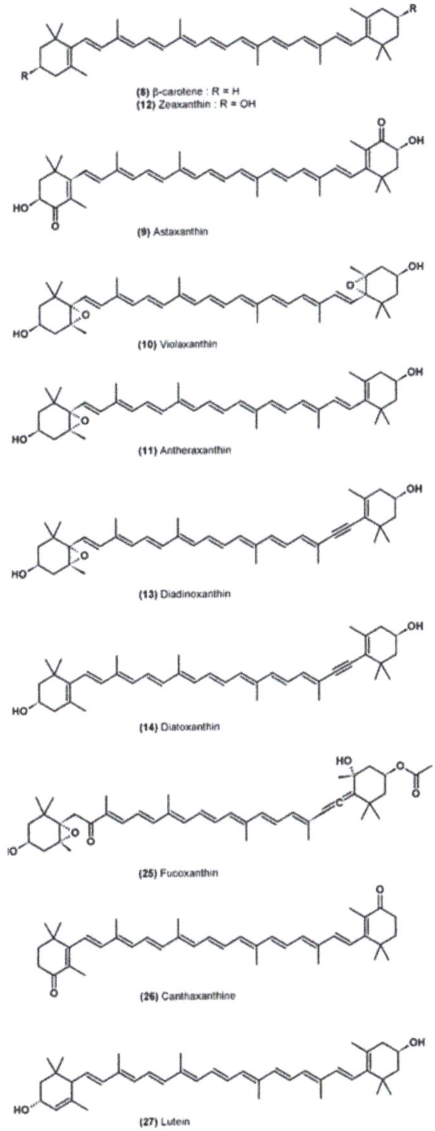

Figure 2. Molecular structure of carotenoids.

Carotenoids are mainly present in the pigment-protein complexes of the thylakoid membrane, but certain species of microalgae can also accumulate carotenoids (β-carotene (**8**) and astaxanthin (**9**)) in lipid globules located in the stroma of the chloroplast or in the cytoplasm [60]. Some carotenoids are only found in specific classes of algae and so be used as chemotaxonomic markers [58].

The role of carotenoids is on the one hand to transfer light energy to chlorophylls and on the other hand to protect the photosynthetic system by deactivating ROS and preventing their formation [61]. The first photoprotection mechanism involves xanthophylls associated with the antennal complexes of the PSII which allows the dissipation of an excess of light energy without damage, according to a series of reactions called the "xanthophyll cycle" [62]. In excess light, violaxanthin (**10**) is converted to antheraxanthin (**11**) and then to zeaxanthin (**12**) by de-epoxidation provided by violaxanthin de-epoxidase, which uses ascorbate as cofactor. This enzyme, bound to thylakoids in the lumen, is activated by an acidic pH, an excess of proton in the lumen signaling that the light energy absorbed exceeds the capacity of the electron transport chain. The de-epoxidation of violaxanthin to zeaxanthin is a very rapid phenomenon on the order of a few minutes and reversible at low light intensity or in darkness by the action of zeaxanthin epoxidase. Zeaxanthin, unlike violaxanthin, can deactivate ^{1}Chl* by dissipating its energy by heat [63]. This nonphotochemical quenching (NPQ) mechanism decreases the lifespan of ^{1}Chl* and therefore prevents the formation of ^{3}Chl* and then singlet oxygen in the PSII. In addition, by dissipating the excess energy, the possibilities of reducing O_2 to the superoxide radical $O_2^{-\bullet}$ in the PSI are minimized (less electron leakage in the transport chain) [32]. The violaxanthin cycle takes place primarily in chlorophytes. There is an alternative xanthophyll cycle, with similar photoprotective functions, in certain classes of microalgae (heterokonts, haptophytes, euglenophytes and dinophyceae) for which diadinoxanthin (**13**) is converted to diatoxanthin (**14**) [62].

At high light intensity, the probability of ^{3}Chl* formation is high despite the action of the xanthophyll cycle [32]. In antenna complexes, carotenoids are located near chlorophylls and can thus quickly neutralize ^{3}Chl* by triplet–triplet transfers before they react with $^3O_2$ to form $^1O_2$ [32]. Carotenoids can also directly deactivate singlet oxygen if it is formed [64]. This ability to deactivate $^1O_2$ is particularly important in the reaction center of PSII where there are no carotenoids in close proximity to the special pair of chlorophylls which can change to the triplet state and then react with the $^3O_2$ without that the reaction is not neutralized beforehand by the carotenoids [32]. Carotenoids therefore deactivate the $^1O_2$ that is formed in the reaction center, thus protecting the photosynthetic system from oxidative damage. The deactivation of ^{3}Chl* and of $^1O_2$ results in the formation of triplet carotenoids (^{3}CAR*) which de-excite without damage by dissipating the excess energy absorbed in the form of heat and can again intervene in a deactivation cycle [32].

Carotenoids are considered to be the most efficient molecules in deactivating $^1O_2$ owing to their system of conjugated double bonds. Thus, the greater the number of conjugated double bonds is, the more effective the carotenoid will be [64]. Carotenoids also have the ability to react with free radicals through three mechanisms: hydrogen atom transfer, monoelectronic transfer and adduct formation [65].

The interactions between carotenoids and free radicals are complex. Indeed, many parameters are involved, such as the nature of the radical, the polarity of the reaction medium, the partial pressure of oxygen, the interactions with other antioxidants, such as ascorbate or tocopherols, and the concentration and structure of the carotenoid (number of conjugated double bonds, presence and types of oxygen functions, presence of end groups, cis- or trans-configuration, etc.) [65]. Carotenoids can, for example, react with a peroxyl radical (ROO•), which is added to the polyene chain of the carotenoid forming an adduct ROO-CAR• which can react with another peroxyl radical forming a nonradical product ROO-CAR-OOR, thus allowing one to break the reaction chain of lipid peroxidation. This phenomenon takes place at low partial pressure of oxygen; however, at higher partial pressure, the ROO-CAR• radical can react with $^3O_2$ to form a ROO-CAR-OO• radical which acts as a pro-oxidant and could in this case contribute to the spread of lipid peroxidation [65,66].

3.6. Miscellaneous Antioxidants

There are other more specific antioxidant molecules produced by certain microalgae:

Mycosporins-like amino acids (MAA) form a family of thirty-five molecules. They are colorless, water-soluble molecules found in a wide variety of marine organisms [67]. In microalgae, the most abundant MAAs are mycosporin-glycine (**15**), porphyra 334 (**16**), shinorin (**17**), asterina-330 (**18**), palythene (**19**) and palythine (**20**) [68,69] (Figure 3). The main function of these molecules is UV protection, but some of them have also been shown to have antioxidant properties. In particular, they can inhibit lipid peroxidation and neutralize singlet oxygen and certain free radicals [67].

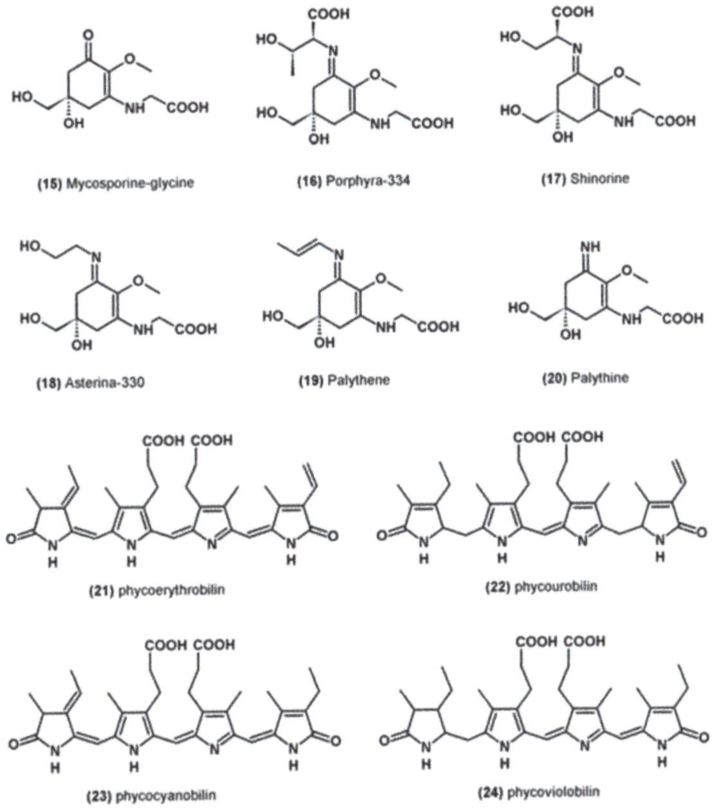

Figure 3. Molecular structure of other miscellaneous molecules with antioxidant activity.

Polysaccharides are polymers composed of osidic units linked to glycosidic bonds attached to the cell wall or released into the medium (exopolysaccharides) [70]. Several polysaccharides derived from microalgae have shown antioxidant activity against free radicals; however, this in vitro activity remains quite low [71–75].

Phycobiliproteins are water-soluble pigments participating in the photosynthesis of certain groups of microalgae. They are composed of a protein and a chromophore called phycobilin particularly effective at absorbing red, orange, yellow and green light, which is not optimally absorbed by chlorophyll a [76]. There are four different structures: phycoerythrobilin (**21**), phycourobilin (**22**), phycocyanobilin (**23**) and phycoviolobilin (**24**) (Figure 3). They can neutralize ROS and chelate or reduce ferrous ions [77].

4. Common and Recent Assays Used to Evaluate Antioxidant Activity of Microalgae

Many antioxidant assays have been developed with different types of reactions to highlight the wide variety of antioxidant molecules and ROS, which act with different mechanisms. It is important to note that there is no single ideal test, and it is necessary to use several tests with different mechanisms of action to evaluate the whole antioxidant capacity of an extract or molecule [7,78–80].

The majority of the assays are based on the two main mechanisms of action of antioxidants (AH) to deactivate radicals ($X^{\bullet}$):

Hydrogen atom transfer (or HAT):

$$AH + X^\bullet \rightarrow A^\bullet + XH$$

These reactions are generally fast; they are completed in seconds to minutes. The effectiveness of the antioxidant is determined by its ability to give a hydrogen atom (homolytic dissociation energy); therefore, the weaker the A-H bond, the more effective the antioxidant is [80].

Single electron transfer (SET):

$$AH + X^\bullet \rightarrow AH^{\bullet+} + X^-$$

$$AH^{\bullet+} \xleftrightarrow{H_2O} A^\bullet + H_3O^+$$

$$X^- + H_3O^+ \rightarrow XH + H_2O$$

These reactions are slower than hydrogen transfer reactions. The reaction is pH dependent, and the effectiveness of the antioxidant is mainly determined by its ionization potential. In general, the ionization potential decreases with increasing pH leading to an increase in the ability to donate an electron by deprotonation [80].

Other methods can also be used to evaluate the capacity of antioxidants to chelate transition metals or to inhibit the lipid peroxidation chain reaction. The most commonly used methods to evaluate the antioxidant activity of microalgae are presented in Table 1, and the most relevant results to assess antioxidant activity of microalgae extracts by in vitro chemical methods are presented in Table 2. Some cell-based antioxidant activity assays are presented, although few results using microalgae are found in the literature (Table 3). In addition, there does not seem to be any specific assays to evaluate antioxidant activity of microalgae on an animal model. Indeed, in most cases, microalgae are administrated to animals by food with a defined period and dosage; the testing animals are then sacrificed, and common in vitro chemical antioxidant activity assays (TBARS mostly) are used on animal tissues or blood by comparing with animals that did not consume microalgae (Table 4).

Table 1. Main methods used for antioxidant activity evaluation of microalgae.

	Name of the method	Principle	Mode of Detection	Ref.
In vitro	ORAC (oxygen radical absorbance capacity) assay	measure the chain breaking capacity against peroxyl radical generated by the thermal decomposition of AAPH (2,2′-azobis (2-amidino-propane) dihydrochloride). The peroxyl radical reacts with fluorescein (fluorescent probe), causing a fluorescence loss over time	fluorimetry	[81]
	β-carotene bleaching assay	measure the inhibition capacity of β-carotene oxidation induced by radical products resulting from the peroxidation of linoleic acid. The discoloration of β-carotene is measured at 434 nm	photocolorimetry	[82]
	TEAC (trolox equivalent antioxidant capacity) assay	measure the scavenging capacity of the blue chromophore ABTS (2,2′-azino-bis (3-éthylbenzothiazoline-6-sulphonique)) radical cation, which is reduced to a colorless compound in the presence of a radical scavenger. The discoloration is followed by absorbance measure at 734 nm	photocolorimetry	[83]
	DPPH (2,2-diphenyl-1-picrylhydrazyl) radical scavenging capacity assay	measure the scavenging capacity of the purple DPPH radical which is reduced to a pale-yellow compound in the presence of a radical scavenger. The absorbance decrease is measured at 515 nm	photocolorimetry	[84]
	Reducing power assay	measure the reduction capacity of potassium ferricyanide to potassium ferrocyanide which produces a ferric ferrocyanide blue complex by reaction with ferric chloride. The absorbance of the complex is measured at 700 nm	photocolorimetry	[85]
	FRAP (ferric-reducing antioxidant power) assay	measure the reduction capacity of ferric-TPTZ (tripyridyltriazine) to ferrous-TPTZ, the latter forming a blue complex at acidic pH which is measured at an absorbance of 593 nm	photocolorimetry	[46]
	TAC (total antioxidant capacity) assay or phosphomolybdenum assay	measure the reduction capacity of molybdenum Mo(vi) to Mo(v), the latter forming a green phosphate-Mo(v) complex at low pH which is followed by absorbance measure at 695 nm	photocolorimetry	[8]
	FCA (ferrous-chelating activity) assay	measure the ferrous-chelating activity by following the formation of a magenta-colored Fe^{2+}-ferrozine complex at an absorbance of 562 nm. Coexisting chelator acts as competing agents results in decrease in the absorbance	photocolorimetry	[86]
	CCA (copper-chelating activity) assay	measure the copper-chelating activity by following the dissociation of the blue complex of pyrocatechol violet (PV) with $CuSO_4$. The color turned to yellow when PV dissociated a Cu ion in the presence of chelating agents. The change in color is measured at 632 nm.	photocolorimetry	[45]

Table 1. *Cont.*

	Name of the method	Principle	Mode of Detection	Ref.
In vitro	TBARS (thiobarbituric acid reactive substances) assay	measure of the end-product of lipid peroxidation which formed a pink complex with thiobarbituric acid at 100 °C in acidic condition. The formation of the complex is measured at an absorbance of 534 nm	photocolorimetry	[84]
	Superoxide radical scavenging activity assay	measure the scavenging capacity of superoxide radical generated by the reaction of NADH with phenazine methosulfate or by the oxidation of hypoxanthine by the xanthine oxidase. The inhibition of the reduction of nitroblue tetrazolium in blue-colored formazan by superoxide radical is followed at an absorbance of 560 nm.	photocolorimetry	[49, 87]
	Hydrogen peroxide scavenging activity by FOX (ferrous ion oxidation–xylenol orange) assay	measure the scavenging capacity of hydrogen peroxide. Hydrogen peroxide oxidizes ferrous ion to ferric ion, which then forms a blue-purple complex with xylenol orange. The decrease in absorbance in presence of scavenger is read at 560 nm	photocolorimetry	[49]
	Hydroxyl radical scavenging activity assay	measure the scavenging capacity of hydroxyl radical which is generated by the Fenton reaction. 2-deoxyribose is oxidized by hydroxyl radical and degraded to malondialdehyde. It forms a pink complex with thiobarbituric acid at 100 °C in acidic condition which is measured at an absorbance of 532 nm.	photocolorimetry	[87]
In vitro or on cell	Nitric oxide scavenging activity assay	measure the scavenging capacity of nitric oxide (NO), generated from sodium nitroprusside. NO reacts with oxygen to produce nitrite which can be estimated by use of Griess reagent (mix of sulphanilamide, phosphoric acid and naphthylethylenediamine dihydrochloride). Scavengers of NO compete with oxygen leading to reduced production of nitrite. The absorbance of the chromophore formed by the reaction of Griess reagent and nitrite was read at 546 nm. Nitrite oxide scavenging capacity could also be evaluated with a cellular-based assay. NO release by cells is determined by measurement of nitrite concentration in culture supernatant using the Griess reagent.	photocolorimetry	[88, 89]
On cell	ROS (reactive oxygen species) assay	measure the decrease in ROS produced by cells after stress induction in presence of antioxidant. The cells are incubated with the fluorescent dye CM-DCFDA (5-(e-6)-clorometil-2,7-dichloro dihydrofluorescein diacetate), and the fluorescence of the sample is measured at 535 nm (excitation 490 nm) to follow ROS production.	fluorimetry	[90]
	CLPAA (cellular lipid peroxidation antioxidant activity) assay	measure inhibition of lipid peroxidation in cellular membranes by monitoring red (590/632 nm) and green (485/520 nm) fluorescent products generated by the lipophilic probe C-11-BODIPY after addition of cumene hydroperoxide.	fluorimetry	[91]

Table 1. *Cont.*

	Name of the method	Principle	Mode of Detection	Ref.
On cell	CAA (cellular antioxidant activity) assay	measure the inhibition of oxidation of a fluorescent probe. The nonfluorescent DCFH (2′,7′-dichlorofluorescein) is entrapped in cell and oxidized by peroxyl radical derived from ABAP (2,2′-azobis(2-amidopropane)) or AAPH decomposition producing fluorescent DCF (dichlorofluorescein). Antioxidant prevent oxidation of the probe and attenuate cellular fluorescence (excitation and emission at 485 and 520 nm)	fluorimetry	[91]
	Comet assay (single-cell gel electrophoresis)	measure the nuclear DNA protection by an antioxidant after applying hydrogen peroxide oxidative stress on cells. Treated cells are embedded in agarose and are lysed to form nucleoids containing supercoiled loops of DNA linked to the nuclear matrix. After electrophoresis, the DNA is stained with a fluorescent dye and results in structures resembling comets observed by fluorescence microscopy; the intensity of the comet tail relative to the head reflects the number of DNA breaks.	fluorescence microscopy	[92]

Table 2. Antioxidant activity evaluation of microalgae extracts by in vitro chemical methods (AA: ascorbic acid, AAE: ascorbic acid equivalent, ABS: absorbance, Ac: acetone, AcOH: acetic acid, AIOLA: AAPH induced oxidation of linoleic acid, BHA: butylated hydroxyanisole, BHT: butylated hydroxytoluene, CCA: copper-chelating activity, $CHCl_3$: chloroform, conc.: concentration, Co-Q10: co-enzyme Q10, DCM: dichloromethane, DPPH: 2,2-diphenyl-1-picrylhydrazyl, DW: dry weight, Eq: equivalent, EtOAC: ethyl acetate, EtOH: ethanol, FA: fatty acid, FCA: ferrous-chelating activity, FRAP: ferric-reducing antioxidant power, FTC: ferric thiocyanate assay, FW: fresh weight, GC-MS: gas chromatography–mass spectroscopy, Hex: hexane, IC_{50}: inhibition concentration 50, inhib.: inhibition, i-PrOH: isopropanol, MeOH: methanol, ORAC: oxygen radical absorbance capacity, PBS: phosphate buffer saline, PE: petroleum ether, PLE: pressurized liquid extraction, PUFA: polyunsaturated fatty acid, TAC: total antioxidant capacity, TBARS: thiobarbituric acid reactive substance, TE: trolox equivalent, TEAC: trolox equivalent antioxidant capacity, temp.: temperature, TPC: total phenolic compounds, US: ultrasounds, α-toco.: α-tocopherol).

Microalgae Species	Antioxidant Assay	Composition Analyses	Antioxidant Activity	Positive Control	Molecules Involved in Antioxidant Activity	Method of Extraction	Ref.
Grammatophora marina	(i) DPPH (ii) FCA (iii) hydrogen peroxide scavenging activity (iv) superoxide radical scavenging activity (v) hydroxyl radical scavenging activity (vi) nitric oxide scavenging activity	-	extracts at 2000 µg mL^{-1} (i) 41–86% inhib. (ii) 21–81% inhib. (iii) 14–25% inhib. (iv) 24–45% inhib. (v) 10–35% inhib. (vi) 12–33% inhib.	α-toco. and BHT at 2000 µg mL^{-1} (i) 70 and 72% inhib (ii) 10 and 11% inhib. (iii) 74 and 67% inhib. (iv) 33 and 64% inhib (v) 79 and 77% inhib. (vi) 43 and 56% inhib.	-	maceration 80% MeOH or enzymatic lysis (5 carbohydrases and 5 proteases tested)	[93]

Table 2. *Cont.*

Microalgae Species	Antioxidant Assay	Composition Analyses	Antioxidant Activity	Positive Control	Molecules Involved in Antioxidant Activity	Method of Extraction	Ref.
Chlorella vulgaris	(i) DPPH (ii) TEAC (iii) ORAC (iv) FRAP	TPC	(i) 0.8 μmol TE g^{-1} DW (ii) 15 μmol TE g^{-1} DW (iii) 31 μmol TE g^{-1} DW (iv) 0.6 μmol TE g^{-1} DW	-	phenolic compounds	US (30 min, room temp.) EtOH 50%	[94]
Dunaliella salina, Dunaliella tertiolecta, Phaeodactylum tricornutum, Chaetoceros muelleri, Pavlova salina, Pavlova lutheri, Tetraselmis suecica, Tetraselmis sp., *Tetraselmis chui, Nannochloropsis* sp., *Isochrysis galbana*	ORAC	TPC, total carotenoids	45–577 μmol TE g^{-1} DW	-	-	maceration + EtOAC, Hex or H_2O	[95]
Scenedesmus obliquus	(i) DPPH (ii) TEAC (iii) superoxide radical scavenging activity (iv) nitric oxide scavenging activity	carotenoids, PUFA	(i) IC_{50}: 412–878 μg mL^{-1} (ii) IC_{50}: 41–648 μg mL^{-1} (iii) IC_{50}: 520–1236 μg mL^{-1} (iv) IC_{50} = 60 μg mL^{-1}	-	-	maceration (20 min 40 °C) + EtOH, Ac, ethyl lactate or Hex/i-PrOH (3/2)	[96]
Scenedesmus sp. + 4 *Scenedesmus quadricauda strains*	(i) DPPH (ii) β-carotene bleaching	TPC, tannins, iridoids	(i) 6–70% inhib. (extracts at 200 μg mL^{-1}) (ii) 24–92% inhib. (extracts at 400 μg mL^{-1})	(i) AA: 98% inhib. at200 μg mL^{-1} (ii) BHT: 70% inhib. at 400 μg mL^{-1}	phenolic compounds	maceration + US (30 min, in ice) + MeOH 50%, PE or DCM	[82]
Chlorella minutissima	(i) DPPH (ii) β-carotene bleaching	TPC, tannins, iridoids, pigments	(i) 10–70% inhib. (extracts at 200 μg mL^{-1}) (ii) IC_{50}: 75–600 μg mL^{-1}	(i) AA: 97% inhib. at 200 μg mL^{-1} (ii) IC_{50} BHT = 60.7 μg mL^{-1}	carotenoids, phenolic compounds	maceration (1 night) + US (30 min, in ice) + MeOH, PE or DCM	[44]
Chlorella minutissima + 2 *Chlorella* sp. *strains.*	(i) DPPH (ii) β-carotene bleaching	TPC, tannins, flavonoids, iridoids	(i) 25–100% inhib. (extracts at 200 μg mL^{-1}) (ii) IC_{50}: 25–450 μg mL^{-1}	(i) AA: 97% inhib. at 200 μg mL^{-1} (ii) IC_{50} BHT = 61 μg mL^{-1}	-	maceration (1 night) + US (30 min, in ice) MeOH, PE or DCM	[48]

Table 2. *Cont.*

Microalgae Species	Antioxidant Assay	Composition Analyses	Antioxidant Activity	Positive Control	Molecules Involved in Antioxidant Activity	Method of Extraction	Ref.
Ammatoidea normanii, Ruttnera lamellose, Pavlova granifera, Apistonema sp., 2 *Cryptomonas pyrenoidifera strains, Porphyridium aerugineum, Porphyridium sordidum, Audorinella* sp., *Phragmonema sordidum,* 3 *Characiopsis aquilonaris strains, Characiopsis ovalis,* 2 *Characiopsis* sp. *strains, Characiopsis minima, Pseudostaurastrum enorme, Goniochloris sculpta, Eustigmatos* sp., *Vischeria helvetica, Chlorobotrys gloeothece, Chlorobotrys* sp., *Dioxys* sp., *Coronastrum aestivale, Chlorella vulgaris, Mychonastes homosphaera, Gloeococcus minor, Pectodyction cubicum, Jaagiella apicola, Schizomeris leibleinii, Interfilum paradoxum, Micrasterias radiosa var. elegantior, Haematococcus pluvialis, Lobomonas* sp., *Stephanosphaera pluvialis, Bumilleria sicula, Euglena cantabrica*	(i) DPPH (ii) TEAC	-	(i) IC_{50}: 44–1421 mg FW mL^{-1} (ii) 5–195 mg AAE 100 g^{-1} FW and 17–258 mg TE 100 g^{-1} FW	-	-	US (30 min, dark) + maceration (1 night, −4 °C) + EtOH	[6]
Botryococcus braunii, Chlorella sorokiniana, Nannochloropsis granulata, Neochloropsis oleabundans, Phaeodactylum tricornutum, Porphyridium aerugineum, Scenedesmus obliquus, Scenedesmus sp., *Tetraselmis chuii*	(i) DPPH (ii) ORAC	TPC, carotenoids, lipids, FA	(i) <50% inhib. (extracts at 200 μg mL^{-1}) (ii) 7–53 μmol TE g^{-1} DW	-	phenolic compounds and lipids	maceration MeOH (DPPH) or PLE Hex/DCM (50/50)(70 °C) and then Ac/H_2O/AcOH (70/29.5/0.5) (80 °C) (ORAC)	[97]
Chlorella kessleri	(i) DPPH (ii) TEAC (iii) reducing power	total carotenoids, chlorophylls *a* and *b*	(i) 1–4% inhib. (extracts at 2500 μg mL^{-1}) (ii) 196–346 μmol TE g^{-1} extract (iii) ABS_{700}: 0,266–0,473 (extracts at 2500 μg mL^{-1})	-	-	maceration MeOH	[98]

Table 2. *Cont.*

Microalgae Species	Antioxidant Assay	Composition Analyses	Antioxidant Activity	Positive Control	Molecules Involved in Antioxidant Activity	Method of Extraction	Ref.
Scenedesmus sp.	(i) DPPH (ii) FRAP	TPC, flavonoids, carotenoids	(i) 0.6–3.7 μmol TE g^{-1} DW (ii) 2.8–47.0 μmol TE g^{-1} DW	-	-	US (20 min) + maceration (1h) $EtOH/H_2O$ (3:1), Hex, EtOAc, or H_2O	[99]
Botryococcus braunii	ORAC	-	43 μmol TE g^{-1} extract	-	-	grinding + PBS	[90]
Euglena tuba	(i) DPPH (ii) TBARS (iii) superoxide radical scavenging activity (iv) hydrogen peroxide scavenging activity (v) peroxynitrite scavenging activity (vi) singlet oxygen scavenging activity (vii) hypochlorous acid scavenging activity	TPC, flavonoids, tannins, alkaloids, AA	(i) IC_{50} = 146 μg mL^{-1} (ii) IC_{50} = 42 μg mL^{-1} (iii) IC_{50} = 5.8 μg mL^{-1} (iv) IC_{50} = 47340 μg mL^{-1} (v) IC_{50} = 278 μg mL^{-1} (vi) IC_{50} = 2821 μg mL^{-1} (vii) IC_{50} = 879 μg mL^{-1} (viii) IC_{50} = 223 μg mL^{-1}	(i) IC_{50} AA = 5.3 μg mL^{-1} (ii) IC_{50} mannitol = 571.4 μg mL^{-1} (ii) IC_{50} quercetin = 42.1 μg mL^{-1} (iv) IC_{50} sodium pyruvate = 3.2 mg mL^{-1} (v) IC_{50} curcumin = 90.8 μg mL^{-1} (vi) IC_{50} gallic acid = 0.88 mg mL^{-1} (vii) IC_{50} lipoic acid = 0.05 mg mL^{-1} (viii) IC_{50} AA = 236.0 μg mL^{-1}	-	maceration (15h) + MeOH 70%	[49]
3 *Chlorella* sp. strain	(i) DPPH (ii) FCA (iii) TBARS	TPC	(i) IC_{50}: 810–1400 μg mL^{-1} (ii) IC_{50}: 1220–1500 μg mL^{-1} (iii) 5.9–88% inhib. (extracts at 4000 μg mL^{-1})	(i) IC_{50} BHT = 50 μg mL^{-1} (ii) IC_{50} EDTA = 28 μg mL^{-1} (iii) BHT 94% inhib. (conc. not specified)	-	grinding (20 min) + H_2O 80 °C 20 min or maceration (24h) + EtOH 95%	[100]
Nephroselmis sp., *Tetraselmis* sp., *Dunaliella* sp., *Picochlorum* sp., *Schizochlamydella* sp., 2 *Nitzschia* sp. strain, *Thalassiosira weissflogi*, *Entomoneis punctulata*, *Cylindrotheca closterium*, *Chaetoceros* sp., *Bacillaria* sp.	(i) DPPH (ii) TEAC (iii) ORAC (iv) TBARS	carotenoids composition	(i) IC_{50} from 484 to >1000 μg mL^{-1} (ii) IC_{50} from 193 to >1000 μg mL^{-1} (iii) 0–190 μg TE mg^{-1} extract (iv) IC_{50}: 15.4–473.6 μg mL^{-1} extract	(i) IC_{50} trolox = 4.7 μg mL^{-1}, α-toco. = 6.2 μg mL^{-1}, AA = 8.7 μg mL^{-1}, β-carotene = 257.3 μg mL^{-1}, astaxanthin = 228.6 μg mL^{-1} (ii) IC_{50} trolox = 6.4 μg mL^{-1}, α-toco. = 10.8 μg mL^{-1}, AA = 6.1 μg mL^{-1}, β-carotene = 37.0 μg mL^{-1}, astaxanthin = 98.5 μg mL^{-1} (iv) IC_{50} trolox = 0.2 μg mL^{-1}, α-toco. = 1.3 μg mL^{-1}	carotenoids	US (60 min) + MeOH/DCM (50/50)	[7]
Nephroselmis sp.	ORAC	carotenoids composition	63.6–154.9 μmol TE g^{-1} DW	-	carotenoids	grinding + maceration (30 min, room temp., dark) + EtOH	[22]
Tetraselmis sp.	TBARS	-	IC_{50}: 3.4–11.3 μg mL^{-1} extract	IC_{50} trolox = 0.2 μg mL^{-1}, IC_{50} α-toco. = 1.3 μg mL^{-1}	-	grinding + US (10 min., ice bath, dark) + MeOH/DCM (50/50)	[21]
Tetraselmis chuii, *Nannochloropsis oculata*, *Chlorella minutissima*, *Rhodomonas salina*	(i) DPPH (ii) FCA (iii) CCA	TPC	extracts at 1000 μg mL^{-1} (i) 0–21% inhib. (ii) 12–98% inhib. (iii) 12–22% inhib.	conc. at 1000 μg mL^{-1} (i) BHT: 88% inhib. (ii) EDTA: 95% inhib. (iii) EDTA: 74% inhib.	-	grinding + maceration (1 nuit) + Hex or MeOH	[45]

Table 2. *Cont.*

Microalgae Species	Antioxidant Assay	Composition Analyses	Antioxidant Activity	Positive Control	Molecules Involved in Antioxidant Activity	Method of Extraction	Ref.
Isochrysis galbana T-iso, Tetraselmis sp., *Scenedesmus* sp.	(i) DPPH (ii) FCA (iii) CCA	TPC, FA	(i) $IC_{50} > 1000$ μg mL^{-1} (ii) IC_{50}: 730–4110 μg mL^{-1} (iii) IC_{50}: 900 μg mL^{-1} to >10000 μg mL^{-1}	(i) IC_{50} BHT = 70 μg mL^{-1} (ii) IC_{50} EDTA = 100 μg mL^{-1} (iii) IC_{50} EDTA = 280 μg mL^{-1}	-	grinding + Hex, and, Ac and H_2O in sequential order	[101]
Chlorococcum minutum	(i) TAC (ii) reducing power	TPC	(i) 2.5–10 mg AAE g^{-1} extract (ii) 1–4 mg AAE g^{-1} extract	-	phenolic compounds	maceration (72 h) EtOH, MeOH, or Ac	[102]
Chaetoceros calcitrans	(i) DPPH (ii) TEAC (iii) FCA	TPC, major phenolic compounds, total carotenoids totaux, fucoxanthin	(i) 0.1–1.4 mg TE g^{-1} DW (ii) 1.2–10.6 mg TE g^{-1} DW (iii) 0.3–18.5 mg Na-EDTA Eq g^{-1} DW	-	carotenoids and phenolic compounds	grinding + US (30 min, room temp.) + MeOH, EtOH, Ac, Ac 90%, Ac/$CHCl_3$ (90/10) or Ac/$CHCl_3$/MeOH (80/10/10)	[103]
Chaetoceros calcitrans, Isochrysis galbana, Skeletonema costatum, Odontella sinensis, Phaedactylum tricornatum	(i) TEAC (ii) FRAP (iii) FCA (iv) β-carotene bleaching	TPC, major phenolic compounds, total carotenoids totaux, fucoxanthin	(i) 2.0–21.5 mg TE g^{-1} DW (ii) 0.2–2.0 mg TE g^{-1} DW (iii) 1.5–13.4 mg EDTA eq g^{-1} DW (iv) 0.1–1.4 mg TE g^{-1} DW	-	carotenoids and phenolic compounds	grinding + MeOH	[104]
Chaetoceros sp., *Nannochloropsis* sp.	(i) DPPH (ii) FRAP (iii) FCA (iv) superoxide radical scavenging activity	TPC	(i) 14.0–106.7 μmol TE g^{-1} extract (ii) 171.5–609.8 μmol TE g^{-1} extract (iii) 3.2–82.4 μmol EDTA Eq g^{-1} extract (iv) 227.9–3224.5 μmol TE g^{-1} extract	-	-	maceration (24h) + Hex, DCM, $CHCl_3$ or MeOH	[86]
Nannochloropsis oculata, Nannochloropsis sp., *Isochrysis* sp., *Isochrysis* ISO-T, *Tetraselmis* sp., *Tetraselmis suecica, Botryococcus braunii, Porphyridium cruentum, Neochloris oleabundans, Chaetoceros calcitrans, Chlorella vulgaris, Haematococcus pluvialis* (red and green phase), *Parachlorella kessleri, Phaeodactylum tricornutum, Schizochytrium* sp.	(i) TEAC (ii) FRAP (iii) AIOLA	TPC, total carotenoids	(i) 0–69 μmol TE g^{-1} DW (ii) 3.3–90 μmol TE g^{-1} DW (iii) 1.8–89.7 μmol TE g^{-1} DW	-	carotenoids and phenolic compounds	grinding + maceration (30 min) + EtOH/H_2O (3/1) or Hex, EtOAc and H_2O (80 °C) in sequential order	[10]

Table 2. *Cont.*

Microalgae Species	Antioxidant Assay	Composition Analyses	Antioxidant Activity	Positive Control	Molecules Involved in Antioxidant Activity	Method of Extraction	Ref.
Phaeodactylum tricurnutum, 2 *Chlorella vulgaris strains*, *Haematococcus pluvialis*, *Scenedesmus maximus*, *Scenedesmus obliquus*, *Scenedesmus quadricauda*, *Desmodesmus pleimorphus*, *Nannochloropsis* sp., *Pavlova lutheri*, *Porphyridium aerugineum*	TEAC	carotenoids	0.8–149 mg L^{-1} AAE μg^{-1} chlorophyll *a*	-	-	grinding + EtOH 50%	[105]
Galdieria sulphuraria, *Ettlia carotinosa*, *Neochloris texensis*, *Chlorella minutissima*, *Stichococcus bacillaris*, *Schizochytrium limacinum*, *Crypthecodinium cohnii*, *Chlorella vulgaris*	DPPH	TPC	89–95% inhib. (extracts at 250 μg mL^{-1})	BHT: 98% inhib. at 250 μg mL^{-1}	TPC	US (20 min) MeOH or maceration H_2O (100 °C, 30 min)	[106]
Chlorella stigmatophora, *Phaeodactylum tricornutum*	(i) Superoxide radical scavenging activity (ii) hydroxyl radical scavenging activity (iii) hypochlorous acid scavenging activity	-	(i) IC_{50}: 48–170 μg mL^{-1} (ii) IC_{50}: 180–250 μg mL^{-1} (iii) IC_{50} > 1000 μg mL^{-1}	-	-	US + H_2O then soxhlet + DCM and MeOH on extraction residue	[87]
Chlorella vulgaris	FRAP	TPC	0.01–58.2 μmol TE g^{-1} DW	-	phenolic compounds	maceration + Hex, EtOAc + H_2O (80 °C) in sequential order	[46]
Phaeodactylum tricornutum, *Nannochloropsis gaditana*, *Nannochloris* sp., *Tetraselmis suecica*	(i) DPPH (ii) reducing power (iii) FCA	TPC, flavonoids, carotenoids	(i) IC_{50}: 356–400 μg mL^{-1} (ii) 24–33 AAE mL^{-1} (iii) IC_{50}: 2810–12820 μg mL^{-1}	(i) IC_{50} AA = 3,7 μg mL^{-1} (ii) BHT = 1,4 AAE mg^{-1} (iii) IC_{50} EDTA = 10 μg mL^{-1}	-	Not specified	[51]
Dunaliella salina	TEAC	carotenoids	11–1118 μmol TE g^{-1} extract	-	carotenoids	PLE Hex, EtOH or H_2O	[83]
Dunaliella salina	TEAC	carotenoids	115–452 μmol TE g^{-1} extract	-	carotenoids	sub- and super-critical CO_2	[107]
Chlorella vulgaris, *Chlamydomonas reinhardtii*	(i) DPPH (ii) TAC (iii) FRAP	TPC, flavonoids	(i) IC_{50}: 397–423 μg mL^{-1} (ii) IC_{50}: 55–73 μg mL^{-1} (iii) ABS_{700}: 0.136 to 0.124 (extracts at 250 μg mL^{-1})	(ii) IC_{50} AA = 127.5 μg mL^{-1} (iii) ABS_{700} AA = 0.423 at 250 μg mL^{-1}	flavonoids	maceration MeOH	[47]
Ankistrodesmus sp., *Euglena cantabrica*	DPPH	-	8–71% inhib. (extracts at 1000 μg mL^{-1})	conc. at 1000 μg mL^{-1} BHT: 26% inhib., BHA: 91% inhib.	-	maceration (40 min) + MeOH or H_2O	[108]

Table 2. *Cont.*

Microalgae Species	Antioxidant Assay	Composition Analyses	Antioxidant Activity	Positive Control	Molecules Involved in Antioxidant Activity	Method of Extraction	Ref.
Halochlorococcum porphyrae, *Oltamannsiellopsis unicellularis*	(i) DPPH (ii) FCA (iii) hydrogen peroxide scavenging activity (iv) superoxide radical scavenging activity (v) hydroxyl radical scavenging activity (vi) nitric oxide scavenging activity	TPC	extracts at 2000 μg mL^{-1} (i) 42–95% inhib. (ii) 4–72% inhib. (iii) 5–42% inhib. (iv) 5–58% inhib. (v) 4–31% inhib (vi) 1–51% inhib.	conc. at 2000 μg mL^{-1} (i) BHT and α-toco: 94% inhib. (ii) BHT: 11% inhib., α-toco 10% inhib. (iii) BHT 60% inhib., α-toco 62% inhib. (iv) BHT 63% inhib., α-toco 61% inhib. (v) BHT 77% inhib., α-toco 79% inhib. (vi) BHT 26% inhib., α-toco 25% inhib.	-	80% MeOH then fractionation with Hex, $CHCl_3$ and EtOAc or enzymatic lysis (5 carbohydrases and 5 proteases tested)	[89]
Chlamydomonas nivalis, *Chlorella protothecoides*, *Chlorella pyrenoidosa*, *Chlorella vulgaris*, *Chlorella zofingiensis*, *Crypthecodinium cohnii*, *Nitzschia laevis*, *Schizochytrium* sp., *Schizochytrium mangrovei*, *Thraustochytrium* sp.	TEAC	TPC	0–11.4 μmol TE g^{-1} DW	-	-	maceration (30 min) + Hex, EtOAc and H_2O (80 °C) in sequential order	[13]
Tetraselmis sp., *Dunaliella salina*, *Dunaliella* sp., *Nannochloropsis gaditana*, *Chlorella* sp., *Navicula* sp., *Phaeodactylum tricurnutum*, *Chaetoceros* sp., *Isochrysis* sp.	DPPH	TPC, total carotenoids, PUFA	IC_{50}: 247–464 μg mL^{-1}	IC_{50} BHT = 6.2 μg mL^{-1}, IC_{50} AA = 2.5 μg mL^{-1}	-	maceration (3 h, dark) + EtOH	[109]
Isochrysis galbana	(i) DPPH (ii) TEAC	TPC, β-glucan, Co-Q10, β-carotene, fucoxanthin	(i) 0–17 mg AAE L^{-1} (ii) 52–56 μmol TE g^{-1} DW	-	-	grinding + maceration (18 h) EtOH 96% or H_2O	[110]
Nannochloropsis gaditana	(i) DPPH (ii) β-carotene bleaching (iii) FRAP	carotenoids, tocopherols, FA	(i) 1,1–1,8 μmol TE g^{-1} extract (ii) 64–97% inhib. (extracts at 1000 μg mL^{-1}) (iii) 48–86 μmol Fe(II) g^{-1} extract	-	carotenoids, tocopherols, FA	Supercritical CO_2	[111]
Dunaliella salina, *Oocystis pusilla*, *Scenedesmus rubescens*	DPPH	TPC	0.4–17.5 μmol TE g^{-1}	-	phenolic compounds	maceration (30 min, 25 °C) + Hex, EtOAc and H_2O (80 °C) in sequential order	[112]
Cymbella sp., *Navicula* sp., *Skeletonema costatum*, *Isochrysis galbana*, *Chaetoceros calcitrans*, *Nannochloropsis oculata*, *Tetraselmis tetrathele*, *Scenedesmus quadricauda*, *Chlorella vulgaris*, *Oocystis* sp., *Trachelomonas* sp.	(i) DPPH (ii) FTC (iii)TBARS	-	(i) no activity for extracts at 250–1000 μg L^{-1} (ii) 0–97% inhib.(extracts at 200 μg mL^{-1}) (iii) 0–98% inhib. (extracts at 80 μg mL^{-1})	(i) α-toco: 85% inhib., quercetin: 65% inhib, BHT: 74% inhib. (100 μg L^{-1}) (ii) α-toco: 84% inhib, quercetin: 92% inhib., BHT: 100% inhib. (200 μg mL^{-1}) (iii) α-toco: 71% inhib., quercetin: 90% inhib., BHT: 98% inhib. (80 μg mL^{-1})	-	maceration (4 j) + MeOH	[84]
2 *Nannochloris* sp. *strains*, *Picochlorum* sp., *Desmochloris* sp.	(i) DPPH (ii) FCA (iii) CCA	TPC, pigments	extracts at 1000 μg mL^{-1} (i) <10% inhib. (ii) <25% inhib. (iii) <30% inhib.	conc. at 1000 μg mL^{-1} (i) BHT: 88% inhib. (ii) EDTA: 96% inhib. (iii) EDTA: 76% inhib.	-	grinding + maceration (1 night, 20 °C) + MeOH	[113]

Table 2. *Cont.*

Microalgae Species	Antioxidant Assay	Composition Analyses	Antioxidant Activity	Positive Control	Molecules Involved in Antioxidant Activity	Method of Extraction	Ref.
Chlorella vulgaris	(i) TEAC (ii) ORAC (iii) superoxide radical scavenging activity	TPC	(i) 146–789 µmol TE g^{-1} extract (ii) 243–1008 µmol TE g^{-1} extract (iii) IC_{50}: 8260–10752 µg mL^{-1}	rosemary extract (i) 2805–2811 µmol TE g^{-1} (ii) 4615–4892 µmol TE g^{-1} (iii) IC50: 464–665 µg mL^{-1}	-	supercritical H_2O	[81]
Haematococcus pluvialis	TEAC	GC-MS	366–1974 µmol TE g^{-1} extract	-	α-toco., gallic acid, caramelization products and possible Maillard reaction products	supercritical H_2O	[114]
Phaeodactylum tricornutum, *Nannochloropsis salina*, *Nannochloropsis limnetica*, *Chlorella sorokiniana*, *Dunaliella salina*, *Desmodesmus* sp.	(i) DPPH (ii) TEAC (iii) FCA (iv) FRAP (v) TAC	TPC, flavonoids, phenolic acids, tocopherols, carotenoids composition	(i) 8–14% inhib. (extracts at 250 µg mL^{-1}) (ii) 2.7–24.2 TE g^{-1} (iii) 3–9% chelation (extracts at 250 µg mL^{-1}) (iv) 0.1–0.5 AAE g^{-1} (v) 3.0–8.9 gallic acid Eq g^{-1}	-	phenolic compounds, carotenoids and tocopherols	US (45 min in the dark at room temp.) + MeOH	[8]
Tetraselmis suecica	DPPH	pigment composition	21.1% inhib. (extract at 50 µg mL^{-1})	α-toco: 6% inhib. at 50 µg mL^{-1}	-	maceration (30 min in the dark under nitrogen atmosphere at room temp.) + EtOH/H_2O (3/1)	[115]
Parachlorella kessleri	(i) DPPH (ii) TEAC (iii) FCA (iv) TAC	TPC, chlorophyll *a* and *b*, total carotenoids	(i) 32–69% inhib. (extracts at 100 µg mL^{-1}) (ii) 1.4–3.0 µmol TE g^{-1} extract (iii) 20% inhib. (extracts at 500 µg mL^{-1}) (iv) 2.2–4.3 mg AAE g^{-1} extract	-	-	grinding + maceration MeOH	[116]
Trentepohlia umbrina	(i) DPPH (ii) reducing power (iii) superoxide radical scavenging activity	TPC, flavonoids	(i) IC_{50} = 665.3 µg mL^{-1} (ii) ABS_{700} = 0.0124(extract at 125 µg mL^{-1}) (iii) IC_{50} = 838.8 µg mL^{-1}	(i) IC_{50} AA = 6.4 µg mL^{-1} (ii) ABS_{700} AA = 0.0478 at 125 µg mL^{-1} (iii) IC_{50} AA = 115.6 µg mL^{-1}	-	maceration (72h) + MeOH	[85]
Dunaliella salina	DPPH	chlorophylls, total carotenoid	15–57% inhib. (extract at 250 µg mL^{-1})	AA: 95% inhib. at 250 µg mL^{-1}	-	US (10 min) + maceration (4 j) + EtOH	[117]
Skeletonema marinoi	TEAC	TPC, flavonoids, AA, β-carotene, diatoxanthin	250–1500 fg AAE $cell^{-1}$	-	phenolic compounds, flavonoids, AA	US (1 min, in ice) + maceration (30 min, dark) MeOH	[53]
Chloromonas sp.	(i) DPPH (ii) TEAC	-	(i) IC_{50} = 1.0µg mL^{-1} (ii) IC_{50} = 0.9 µg mL^{-1}	(i) IC_{50} AA = 0.1 µg mL^{-1} (ii) IC_{50} AA = 0.2 µg mL^{-1}	-	maceration (24 h) + EtOH	[118]
Botryidiopsidaceae sp.	(i) DPPH (ii) TEAC	-	(i) IC_{50} = 1.5 µg mL^{-1} (ii) IC_{50} = 1.8 µg mL^{-1}	(i) IC_{50} AA = 0.2 µg mL^{-1} (ii) IC_{50} AA = 0.2 µg mL^{-1}	-	maceration (24 h) + EtOH	[119]
Crypthecodinium cohnii, *Schizochytrium* sp.	(i) DPPH (ii) TAC (iii) FCA (iv) reducing power	TPC, flavonoids	extracts at 500 µg mL^{-1} (i) 15–30% inhib. (ii) ABS_{695}: 0,500–1,000 (iii) 10–60% inhib. (iv) ABS_{700}: 0,050–0,300	(ii) BHT: ABS_{695} = 0,500 at 500 µg mL^{-1} (iii) EDTA: 65% inhib. at 50 µg mL^{-1} (iv) BHT: ABS_{700} = 0,300 at 500 µg mL^{-1}	phenolic compounds	maceration (2 j) EtOH 70%	[54]

Table 3. Antioxidant activity evaluation of microalgae extracts by cellular assays (AA: ascorbic acid, Ac: acetone, CAA: cellular antioxidant activity, CHCl3: chloroform, CLPAA: cellular lipid peroxidation antioxidant activity, EtOH: ethanol, Hex: hexane, IC50: inhibition concentration 50, inhib.: inhibition, MeOH: methanol, NMR: nuclear magnetic resonance, PBS: phosphate buffer saline, ROS: reactive oxygen species, TPC: total phenolic compounds, US: ultrasounds).

Microalgae Species	Antioxidant Assay	Composition Analyses	Antioxidant Activity	Positive Control	Molecules Involved in Antioxidant Activity	Extraction Method	Ref.
Chaetoceros calcitrans	Nitric oxide scavenging activity assay on RAW 264.7 cells (mouse macrophage)	metabolites profiling by ^{1}H NMR + TPC	IC_{50}: 3.5–187.7 μg mL^{-1}	IC_{50} quercetin = 4.7 μg mL^{-1} IC_{50} curcumin = 6.1 μg mL^{-1}	Fucoxanthin (25), astaxanthin, violaxanthin, zeaxanthin, canthaxanthin (26), and lutein (27)	US (30 min, room t °C) + MeoH, 70% EtOH, Ac, $CHCl_3$ or Hex	[88]
Botryococcus braunii	(i) ROS assay and (ii) Comet assay on NIH3T3 cells (mouse embryonic fibroblast cells)	-	extract at 0.1–0.05% (i) reduction of ROS production of 35% over the control (=no microalgae extract) after stress induction (ii) no activity	(i) AA: reduction of ROS production by 64% over the control at 250 μM	-	crushing in PBS + silica sand	[90]
Pediastrum duplex, *Halochlorococcum porphyrae*, *Oltmannsiellopsis unicellularis*, *Achnanthes longipes*, *Navicula* sp., *Amphora coffeaeformis*	Comet assay on L5178 cells (mouse lymphoma cells)	Crude lipid content	extract at 25–100 μg mL^{-1} (i) inhibitory effect to DNA damage until 80% over the control (=no microalgae extract) after stress induction	-	-	Enzymatic extraction by 5 carbohydrases and 5 proteases	[92]
Cylindrotheca closterium, *Coscinodiscus actinocyclus*, *Nitzschia closterium*, 2 *Pseudo-nitzschia pseudodelicatissima* strains, *Tetraselmis suecica*, *Isochrysis galbana*, *Skeletonema costatum*, *Lauderia annulata*, *Leptocylindrus danicus*, *Chaetoceros affinis*, *Odontella mobiliensis*, *Leptocylindrus aporus*, *Thalassiosira rotula*, *Thalassiosira weissflogii*, 2 *Skeletonema marinoi* strains, *Thalassiosira rotula*, *Skeletonema costatum*, *Stephanopyxis turris*, *Bacteriastrum hyalinum*, *Guinardia striata*, *Proboscia alata*, *Guillardia theta*, *Rhodomonas baltica*, *Rhinomonas reticulata*, *Alexandrium tamutum*, *Alexandrium andersonii*, *Ostreopsis ovata*, *Alexandrium minutum*, *Lepidodinium viride*, *Prorocentrum gracile*	(i) CAA and (ii) CLPAA on HepG2 cells (human liver cancer cell line)		extract at 50 μg mL^{-1} (i) 66–70% inhib. for *Ostreopsis ovata* (ii) 61–74% inhib. for *Ostreopsis ovata* and 100% inhib. for *Alexandrium minutum* but both species showed toxicity in cytotoxicity assay	-	-	US (1 min)+ H_2O then addition of Ac + maceration (50 min, room temp.) then fractionation on Amberlite XAD16N resin	[91]

Table 4. Antioxidant activity evaluation of microalgae extracts by in vivo experimentations (CAT: catalase, DNPH: 2,4-dinitrophenyl hydrazine, FA: fatty acid, FRAP: ferric-reducing antioxidant power, GPX: glutathione peroxidase, GSH: reduced glutathione, MDA: malondialdehyde, PX: peroxidase SOD: superoxide dismutase, TAC: total antioxidant capacity, and TBARS: thiobarbituric acid reactive substance).

Microalgae Species	Experimental Animals	Concentration of Microalgae Tested	Experimental Time	Antioxidant Assay	Other Measure	Activity	Ref.
Schizochytrium sp.	Pacific white shrimps (*Litopenaeus vannamei*)	0–75 g of dry microalgae kg^{-1} of feed	12 weeks	TBARS on tail muscle	antioxidant enzymes activity (CAT, SOD), lipid composition of food and muscle	No effect of microalgae	[120]
Chlorella vulgaris and *Amphora coffeaformis*	Chickens (Cobb 500 broiler chick)	1 g of dry microalgae kg^{-1} of feed	32 days	TBARS on breast meat	SOD activity, FA and amino acids profiles of microalgae	28–31% decrease in MDA compared to control group (feed without microalgae)	[121]
Schizochytrium limacinum	Chickens (Arbor Acres chick)	1–2% of dry microalgae in feed	42 days	In breast and thigh muscle (i) TAC (ii) TBARS	antioxidant enzymes activity of serum (SOD, GPX, CAT), FA composition of diet and muscle	Compared to control group (feed without microalgae): (i) 33–81% increase in TAC (ii) 11–35% decrease in MDA content	[122]
Acutodesmus obliquus	Catfish (*Rhamdia quelen*)	1–3% of residual microalgae biomass (after oil extraction) in feed	60 days	(i) TBARS in liver (ii) Comet assay in erythrocytes, liver, and brain	Antioxidant enzymes activity (SOD, CAT), pigment determination of microalgae residual biomass	(i) No effect of microalgae (ii) Decrease in DNA damage with 3% of microalgae in erythrocytes and liver, no effect in brain tissue	[123]
Nannochloropsis gaditana	Normal and diabetic Wistar rats	10% of dry microalgae in feed	8 weeks	(i) TBARS of liver mitochondria and liver tissue (ii) DNPH (protein oxidation) on liver mitochondria and liver tissue	On microalgae biomass: total carotenoids, carbohydrates, total lipids and total protein On liver mitochondria and tissue: antioxidant enzymes activity (SOD, CAT, GSH)	Compared to control group (feed without microalgae): (i) Normal rats: 0–8% decrease in MDA content Diabetic rats: 35% decrease in MDA content (ii) Normal rats: no effect. Diabetic rats: 18–25% decrease in protein oxidation	[124]
Nannochloropsis sp.	Juvenile turbots (*Scophthalmus maximus* L.)	2.5–10% of dry microalgae in feed	10 weeks	(i) TBARS in serum and liver (ii) TAC	Antioxidant enzyme activity (SOD, GPX) in serum and liver	Compared to control group (feed without microalgae): (i) 19–56% decrease in MDA content, (ii) 9–44% increase in TAC	[125]
Tetraselmis chuii	Pacific white shrimps postlarvae (*Litopenaeus vannamei*)	25–100% of dry microalgae in feed	12 days	In shrimp tissue (i) hydrogen peroxide content (ii) TBARS	Proximate analysis and antioxidant activity of the feed	(i) Decrease of about 0–25% of hydrogen peroxide content (ii) No effect of microalgae on lipid peroxidation	[126]
Haematococcus pluvialis, *Botryococcus braunii*	Wistar rats	Administration by intubation to the stomach of a single dose of one of the two microalgae biomass solubilized in olive oil as source of 200 µM equivalent of astaxanthine or lutein	9 h	TBARS in plasma and liver	Analysis of carotenoids from plasma, liver and eyes Antioxidant enzyme activity (SOD, CAT, PX) in plasma and liver	25–61% decrease in MDA content compared to MDA content at t_0	[127]

Table 4. *Cont.*

Microalgae Species	Experimental Animals	Concentration of Microalgae Tested	Experimental Time	Antioxidant Assay	Other Measure	Activity	Ref.
Haematococcus pluvialis, Botryococcus braunii	Wistar rats	Administration of a daily dose of one of the two microalgae biomass solubilized in olive oil as source of 200 μM equivalent of astaxanthine or lutein	15 days	TBARS in plasma and liver	Analysis of carotenoids from plasma, liver and eyes Antioxidant enzyme activity (SOD, CAT, PX) in plasma and liver	45–64% decrease in MDA content compared to MDA content at t_0	[128]
Haematococcus pluvialis	Juvenile rainbow trout (*Oncorhynchus mykiss*)	1–10 g of dry microalgae kg^{-1} of feed	30 days	In serum (i) FRAP (ii) TBARS	alkaline phosphatase, alanine aminotransferase, aspartate and serum total protein, glucose, triglycerides, and cholesterol	Compared to control group (feed without microalgae): (i) 36–75% increase in activity (ii) 44–69% decrease in MDA content	[129]

5. Antioxidant Activity of Microalgae

The studies included in this part have been selected using Scopus and Google Scholar databases, using the terms "microalgae" in combination with "antioxidant", "antioxidant activity", "antioxidant capacity" or "antioxidant properties" as keywords. The research was limited to publication with an impact factor higher than 0.5, published until 2020. The studies have been selected based on these criteria: studies using in vitro (Table 2) or in cellular assays (Table 3) reporting the antioxidant activity of crude extract of eukaryotic microalgae. Studies focusing in the antioxidant activity of specific purified metabolite(s), or antioxidant enzyme activity have not been considered.

In addition, we have included a nonexhaustive selection of studies evaluating the antioxidant activity of microalgae in different animal models (Table 4).

The main publications evaluating the antioxidant activity of crude microalgal extracts by in vitro chemical tests are presented in Table 2. In these studies, more than two hundred *strains* of microalgae were evaluated. The most studied genera are *Chlorella* (29 *strains*), *Scenedesmus* and *Tetraselmis* (14 *strains*) (Supplementary Materials Figure S1).

The most commonly used assays to evaluate the antioxidant activity of microalgae are the DPPH (36 studies out of 52 referenced), ABTS (20 studies) and FCA assays (13 studies) (Supplementary Materials Figure S2). Overall, the results are very heterogeneous depending on the species of microalgae studied and the tests used to measure antioxidant activity. The protocols of the assays vary from one study to another, notably in terms of the extraction method, the solvents used, the reaction time and the concentrations tested. In addition, the results are expressed in different ways, making it difficult to compare the results. For example, for the DPPH assay, the results are expressed in percentage of inhibition for a given concentration (at different concentrations according to the studies), in IC_{50}, in equivalent trolox (per unit of weight of extract or per unit of dry weight) or in equivalent ascorbic acid. Finally, the use of a reference product as a point of comparison is not systematic, and the choice of the reference product is not always relevant according to the assay used.

Nevertheless, several studies highlight the potential of microalgae as a source of antioxidants:

Chloromonas sp. and *Botryidiopsidaceae* sp. (ethanolic extracts) show a strong ability to neutralize DPPH radicals (IC_{50} of 0.97 and 1.53 μg mL^{-1}, respectively) and ABTS (IC_{50} of 0.95 μg mL^{-1} and 1.79 μg mL^{-1}) similar to vitamin C [118,119]. The ABTS assay also revealed interesting activities of *Scenedesmus obliquus* (IC_{50} of 41 μg mL^{-1}, [96]), *Haematococcus pluvialis* (activity up to 1974 μmol TE g^{-1} extract for supercritical H_2O extraction, [114]) and *Dunaliella salina* (activity up to 1118 μmol TE g^{-1} extract with hexane extraction, [83]). Interesting results are also obtained with the DPPH assay for *Galdieria sulphuraria, Ettlia carotinosa, Neochloris texensis, Chlorella minutissima, Chlorella vulgaris, Schizochytrium limacinum, Stichococcus bacillaris* and *Crypthecodinium cohnii* with inhibition percentages between 89% and 95% with aqueous or methanolic extracts at concentrations of 250 μg mL^{-1} [106].

Natrah et al. [84] showed that *Chaetoceros calcitrans, Scenedesmus quadricauta, Isochrysis galbana, Chlorella vulgaris, Nannochloropsis oculata*, and *Tetraselmis tetrahele* had a strong ability to inhibit lipid peroxidation with inhibition percentages ranging from 88% to 98% for methanolic extracts at 80 μg mL^{-1} with the TBARS assay and between 88.4 and 97% for extracts at 200 μg mL^{-1} with the FTC assay (Ferric ThioCyanate assay, indirect measurement of the quantity of hydroperoxides formed during the first stages of lipid oxidation). The ability of the genera *Tetraselmis* to inhibit lipid peroxidation is confirmed by Coulombier et al. [21] who have obtained an IC_{50} up to 3,4 μg mL^{-1} with a methanol-dichloromethane extract. *Euglena tuba* also seems to be an interesting species for its ability to inhibit lipid peroxidation (IC_{50} with TBARS assay = 42 μg mL^{-1}) and to neutralize the superoxide radical (IC_{50} = 5.2 μg mL^{-1}, [49]). Some species show good ability to neutralize superoxide radical such as *Chaetoceros* sp. (1029 μmol TE g^{-1} dichloromethane extract), *Nannochloropsis* sp. (3224 μmol TE g^{-1} methanol extract), *Chlorella stigmatophora* and *Phaeodactylum tricornutum* (IC_{50} of 48.37 and 68.61 µg mL^{-1} with aqueous extracts, [87]). Chloroform and methanol extracts of *Chaetoceros* sp. also show interesting results with the FRAP assay (610 and 492.50 µmol TE g^{-1}, [86]). Good results are also obtained with the TAC assay with IC_{50} below 100 µg mL^{-1} for methanolic extracts of *Chlorella vulgaris* and *Chlamydomonas reinhardtii* [47].

The antioxidant activity of the genus *Chlorella* has been demonstrated by several authors with different antioxidant assays. In addition to the results obtained with the DPPH, TBARS, FTC, TAC, and superoxide radical neutralization assays presented above, Aremu et al. [44,48] obtained IC_{50} up to 25 µg mL^{-1} with the β-carotene bleaching assay for *Chlorella minutissima* and *Chlorella* sp. and Plaza et al. [81] showed activities up to 1008 µmol TE g^{-1} of *Chlorella vulgaris* extract with the ORAC assay.

Overall, few links are made between these antioxidant activities and the metabolites involved. Still, correlations have been shown with carotenoid content [44,83], phenolic compound content [44,106] including flavonoids [47] and gallic acid and vitamin E content [114].

Despite cellular assays potentially giving more biological relevant information, as they take into account the bioavailability and metabolism of the tested compounds, we found only four studies using cellular assays to determine antioxidant activity of microalgae extract (Table 3). Those studies use different antioxidant cellular assays and different cell models (mouse fibroblast, macrophage or lymphoma cells and human liver cancer cell line).

Chloroform, methanol, acetone and 70% ethanol extracts of *Chaetoceros calcitrans* showed high nitric oxide scavenging activity in mouse macrophage with IC_{50} values of 3.46, 3.83, 15.35 and 17.94 µg mL^{-1}, respectively, that is closed to reference compounds (IC_{50} of 4.7 and 6.1 µg mL^{-1} for quercetin and curcumin, [88]). This strong inhibitory activity of nitric oxide was attributed to the carotenoid content of *Chaetoceros calcitrans* (fucoxanthin, astaxanthin, violaxanthin, zeaxanthin, canthaxanthin and lutein). Karawita et al. [92] showed that *Pediastrum duplex* extract has a good protective effect against DNA damage induced by hydrogen peroxyde exposure (Comet assay). Indeed, a decrease of 80% of DNA damage on mouse lymphoma cells was measured with *Pediastrum* extract at 100 µg mL^{-1} compared to control with no microalgae extract. Good antioxidant activity was also measured with CAA (cellular antioxidant activity) and CLPAA (cellular lipid peroxidation antioxidant activity) assays on human liver cancer cell line with *Ostreopsis ovata* and *Alexandrium minutum*; however, both species extracts showed toxicity in cytotoxicity assay [91].

Similarly to cellular assays, the evaluation of the antioxidant activity of microalgae extracts by in vivo experimentations are limited compared to in vitro assays (Table 4). Those studies used different antioxidant in vitro assays couple with other physiological measurement, such as antioxidant enzyme activity, on various animal models (e.g., shrimps, chicken, catfish, rats, turbots or trouts, Table 4) to assess the effect of microalgae. The microalgae (*Schizochytrium* sp., *Chlorella vulgaris*, *Amphora coffeaformis*, *Schizochytrium limacinum*, *Acutodesmus obliquus*, *Nannochloropsis* spp., *Tetraselmis chuii* and *Botryococcus braunii*) were mostly included in the animal feed as dry microalgae with a percentage of inclusion mainly going from 1–10% or as a molecule equivalent of given antioxidant compounds. The results are variable depending on species from no effect of the microalgae tested [120,123] to a decrease in oxidative stress measurements such as the malondialdehyde or hydrogen peroxide content [124–129] or a decrease in DNA damage [123]. In most cases, it seems that the inclusion of microalgae directly in the fed has a positive effect on the animal physiology, which is promising regarding further used of microalgae in the food industry either in human or animal nutrition as functional ingredients. It also raises the question of the bioavailability of an antioxidant compound in the algal matrices and thus of the digestibility of the microalgae tested.

6. Applications in the Food Industry

In the food industry, antioxidants are used for human and animal nutrition as functional ingredients to provide nutritional benefits to a product (e.g., orange juice enriched with vitamin C), and as preservatives to extend the shelf life of foods and beverages to prevent their degradation by oxidation [130,131]. The use of antioxidant ingredients in food products intended for humans is highly regulated by country-specific laws owing to their potential toxicity. In the European Union, there is a list of authorized antioxidant additives, some of which may be of natural origin such as vitamin C (E300-E304), vitamin E (E306-E309), guaiac resin (E314) and rosemary extract (E392). Certain carotenoids are also authorized as dyes but can have an antioxidant role such as β-carotene (E160a), lycopene (E160d), lutein (E161b), violaxanthin (E161e), zeaxanthin (E161h), canthaxanthin (E161g) or astaxanthin (E161j) [130]. For foods and ingredients that were not significantly consumed before 1997, such as most microalgae, the "Novel Food" regulation framework was to be applied in Europe [132]. New microalgae on the market must obtain this authorization; however, to receive it, it has to be demonstrated that the product does not present any risk in terms of safety for human health [133] as some microalgae are known to produce phytotoxins [134–136]. In addition, and beyond the regulatory framework, to be of interest to the food industry, an antioxidant should not affect the color, smell and taste of the food and should be effective at low concentrations (0.001–0.01%), be easily usable, stable during processing and storage and be inexpensive [130,131]. The use of microalgae may thus be regarded as promising additive for human food, livestock feed and shelf life; however, it greatly depends on the microalgae productivity and nutrient compositions in protein, carbohydrates, lipids, vitamins, and antioxidants, which also strongly depend on species, mode of cultivation and culture medium composition e.g., [7,21,22]. Currently, around 10 species of microalgae or microalgae extract are authorized for human consumption in Europe as a food or food ingredient [137].

For livestock feed, antioxidant additives are subject to authorization before going on the market, an authorization that remains only valid ten years. On the other hand, raw materials are not subject to authorization, but a contribution of microalgae as an antioxidant in animal feed could only be considered as a raw material if it also provides proteins, minerals, fats, fibers, energy or carbohydrates [138]. Microalgae presents growth rate and dietary value of interest (e.g., polyunsaturated fatty acids, vitamins, pigments, polysaccharides) for livestock feed or aquaculture feed either fish, live feed and shellfish applications (e.g., in Table 4). Indeed, in aquaculture, the polyunsaturated fatty acids (PUFAs) eicosapentaenoic acid (DHA) and docosahexanoic acid (DHA) are of nutritional importance in aquafeeds and are hitherto ensured by inclusion of fish oil in aquafeeds. However, this resource is limited, and microalgae offer an alternative to fish oil. In addition, microalgae are not only seen as a source of PUFAs but also as source of other metabolites of interest such as pigments, polysaccharides, vitamins (e.g., vitamin E and C) and sterols which are introduced as dietary supplements for dietetic and therapeutic purposes [3,129]. In terms of applications, antioxidant molecules (asthaxanthin, lutein, β-carotene) carotenoids are produced by a wide variety of microalgae (see Table 2).

7. Conclusion

Antioxidant molecules from microalgae are more and more considered as a potential source of natural antioxidant compounds by the food, the cosmetic and nutraceutical industries as they may bring benefits to their products.

However, it is very crucial to assess properly the antioxidant activity of an algal extract owing to the wide diversity of antioxidant compounds and the mode of action combined with the diversity of ROS involved. This review highlights the lack of standardization between extractions procedures used to assess antioxidant activity from microalgae matrices, and more disturbingly, it highlights the inappropriateness between the assay used and the molecules studied. These often hamper the comparison between studies and bring the authors to false or incorrect interpretation of their results.

Therefore, although all the assays have their merits and demerits, the appropriate selection of a given assay was to be made based on the mode of action of a studied molecule in front of the principle and mechanism of an assay, especially in vitro assays. In addition to the need of normalization of the extraction procedures and to the appropriate use of an assay, we conclude that it is crucial to combine many assays to assess microalgae full antioxidant activity.

This review also highlights that microalgae are rich in antioxidant molecules with more or less potent activities, which can be used as an ingredient in food, cosmetic and nutraceutical industries. In addition, research publications are available on modern in vitro chemical methods, but application on cellular assays and in vivo experimentations are still lacking. There is a need to develop models to improve our ability to assess the activity of antioxidant molecule on these kinds of models to further improve industrial adaption and application.

Supplementary Materials: The following are available online at https://www.mdpi.com/article/10.3390/md19100549/s1, Figure S1. Top 15 microalgae genus studied for their antioxidant activity, Figure S2. In vitro chemical assays used to evaluate the antioxidant activity of microalgae crude extracts.

Author Contributions: N.C. built the bibliographic database. N.C., N.L. and T.J. wrote the manuscript. All authors have read and agreed to the published version of the manuscript.

Funding: This work was funded by the Province Nord, the Province Sud, the Government of New Caledonia and the Comité Interministériel de l'Outre-Mer (CIOM) through the AMICAL (Aquaculture of Microalgae in NewCALedonia) 1 and 2 research programs.

Institutional Review Board Statement: Not applicable.

Conflicts of Interest: The authors declare no conflict of interest.

References

1. Mimouni, V.; Ulmann, L.; Pasquet, V.; Mathieu, M.; Picot, L.; Bougaran, G.; Cadoret, J.-P.; Morant-Manceau, A.; Schoefs, B. The Potential of Microalgae for the Production of Bioactive Molecules of Pharmaceutical Interest. *Curr. Pharm. Biotechnol.* **2012**, *13*, 2733–2750. [CrossRef]
2. Wijffels, R.H.; Barbosa, M.J.; Eppink, M.H.M. Microalgae for the Production of Bulk Chemicals and Biofuels. *Biofuels. Bioprod. Biorefining* **2010**, *4*, 287–295. [CrossRef]
3. Aklakur, M. Natural Antioxidants from Sea: A Potential Industrial Perspective in Aquafeed Formulation. *Rev. Aquacult.* **2016**, *10*, 385–399. [CrossRef]
4. Guedes, A.C.; Amaro, H.M.; Malcata, F.X. Microalgae as Sources of High Added-Value Compounds—a Brief Review of Recent Work. *Biotechnol Prog.* **2011**, *27*, 597–613. [CrossRef]
5. Abad, M.J.; Bedoya, L.M.; Bermejo, P. Natural Marine Anti-Inflammatory Products. *Mini-Rev. Med. Chem.* **2008**, *8*, 740–754. [CrossRef]
6. Assunção, M.F.G.; Amaral, R.; Martins, C.B.; Ferreira, J.D.; Ressurreição, S.; Santos, S.D.; Varejão, J.M.T.B.; Santos, L.M.A. Screening Microalgae as Potential Sources of Antioxidants. *J. Appl. Phycol.* **2016**, *29*, 865–877. [CrossRef]
7. Coulombier, N.; Nicolau, E.; Le Déan, L.; Antheaume, C.; Jauffrais, T.; Lebouvier, N. Impact of Light Intensity on Antioxidant Activity of Tropical Microalgae. *Mar. Drugs* **2020**, *18*, 122. [CrossRef]
8. Safafar, H.; van Wagenen, J.; Møller, P.; Jacobsen, C. Carotenoids, Phenolic Compounds and Tocopherols Contribute to the Antioxidative Properties of Some Microalgae Species Grown on Industrial Wastewater. *Mar. Drugs* **2015**, *13*, 7339–7356. [CrossRef]
9. Sathasivam, R.; Ki, J.-S. A Review of the Biological Activities of Microalgal Carotenoids and Their Potential Use in Healthcare and Cosmetic Industries. *Mar. Drugs* **2018**, *16*, 26. [CrossRef]
10. Goiris, K.; Muylaert, K.; Fraeye, I.; Foubert, I.; Brabanter, J.D.; Cooman, L.D. Antioxidant Potential of Microalgae in Relation to Their Phenolic and Carotenoid Content. *J. Appl. Phycol.* **2012**, *24*, 1477–1486. [CrossRef]
11. Cezare-Gomes, E.A.; del Carmen Mejia-da-Silva, L.; Pérez-Mora, L.S.; Matsudo, M.C.; Ferreira-Camargo, L.S.; Singh, A.K.; de Carvalho, J.C.M. Potential of Microalgae Carotenoids for Industrial Application. *Appl. Biochem. Biotechnol.* **2019**, *188*, 602–634. [CrossRef]
12. Ito, N.; Hirose, M.; Fukushima, S.; Tsuda, H.; Shirai, T.; Tatematsu, M. Studies on Antioxidants: Their Carcinogenic and Modifying Effects on Chemical Carcinogenesis. *Food Chem. Toxicol.* **1986**, *24*, 1071–1082. [CrossRef]
13. Li, H.-B.; Cheng, K.-W.; Wong, C.-C.; Fan, K.-W.; Chen, F.; Jiang, Y. Evaluation of Antioxidant Capacity and Total Phenolic Content of Different Fractions of Selected Microalgae. *Food Chem.* **2007**, *102*, 771–776. [CrossRef]

14. Safer, A.M.; Al-Nughamish, A.J. Hepatotoxicity Induced by the Anti-Oxidant Food Additive, Butylated Hydroxytoluene (BHT), in Rats: An Electron Microscopical Study. *Histol. Histopathol.* **1999**, *14*, 391–406.
15. Batista, A.P.; Niccolai, A.; Fradinho, P.; Fragoso, S.; Bursic, I.; Rodolfi, L.; Biondi, N.; Tredici, M.R.; Sousa, I.; Raymundo, A. Microalgae Biomass as an Alternative Ingredient in Cookies: Sensory, Physical and Chemical Properties, Antioxidant Activity and in Vitro Digestibility. *Algal Res.* **2017**, *26*, 161–171. [CrossRef]
16. Batista, A.P.; Niccolai, A.; Bursic, I.; Sousa, I.; Raymundo, A.; Rodolfi, L.; Biondi, N.; Tredici, M.R. Microalgae as Functional Ingredients in Savory Food Products: Application to Wheat Crackers. *Foods* **2019**, *8*, 611. [CrossRef]
17. Goiris, K.; Muylaert, K.; De Cooman, L. Microalgae as a Novel Source of Antioxidants for Nutritional Applications. In *Handbook of Marine Microalgae*; Kim, S.-K., Ed.; Academic Press: Boston, MA, USA, 2015; pp. 269–280; ISBN 978-0-12-800776-1.
18. Sansone, C.; Brunet, C. Promises and Challenges of Microalgal Antioxidant Production. *Antioxidants* **2019**, *8*, 199. [CrossRef]
19. Halliwell, B. How to Characterize an Antioxidant: An Update. *Biochem. Soc. Symp.* **1995**, *61*, 73–101. [CrossRef]
20. Chen, B.; Wan, C.; Mehmood, M.A.; Chang, J.-S.; Bai, F.; Zhao, X. Manipulating Environmental Stresses and Stress Tolerance of Microalgae for Enhanced Production of Lipids and Value-Added Products–A Review. *Bioresour. Technol.* **2017**, *244*, 1198–1206. [CrossRef]
21. Coulombier, N.; Blanchier, P.; Le Dean, L.; Barthelemy, V.; Lebouvier, N.; Jauffrais, T. The Effects of CO_2-Induced Acidification on *Tetraselmis* Biomass Production, Photophysiology and Antioxidant Activity: A Comparison Using Batch and Continuous Culture. *J. Biotechnol.* **2021**, *325*, 312–324. [CrossRef]
22. Coulombier, N.; Nicolau, E.; Le Déan, L.; Barthelemy, V.; Schreiber, N.; Brun, P.; Lebouvier, N.; Jauffrais, T. Effects of Nitrogen Availability on the Antioxidant Activity and Carotenoid Content of the Microalgae *Nephroselmis* sp. *Mar. Drugs* **2020**, *18*, 453. [CrossRef]
23. Gauthier, M.R.; Senhorinho, G.N.A.; Scott, J.A. Microalgae under Environmental Stress as a Source of Antioxidants. *Algal Res.* **2020**, *52*, 102104. [CrossRef]
24. Goiris, K.; Van Colen, W.; Wilches, I.; León-Tamariz, F.; De Cooman, L.; Muylaert, K. Impact of Nutrient Stress on Antioxidant Production in Three Species of Microalgae. *Algal Res.* **2015**, *7*, 51–57. [CrossRef]
25. Guedes, A.C.; Amaro, H.M.; Pereira, R.D.; Malcata, F.X. Effects of Temperature and PH on Growth and Antioxidant Content of the Microalga *Scenedesmus obliquus*. *Biotechnol. Prog.* **2011**, *27*, 1218–1224. [CrossRef]
26. Paliwal, C.; Mitra, M.; Bhayani, K.; Bharadwaj, V.S.V.; Ghosh, T.; Dubey, S.; Mishra, S. Abiotic Stresses as Tools for Metabolites in Microalgae. *Bioresour. Technol.* **2017**, *244*, 1216–1226. [CrossRef]
27. Rice-Evans, C.A.; Diplock, A.T. Current Status of Antioxidant Therapy. *Free Radic. Biol. Med.* **1993**, *15*, 77–96. [CrossRef]
28. Saed-Moucheshi, A.; Shekoofa, A.; Pessarakli, M. Reactive Oxygen Species (ROS) Generation and Detoxifying in Plants. *J. Plant. Nutr.* **2014**, *37*, 1573–1585. [CrossRef]
29. Cirulis, J.T.; Scott, J.A.; Ross, G.M. Management of Oxidative Stress by Microalgae. *Can. J. Physiol. Pharmacol.* **2013**, *91*, 15–21. [CrossRef]
30. Demidchik, V. Mechanisms of Oxidative Stress in Plants: From Classical Chemistry to Cell Biology. *Environ. Exp. Bot.* **2015**, *109*, 212–228. [CrossRef]
31. Triantaphylidès, C.; Havaux, M. Singlet Oxygen in Plants: Production, Detoxification and Signaling. *Trends Plant. Sci.* **2009**, *14*, 219–228. [CrossRef]
32. Telfer, A.; Pascal, A.; Gall, A. Carotenoids in Photosynthesis. In *Carotenoids: Volume 4: Natural Functions*; Britton, G., Liaaen-Jensen, S., Pfander, H., Eds.; Birkhäuser: Basel, Switzerland, 2008; pp. 265–308; ISBN 978-3-7643-7499-0.
33. Gill, S.S.; Tuteja, N. Reactive Oxygen Species and Antioxidant Machinery in Abiotic Stress Tolerance in Crop Plants. *Plant. Physiol. Biochem.* **2010**, *48*, 909–930. [CrossRef]
34. Ledford, H.K.; Niyogi, K.K. Singlet Oxygen and Photo-Oxidative Stress Management in Plants and Algae. *Plant. Cell Environ.* **2005**, *28*, 1037–1045. [CrossRef]
35. Pospíšil, P.; Yamamoto, Y. Damage to Photosystem II by Lipid Peroxidation Products. *Biochim. Biophys. Acta Gen. Subj.* **2017**, *1861*, 457–466. [CrossRef]
36. Halliwell, B.; Gutteridge, J.M.C. *Free Radicals in Biology and Medicine*, 5th ed.; Oxford University Press: Oxford, UK, 2015.
37. Sharma, P.; Jha, A.B.; Dubey, R.S.; Pessarakli, M. Reactive Oxygen Species, Oxidative Damage, and Antioxidative Defense Mechanism in Plants under Stressful Conditions. *J. Bot.* **2012**, *2012*, 21703. [CrossRef]
38. Richards, S.L.; Wilkins, K.A.; Swarbreck, S.M.; Anderson, A.A.; Habib, N.; Smith, A.G.; McAinsh, M.; Davies, J.M. The Hydroxyl Radical in Plants: From Seed to Seed. *J. Exp. Bot.* **2015**, *66*, 37–46. [CrossRef]
39. Gulcin, İ. Antioxidants and Antioxidant Methods: An Updated Overview. *Arch. Toxicol.* **2020**, *94*, 651–715. [CrossRef]
40. Duarte, T.L.; Lunec, J. When Is an Antioxidant Not an Antioxidant? A Review of Novel Actions and Reactions of Vitamin C. *Free Radic. Res.* **2005**, *39*, 671–686. [CrossRef]
41. Herrera, E.; Barbas, C. Vitamin E: Action, Metabolism and Perspectives. *J. Physiol. Biochem.* **2001**, *57*, 43–56. [CrossRef]
42. Munné-Bosch, S.; Alegre, L. The Function of Tocopherols and Tocotrienols in Plants. *Crit. Rev. Plant. Sci.* **2002**, *21*, 31–57. [CrossRef]
43. Dai, J.; Mumper, R.J. Plant Phenolics: Extraction, Analysis and Their Antioxidant and Anticancer Properties. *Molecules* **2010**, *15*, 7313–7352. [CrossRef]

44. Aremu, A.O.; Neményi, M.; Stirk, W.A.; Ördög, V.; van Staden, S. Manipulation of Nitrogen Levels and Mode of Cultivation Are Viable Methods to Improve the Lipid, Fatty Acids, Phytochemical Content, and Bioactivities in *Chlorella minutissima*. *J. Phycol.* **2015**, *51*, 659–669. [CrossRef]
45. Custódio, L.; Justo, T.; Silvestre, L.; Barradas, A.; Duarte, C.V.; Pereira, H.; Barreira, L.; Rauter, A.P.; Alberício, F.; Varela, J. Microalgae of Different Phyla Display Antioxidant, Metal Chelating and Acetylcholinesterase Inhibitory Activities. *Food Chem.* **2012**, *131*, 134–140. [CrossRef]
46. Hajimahmoodi, M.; Faramarzi, M.A.; Mohammadi, N.; Soltani, N.; Oveisi, M.R.; Nafissi-Varcheh, N. Evaluation of Antioxidant Properties and Total Phenolic Contents of Some Strains of Microalgae. *J. Appl. Phycol.* **2010**, *22*, 43–50. [CrossRef]
47. Jayshree, A.; Jayashree, S.; Thangaraju, N. *Chlorella vulgaris* and *Chlamydomonas reinhardtii*: Effective Antioxidant, Antibacterial and Anticancer Mediators. *Indian J. Pharm. Sci.* **2016**, *78*, 575–581. [CrossRef]
48. Aremu, A.O.; Masondo, N.A.; Molnár, Z.; Stirk, W.A.; Ördög, V.; Van Staden, S. Changes in Phytochemical Content and Pharmacological Activities of Three *Chlorella* Strains Grown in Different Nitrogen Conditions. *J. Appl. Phycol.* **2016**, *28*, 149–159. [CrossRef]
49. Chaudhuri, D.; Ghate, N.B.; Deb, S.; Panja, S.; Sarkar, R.; Rout, J.; Mandal, N. Assessment of the Phytochemical Constituents and Antioxidant Activity of a Bloom Forming Microalgae *Euglena tuba*. *Biol. Res.* **2014**, *47*, 24. [CrossRef] [PubMed]
50. Goiris, K.; Muylaert, K.; Voorspoels, S.; Noten, B.; Paepe, D.D.; Baart, G.J.E.; Cooman, L.D. Detection of Flavonoids in Microalgae from Different Evolutionary Lineages. *J. Phycol.* **2014**, *50*, 483–492. [CrossRef]
51. Haoujar, I.; Cacciola, F.; Abrini, J.; Mangraviti, D.; Giuffrida, D.; Oulad El Majdoub, Y.; Kounnoun, A.; Miceli, N.; Fernanda Taviano, M.; Mondello, L.; et al. The Contribution of Carotenoids, Phenolic Compounds, and Flavonoids to the Antioxidative Properties of Marine Microalgae Isolated from Mediterranean Morocco. *Molecules* **2019**, *24*, 4037. [CrossRef] [PubMed]
52. López, A.; Rico, M.; Santana-Casiano, J.M.; González, A.G.; González-Dávila, M. Phenolic Profile of *Dunaliella Tertiolecta* Growing under High Levels of Copper and Iron. *Environ. Sci. Pollut. Res.* **2015**, *22*, 14820–14828. [CrossRef]
53. Smerilli, A.; Balzano, S.; Maselli, M.; Blasio, M.; Orefice, I.; Galasso, C.; Sansone, C.; Brunet, C. Antioxidant and Photoprotection Networking in the Coastal Diatom *Skeletonema marinoi*. *Antioxidants* **2019**, *8*, 154. [CrossRef] [PubMed]
54. Yu, J.; Wang, Y.; Sun, J.; Bian, F.; Chen, G.; Zhang, Y.; Bi, Y.; Wu, Y. Antioxidant Activity of Alcohol Aqueous Extracts of *Crypthecodinium cohnii* and *Schizochytrium* sp. *J. Zhejiang Univ.-Sci. B* **2017**, *18*, 797–806. [CrossRef]
55. Shahidi, F.; Ambigaipalan, P. Phenolics and Polyphenolics in Foods, Beverages and Spices: Antioxidant Activity and Health Effects—A Review. *J. Funct. Foods* **2015**, *18*, 820–897. [CrossRef]
56. Leopoldini, M.; Russo, N.; Toscano, M. The Molecular Basis of Working Mechanism of Natural Polyphenolic Antioxidants. *Food Chem.* **2011**, *125*, 288–306. [CrossRef]
57. Pouvreau, J.-B.; Morançais, M.; Taran, F.; Rosa, P.; Dufossé, L.; Guérard, F.; Pin, S.; Fleurence, J.; Pondaven, P. Antioxidant and Free Radical Scavenging Properties of Marennine, a Blue-Green Polyphenols Pigment from the Diatom *Haslea ostrearia* (Gaillon/Bory) Simonsen Responsible for the Natural Greening of Cultured Oysters. *J. Agric. Food Chem.* **2008**, *56*, 6278–6286. [CrossRef]
58. Takaichi, S. Carotenoids in Algae: Distributions, Biosyntheses and Functions. *Mar. Drugs* **2011**, *9*, 1101–1118. [CrossRef] [PubMed]
59. Stahl, W.; Sies, H. Antioxidant Activity of Carotenoids. *Mol. Asp. Med.* **2003**, *24*, 345–351. [CrossRef]
60. Solovchenko, A.E. Physiology and Adaptive Significance of Secondary Carotenogenesis in Green Microalgae. *Russ. J. Plant. Physiol.* **2013**, *60*, 167–176. [CrossRef]
61. Huang, J.J.; Lin, S.; Xu, W.; Cheung, P.C.K. Occurrence and Biosynthesis of Carotenoids in Phytoplankton. *Biotechnol. Adv.* **2017**, *35*, 597–618. [CrossRef] [PubMed]
62. Goss, R.; Jakob, T. Regulation and Function of Xanthophyll Cycle-Dependent Photoprotection in Algae. *Photosynth. Res.* **2010**, *106*, 103–122. [CrossRef] [PubMed]
63. Jahns, P.; Holzwarth, A.R. The Role of the Xanthophyll Cycle and of Lutein in Photoprotection of Photosystem II. *Biochim. Biophys. Acta* **2012**, *1817*, 182–193. [CrossRef]
64. Edge, R.; McGarvey, D.J.; Truscott, T.G. The Carotenoids as Antioxidants—A Review. *J. Photochem. Photobiol. B Biol.* **1997**, *41*, 189–200. [CrossRef]
65. El-Agamey, A.; McGarvey, D.J. Carotenoid Radicals and Radical Ions. In *Carotenoids*; Britton, G., Liaaen-Jensen, S., Pfander, H., Eds.; Birkhäuser: Basel, Switzerland, 2008; Volume 4, pp. 119–154. ISBN 978-3-7643-7498-3.
66. Ribeiro, D.; Freitas, M.; Silva, A.M.S.; Carvalho, F.; Fernandes, E. Antioxidant and Pro-Oxidant Activities of Carotenoids and Their Oxidation Products. *Food Chem. Toxicol.* **2018**, *120*, 681–699. [CrossRef]
67. Wada, N.; Sakamoto, T.; Matsugo, S. Mycosporine-Like Amino Acids and Their Derivatives as Natural Antioxidants. *Antioxidants* **2015**, *4*, 603–646. [CrossRef]
68. Llewellyn, C.A.; Airs, R.L. Distribution and Abundance of MAAs in 33 Species of Microalgae across 13 Classes. *Mar. Drugs* **2010**, *8*, 1273–1291. [CrossRef]
69. Sinha, R.P.; Singh, S.P.; Häder, D.-P. Database on Mycosporines and Mycosporine-like Amino Acids (MAAs) in Fungi, Cyanobacteria, Macroalgae, Phytoplankton and Animals. *J. Photochem. Photobiol. B Biol.* **2007**, *89*, 29–35. [CrossRef]
70. Raposo, M.F.D.J.; De Morais, R.M.S.C.; de Morais, B.A.M.M. Bioactivity and Applications of Sulphated Polysaccharides from Marine Microalgae. *Mar. Drugs* **2013**, *11*, 233–252. [CrossRef]
71. Balavigneswaran, C.K.; Sujin Jeba Kumar, T.; Moses Packiaraj, R.; Veeraraj, A.; Prakash, S. Anti-Oxidant Activity of Polysaccharides Extracted from *Isochrysis galbana* Using RSM Optimized Conditions. *Int. J. Biol. Macromol.* **2013**, *60*, 100–108. [CrossRef]

72. Sun, Y.; Wang, H.; Guo, G.; Pu, Y.; Yan, B. The Isolation and Antioxidant Activity of Polysaccharides from the Marine Microalgae *Isochrysis galbana*. *Carbohydr. Polym.* **2014**, *113*, 22–31. [CrossRef]
73. Tannin-Spitz, T.; Bergman, M.; Van-Moppes, D.; Grossman, S.; Arad, S. Antioxidant Activity of the Polysaccharide of the Red Microalga *Porphyridium* sp. *J. Appl. Phycol.* **2005**, *17*, 215–222. [CrossRef]
74. Yu, M.; Chen, M.; Gui, J.; Huang, S.; Liu, Y.; Shentu, H.; He, J.; Fang, Z.; Wang, W.; Zhang, Y. Preparation of *Chlorella vulgaris* Polysaccharides and Their Antioxidant Activity in Vitro and in Vivo. *Int. J. Biol. Macromol.* **2019**, *137*, 139–150. [CrossRef]
75. Zhong, Q.; Wei, B.; Wang, S.; Ke, S.; Chen, J.; Zhang, H.; Wang, H. The Antioxidant Activity of Polysaccharides Derived from Marine Organisms: An Overview. *Mar. Drugs* **2019**, *17*, 674. [CrossRef] [PubMed]
76. Glazer, A.N. Phycobiliproteins—A Family of Valuable, Widely Used Fluorophores. *J. Appl. Phycol.* **1994**, *6*, 105–112. [CrossRef]
77. Sonani, R.R.; Rastogi, R.P.; Madamwar, D. Antioxidant Potential of Phycobiliproteins: Role in Anti-Aging Research. *Biochem. Anal. Biochem.* **2015**, *4*, 1009–2161. [CrossRef]
78. Carocho, M.; Ferreira, I.C.F.R. A Review on Antioxidants, Prooxidants and Related Controversy: Natural and Synthetic Compounds, Screening and Analysis Methodologies and Future Perspectives. *Food Chem. Toxicol.* **2013**, *51*, 15–25. [CrossRef]
79. Frankel, E.N.; Meyer, A.S. The Problems of Using One-Dimensional Methods to Evaluate Multifunctional Food and Biological Antioxidants. *J. Sci. Food Agric.* **2000**, *80*, 1925–1941. [CrossRef]
80. Prior, R.L.; Wu, X.; Schaich, K. Standardized Methods for the Determination of Antioxidant Capacity and Phenolics in Foods and Dietary Supplements. *J. Agric. Food Chem.* **2005**, *53*, 4290–4302. [CrossRef]
81. Plaza, M.; Santoyo, S.; Jaime, L.; García-Blairsy Reina, G.; Herrero, M.; Señoráns, F.J.; Ibáñez, E. Screening for Bioactive Compounds from Algae. *J. Pharm. Biomed. Anal.* **2010**, *51*, 450–455. [CrossRef]
82. Aremu, A.O.; Masondo, N.A.; Stirk, W.A.; Ördög, V.; Staden, J.V. Influence of Culture Age on the Phytochemical Content and Pharmacological Activities of Five *Scenedesmus* Strains. *J. Appl. Phycol.* **2014**, *26*, 407–415. [CrossRef]
83. Herrero, M.; Jaime, L.; Martín-Alvarez, P.J.; Cifuentes, A.; Ibáñez, E. Optimization of the Extraction of Antioxidants from *Dunaliella salina* Microalga by Pressurized Liquids. *J. Agric. Food Chem.* **2006**, *54*, 5597–5603. [CrossRef]
84. Natrah, F.M.I.; Yusoff, F.M.; Shariff, M.; Abas, F.; Mariana, N.S. Screening of Malaysian Indigenous Microalgae for Antioxidant Properties and Nutritional Value. *J. Appl. Phycol.* **2007**, *19*, 711–718. [CrossRef]
85. Simic, S.; Kosanic, M.; Rankovic, B. Evaluation of In Vitro Antioxidant and Antimicrobial Activities of Green Microalgae *Trentepohlia umbrina*. *Not. Bot. Horti Agrobot.* **2012**, *40*, 86–91. [CrossRef]
86. Goh, S.-H.; Yusoff, F.M.; Loh, S.P. A Comparison of the Antioxidant Properties and Total Phenolic Content in a Diatom, *Chaetoceros* sp. and a Green Microalga, *Nannochloropsis* sp. *J. Agric. Sci.* **2010**, *2*, 123. [CrossRef]
87. Guzmán, S.; Gato, A.; Calleja, J.M. Antiinflammatory, Analgesic and Free Radical Scavenging Activities of the Marine Microalgae *Chlorella stigmatophora* and *Phaeodactylum tricornutum*. *Phytother. Res.* **2001**, *15*, 224–230. [CrossRef] [PubMed]
88. Azizan, A.; Ahamad Bustamam, M.S.; Maulidiani, M.; Shaari, K.; Ismail, I.S.; Nagao, N.; Abas, F. Metabolite Profiling of the Microalgal Diatom *Chaetoceros calcitrans* and Correlation with Antioxidant and Nitric Oxide Inhibitory Activities via ^{1}H NMR-Based Metabolomics. *Mar. Drugs* **2018**, *16*, 154. [CrossRef] [PubMed]
89. Lee, S.-H.; Lee, J.-B.; Lee, K.-W.; Jeon, Y.-J. Antioxidant Properties of Tidal Pool Microalgae, *Halochlorococcum porphyrae* and *Oltamannsiellopsis unicellularis* from Jeju Island, Korea. *Algae* **2010**, *25*, 45–56. [CrossRef]
90. Buono, S.; Langellotti, A.L.; Martello, A.; Bimonte, M.; Tito, A.; Carola, A.; Apone, F.; Colucci, G.; Fogliano, V. Biological Activities of Dermatological Interest by the Water Extract of the Microalga *Botryococcus braunii*. *Arch. Derm. Res.* **2012**, *304*, 755–764. [CrossRef] [PubMed]
91. Lauritano, C.; Andersen, J.H.; Hansen, E.; Albrigtsen, M.; Escalera, L.; Esposito, F.; Helland, K.; Hanssen, K.Ø.; Romano, G.; Ianora, A. Bioactivity Screening of Microalgae for Antioxidant, Anti-Inflammatory, Anticancer, Anti-Diabetes, and Antibacterial Activities. *Front. Mar. Sci.* **2016**, *3*, 68. [CrossRef]
92. Karawita, R.; Senevirathne, M.; Athukorala, Y.; Affan, A.; Lee, Y.-J.; Kim, S.-K.; Lee, J.-B.; Jeon, Y.-J. Protective Effect of Enzymatic Extracts from Microalgae against DNA Damage Induced by H2O2. *Mar. Biotechnol.* **2007**, *9*, 479–490. [CrossRef]
93. Affan, A.; Karawita, R.; Jeon, Y.-J.; Kim, B.-Y.; Lee, J.-B. Growth Characteristics, Bio-Chemical Composition and Antioxidant Activities of Benthic Diatom *Grammatophora marina* from Jeju Coast, Korea. *Algae* **2006**, *21*, 141–148. [CrossRef]
94. Agregán, R.; Munekata, P.E.S.; Franco, D.; Carballo, J.; Barba, F.J.; Lorenzo, J.M. Antioxidant Potential of Extracts Obtained from Macro- (*Ascophyllum nodosum*, *Fucus vesiculosus* and *Bifurcaria bifurcata*) and Micro-Algae (*Chlorella vulgaris* and *Spirulina platensis*) Assisted by Ultrasound. *Medicines* **2018**, *5*, 33. [CrossRef]
95. Ahmed, F.; Fanning, K.; Netzel, M.; Turner, W.; Li, Y.; Schenk, P.M. Profiling of Carotenoids and Antioxidant Capacity of Microalgae from Subtropical Coastal and Brackish Waters. *Food Chem.* **2014**, *165*, 300–306. [CrossRef] [PubMed]
96. Amaro, H.M.; Fernandes, F.; Valentão, P.; Andrade, P.B.; Sousa-Pinto, I.; Malcata, F.X.; Guedes, A.C. Effect of Solvent System on Extractability of Lipidic Components of *Scenedesmus obliquus* (M2-1) and *Gloeothece* sp. on Antioxidant Scavenging Capacity Thereof. *Mar. Drugs* **2015**, *13*, 6453–6471. [CrossRef] [PubMed]
97. Banskota, A.H.; Sperker, S.; Stefanova, R.; McGinn, P.J.; O'Leary, S.J.B. Antioxidant Properties and Lipid Composition of Selected Microalgae. *J. Appl. Phycol.* **2018**, *31*, 309–318. [CrossRef]
98. Bauer, L.M.; Costa, J.A.V.; da Rosa, A.P.C.; Santos, L.O. Growth Stimulation and Synthesis of Lipids, Pigments and Antioxidants with Magnetic Fields in Chlorella Kessleri Cultivations. *Bioresour. Technol.* **2017**, *244*, 1425–1432. [CrossRef] [PubMed]

99. Bulut, O.; Akın, D.; Sönmez, Ç.; Öktem, A.; Yücel, M.; Öktem, H.A. Phenolic Compounds, Carotenoids, and Antioxidant Capacities of a Thermo-Tolerant *Scenedesmus* sp. (Chlorophyta) Extracted with Different Solvents. *J. Appl. Phycol.* **2019**, *31*, 1675–1683. [CrossRef]
100. Choochote, W.; Suklampoo, L.; Ochaikul, D. Evaluation of Antioxidant Capacities of Green Microalgae. *J. Appl. Phycol.* **2014**, *26*, 43–48. [CrossRef]
101. Custódio, L.; Soares, F.; Pereira, H.; Barreira, L.; Vizetto-Duarte, C.; Rodrigues, M.J.; Rauter, A.P.; Alberício, F.; Varela, J. Fatty Acid Composition and Biological Activities of *Isochrysis galbana* T-ISO, *Tetraselmis* sp. and *Scenedesmus* sp.: Possible Application in the Pharmaceutical and Functional Food Industries. *J. Appl. Phycol.* **2014**, *26*, 151–161. [CrossRef]
102. Elshobary, M.E.; El-Shenody, R.A.; Ashour, M.; Zabed, H.M.; Qi, X. Antimicrobial and Antioxidant Characterization of Bioactive Components from *Chlorococcum minutum*. *Food Biosci.* **2020**, *35*, 100567. [CrossRef]
103. Foo, S.C.; Yusoff, F.M.; Ismail, M.; Basri, M.; Khong, N.M.H.; Chan, K.W.; Yau, S.K. Efficient Solvent Extraction of Antioxidant-Rich Extract from a Tropical Diatom, *Chaetoceros calcitrans* (Paulsen) Takano 1968. *Asian Pac. J. Trop. Biomed.* **2015**, *5*, 834–840. [CrossRef]
104. Foo, S.C.; Yusoff, F.M.; Ismail, M.; Basri, M.; Yau, S.K.; Khong, N.M.H.; Chan, K.W.; Ebrahimi, M. Antioxidant Capacities of Fucoxanthin-Producing Algae as Influenced by Their Carotenoid and Phenolic Contents. *J. Biotechnol.* **2017**, *241*, 175–183. [CrossRef]
105. Guedes, A.C.; Gião, M.S.; Seabra, R.; Ferreira, A.C.S.; Tamagnini, P.; Moradas-Ferreira, P.; Malcata, F.X. Evaluation of the Antioxidant Activity of Cell Extracts from Microalgae. *Mar. Drugs* **2013**, *11*, 1256–1270. [CrossRef]
106. Gürlek, C.; Yarkent, Ç.; Köse, A.; Tuğcu, B.; Gebeloğlu, I.K.; Öncel, S.Ş.; Elibol, M. Screening of Antioxidant and Cytotoxic Activities of Several Microalgal Extracts with Pharmaceutical Potential. *Health Technol.* **2020**, *10*, 111–117. [CrossRef]
107. Jaime, L.; Mendiola, J.A.; Ibáñez, E.; Martin-Álvarez, P.J.; Cifuentes, A.; Reglero, G.; Señoráns, F.J. β-Carotene Isomer Composition of Sub- and Supercritical Carbon Dioxide Extracts. Antioxidant Activity Measurement. *J. Agric. Food Chem.* **2007**, *55*, 10585–10590. [CrossRef]
108. Jerez-Martel, I.; García-Poza, S.; Rodríguez-Martel, G.; Rico, M.; Afonso-Olivares, C.; Gómez-Pinchetti, J.L. Phenolic Profile and Antioxidant Activity of Crude Extracts from Microalgae and Cyanobacteria Strains. *J. Food Qual.* **2017**, *2017*, 2924508. [CrossRef]
109. Maadane, A.; Merghoub, N.; Ainane, T.; El Arroussi, H.; Benhima, R.; Amzazi, S.; Bakri, Y.; Wahby, I. Antioxidant Activity of Some Moroccan Marine Microalgae: Pufa Profiles, Carotenoids and Phenolic Content. *J. Biotechnol.* **2015**, *215*, 13–19. [CrossRef] [PubMed]
110. Matos, J.; Cardoso, C.; Gomes, A.; Campos, A.M.; Falé, P.; Afonso, C.; Bandarra, N.M. Bioprospection of *Isochrysis galbana* and Its Potential as a Nutraceutical. *Food Funct.* **2019**, *10*, 7333–7342. [CrossRef] [PubMed]
111. Millao, S.; Uquiche, E. Antioxidant Activity of Supercritical Extracts from *Nannochloropsis Gaditana*: Correlation with Its Content of Carotenoids and Tocopherols. *J. Supercrit. Fluids* **2016**, *111*, 143–150. [CrossRef]
112. Morowvat, M.H.; Ghasemi, Y. Evaluation of Antioxidant Properties of Some Naturally Isolated Microalgae: Identification and Characterization of the Most Efficient Strain. *Biocatal. Agric. Biotechnol.* **2016**, *8*, 263–269. [CrossRef]
113. Pereira, H.; Custódio, L.; Rodrigues, M.J.; De, S.; Oliveira, M.; Barreira, L.; Neng, N.D.R.; Nogueira, J.M.F.; Alrokayan, S.A.; Mouffouk, F.; et al. Biological Activities and Chemical Composition of Methanolic Extracts of Selected Autochthonous Microalgae Strains from the Red Sea. *Mar. Drugs* **2015**, *13*, 3531–3549. [CrossRef] [PubMed]
114. Rodríguez-Meizoso, I.; Jaime, L.; Santoyo, S.; Señoráns, F.J.; Cifuentes, A.; Ibáñez, E. Subcritical Water Extraction and Characterization of Bioactive Compounds from *Haematococcus pluvialis* Microalga. *J. Pharm. Biomed. Anal.* **2010**, *51*, 456–463. [CrossRef]
115. Sansone, C.; Galasso, C.; Orefice, I.; Nuzzo, G.; Luongo, E.; Cutignano, A.; Romano, G.; Brunet, C.; Fontana, A.; Esposito, F.; et al. The Green Microalga *Tetraselmis suecica* Reduces Oxidative Stress and Induces Repairing Mechanisms in Human Cells. *Sci. Rep.* **2017**, *7*, 41215. [CrossRef]
116. Sharma, A.K.; General, T. Variation of Both Chemical Composition and Antioxidant Properties of Newly Isolated *Parachlorella kessleri* GB1, by Growing in Different Culture Conditions. *LWT* **2019**, *112*, 108205. [CrossRef]
117. Singh, P.; Baranwal, M.; Reddy, S.M. Antioxidant and Cytotoxic Activity of Carotenes Produced by *Dunaliella Salina* under Stress. *Pharm. Biol.* **2016**, *54*, 2269–2275. [CrossRef] [PubMed]
118. Suh, S.-S.; Yang, E.J.; Lee, S.G.; Youn, U.J.; Han, S.J.; Kim, I.-C.; Kim, S. Bioactivities of Ethanol Extract from the Antarctic Freshwater Microalga, *Chloromonas* sp. *Int. J. Med. Sci.* **2017**, *14*, 560–569. [CrossRef] [PubMed]
119. Suh, S.-S.; Kim, S.-M.; Kim, J.E.; Hong, J.-M.; Lee, S.G.; Youn, U.J.; Han, S.J.; Kim, I.-C.; Kim, S. Anticancer Activities of Ethanol Extract from the Antarctic Freshwater Microalga, *Botryidiopsidaceae* sp. *BMC Complemen. Altern. Med.* **2017**, *17*, 509. [CrossRef] [PubMed]
120. Allen, K.M.; Habte-Tsion, H.-M.; Thompson, K.R.; Filer, K.; Tidwell, J.H.; Kumar, V. Freshwater Microalgae (*Schizochytrium* sp.) as a Substitute to Fish Oil for Shrimp Feed. *Sci. Rep.* **2019**, *9*, 6178. [CrossRef]
121. El-Bahr, S.; Shousha, S.; Shehab, A.; Khattab, W.; Ahmed-Farid, O.; Sabike, I.; El-Garhy, O.; Albokhadaim, I.; Albosadah, K. Effect of Dietary Microalgae on Growth Performance, Profiles of Amino and Fatty Acids, Antioxidant Status, and Meat Quality of Broiler Chickens. *Animals* **2020**, *10*, 761. [CrossRef]
122. Long, S.F.; Kang, S.; Wang, Q.Q.; Xu, Y.T.; Pan, L.; Hu, J.X.; Li, M.; Piao, X.S. Dietary Supplementation with DHA-Rich Microalgae Improves Performance, Serum Composition, Carcass Trait, Antioxidant Status, and Fatty Acid Profile of Broilers. *Poult. Sci.* **2018**, *97*, 1881–1890. [CrossRef]

123. Marques, A.E.M.L.; Balen, R.E.; da Silva Pereira Fernandes, L.; Motta, C.M.; de Assis, H.C.S.; Taher, D.M.; Meurer, F.; Vargas, J.V.C.; Mariano, A.B.; Cestari, M.M. Diets Containing Residual Microalgae Biomass Protect Fishes against Oxidative Stress and DNA Damage. *J. Appl. Phycol.* **2019**, *31*, 2933–2940. [CrossRef]
124. Nacer, W.; Baba Ahmed, F.Z.; Merzouk, H.; Benyagoub, O.; Bouanane, S. Evaluation of the Anti-Inflammatory and Antioxidant Effects of the Microalgae *Nannochloropsis gaditana* in Streptozotocin-Induced Diabetic Rats. *J. Diabetes Metab. Disord.* **2020**, *19*, 1483–1490. [CrossRef]
125. Qiao, H.; Hu, D.; Ma, J.; Wang, X.; Wu, H.; Wang, J. Feeding Effects of the Microalga *Nannochloropsis* sp. on Juvenile Turbot (*Scophthalmus maximus* L.). *Algal Res.* **2019**, *41*, 101540. [CrossRef]
126. Rahman, N.A.; Khatoon, H.; Yusuf, N.; Banerjee, S.; Haris, N.A.; Lananan, F.; Tomoyo, K. Tetraselmis Chuii Biomass as a Potential Feed Additive to Improve Survival and Oxidative Stress Status of Pacific White-Leg Shrimp Litopenaeus Vannamei Postlarvae. *Int. Aquat. Res.* **2017**, *9*, 235–247. [CrossRef]
127. Ranga Rao, A.; Raghunath Reddy, R.L.; Baskaran, V.; Sarada, R.; Ravishankar, G.A. Characterization of Microalgal Carotenoids by Mass Spectrometry and Their Bioavailability and Antioxidant Properties Elucidated in Rat Model. *J. Agric. Food Chem.* **2010**, *58*, 8553–8559. [CrossRef] [PubMed]
128. Ranga Rao, A.; Baskaran, V.; Sarada, R.; Ravishankar, G.A. In Vivo Bioavailability and Antioxidant Activity of Carotenoids from Microalgal Biomass—A Repeated Dose Study. *Food Res. Int.* **2013**, *54*, 711–717. [CrossRef]
129. Sheikhzadeh, N.; Tayefi-Nasrabadi, H.; Khani Oushani, A.; Najafi Enferadi, M.H. Effects of *Haematococcus pluvialis* Supplementation on Antioxidant System and Metabolism in Rainbow Trout (*Oncorhynchus mykiss*). *Fish. Physiol. Biochem.* **2012**, *38*, 413–419. [CrossRef]
130. Carocho, M.; Morales, P.; Ferreira, I.C.F.R. Antioxidants: Reviewing the Chemistry, Food Applications, Legislation and Role as Preservatives. *Trends Food Sci. Technol.* **2018**, *71*, 107–120. [CrossRef]
131. Lourenço, S.C.; Moldão-Martins, M.; Alves, V.D. Antioxidants of Natural Plant Origins: From Sources to Food Industry Applications. *Molecules* **2019**, *24*, 4132. [CrossRef] [PubMed]
132. European Commission Règlement (UE). 2015/2283 du Parlement Européen et du Conseil du 25 Novembre 2015 Relatif aux Nouveaux Aliments, Modifiant le Règlement (UE) No. *1169/2011 du Parlement Européen et du Conseil et Abrogeant le Règlement (CE) n° 258/97 du Parlement Européen et du Conseil et le Règlement (CE) No. 1852/2001 de la Commission (Texte Présentant de L'intérêt pour l'EEE)*. Available online: http://data.europa.eu/eli/reg/2015/2283/oj/fra (accessed on 15 May 2020).
133. Turck, D.; Bresson, J.-L.; Burlingame, B.; Dean, T.; Fairweather-Tait, S.; Heinonen, M.; Hirsch-Ernst, K.I.; Mangelsdorf, I.; McArdle, H.; Naska, A.; et al. Guidance on the Preparation and Presentation of an Application for Authorisation of a Novel Food in the Context of Regulation (EU) 2015/2283. *EFSA J.* **2016**, *14*, e04594. [CrossRef]
134. Lefebvre, K.A.; Robertson, A. Domoic Acid and Human Exposure Risks: A Review. *Toxicon* **2010**, *56*, 218–230. [CrossRef] [PubMed]
135. McKenzie, C.H.; Bates, S.S.; Martin, J.L.; Haigh, N.; Howland, K.L.; Lewis, N.I.; Locke, A.; Peña, A.; Poulin, M.; Rochon, A.; et al. Three Decades of Canadian Marine Harmful Algal Events: Phytoplankton and Phycotoxins of Concern to Human and Ecosystem Health. *Harmful Algae* **2021**, *102*, 101852. [CrossRef]
136. Kilcoyne, J.; Nulty, C.; Jauffrais, T.; McCarron, P.; Herve, F.; Foley, B.; Rise, F.; Crain, S.; Wilkins, A.L.; Twiner, M.J.; et al. Isolation, Structure Elucidation, Relative LC-MS Response, and in Vitro Toxicity of Azaspiracids from the Dinoflagellate Azadinium Spinosum. *J. Nat. Prod.* **2014**, *77*, 2465–2474. [CrossRef] [PubMed]
137. CEVA. *Macroalgues et Microalgues Alimentaires—Statut Règlementaire en France et en Europe*; CEVA: Marseille, France, 2019; p. 15.
138. European Commission Règlement (UE). No. 68/2013 de la Commission du 16 Janvier 2013 Relatif au Catalogue des Matières Premières pour Aliments des Animaux Texte Présentant de L'intérêt pour l'EEE. Available online: https://eur-lex.europa.eu/legal-content/FR/ALL/?uri=CELEX%3A32013R0068 (accessed on 15 May 2020).

MDPI
St. Alban-Anlage 66
4052 Basel
Switzerland
Tel. +41 61 683 77 34
Fax +41 61 302 89 18
www.mdpi.com

Marine Drugs Editorial Office
E-mail: marinedrugs@mdpi.com
www.mdpi.com/journal/marinedrugs

www.ingramcontent.com/pod-product-compliance
Lightning Source LLC
Chambersburg PA
CBHW061611220926
51850CB00005B/480
9783036567723